BIOTECHNOLOGY IN THE SUSTAINABLE ENVIRONMENT

ENVIRONMENTAL SCIENCE RESEARCH

Recent Volumes in this Series

Volume 44 — SECONDARY-METABOLITE BIOSYNTHESIS AND METABOLISM
Edited by Richard J. Petroski and Susan P. McCormick

Volume 45 — GLOBAL CLIMATE CHANGE: Linking Energy, Environment, Economy, and Equity
Edited by James C. White

Volume 46 — PRODUCED WATER: Technological/Environmental Issues and Solutions
Edited by James P. Ray and F. Rainer Engelhardt

Volume 47 — GLOBAL ENERGY STRATEGIES: Living with Restricted Greenhouse Gas Emissions
Edited by James C. White

Volume 48 — GLOBAL ATMOSPHERIC–BIOSPHERIC CHEMISTRY
Ronald G. Prinn

Volume 49 — BIODEGRADATION OF NITROAROMATIC COMPOUNDS
Edited by Jim C. Spain

Volume 50 — BIOMONITORS AND BIOMARKERS AS INDICATORS OF ENVIRONMENTAL CHANGE: A Handbook
Edited by Frank M. Butterworth, Lynda D. Corkum, and Judith Guzmán-Rincón

Volume 51 — CHEMISTRY FOR THE PROTECTION OF THE ENVIRONMENT 2
Edited by Lucjan Pawłowski, William J. Lacy, Christopher G. Uchrin, and Marzenna R. Dudzińska

Volume 52 — PRODUCED WATER 2: Environmental Issues and Mitigation Technologies
Edited by Mark Reed and Ståle Johnsen

Volume 53 — EVALUATING CLIMATE CHANGE ACTION PLANS: National Actions for International Commitment
Edited by James C. White

Volume 54 — BIOTECHNOLOGY IN THE SUSTAINABLE ENVIRONMENT
Edited by Gary S. Sayler, John Sanseverino, and Kimberly L. Davis

---

BIOTECHNOLOGY IN THE SUSTAINABLE ENVIRONMENT

Edited by

Gary S. Sayler
John Sanseverino
and
Kimberly L. Davis

The University of Tennessee
Knoxville, Tennessee

Plenum Press • New York and London

Library of Congress Cataloging-in-Publication Data

Biotechnology in the sustainable environment / edited by Gary S.
Sayler, John Sanseverino, and Kimberly L. Davis.
p. cm. -- (Environmental science research ; v. 54)
"Proceedings of a conference on Biotechnology in the Sustainable
Environment, held April 14-17, 1996, in Knoxville, Tennessee"--T.p.
verso.
Includes bibliographical references and index.
ISBN 0-306-45717-2
1. Bioremediation--Congresses. I. Sayler, Gary S., 1949- .
II. Sanseverino, John. III. Davis, Kimberly L. IV. Conference on
Biotechnology in the Sustainable Environment (1996 : Knoxville,
Tenn.) V. Series.
TD192.5.B57 1997
628.5--dc21 97-33246
CIP

Proceedings of a conference on Biotechnology in the Sustainable Environment,
held April 14 – 17, 1996, in Knoxville, Tennessee

ISBN 0-306-45717-2

A Division of Plenum Publishing Corporation
233 Spring Street, New York, N. Y. 10013

http://www.plenum.com

10 9 8 7 6 5 4 3 2 1

Printed in the United States of America

PREFACE

The purpose of these proceedings was to address the technological, economic, institutional and social questions raised by the rapidly evolving use of biotechnology for the protection, restoration, and sustainability of our environment. The symposium, "Biotechnology in the Sustainable Environment" emphasized the future-oriented nature of biotechnology, that is, the potential use of biotechnology in upstream decisions about materials management, in addition to the status and directions of current applications. This symposium incorporated state-of-the-science, as well as risk and policy issues into the topics of biotechnology and remediation, waste treatment, environmental evaluation and monitoring, and versatility and future directions in biotechnology and sustainability. Achieving a sustainable environment cannot be accomplished by one discipline or one sector of society. Likewise, long-term environmental sustainability cannot be accomplished by one government. Sustainability must be a global effort as is evident from the multinational participation in these proceedings.

The importance of biotechnology, sustainability, and the environment is highlighted by the combined support that this discipline receives from industry and the government. In this regard, we acknowledge the foresight of the sponsors and supporters of this symposium: Eastman Chemical Company, E.I. DuPont De Nemours and Company, Inc., Procter and Gamble, IT Corporation, Dow Chemical Company, Science Applications International Corporation, Oak Ridge National Laboratories, General Electric, the Waste Management Research and Education Institute of the University of Tennessee, Biotreatment News, and the Bio-Cleanup Report. The efforts of the planning and steering committee as well as the symposium participants were essential for the execution of this successful symposium.

John Sanseverino
UT Center for Environmental Biotechnology

Kimberly L. Davis
UT Waste Management Research and Education Institute

CONTENTS

1

CHALLENGES AND OPPORTUNITIES IN THE AREA OF ENVIRONMENTAL BIOTECHNOLOGY

G. S. Sayler

Center for Environmental Biotechnology
Waste Management Research and Education Institute
University of Tennessee
Knoxville, Tennessee 37996

1. INTRODUCTION

The explosive growth of biotechnology over the past decade has been fueled by ever expanding development of the tools of molecular biology, permitting unpredicted ability to analyze, manipulate and apply living organisms and biological processes across the spectrum of health, industrial and environmental challenges facing society. Biomedical applications in particular have been driven by major federal initiatives such as whole genome sequencing, as well as potential multibillion dollar human and animal health markets for the products and knowledge resulting from biotechnical research. Environmentally, biotechnology continues to grow and develop but it is apparent that the driving forces behind these applications wll be changing in the coming decades.

While environmental biotechnology is broadly based in a variety of environmental protection, restoration, agriculture and industrial practices, for over a decade it has been most commonly linked with bioremediation and waste treatment technology. Research and applications have thus been largely driven by environmental policy and federal regulation and, too a much lesser extent, economics. However, in the last half decade, there has developed a major shift in these driving forces. Clearly, economics and cost benefit issues are replacing regulatory drivers as primary considerations in environmental waste management and restoration with both the potential positive and negative effects for environmental biotechnology R&D. Nowhere is this more evident than in approaches to natural attenuation of environmental contamination where intrinsic or native bioremediation is viewed as a major co-contributor along with physical-chemical processes in maintaining chemical exposure risks below a critical threshold, a threshold that would require aggressive and expensive remediation and cleanup.

The movement toward cost benefit and risk based approaches in environment protection and restoration, over contaminant concentration dependent cleanup standards, may have important implications for bioremediation. Such implications may include the use of

Biotechnology in the Sustainable Environment, edited by Sayler *et al.*
Plenum Press, New York, 1997

bioavailability as a determinant in cleanup, which could permit applications for bioremediation processes, which heretofore were noncompetitive because they could not achieve mandated cleanup levels. Likewise, natural attenuation as a cleanup option may produce applications for environmental biotechnology in the diagnosis, detection and monitoring of the progress of the cleanup action.

What is clear is that fundamental change is occurring with respect to environmental biotechnology. Rather than view the technology as only a way to reduce cost and liabilities associated with past environmental problems, we can now view the technology in a much broader market including those of value added products.

The global market for environmental technology in the broadest terms is large, measured in the billions of dollars, with some estimates approaching $500 billion annually early in the next century. The proportionate share of such a market for environmental biotechnology is not clear. However, if the breadth of environmental biotechnology is fully exploited, it is anticipated that a very significant fraction of this market is available. Conceptually, the field is in the midst of a significant transition. This transition is summarized as a move from a bioremediation focus to a greater emphasis on pollution abatement, and pollution prevention to environmental sustainability. Figure 1 represents this conceptual shift in terms of either research dollar investments or market potential over a time interval of 40 years. While absolute dollar values are difficult, if not impossible, to assess, the general trends are illustrative of the changing landscapes driving environmental biotechnology.

2. ENVIRONMENTAL BIOTECHNOLOGY AND WASTE MANAGEMENT

Research into management of past environmental contamination and remediation of those contamination problems has likely plateaued. As societies redefine what are acceptable endpoints of cleanup relative to land use objectives and the relative degree of risk

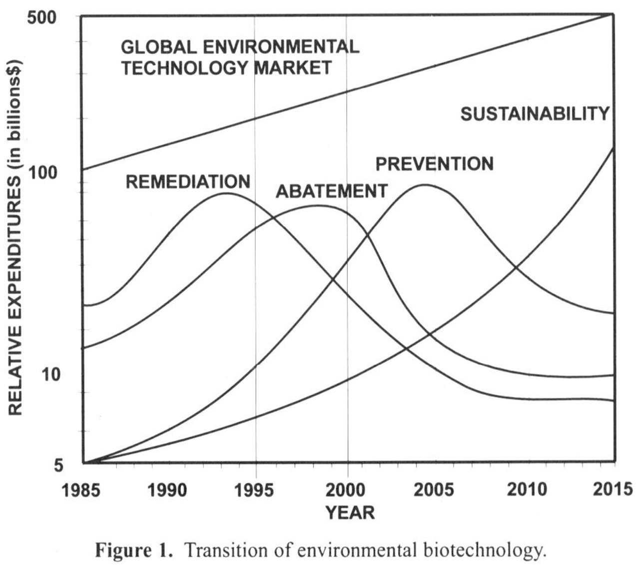

Figure 1. Transition of environmental biotechnology.

they are willing to accept per capital, it is anticipated that remediation will have a finite life. While future environmental contamination will doubtlessly occur, the frequency and scale of such contamination will hopefully lessen, at least for the industrialized world. Opportunities will continue to exist for biotechnology in remediation of environmental contamination problems and it is likely that these opportunities will be driven by cost and risk rather than by regulated cleanup endpoints. Consequently, environmental biotechnology may experience the growth of a competitive advantage where cleanup endpoints are defined relative to biological availability and exposure risk. In addition, environmental diagnostics and monitoring of long term remediations associated with natural attenuation may be an additional area of research and market opportunity.

While remediation technology may experience somewhat of a lessening imperative as we move into the next century, there remains a strong driving force for understanding natural biodegradation, transformation and immobilization of environmental contaminants, either from the perspective of predictive risk assessment or from cost effect non-aggressive treatment technology such as phytoremediation.

We may anticipate somewhat of a resurgence and growth in research and applications of biotechnology with respect to waste treatment and pollution abatement. These developments are broad ranging and may include a greater emphasis on monitoring and control of conventional technology such as activated sludge processes, development of niche specific biotechnology for individual waste streams, fugative emissions or stack gases; re-engineering of waste treatment facilities to deal with new component mixtures of domestic or industrial waste; or greater exploitation of biotechnology in solid waste management and anaerobic treatment technology.

3. POLLUTION PREVENTION AND RECYCLING

The environmental sector dealing with pollution prevention technology appears to be experiencing more rapid growth than generally acknowledged by the research and development community. This growth is due to awareness that pollution and generation of waste is bad business. Economically, generating waste and waste treatment are costs that need not be incurred in many cases. In addition, waste generation is often viewed socially irresponsible and unacceptable. The implication for environmental biotechnology is broad ranging. To some degree, pollution prevention can be considered an extension of waste treatment, such as power plant stack gas emission and recovery of valuable products e.g., sulfur, or silver recovery from photographic emulsion waste. However, in a much broader sense pollution prevention extends into the very essence of manufacturing and chemical processing to ultimately change product cycles and feedstocks to avoid the use of materials that generate problematic wastes or to create products that can largely be recycled or reused once the serviceable life of the product has been met.

When heavy industry such as automobile production seeks to make vehicles that are 90% or greater recyclable, it should not be a surprise that other direct consumer based companies are seeking compostable diapers or biochemical replacements for synthetic chemicals in their dishwasher and washing machine detergents, or that major chemical companies are seeking biologically based feedstocks and processes to replace petroleum and fossil fuel based feedstocks in the 21st century. Unlimited opportunities for environmental biotechnology are offered by such fundamental industrial changes. These opportunities range from bioprocess development for product recovery from waste streams, isolation and development of organisms and processes for chemical replacements and bioconversions of feedstock chemicals, environmentally stable enzyme technology, and chaperon organisms, enzymes and inhibitors for facilitating or modifying composting and product degradation rates in the environment.

4. ENVIRONMENTAL SUSTAINABILITY

All of these new driving forces effecting environmental biotechnology are ultimately encompassed by environmental sustainability. Environmental sustainability is operationally defined as utilization and management of natural resources in a manner that is consistent with their steady state availability for future generations.

With this definition in mind, it would appear self evident that applications for environmental biotechnology in bioremediation, pollution abatement and prevention are well-suited, if not fundamental, in sustainable resources management and protection of soil, water and air. Moreover as biological resources, the living organisms and genetic elements exploitable by environmental biotechnology are agents of and targets for sustainability. As genetic resources for future food, fiber, and feedstocks, microorganisms represent an untapped wealth of biological diversity. Based on an expanding database of phylogenetic diversity, the diversity of the microbial world exceeds that of all other living organisms combined. The extent that this phylogenetic diversity is reflected in gene sequence diversity and genomic complexity, to be exploited to serve society's needs, remains an open question. However, as the genomes of microbial species are sequenced and compared to ever expanding databases, the diversity of this biological resource will become more fully defined. The prospects for exploiting this resource, by recovery of organisms from the environment or by direct environmental recovery of genes, looms as a major development now and into the next century. Will microbial and genetic biodiversity be an inexhaustible source for fueling new development of environmental biotechnology? Or, is it a depletable resource that may be lost in part due to depletion of higher order biodiversity (plant and animals) to which it is undoubtedly linked? Such questions will likely be answered only by developments in analytical methods, such as gene probe arrays, which can be usurped for their direct applications in environmental analysis.

5. SUMMARY

Environmental biotechnology is in full-fledged research and development across a broad spectrum of applications contributing to environmental sustainability. These advances extend from bioremediation and protection and restoration of natural resources, using genetically engineered organisms, to exploiting biological diversity to recover microorganisms and products responsive to the changing driving forces of society and industry.

Developing analytical methods suitable for use in environmental biotechnology now permit recovery of genes expressed under field conditions, and soon will allow facilitated analysis of the vast genetic information available to us in the natural environment. These analytical tools will further facilitate improvements in environmental process monitoring and control to refine and create more efficient and effective waste treatment technology, as well as to diagnose the performance of natural and engineered remediation processes. With awareness of the broad application potential for environmental biotechnology, and growing markets available for the technology in the next century, it would appear that there is virtually unlimited opportunity in both research and enterprise.

2

GREEN TECHNOLOGY TRENDS

The Changing Context of the Environmental Technology Industry

Doug Miller

International Environmental Monitor Ltd.
Toronto, Canada

1. INTRODUCTION

The environmental technology business, like many others in these latter days of the 20th century, is vulnerable to contextual forces that are changing at an unprecedented rate. Without understanding the trends in the immediate social and political context, our success will be limited and environmental progress will be slowed.

International Environmental Monitor Limited's expertise is in applying quantitative survey research techniques to reliably monitor changes in both public opinion and expert views related to environmental progress. A deep understanding of public opinion serves to anticipate the evolving social context and predict the "pull" part of the equation for environmental progress—the extent to which citizens will "vote for green" with their ballots and their wallets. Expert views are the "push" part of the change equation in that they are key influencers on those making decisions in both public and private organizations. A good understanding of experts' changing view of priorities, drivers, and markets is perhaps the best predictor of the immediate future. Green technology companies can then develop business strategies that take full advantage of their evolving business context.

This paper outlines some of the trends effecting the environmental biotechnology industry.

2. THE PUBLIC ENVIRONMENT

Public expectation for continuous progress towards achieving environmental sustainability continues to be strong across all industrialized countries.

Environmental concerns remain high despite ongoing political preoccupation with employment and the economy. Large numbers of people reject the suggestion that environmental protection must be traded-off for economic gains. A 1993 public opinion survey by the University of Chicago showed that in developed countries like Canada, Norway,

New Zealand, West Germany, Israel, the Netherlands and Russia, over 50 percent of the public disagreed that "we worry too much about the future of the environment and not enough about prices and jobs today," with near-majorities in other industrialized countries disagreeing with the statement.

The key reason for this enduring concern appears to be ongoing fears over the impacts of pollution on human health. In 1992, a Gallup survey of 23 countries around the world found that majorities in every country felt that their health had been affected by pollution. In Canada, recent Environmental Monitor surveys have found that concern about the health effects of pollution has continued to increase steadily since 1992. Today, almost two in three Canadians believe their health has already been affected by pollution.

Increasing health concerns are seen as important precursors of a resurgence in top-of-mind environmental concerns in developed countries. Combined with recently observed increases in criticism of the environmental performance of industries and governments, leads International Environmental Monitor Limited to predict another "green wave" of public concern to emerge in many countries within the next five years.

In summary, the growing body of international social research reveals that the environmental concerns of citizens are linked to deep worries about the health of current and future generations, as well as the long-term viability of global economic and ecological systems. It is clear that the depth and strength of these perceptions will continue to press governments and industries for continuous environmental improvement to ensure sustainability. Since governments and industries will need to respond to these societal pressures, green technology companies can expect continuing overall market growth.

3. PROGRESS TOWARDS SUSTAINABILITY

Our twice-yearly GlobeScan surveys of sustainable development experts across Organization of Economic Cooperation and Development (OECD) countries reveal expert views that mirror public expectations for continued environmental progress. Almost half the experts (drawn equally from industry, government, universities, environmental groups

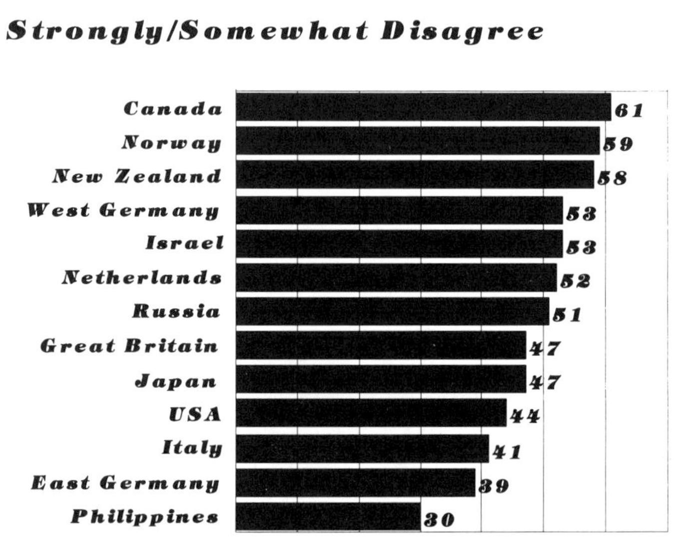

Figure 1. Prices/jobs more important than environment?

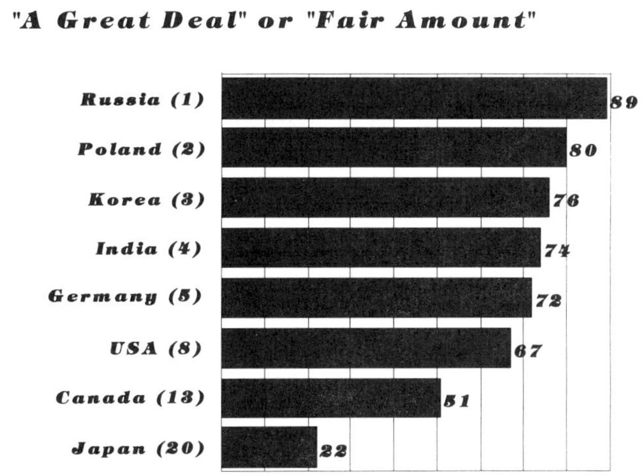

Figure 2. Our own health affected by pollution?

and consultants in the 26 most industrialized countries) say that the most concentrated progress towards sustainability will occur over the next decade (mostly between 5 to 10 years from now), with another one in four expecting to see steady progress over the next twenty years. These findings suggest the potential for a major expansion in the green technology market in the years just following 2000, for those able to successfully navigate the intervening years.

In fact, experts predict mega-changes to occur over the next decade. Fully two in three predict the widespread adoption of carbon/energy taxes, and almost as many expect sustainable forestry practices and tradable emission permits in their countries in ten years. Pluralities expect even more aggressive measures to be in place in the same time frame, including sustainable agriculture practices, a doubling of the average auto fleet efficiency, a ban of chlorine compounds, a 50% reduction in farm chemical use, and even a replacement measure for the GNP/GDP economic indicator. If these expert predictions come true,

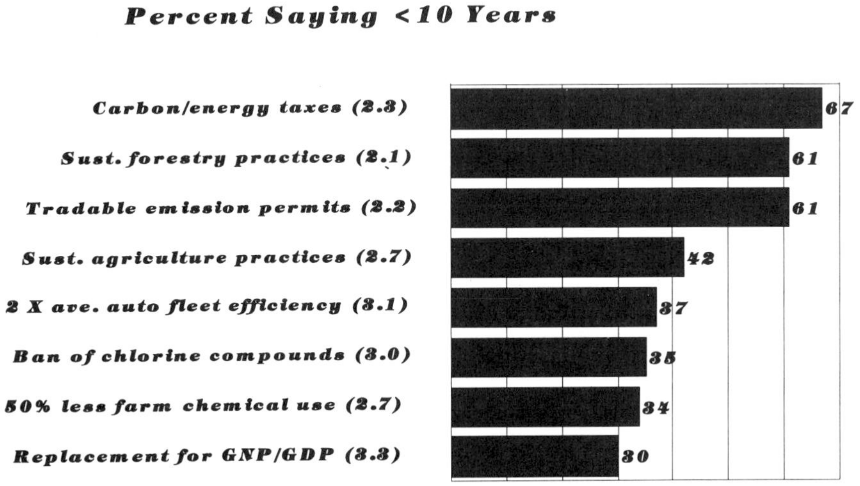

Figure 3. Predicted adoption of sustainability measures.

the next decade offers strong promise for green technology companies who are positioned to take advantage of upcoming changes.

4. CHANGING GOVERNANCE

While most experts and members of the public expect continued progress towards sustainable development, actual environmental progress and market opportunities for new green technologies require changes in societal and corporate behavior. The "mechanisms" used by governments to shape behavior around sustainable development have evolved significantly over the past decade. While ten years ago, "regulation" would have been the only mechanism mentioned by experts, the number of such tools that governments are applying has expanded significantly since then.

In fact, experts now see non-regulatory approaches like economic instruments, public environmental reporting, and voluntary or negotiated agreements as playing an even greater role than regulation in driving progress toward sustainability over the next five years.

Programs like the USEPA Toxic Release Inventory (TRI) program, which require companies to report publicly their emissions of certain toxic chemicals, are broadly recognized as being effective. Public pressure and embarrassment resulting from TRI reporting drove U.S. companies to reduce their toxic emissions by 33 percent within three years, and 50 percent within five years. One of the most recent adherents to public reporting is the Netherlands, which is advancing a law requiring its top 300 polluters to publicly report a whole range of emissions every year.

An equally potent industry-led driver of sustainability is ISO 14000. While the ISO 9000 series was about quality, the ISO 14000 series is about environmental stewardship and environmental management systems. The number of GlobeScan experts recognizing the significance of ISO 14000 as a driver of progress has increased notably over the past year.

The use of non-regulatory approaches to environmental governance are expected to increase over the next five years. This change suggests that environmental technology companies looking to the next new government regulation to provide market opportunities are missing important other markets. In order to get in on the "ground floor" of these new markets, technology companies are advised to develop strategic partnerships with multinational corporations and industry sectors that are on the leading edge of sustainability. Products engineered to fit the niche needs of a particular industry will stand the best chance of capturing a global market. Yet how does one pick the winners?

5. LEAD INDUSTRIES

GlobeScan experts were asked to rate how well different industry sectors were managing their transition toward sustainability. The chemical sector is viewed most positively, with 40 percent of experts rating its performance as "good." The next most positively rated sectors are electronics and packaging. Rated most poorly are oil and gas, agri-food, and automotive, with only one in ten experts viewing them in a positive light. The chemical and electronics sectors are rated highly because they have taken the lead and exceeded government regulations in their sustainability efforts. The chemical industry's Responsible Care Program, underway in 42 countries, applies comprehensive standards to the full life-cycle of chemical products. The electronics industry surpassed regulatory requirements by

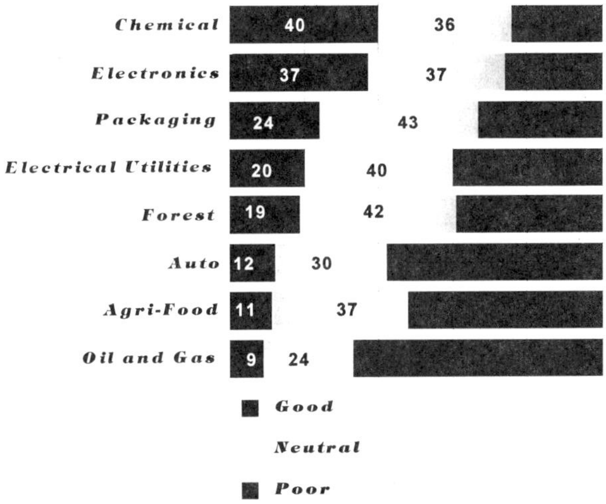

Figure 4. Industry sector performance: making transition to sustainability.

eliminating the use of ozone-damaging CFCs well ahead of the schedule specified in international agreements.

6. GREEN TECHNOLOGY DRIVERS

GlobeScan experts predict that the ongoing drive for "cost reduction" will continue to play a major role in industry's adoption of green technologies, requiring suppliers to demonstrate how their technologies lower overall system costs. Market demand, especially public and private sector green procurement policies, is also viewed as an important driver of green technology adoption as firms work to reduce their exposure to environmental liability, or to qualify for ISO 14000, or to turn environmental leadership into their competitive advantage. Experts also see full cost pricing of natural resources as a possible future driver. On the other hand, experts do not see international agreements and government-led green technology initiatives as significant drivers of green technology adoption and application by industry.

7. STAGES OF TECHNOLOGY

The biggest challenge for those in the green technology business is posed by the fact that, by its very nature, technology is constantly advancing. And may it ever be so! Only one in five of our experts believe that existing "off-the-shelf" technology will get us to a sustainable future. One of the challenges they point to is the successful and rapid commercialization of promising technologies currently at the research and development (R&D) stage. It is perhaps at this phase that governments can be most useful, either by helping to remove regulatory and other institutional barriers, facilitating the exchange of market information, or through promoting partnerships.

But to understand the most important dynamics of the green technology market, one needs to be introduced to a practical framework developed by the International Institute for Sustainable Development (IISD). This framework identifies four successive generations or stages of green technology development:

Stage 1: Remediation
Stage 2: Abatement
Stage 3: Pollution Prevention
Stage 4: Sustainable Technologies

Stage One technologies are focused on remediation, or cleaning up pollution after we have let it enter the environment. Because of ongoing problems with existing contaminated sites and the advantages offered by brownfields development, these technologies are going to have continuing importance in the near future. In particular, the remediation market in the Pacific Rim represents huge growth potential. However, since remediation technologies represent a rearguard approach to the environment, they represent more a near-term, than a long-term growth market. They are already losing ground as the major growth segment of the green technology market.

Stage Two technologies are focused on abatement or end-of-pipe solutions. Such technologies include scrubbers on smokestacks, catalytic converters on cars and liquid effluent treatment systems. The market for this generation of technologies will be increasingly limited to the "back of the wave" markets, which are willing to continue to incur the production cost of wastes captured at end-of-pipe. These markets are increasingly located offshore in developing countries and raise questions about the ethics of exporting old technologies that prevent emerging economies from leap-frogging past the dirtiest stages of industrialization.

Stage Three or pollution prevention technologies do not take industrial processes as a given, and focus on optimizing these internal processes to eliminate waste. In North America and Europe, most emphasis is currently on developing and adopting pollution prevention technologies. This generation of technology is typified in the forest sector, where old chlorine-based bleaching processes are being replaced with non-chlorine-based production methods that no longer require end-of-pipe chlorine abatement because there is no need to clean up pollution that is not created in the first place.

Stage Four represents sustainable technologies, the market opportunity of tomorrow. Just as pollution prevention technologies focus on optimizing internal processes, sustainable technologies seek to optimize the full life-cycle of a product from raw material to final resting place, not only from an environmental but from a social and economic perspective. Sustainable technologies grapple with what tend to be viewed as the externalities of production and represent a major paradigm shift. For example, there is a pilot plant in Canada that produces paper from grass. The grass is produced on small-scale farms using sustainable agriculture and natural organic systems to produce grass at a higher yield of paper per acre than trees. Also, the raw material has a lower lignum content so it requires much less bleaching. Production waste is plowed back into the fields to help produce more grass.

The great challenge of the green technology industry stems from the fact that each progressive stage of technology eliminates the need for the previous stages in similar applications. And that makes for very vulnerable and relatively short-lived markets. The answer, if you are a green technology company, is to continually repackage your technology offerings to keep up with the rapidly evolving marketplace, and focus your R&D efforts on more advanced stages of technology.

When GlobeScan experts in OECD countries are asked which of the four stages of environmental technologies will enjoy the most rapid market growth in their respective

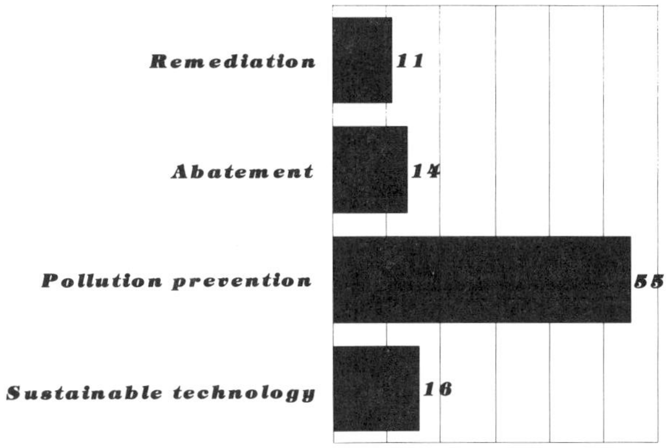

Figure 5. Fastest growing green technology markets over the next five years.

countries over the next five years, most point to pollution prevention. This means that a lot of abatement technologies will soon lose their markets.

While less than one in five experts believe sustainable technologies will grow the most over the next five years, it is interesting to note that this was a much more prevalent choice among experts from multinational corporations, indicating this type of whole system thinking has already gained currency among leading corporate thinkers, and represents a significant area of market opportunity for green technologies in the future.

8. CHALLENGES AND OPPORTUNITIES

While the worldwide environmental technology industry will continue to grow over the next decade, individual technologies are increasingly vulnerable to extinction as rapid technological advances, changes in governance and market opportunities push multinational corporations and other customers to embrace whole system thinking and leap-frog change. Pollution prevention and sustainable technologies are already making a number of abatement technologies obsolete; and expensive site remediation projects are increasingly being done in Asia (as the worst of the North American sites are cleaned up).

Successful green technology companies will be ones that constantly re-invent themselves and re-package their proprietary expertise to meet the changing needs of their customers. They will have the capacity to think "outside the envelope," seeing their technology as part of a system that includes their customers' supply chain through to waste disposal. Those able to match the whole-system thinking of leading multinational companies will be able to develop global strategic partnerships that will ensure their mutual success. These partnerships will focus not only on "draining the swamp" but on taking environmental protection from the cost side of the business equation, over to the opportunity side.

These fundamentally different end-points can be achieved in different ways. Rather than focusing on treating contaminated sites to produce stable, non-toxic residues in the environment, a remediation company may come up with a way of producing a marketable raw material from a common contaminant. Another firm might reverse the polarity of its knowledge of how best to break down a particular chemical compound, and develop a less damaging way to produce it in the first place, or develop a more benign formulation of the compound.

Sectors providing the greatest opportunity for the application of systems thinking include those already most advanced (like the chemical, electronics and packaging industries) and those faced with the greatest pressure for change. Some chemical companies are seriously considering going out of the business of selling chemicals and into the business of leasing molecules, thereby taking responsibility for the full life cycle. They need help to close some of their cycles so they can reprocess used chemistry into useful by-products and raw materials. The automotive sector in North America is under a great deal of pressure because of air quality and fueling issues. The packaging industry continues to be targeted to live up to its waste reduction targets. Agricultural systems are being reinvented.

There are many opportunities for environmental biotechnology companies to carve sustainable market niches in addition to their short-term applications in the bio-remediation field.

3

GREEN CHEMISTRY

Using Enzymes as Benign Substitutes for Synthetic Chemicals and Harsh Conditions in Industrial Processes

Glenn E. Nedwin

Novo Nordisk Biotech, Inc.
1445 Drew Avenue
Davis, California 95616

1. INTRODUCTION

Chemophobia is on the rise in the general public, and federal and state regulators continue to put pressure and controls on the industrial use of chemicals. From a more positive angle, industry itself has shown an interest in "sustainable industrial development." Sustainable industrial development refers to the concept that our future economic health depends on industry cutting pollution and using fewer natural resources. As a result, public, regulatory and private industrial forces are all fueling a conversion to greener alternatives by chemical producers and industrial manufacturers (Illman, 1994).

Since the beginning of the Industrial Revolution in the 1700s, industry has not consistently paid attention to the environmental impact of its operations. This inconsistency has grossly intensified since the early 1900s due to a number of complicating factors, including competition driven by capitalism, technological advancements and the need to sustain an increasingly rapid growth in population. Today we are faced with an inordinate number of toxic waste sites due to industrial carelessness and little attention has been paid to energy and resource efficiency. Historically, most known synthetic reactions were developed primarily on the basis of product yield, with little or no regard for the toxic nature of the starting materials, catalysts, solvents, reagents, byproducts or impurities. Further, in many instances there was an incomplete technical understanding of the process, no stated set of emissions or waste parameters for operating, lack of detection systems, and in some cases, instances where arrogance, neglect and greed took over.

As a result, there are enormous technical challenges to correct the past environmental disasters. Moving ahead in an environmentally responsible manner, industry on a world-wide basis must meet these challenges. As the world population continues to grow at an ever rapid pace, industry must produce enough goods, service and employment to meet these needs, but the dilemma is that at the same time industry must preserve the en-

Biotechnology in the Sustainable Environment, edited by Sayler *et al.*
Plenum Press, New York, 1997

vironment. Simply stated: we have only one environment. There must be a conversion to using safer, cleaner and more resource efficient industrial processes. Furthermore, industry must produce products that are more energy efficient, less polluting and less hazardous. Only through a serious commitment to sustainable industrial development will industry be able to operate and continue to meet the needs of society.

Our problems cannot be solved by technology alone. However, technology, and more specifically, biotechnology, does have an important role to play. For example, in a number of instances, microorganisms (through enzymatic degradation or conversion) are being used to clean up hazardous waste sites. The frequent heterogenous nature of toxic waste sites (and therefore the heterogeneous nature of the enzymatic substrates) requires the use of whole microorganisms instead of pure or monocomponent enzymes. In addition, unfavorable economics prevent the use of pure enzymes in bioremediation. However, enzymes do have an important and growing impact on a number of environmentally important issues regarding sustainable industrial development. Enzymes are being used in industrial process development and manufacturing, thereby circumventing the harsh chemical and toxic loads normally put into the environment. Too often water has been viewed as an unlimited and inexpensive resource and as a convenient diluent for discharge of toxic byproducts, instead of a precious limited and vital natural resource. In the future, industrial manufacturers must pay greater attention to accounting for all of its waste products and where possible, recycling its water. Enzymes in many different applications can mitigate this problem, being used in lieu of harsh and toxic chemicals. The beauty of enzymes is that they do their work, become inactive, and break down into simpler, non-toxic, natural components.

Enzymes are used today in many manufactured goods we use and consume (Figure 1). Most catalytic industrial processes operate efficiently only at high temperatures, high pressures or in highly acidic or alkaline conditions. Safety and environmental problems are often the result under such extreme physical conditions. Enzymes can replace harsh chemicals and harsh conditions (Figure 2), working best at mild temperatures, mild pressures and neutral conditions. Enzymes are highly specific, resulting in more pure products with fewer side reactions, so harmful processes which endanger the environment can be replaced by environmentally safe biological enzymes. Replacing harmful chemicals with environmentally safe biological enzymes is making many industrial processes cleaner and less expensive. The following is an overview of how enzymes are reducing the chemical loads in industrial processes and how enzymes can be used to eliminate the strain and toxicity on the environment, thereby contributing to sustainable industrial development.

2. ENZYME MARKETS

Today, the estimated worldwide industrial enzyme market is valued at approximately $1.1 billion. The industry segments (approximate market share) where enzymes are used today are the detergent (45%), textile (14%), starch (13%), baking, brewing, wine and juice, alcohol, food functionality, dairy (combined food and beverage, 18%), animal feed, personal care, pulp and paper, leather, fine chemicals and the fats and oil industries (all others combined, 10%) (Figure 3). The largest enzyme segments consist of proteases, amylases, and cellulases.

While enzymes are widely used in industry, they make up only a small percent of the overall chemical market. This is due to a number of reasons including, a) the economics of the enzyme manufacture - where price competition with the competing chemical technology requires enzymes to be made in the multi-gram per liter amounts in the fermentation

Detergent Industry
- Degradation of protein, starch, and fatty stains in laundry
- Color clarification and softening of cotton laundry
- Automatic dishwashing
- Surfactant production

Textile Industry
- Polishing of cotton fabrics
- Stonewashing of denim garments
- Degumming of silk
- Bleach clean-up
- Removal of starch from woven materials

Starch Industry
- Production of dextrose, fructose and special syrups for the baking, confectionery and soft drink industries, among others

Baking Industry
- Starch, pentosan, and protein modification for improvement of the baking properties

Brewing Industry
- Degradation of starch, protein, and glucans when brewing with a combination of malt and unmalted raw materials, e.g. barley, corn and rice

Wine and Juice Industry
- Degradation of pectin when manufacturing fruit juices, wine, etc.

Alcohol Industry
- Degradation of starch into sugars which are converted to alcohol through fermentation

Food Functionality Industry
- Improvement of nutritional and functional properties of animal and vegetable proteins
- Process of optimization, e.g. energy savings by lowering of viscosity

Dairy Industry
- Curdling of milk
- Conversion of lactose in milk and whey into sweeter, more easily digestible sugars
- Flavor development in specialty cheeses

Animal Feed Industry
- Degradation of feed components for improvement of feed utilization and nutrient digestion

Personal Care Industry
- Biotechnological ingredients for personal care products

Pulp and Paper Industry
- Control of pitch problems caused by the use of mechanical pulps
- Reduction of chlorine consumption in pulp bleaching process
- Viscosity control in starch-based coatings

Leather Industry
- Soaking of hides and skins, unhairing, bating and defatting

Fats and Oils Industry
- Modifications of fats and lecithins, and synthesis of esters

Biocatalysis
- Synthesis of organic compounds

Figure 1. Enzyme market segments/applications.

Industry Segment	Enzymes	Chemicals / Process Replaced
• Detergents	• Lipases, proteases, cellulases, amylases	• Phosphates, silicates, high temperatures
• Textile	• Amylases, cellulases, catalases	• Acid, alkali, oxidizing agents, reducing agents, water, pumice, energy, new garment manufacture
• Starch	• Amylases, pullulanases	• Acids, high temperatures
• Baking	• Amylases, proteases, xylanases	• Emulsifying agents, sodium bisulfate
• Pulp & paper	• Xylanases, mannanases	• Chlorine, toxic waste
• Leather	• Proteases, lipases	• Sulfides, high temperature
• Biocatalysis	• Isomerases, lipases, reductases, acylases	• Acids, organic solvents, high temperatures

Figure 2. Enzymes contributing to sustainable industrial development.

Enzyme Markets

Segment	Market Share
Detergent Laundry Automatic Dishwashing Industrial Cleaning	45%
Textile Desizing Biopolishing Biostoning	14%
Starch	13%
Food & Beverage Baking Brewing Wine Juice Alcohol Dairy Food Modification	18%
Other Industries Animal Feed Personal Care Pulp & Paper Leather Fine Chemicals Fats & Oils	10%

Figure 3. Enzyme markets.

process, b) the lack of understanding, in many industry segments, of enzymology - knowledge about basic chemistry mechanisms and how enzymes might be used to solve the problem is limiting in many applications-oriented fields, c) the difficulties in incorporating the enzyme into many "older" manufacturing processes - capital intensive equipment and "the old way of doing things" may become a hindrance, and, d) finding the right enzyme - each industrial process has its own conditions for operation and it is important to find the best enzyme for each application. With this background, several examples will be discussed where enzymes are being used, and where they might be used in the near future, to provide a more environmentally friendly solution to industrial processes.

3. ENZYMES CURRENTLY CONTRIBUTING TO SUSTAINABLE INDUSTRIAL DEVELOPMENT

An overview of a number of enzymes contributing to sustainable industrial development is given in Figure 2.

3.1. Starch Processing

One of the most well known applications of enzymes is in the starch processing industry (Lloyd, 1984). Enzymes replaced acids and heat as processing aids in starch liquefaction and saccharification about 50 years ago. In the 70s, enzymes also replaced alkali in the isomerization of glucose to fructose. Converting starch to dextrose, in the traditional process, was achieved by acid treatment at pH 2, followed by 140°C heat treatment (Figure 4). A heat stable alpha-amylase (Novo BAN®, a bacterial amylase), operating at up to 90°C was first used to replace the acid treatment. This enzyme, which acts randomly on

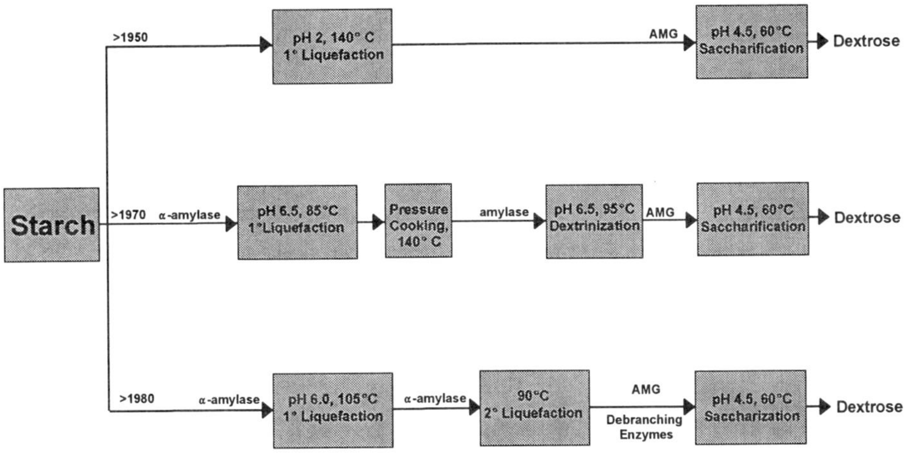

Figure 4. Enzymatic starch processing.

α-1,4 linkages in amylose and amylopectin, required multiple dosing as the enzyme was inactivated at the higher temperatures. Later developments included the use of even more thermostable amylases (Novo Termamyl®) derived from *B. licheniformis* enabling reaction temperatures to reach 110°C for short periods. Today, the process may involve several different enzymes, depending on the final starch product required, including fungal and bacterial amyloglucosidases and debranching enzymes, such as pullulanase, which acts on α-1,6 linkages in the branched dextrin chain. The end result is a higher yield, coupled with a more energy efficient process and less toxic waste stream.

3.2. Detergents — Laundry

Detergents represent the single largest industry segment for the use of today's enzymes, representing about 45% of the total enzyme market. Enzymes used in detergents today include both bacterial- and fungal-derived proteases, amylases, cellulases, lipases (Kochavi, 1993; Malmos, 1990); others are in the R&D pipeline. The single largest selling industrial enzymes are the detergent proteases, which are the bacterial-derived subtilisin type proteases. In addition to the wild-type enzyme, a number of protein engineered variants have entered the market place, with such properties as increased stability towards oxidation and increased cleaning performance. Alpha-amylases have proven useful in catalyzing the degradation of starch stains and in improving overall soil removal by hydrolyzing the starch which binds particulate soil to the fabric. Using both proteases and amylases, the overall performance of the detergents is increased at lower washing temperatures with a decrease in the detergent's overall chemical makeup. Cellulases can be used in lieu of cationic surfactants (fabric softeners) as they improve the softness of cotton fabrics. Cellulases are also active in removal of particulate soil by removing microfibrils from the cotton fibers. As a result there is a color brightening effect of the cellulase treatment. Lipases catalyze the hydrolysis of triglycerides present in fatty stains, making them more hydrophilic and more easily removable during the wash.

Enzymes used in detergents have a positive environmental impact, as they save energy by reducing washing temperatures, allow the content of undesirable chemicals in detergents to be reduced, are totally biodegradable (leaving no harmful residue), have no negative impact on sewage treatment processes, and most importantly do not present a risk to aquatic life. Chemicals eliminated or reduced by the use of enzymes in detergents include phosphates, surfactants, carbonates, silicates and chlorine bleach.

From an environmental viewpoint there are a number of good reasons to use enzymes in detergents. One of the most important is that enzymes save energy. Calculations show that if a population of 5 million people (Denmark's total population) were to reduce their wash temperatures by 20°C, from 60°C to 40°C, the energy saving on heating the wash water would correspond to around 37,000 tons of coal, while the energy used to make the enzymes would account for less than 250 tons of coal (Novo Nordisk, 1991). These savings figures are geographically specific as average washing temperatures differ culturally.

3.3. Textiles — Desizing

Starch is usually coated onto cotton to give it strength in the mechanical weaving process. The starch must be removed before finishing the fabric, having been traditionally been carried out using acid treatment. Enzymatic desizing using amylases, eliminating the acid reaction, dates back to the early 1900s. Amylases work efficiently in the desizing process and without the environmentally hazardous waste treatment problems associated with harsh chemicals such as acids, base and oxidizing agents.

3.4. Leather Processing

The production of leather is one of the oldest applications of industrial enzymes. Traditionally, steps include curing, soaking, dehairing and dewooling, bating and tanning. The process involves the use of numerous chemicals and unpleasant working environments, in order to remove hair, fat and unwanted protein such as elastin, keratin, albumins and globulins, leaving the collagen fibers intact prior to the tanning of the hides and skins. Traditionally, dog or pigeon feces was used in the early bating steps to make the leather more pliable. The feces bating owed its softening effects to the action of proteases. In 1908, the first standardized enzymatic bating process was patented and was based on using animal pancreatic enzymes. Today, bacterial and fungal proteases, mammalian trypsin (a by-product of insulin production) and lipases are used in leather processing (Muthukumaran, 1982). The results are a reduction in the use of sulfides, lime, non-ionic surfactant compounds and organic solvents. The benefits are in a more standardized, well defined process, with advantages with respect to the final characteristics and properties of the leather itself.

Enzymes are being used more and more in industries that previously applied only chemical processes.

3.5. Pulp and Paper Manufacturing

Paper is comprised of wood fibers derived from pulp produced by chemical and/or mechanical means. The world-wide production of pulp raw materials is approximately 70 million tons per year, corresponding to a market value of pulp and paper chemicals of more than $5 billion. In the manufacture of white paper derived from pulp there is a strong desire to reduce (and eliminate) the use of chlorine and other chlorinated organic compounds as bleaching agents. This desire stems from tougher and tougher federal, state and local regulatory and public environmental pressure to create fewer pollutants (Thayer, 1993; *Chemical & Engineering News,* 1993). A key concern in the reduction of chemical use is to provide the same high quality of brightness of the final paper product without losing paper strength. Enzymes are being used to enhance the bleachability of the chemically derived pulp.

The purpose of processing wood and bleaching the pulp is to remove lignin (delignification). Lignin is a natural resinous adhesive, which causes the yellow-brownish color. The goal is to reduce wood to cellulose fibers and to do so with as little damage as possible to the other wood constituents such as hemicellulose and, most importantly, the cellulose itself. Figure 5 shows the chemical pulping process with the enzymatically aided bleach boosting process. Pulp production begins with debarking and chipping wood logs. This is followed by a caustic sodium hydroxide cooking wherein the main part of the lignin is dissolved and then washed away. The result is a darkly colored pulp which has to be bleached in order to obtain white pulp for paper production. Chlorine and chlorine dioxide are the most common bleaching chemicals used in multiple stages of pulp bleaching, coupled with alkaline extraction of the dissolved lignin.

Environmental pressures have been mounting over the past 15 years on the control of toxic dioxins and chlorinated organic compounds in paper mill effluents. Regulators have been pressuring industry to move away from using chlorine based chemicals and towards other alternatives, such as the use of hydrogen peroxide, oxygen, ozone, and more recently, the use of enzymes. Hydrogen peroxide, oxygen, and ozone have the disadvantages of being too expensive, requiring costly capital investments and resulting in the loss of paper strength, respectively. A lot of work has been devoted to finding alternative bleaching agents and alternatives have been examined to supplement the delignification

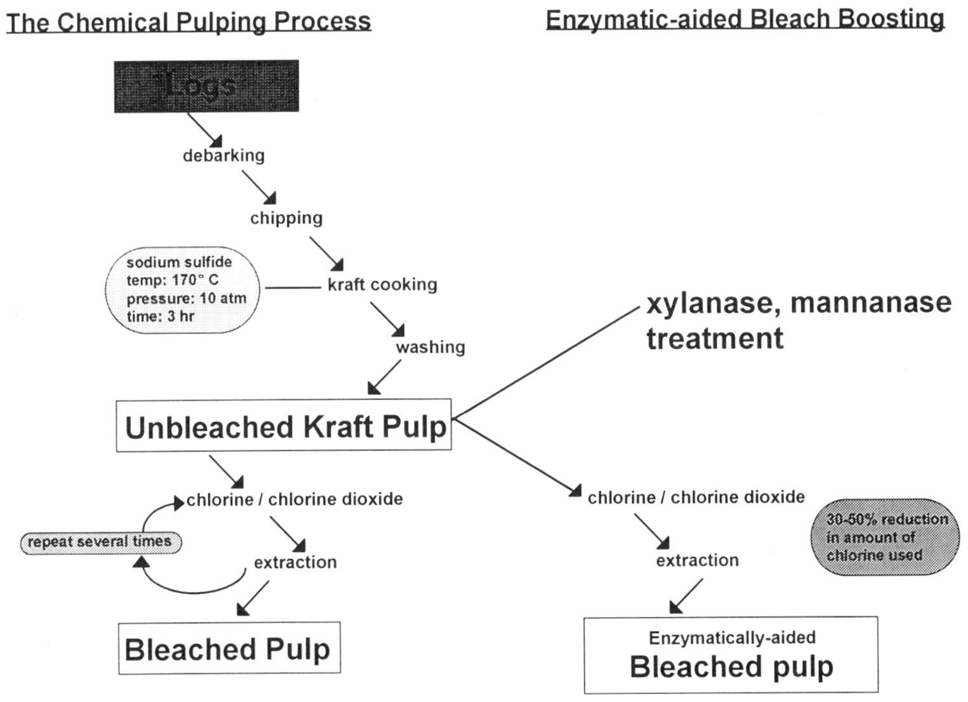

Figure 5. The pulping process.

stage. The application of xylanases and mannanases has turned out to be useful in removing part of the lignin after the alkaline cooking process and before the bleaching step (Patel, 1993; Elegir, 1995).

In a typical xylanase procedure, a pH adjustment of the pulp from 10–11 to 7–9 is necessary. (The temperature is about 55–65 °C). After the pH adjustment, xylanase is mixed into the pulp which is kept in a holding tank for about 2 hours. Mannanases may also be added in some cases for additional bleach boosting effects. Due to the significant lignin removal from the enzyme treatment, much lower consumption of chlorine (30–50% reduction) is required to bleach the pulp further (depending on the type of wood and particular enzyme(s)). In addition, the enzyme treatment offers improved bleachability, reduction in the bleaching costs, less chlorinated organic compounds in the effluent and higher brightness in the paper. The economic benefits include very simple equipment needs and savings in chlorine compounds. These benefits can easily pay for the enzyme treatment; not to mention the supplementary beneficial effect on the environment. The cost of chlorine bleaching is about $16 per ton, whereas enzyme treatment costs are about $7 per ton.

Although the mechanism of enzymatic bleach boosting is not fully understood, it appears that xylan is solubilized during the pressurized alkaline cooking process. As the pressure, temperature and pH drops, xylan precipitates on the wood fiber surfaces. Some lignin is chemically bound to the xylan, which may become released by the action of the xylanase. The precipitated xylan may create a physical barrier for the bleaching chemicals to get into the fibers and block the extraction of lignin and other chromophores. The paper fibers become more susceptible to bleaching after xylanases selectively remove xylan from the surface and the pores of the fibers, releasing the darkly-colored chromophores.

The use of xylanases in bleach boosting represent an example of the need to develop the right enzyme, which has taken time, as well as collaborative efforts with several pulp mills. It took some time for microbial screening efforts to identify a xylanase that most closely met the conditions of the pulp mill. This experimentation took into account both the physical constraints of the pulp capital equipment and the operating conditions of the pulp process. The pH optima curves for three different xylanases is shown in Figure 6. A first fungal xylanase was identified from *Trichoderma reesei*, named Pulpzyme HA® (Novo Nordisk A/S). One shortcoming of Pulpzyme HA® was that it worked only up to pH 7 and at temperatures below 55 °C (131 °F). These reaction requirements made it difficult to adjust the mill conditions to suit the enzyme. The Pulpzyme HA® preparation also had minor amounts of cellulytic activity which was a further shortcoming of widespread use of this particular enzyme. A second xylanase was then developed in 1992, Pulpzyme HB®, a bacterial enzyme derived from *Bacillus pumilus*, which was an improvement in its pH and temperature activity optimums. Its pH optimum is around pH 7. Further development work was focused on producing xylanases that are better adapted to overall mill conditions. Pulpzyme HC®, a xylanase derived from an alkalophilic *Bacillus sp.*, is the outcome of research and development work and was launched as a product in August, 1994. Pulpzyme HC® is an alkaline, thermo-stable xylanase that works best at pH 6.5–9.5 and 40–65 °C (104–149 °F). The ability to operate at higher pH and temperature also makes Pulpzyme HC® more capable of withstanding fluctuations in process conditions (Novo Nordisk, 1995).

The benefits of using xylanases, and mannanases, in paper manufacture include reduced need for bleaching chemicals, giving economical and environmental benefits. In addition, they reduce the amount of chlorinated organic substances in the paper mill's effluent based on a low capital investment compared with alternative oxygen and/or extended delignification. Further, by making better use of a limited amount of bleaching

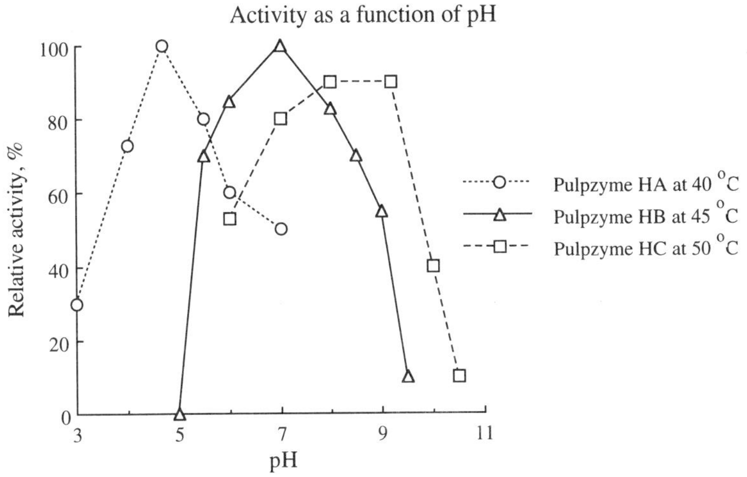

Figure 6. The development of xylanases for bleach boosting.

chemicals, production throughput can be increased. The longer term goal is to define an enzymatic process that completely eliminates the need for any added chlorine bleaching.

3.6. Baking

Six billion loaves of bread are baked in the U.S. alone each year. Approximately 600 million to over 1 billion of these loaves become unsalable due to staling and mold spoilage. Research into bread staling has had a long history, with no good answers, until recent developments in the use of enzymes, such as amylases, to retard the process.

Many bakers employ the use of chemical emulsifiers, such as monoglycerides to retard staling. The emulsifying agents make a softer bread and reduce the rate at which bread becomes firm (stale). Enzymes offer a more natural way to prevent staling and eliminate the need for emulsifiers (see also Section 4.3). Amylases can be used as natural anti-staling baking ingredients, as they reduce the need for chemical emulsifier additives (Novo Nordisk, 1995). Amylases have been shown to significantly increase the freshness of bread compared to results using emulsifying agents. Figure 7a shows the freshness of untreated bread drops in 48 hours to about 20% of its original level. However, the bread retained greater than 80% of its original freshness at the end of 48 hours when the dough was made using Novamyl® (Novo Nordisk A/S), a bacterial alpha-amylase. Figure 7a also shows the synergistic effect of Novamyl® plus emulsifiers such as monoglycerides. Figure 7b shows that the freshness of the bread increases with increasing dose of Novamyl®. In fact, bread made with the Novamyl® is more likely to spoil from mold than from staling. Novamyl® retards the staling process without affecting the handling of the dough. The use of amylases can also reduce wastage, reduce stale returns, expand sales-distribution ranges, reduce production costs and increase the baker's profitability.

Several theories exist regarding why bread stales. Bread is primarily comprised of starch (amylose and amylopectin) and a heterogeneous mixture of gluten proteins. In fresh bread, starch is mostly in an amorphous state. A modification in starch structure plays a major role in the staling of bread (*Food Engineering*, 1989). Starch molecules tend to revert from an amorphous state to a less hydrated, crystalline state. As this happens, water is

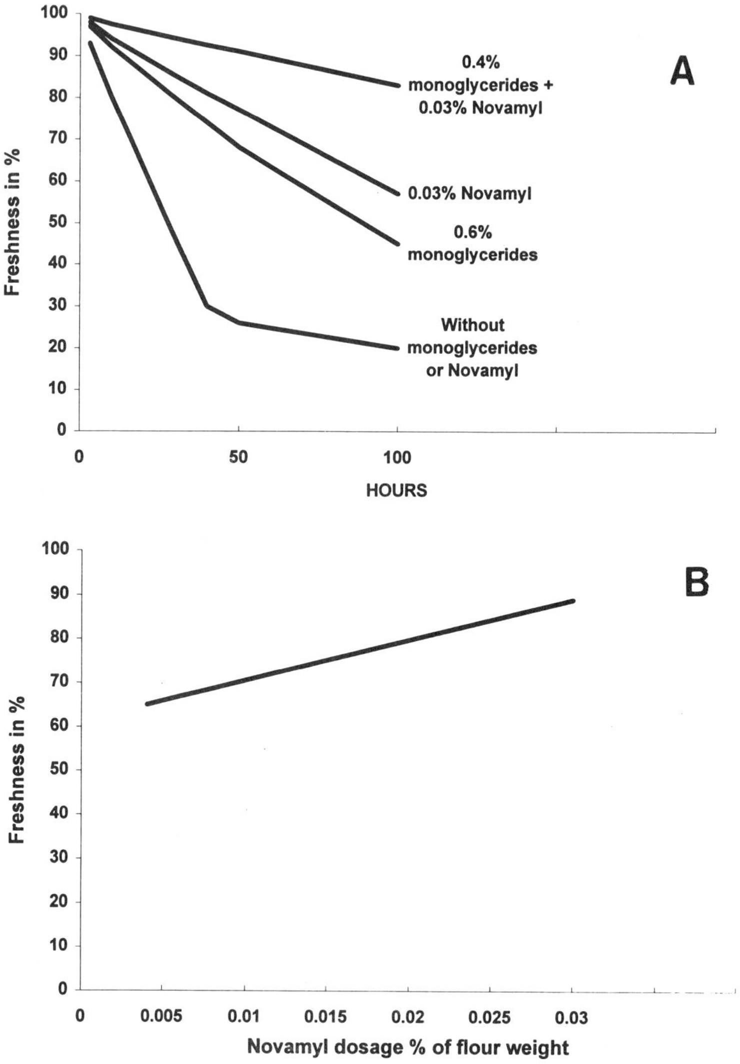

Figure 7. (A) Effects of Novamyl® on freshness of white bread made with and without emulsifiers. (B) Bread freshness: dose response of Novamyl®.

released from the starch and is presumably absorbed by the gluten proteins. Long linear amylose chains, extending through several amorphous regions of the bread, interact with the crystalline regions (which may involve as much as 15% of the starch) making it rigid, affecting the entire bread.

Novamyl® is a very pure amylase and it is believed this amylase interferes with the ability of starch to crystallize. Novamyl® produces maltotriose and maltotetraose (without producing the intermediate dextrins that cause gumminess), which may also retain part of the water, keeping the bread moist. Besides reducing wastage and the need for additives, a longer shelf life helps bakeries cut costs by increasing the interval between deliveries. Considerable transportation and labor costs involved in rotating bread stocks in the vast number of supermarkets and thrift stores are therefore reduced.

3.7. Detergents — Automatic Dishwashing

Automatic dishwashing detergents have traditionally contained high concentrations of phosphate builders (sodium tripolyphosphate), silicates (sodium metasilicate) as high pH buffers, and chlorine (sodium dichloroisocyanurate) as bleach. Safety, environmental and performance concerns have shifted manufacturers to begin using enzymes in automatic dishwashing detergents. Phosphates were causing eutrophication of surface water; silicates, while being added to minimize corrosion of metals and ceramic dish surfaces, were too alkaline and posed a health threat in some European countries if ingested; and chlorine-based chemicals were reacting with organic compounds. Enzyme use, such as with proteases, lipases and amylases, has thus lowered the requirement of these chemicals, generating "greener," safer and higher performing automatic dishwashing detergents.

3.8. Textiles — Fabric Dyeing

In the textile and fashion industries, product appearance is everything, so fabric dyeing must be done right. Prior to textile dyeing, cotton fibers are bleached in order to provide a uniform whiteness. The bleach must then be completely removed from the fabric, otherwise it will interfere with subsequent dyeing steps. Bleach removal is achieved by either several rinses in water or by neutralization of the bleaching chemical by a reducing agent, followed by rinsing in water. For example, in some situations approximately 40 liters of water are needed to rinse the bleach from one kilogram of fabric. When an enzyme such as catalase is used, the fabric requires no rinsing with water at all (Novo Nordisk, 1994). The enzyme can be added and after about 15 minutes the dye can be added to the fabric. In the case of the non-enzymatic reaction, the textile manufacturer will often overdose the reducing agents, such as sodium thiosulfate or sodium bisulfate, to be sure of neutralizing the bleach. These reducing agents can also interfere with the subsequent dyeing steps and must also be rinsed away. In the traditional process, energy consumption is also required as high temperatures are needed in order for the reducing agents to react efficiently with the hydrogen peroxide. Catalase treatment therefore has a number of environmental advantages including lower water and energy consumption, elimination of sulphate and nitrogen salt compounds in effluents and the breakdown of hydrogen peroxide into its natural components of water and oxygen. In addition, the process is more controllable and shorter in duration.

3.9. Textiles — Denim Processing — Biostoning Jeans and Other Denim Garments

About 1 billion pairs of jeans are produced each year, as well as a significant number of other denim garments (shirts, jackets). A large percent of these denim garments are abraded and faded, previously by such methods of stonewashing, to give a "washed out" and/or softer look to the garment. Stone washing has traditionally involved "beating up" the garments in large washing machines in the presence of pumice stones, typically 1–2 kilograms of stones per pair of jeans. Bleach is often added when a lighter color is the intent. Neutral or acid cellulases have been an excellent replacement for stonewashing of denim garments, a process referred to as "biostoning" (Pedersen, 1992). A cupful of enzyme can replace about 100 times its weight of pumice stones, thereby preventing pumice particle dust in the manufacturing plant and preventing clogged drains by spent pumice sand.

Using cellulases as a substitute for pumice stones also prevents the damage to wash-machines and the garments, eliminates the need for disposal of the used stones, improves the quality of the wastewater and eliminates the need for labor-intensive removal of dust from the finished garments. Without the use of stones, the garment load can be increased by as much as 50%, thereby increasing throughput. Neutral cellulases have an advantage in the process of biostoning because of the reduction in backstaining and a broader pH profile. Backstaining is a result of the released indigo dye, from the denim, being re-deposited onto the surface of the fibers.

3.10. Surfactants

Surfactants are amphipathic molecules with hydrophobic and hydrophilic moieties. They are used in a number of cleaning applications including washing powders, fabric softeners, shampoos and laundry detergents. They are also used in foods as emulsifiers, personal care products (lotions), specialty chemicals and in other products. Their structural characteristics, enabling them to adhere to surfaces and thereby reduce surface tension, allows them to solubilize or disperse otherwise insoluble substances. They also have the capacity to alter the foaming properties of solutions. Surfactants until recently, were primarily derived from hydrocarbons from petroleum products, e.g., fatty alcohol carboxylates, sulfonates, and sulfuric acid esters. Their manufacture has typically relied on organic chemistry requiring solvents such as pyridine or dimethylsulfoxide. Enzymes have not done well in the more traditional methods of hydrocarbon based syntheses due to low yields. More recently, enzymology has shown that lipases can be used on commercial scale to make a particular type of surfactants, namely glycolipids (Sarney, 1995; Kirk, 1991). Glycolipids are fatty acid esters of carbohydrates. Glycolipids, such as ethyl glucoside fatty acid esters, may replace alcohol ethoxylates in household detergents, resulting in better performance at lower temperatures. Glycolipids also show superior performance to petroleum based surfactants in oil removal. Glycolipid surfactants are quickly biodegraded into natural sugars and fatty acids in the environment and are non-toxic. The enzymatic synthesis of glycolipid surfactants has shown to be an environmentally-friendly, commercially viable, solvent-free process.

3.11. Biocatalysis

The role of enzymes in the manufacture of organic chemicals is becoming more established, (Turner, 1995). Examples include the isomerization of glucose to isoglucose by glucose isomerase, beta-cyclodextrin production using cyclodextrin glucanotransferase, carnitine production using beta-ketoreductase and Aspartame production using the protease, Thermolysin. The use of enzymes is particularly important when chiral molecules must be distinguished. In the production of amino acids, the conversion of racemic mixtures to the chirally-pure desired product is being carried out using amidohydrolases and hydantoinases. The production processes of medically important antibiotics, penicillin and ampicillin, both benefit from enzymatically-aided synthesis in lowered chemicals (inorganic and organic solvents) usage, waste and reduction in energy. Biocatalysis offers significant advantages over equivalent chemical processes in allowing more benign pHs, temperatures and pressures to by used. In addition, the enzymatic route of synthesis results in fewer undesirable byproducts and less waste products.

Industry Segment	Enzymes	Chemicals / Process Replaced
· Paper recycling	· Cellulases	· Sodium hydroxide, sodium silicates, chelating agents, organic solvents, high temperatures
· Pulp & Paper	· Xylanases, mannanases	· Chlorine
· Wood / Fiberboard	· Laccases, peroxidases	· Formaldehyde glues
· Food Functionality	· Lipases	· Organic solvents, emulsifiers, high energy, toxic waste
· Baking	· Oxidases	· Bromates, azodicarbonamide
· Personal Care	· Isomerases	· Sulfur chemicals
· Coal & Oil	· Oxidases	· Acids, high energy desulfurization methods
· Waste Water Treatment	· Laccases	· Toxic burial
· Bioremediation	· Oxidases	· Toxic burial

Figure 8. Enzyme applications in development.

4. NEW AREAS FOR ENZYMATIC PROCESSES

Figure 8 shows a number of selected enzyme applications currently in the research phase or under development.

4.1. Paper Recycling — De-Inking

Paper recycling is increasing rapidly on a worldwide basis (*Recovered Paper Statistical Highlights,* 1991). Today more than 20 million tons of paper is de-inked every year and the increase in total de-inking of recycled paper is expected to increase to over 30 million tons by the year 2000. Recycled paper represents about 30% of total paper consumption. Of the paper that is recycled, 60% represents old newspaper and 40% represents mixed office wastepaper and tissue paper. Municipal waste is comprised of about 35–40% paper and with the current rate of usage, by the year 2000 there will not be enough landfill to bury all the municipal waste. Recycling paper utilizing de-inking technology saves land (from municipal waste burial), trees and energy. Efficient ink removal is a vital requirement before paper can be recycled.

Current de-inking methods involve a highly alkaline solution containing sodium hydroxide, sodium silicates, hydrogen peroxide, chelating agents and synthetic surfactants (Borchardt, 1993). These chemicals provide fiber swelling, ink removal and the disaggregation of the paper into fibers. As the de-inking process initiates, the paper has a tendency to darken and therefore large amounts of hydrogen peroxide are required to keep the pulp white. Often harsh chlorine-based bleaching chemicals are used as a final whitening step.

In addition to the chemicals, the process also involves a combination of thermal and mechanical energy.

Pilot plant de-inking trials using enzymes has shown that fungal cellulases are effective in de-inking newspaper (Prasad, 1992) and most importantly in aiding toner removal from xerographic and laser-printed paper, largely comprising mixed office waste paper (Franks, 1994; Jeffries, 1994). Toner used in office copiers are made of resilient plastic polymers that are tightly bound to the paper fibers. Their removal is difficult with the chemical process, but cellulases are able to release toners from fiber surfaces, presumably by the enzymatic removal of the tiny strands of cellulose that protrude from the surface of the paper fibers, dislodging the cellulose that holds the particle of ink to the paper. De-inking mixed office waste paper can be achieved using low doses of alkaline cellulases without any added sodium hydroxide. The alkaline cellulases are required as the pulp already has a high pH due to calcium carbonate used as a filler in the initial paper product. It is expected that enzyme technology will be used on a commercial scale in the near future to modify and/or replace the current de-inking methods.

4.2. Wood Products — Fiberboards

During the processes of making wood composites such as fiberboards, particle boards, cardboards, etc., the wood is first disintegrated into small pieces which are afterwards put together into larger boards using high energy and petrochemical adhesives like urea-formaldehyde and phenol-formaldehyde glues. There is a growing interest in eliminating the use of such toxic and carcinogenic chemicals in the process. In addition, boards made with these adhesives give off small emissions of formaldehyde for months after manufacture.

Enzyme technology may provide a natural alternative to formaldehyde glues for board manufacturers. Lignin, a heterogenous polyaromatic polymer, is a naturally resinous adhesive material. Non-sulfonated and sulfonated lignin can be polymerized by oxidase reaction. Enzymatic treatment of the wood particles with oxidases, such as peroxidases and laccases, leads to a rough surface of the fiber, rendering them very susceptible to interfibrous bonding. Fiberboards prepared from enzymatically treated fibers by simple pressing under normal production conditions has shown, in preliminary trials, to have very good technical properties (Kharazipour, 1995). Both the tensile strength and the swelling properties were meeting the required technical standards without the use of the formaldehyde-based adhesives. While a commercial process is still being developed, it should be possible to completely avoid the use of the hazardous formaldehyde based adhesives. Enzymatic gluing of wood could significantly reduce the world-wide consumption of the approximately 300,000 tons of phenyl–formaldehyde and the 2,500,000 tons of urea–formaldehyde, used in the manufacture of today's fiber and chip boards.

4.3. Food Functionality

There is considerable interest in using enzymes in the food industry due to their mild and more controllable reaction conditions. Lipases have proven to be very useful in food technology as the hydrolysis and esterification reactions they carry out can be done under mild conditions and without toxic solvents and other catalysts (Mukherjee, 1990; Nagao, 1990). Research is being conducted in the emulsifier segment of the food industry using lipase catalyzed esterification to produce monoacylglycerols. Monoglycerides are produced via base catalyzed transesterification reactions between glycerol and triglycerides.

The process takes place at about 240°C, followed by distillation. Commercial drawbacks are the high energy requirements, lack of specificity and distillation expense. Lipase catalyzed synthesis of monoglycerides represents a reasonable alternative to high temperature chemical synthesis. Lipases are also being used to replace inorganic chemicals used to hydrolyze fatty acids from triglycerides. Polyunsaturated fatty acids used in dietetic products are prepared under mild conditions of hydrolysis of marine and plant oils with triacylglycerol lipases. More specific lipases (e.g., *sn*-1,3, specific lipase) can be used to hydrolyze cruciferous oils resulting in the production of very long chain monounsaturated fatty acids (gadoleic, erucic, and nervonic), of value to the oleochemcial industry. The results are improved fatty acid selectivity, energy minimization and generation of more environmentally sound production waste streams.

4.4. Baking

As discussed in section 3.6, bread is comprised of starch and gluten proteins. During baking, the gluten proteins orient, align, partially unfold, and interact with one another through hydrophobic bonds and disulfide linkages. This protein network transforms into a thin film in the bread, entrapping the starch granules and other flour components. The addition of oxidizing agents such as potassium bromate, azodicarbonamide and ascorbic acid enhance the cohesive structures of hydrated gluten proteins and bread dough, causing an increase in the toughness and elasticity of the bread. The effect is an overall increase in the volume of the bread, better crumb structure and dough handling. Bromates and azocarbonamide have come under regulatory fire for health reasons and are restricted and/or banned in some countries. Animal feeding studies have shown potassium bromate to be carcinogenic (Kurokawa, 1983). In the U.K. potassium bromate as a bread improver was banned in 1990. Oxidases, such as glucose oxidase and sulfhydryloxidase, have dough strengthening effects and may become important players in the replacement of these oxidizing chemicals.

5. ENZYME MANUFACTURING AND INDUSTRIAL ECOLOGY: AN ENVIRONMENTAL ROLE TO PLAY

5.1. Recombinant DNA (rDNA) Production Technology

A revolution has taken place in biotechnology over the past two decades with the growing understanding of rDNA technology. rDNA has enabled enzyme manufacturers to produce large quantities (thousands of tons) of almost any enzyme no matter what the source (*Gene Expression in Recombinant Microorganisms*, 1995). Figure 9 depicts the rDNA industrial biotechnology development process.

Almost all industrial enzymes used today are derived from either bacteria or fungi. The genes encoding these enzymes are either cloned or synthesized, and expressed in bacterial or fungal production strains. Such production strains are tested for their high level secretion capacity (gram per liter range), then the enzymes are purified and analyzed for their biochemical properties. The strains and enzyme products are tested for safety, scaled up and the enzyme product is recovered, formulated and readied for sale. Enzymes produced by recombinant DNA technology offers a number of advantages over the wild-type microorganism fermentation process:

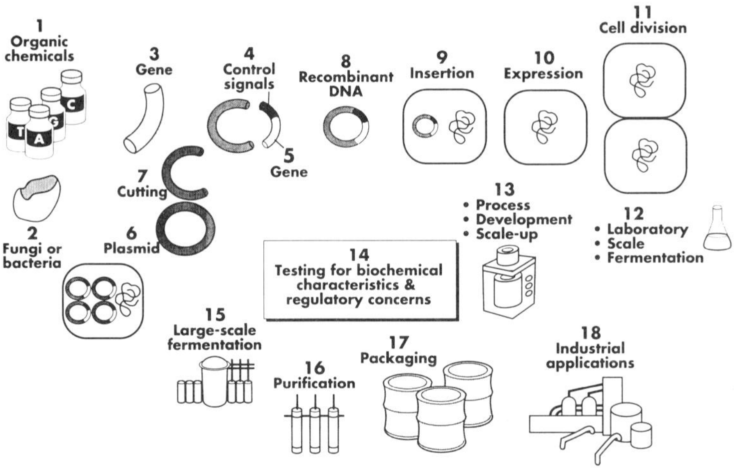

Figure 9. Recombinant DNA technology: the industrial biotechnology development process.

a. Utilization of production strains with a history of safety and fermentation knowledge are utilized.
b. Potential for higher yields in the fermentation process enabling less expensive pricing, resulting in greater usage by industry.
c. Lower consumption of energy and raw materials.
d. More controllable fermentation processes.
e. Ability to manipulate production strains to produce more pure products with less undesirable by products.
f. Greater stability of production organisms.

Protein engineering, using a number of site-specific and random mutagenesis approaches, is being used to adjust the biochemical properties of the enzyme to fine tune it for particular applications.

Important considerations for production and use of industrial enzymes include the number one concern of cost. Enzymes must be inexpensive (and most often cheaper than the chemical alternative), and must be produced in high fermentation yield, and with low recovery/purification costs. The production organisms must be safe, non-pathogenic and non-toxigenic. Today almost all enzymes are manufactured from fermentation of either *Bacillus* spp or fungal strains such as *Aspergillus* spp. or *Trichoderma* spp., microorganisms with a long history of safety.

5.2. Industrial Ecology

A novel approach taken by Novo Nordisk, the world's largest producer of industrial enzymes, in dealing with spent enzyme fermentation, is their conversion of fermentation waste into valuable fertilizer, which adheres to an environmentally sound industrial ecology (Larsen, 1991). Enzyme manufacture involves microbial fermentation at approximately the 100,000 liter scale. The production of thousands of tons of enzymes results in large volumes of spent fermentation waste products, which, in the case of Novo Nordisk, are converted into a valuable by-product for use as a fertilizer. The waste from the fermentation process is non-toxic, but if directly discharged into a marine environment, may cause eutrophication due to its high content of nitrogen and phosphate. The raw materials used in fermentation feed stocks include agricultural products such as wheat and potato starch, corn steep liquor and soy sauce. The spent fermentation is the fermentation broth after product (enzyme) recovery, which contains microorganisms, nutrient surplus and filter aid components. The conversion of the spent fermentation waste into fertilizer involves inactivation of the microorganisms with lime and heat reaction at 90°C for at least 1 hour. After 10 minutes of this treatment, no viable microorganisms can be detected. The product after cooling can be stored and transported to local farms.

At the site of its largest enzyme manufacturing plant, in Kalundborg, Denmark, Novo Nordisk distributes over 1,200,000 liters per year of the converted spent fermentation waste/fertilizer to approximately 200 farms, free of charge. This fertilizer can be used to cover about 36,000 acres of farmland. The fermenter fertilizer, which differs in its content of major plant nutrients (N:P:K; 10:4:3) from commercial inorganic fertilizers, has proven to be valuable for use with many crops including, grass, corn, sugar beet, wheat, barley and others.

Taking a cue from the interactive networks inherent in natural biological ecosystems, industrial symbiosis offers a way to reduce the trade off between industrial process and product formation and waste management - thereby reducing potential negative environmental impacts. The town of Kalundborg, Denmark (noted above) involves such an industrial symbiosis, where several different enterprises utilize each others residual products in a network (Gentler, 1996) (Figure 10). The site includes the participation of Novo Nordisk. For example, at this site the Asnæs Power Station supplies heating to the town of Kalundborg and process steam to Novo Nordisk and Statoil Refinery (Denmark's largest

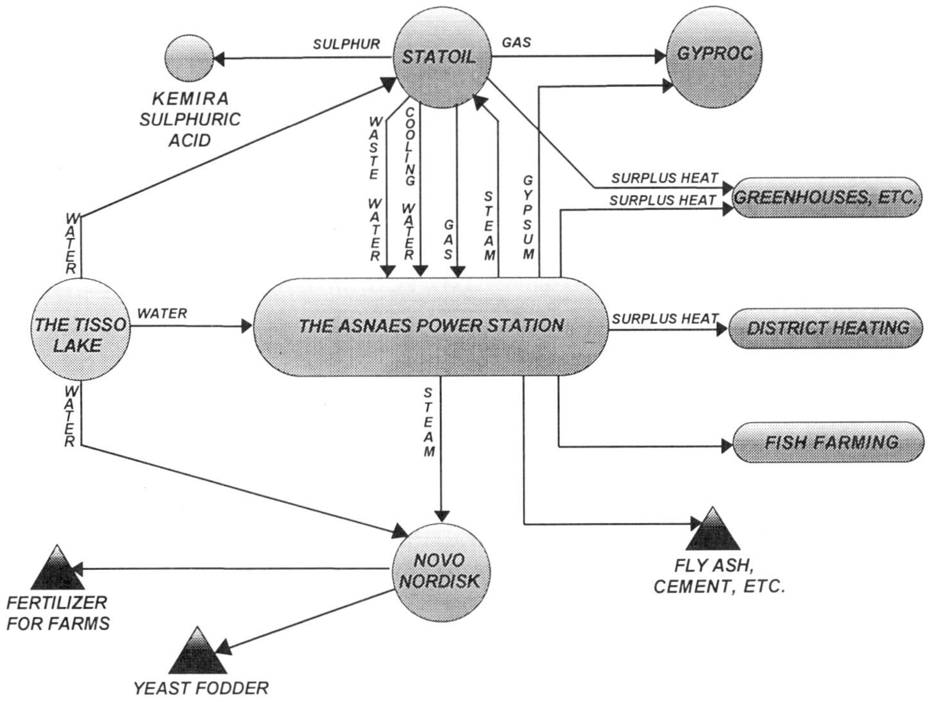

Figure 10. Industrial symbiosis.

oil refinery). As discussed above, Novo Nordisk converts its spent microbial fermentation waste to a high value fertilizer and provides and distributes it free of charge to local area farmers. The power station sells about 100,000 tons of gypsum annually from its gas desulphurization plant to Gyproc (Scandinavia's largest producer of plasterboard products), while the Power Station receives part of its cooling water as recycled water from Statoil. The result of this industrial symbiosis is both reduced resource consumption and burden on the environment, as well as economical use of by-products while complying with ever-stricter environmental regulations.

6. SUMMARY

The advantage of using enzymes in industrial processes comes from their specificity (high or low) to suit the desired end product. In addition, little to no byproducts are formed and the optimal enzymatic activity occurs under defined reaction conditions, which are most often relatively mild compared to more traditional chemistries. Enzymes are cost effective, non-polluting process aids.

How about the future? A number of new applications are being researched and developed today. More efficient enzymes tailored for specific and often extreme process conditions are being sought, i.e., hemicellulytic systems being active at the conditions existing after the kraft pulp cooking, e.g., pH 10–11 and temperatures higher than 70°C, thus further simplifying the process by avoiding any adjustments of the pulp. Systems to eliminate chlorine use completely in the bleaching of pulp and enzymes to be used in delignification steps are also of major interest. It is likely that we will see a more broad use of enzymes in organic chemical synthesis (e.g., lipases), for bioremediation (e.g., peroxidases, laccases), coal and oil desulfurization (e.g., peroxidases), and in the personal care sector (e.g., isomerases to replace sulfur chemicals in the permanent waving of hair).

Identifying commercially useful enzymes requires a good biochemical understanding of the process operating conditions and chemistry of the particular process reactions. It is not enough to isolate microorganisms from extreme environments (e.g., growing at 100°C or at pH 2) and their enzyme encoding genes, hoping to identify industrially useful enzymes. Each enzyme is unique with respect to its biochemical properties (e.g., substrate specificity, K_m, K_{cat}, etc.) and its optimal operating conditions (e.g., pH, temperature, stability). It is crucial that such parameters are carefully evaluated for each enzyme. Of course, the next step is that the enzyme encoding gene must be amenable to very high level expression in a recombinant host. Otherwise the product is unlikely to ever achieve commercial economic viability. Many of the industrially useful enzymes will not be found existing in nature, but will rather be a result of protein engineering efforts on the wild-type enzyme backbone.

The revolution in industrial enzymology has been a quiet one. The widespread use of enzymes today is a direct result of recombinant DNA technology and it is expected that we will see a broader dissemination of enzyme use into many other segments of industry. Greater awareness of the environment has brought about a more intensive search for alternative cleaner technology. The balance must be shifted from an exclusive focus on yield to one that places a value, both economic and environmental, on minimizing or eliminating wasteful and/or toxic byproducts. Enzyme technology, it is believed, will be able to provide some of these alternatives and will gradually replace many chemical industrial processes.

ACKNOWLEDGMENTS

The data from Figures 6, 7A and 7B were derived respectively from the March 1995 and October 1995 Novo Nordisk publication *Biotimes*. The author would like to thank Ms. Marta L. Hillesheim for critical reading and help in editing this manuscript.

REFERENCES

American Paper Institute, Inc., 1991. *Recovered Paper Statistical Highlights*. Washington, D.C.

Borchardt, J.K., 1993. Paper De-Inking Technology, *Chemistry & Industry*, April 19, 273–276.

Chemical & Engineering News, 1993. Chlorine Based Bleaching: EPA Rule to Curb Paper Industry Use, 1993, November 8, 5–8.

Dalgård, L.H., Kochavi, D., and Thelleisen, M., 1991, Dishwashing — a New Area for Enzymes, *INFORM*, 2:(533–536) .

Elegir, G., Sykes, M., Jeffries, T., 1995. Differential and Synergistic Action of *Streptomyces* Endoxylanases in Prebleaching of Kraft Pulps, *Enzyme and Microbial Technology* 17:954–959.

Enzyme-Enhanced with Conventional Deinking of Xerographic and Laser-Printed Paper, *Tappi Journal*, 77:173–179.

Franks, N., 1994. Enzyme Facilitated Deinking of Mixed Office Waste: the Use of Alkaline Cellulases, Abstract Presented at the Paper Recycling '94 Conference, London, UK.

Food Engineering, 1989. Shelf Life of Bread Increased With... Enzyme Systems, 61(3): 57–59.

Gene Expression in Recombinant Microorganisms, (A. Smith, ed.), New York: Marcel Dekker, Inc., 1995.

Gentler, N., 1996. A Down-to-Earth Approach to Clean Production, *Technology Review*, January/February, 48–54.

Illman, D. L., 1994. Environmentally Benign Chemistry Aims for Processes that Don't Pollutc, *Chemical & Engineering News*, September 5: 22–27.

Jeffries, T., Klungness, J., Sykes, M., and Rutledge-Cropsey, K.R., 1994. Comparison of Kharazipour, A., Hütterman, A., 1995. Enzymatic Activation of the Middle Lamella-Lignin of Wood Fibres as a Means for the Production of Binder-Free Fibre Boards, The 6th International Conference on Biotechnology in the Pulp and Paper Industry, Vienna, Austria.

Kirk, O., et al., 1991. Monoesters of Glycosides and a Process for Enzymatic Preparation Thereof , United States Patent #5,191,071.

Kochavi, D., 1993. Enzymes for Household Detergents, *INFORM*, 4:990–995.

Kurokawa, Y., Hayashi, Y., Maekawa, A., Takahashi, T., Kokubo, T., and Odashima, S., 1983. Carcinogenicity of Potassium Bromate Administered Orally to F344 Rats, *Journal of the National Cancer Institute*, 71:965–970.

Larsen, A., Funch, F., 1991. The Use of Fermentation Sludge as a Fertilizer in Agriculture, *Water Scientific Technology*, 24(12), 33–42.

Lloyd, N.E., and Nelson, W.J., 1984. Glucose and Fructose-Containing Sweeteners from Starch, *Starch: Chemistry and Technology*, 2nd ed., (R.L. Whistler, J.N. Bemiller, and E.F. Paschal, eds.) Academic Press, New York, 611–660.

Malmos, H., 1990. Enzymes for Detergents, *Chemistry & Industry*, No.6, March, 183–186.

Mukherjee, K.D., 1990. Lipase-Catalyzed Reactions for Modifications of Fats and Other Lipids, *Biocatalysis*, 3:277–293.

Muthukumaran, N., Dhar, S.C., 1982. *Leather Science*, 29:417–424.

Nagao, A., Kito, M., 1990. Lipase-Catalyzed Synthesis of Fatty Acid Esters Useful in the Food Industry, *Biocatalysis*, 3:295–305.

Novo Nordisk, 1991. Cool Down with Enzymes, *Biotimes*, (L.L. Bundesen, ed.), a Quarterly Bioindustrial Magazine from Novo Nordisk A/S, Bagsværd, Denmark, VI, No.1, 10–11.

Novo Nordisk, 1994, Terminox® Saves Water in the Dyehouse, 1994. *Biotimes* (D. Hansen, ed.)a Quarterly Bioindustrial Magazine from Novo Nordisk A/S, Bagsværd, Denmark, September, 4.

Novo Nordisk, 1995. Bread Stays Fresh Longer with Novamyl, *Biotimes* (D. Hansen, ed.), a Quarterly Bioindustrial Magazine from Novo Nordisk A/S, Bagsværd, Denmark, V, No.4, 2–3.

Novo Nordisk, 1995. The High-Performance Enzyme for Bleach Boosting, *Biotimes* (D. Hansen, ed.), a Quarterly Bioindustrial Magazine from Novo Nordisk A/S, March , No. 1, 6–7.

Patel, R.N., Grabski, A.C., and Jeffries, T.W., 1993. Chromophore Release from Kraft Pulp by Purified *Streptomyces roseiscleroticus* Xylanases, *Applied Microbiology Biotechnology*, 39:405–412.

Pedersen, G.L., Screws, Jr., G.A., and Cedroni, D.M., 1992. Biopolishing of Cellulosic Fibers, *Canadian Textile Journal,* December, 31–35.

Prasad, D.Y., Heitmann, J.A., and Joyce, T.W., 1992. Progress in Paper Recycling 1(3):21.

Sarney, D., Vulfson, E., 1995. Application of Enzymes to the Synthesis of Surfactants, *Trends in Biotechnology,* 13: 164–176.

Thayer, A., 1993. Paper Chemicals, *Chemical & Engineering News,* November 1, 28–41.

Turner, M., 1995. Biocatalysis in Organic Chemistry (Part II): Present and Future, *Trends in Biotechnology*, 13, 253–258.

4

FUNGAL DEGRADATION OF AZO DYES AND ITS RELATIONSHIP TO THEIR STRUCTURE*

Andrzej Paszczynski, Stefan Goszczynski, and Ronald L. Crawford

Center for Hazardous Waste Remediation Research, and Department of Microbiology, Molecular Biology, and Biochemistry
University of Idaho
Moscow, Idaho 83844-3052

1. INTRODUCTION

Organic chemists add approximately 200,000 new chemicals per year to the millions already used by the USA, Japan, and the advanced industrial nations of Europe. About 100,000 of the chemicals in use are synthetic dyes (Meyer, 1981). Azo dyes are the most numerous and widely manufactured of the synthetic dyes, having a great variety of uses ranging from food dyes to gasoline additives (solvent dyes). Highly water soluble azo dyes are widely used in the fiber dyeing industry, and azo pigments that are highly insoluble in water and in organic solvents are used in paint products (Hunger et al., 1985). The characteristic feature of azo dyes is the presence of one or more chromophoric azo group(s) R–N=N–R, where at least one nitrogen atom is linked to a carbon atom belonging to an aromatic carbocycle.

Naturally occurring azo dyes are unknown; consequently, these compounds are resistant to degradation by microorganisms and can become persistent environmental pollutants. Because of the widespread use of azo dyes in textiles, printing, cosmetics, drugs, and foods, the toxicity and mutagenicity of this group of chemicals are studied in many laboratories (Chung, 1983; Reid et al., 1984; Joachim et al., 1985; Lin and Solodar, 1988). Toxicity studies usually involve standard tests such as the Ames test, which quantifies reversions of *Salmonella* mutants as indicators of mutagenicity. The mutagenicity of azo dyes has been evaluated by studying the relationship between structure and activity (Shahin, 1989). However, these studies of structure activity relationships are difficult to interpret because of the complexity of the azo dye molecule and the variety of reactive degradation products forming during their biotransformation.

A computerized method using *A*utomated *D*ata *A*nalysis by *P*attern Recognition *T*echnique (ADAPT) (Claxton et al., 1990) has been used for analyzing the mutagenic activity of azo dyes and related compounds. When the ADAPT technique was used, the

---

* Publication no. 96502 of the Idaho Agricultural Experiment Station

Ames mutagenicity test indicated that the heteroatoms N, O, S increased mutagenicity, while decreased mutagenic potential was shown after increasing the size of the dye molecule. A recent review of the literature regarding the toxicity and mutagenicity of azo dyes (Brown and de Vito, 1993) concluded that azo dyes are toxic and carcinogenic only after reduction and cleavage to aromatic amines. After oxidation to reactive species, the aromatic amines can bind to DNA. Azo dyes containing aromatic amines can be activated without azo linkage reduction, and the azo linkage of azo dyes can also be activated by oxidation to the highly reactive diazonium salt. Aromatic amine pollutants undergo many complex transformations in the environment: they can be mineralized, acylated, or polymerized by microorganisms, but they can also photooxidize, autooxidize, and chemically bind to soil mineral or organic matter. Biological and physicochemical processes can interact in the environment to further complicate the picture (Lyons et al., 1984).

Our studies have shown that during oxidation of azo dyes by fungal peroxidases, reactive diazene and nitroso derivatives can be formed (Goszczynski et al., 1994). Other studies have suggested the formation by peroxidative oxidation of highly unstable sulfophenyl hydroperoxide moieties (Chivukula et al., 1995), which might be toxic to living cells.

Another method of evaluating the environmental fate of azo dyes is to examine the correlation between structure and biodegradability by use of sewage sludge enrichment cultures. Biodegradability is measured as the cell protein yield, with organic compounds serving as the sole source of carbon and/or nitrogen and energy in a mineral solution (Rothkopf and Bartha, 1984). Often, only partial degradations are seen. Such partial degradation, e.g., reduction or oxidation, of azo dyes is not an outcome that will protect the environment. Only total degradation, i.e., mineralization, will guarantee that potentially toxic intermediates are removed.

We believe that the construction of nontoxic and nonmutagenic azo compounds with biodegradability designed into their structures will make them safer to use. This work can also serve as a model for research on designing biodegradability into other types of industrial and agricultural chemicals.

2. DEGRADATION PATHWAYS AND STRUCTURE-BIODEGRADABILITY CORRELATIONS IN AZO DYES

Although azo dyes appear to be resistant to aerobic degradation by bacteria (Kulla, 1981; Kulla et al., 1983,1984; Haug et al., 1991), a bacterial consortium was able to largely degrade several azo dyes during sequential anaerobic/aerobic growth (Haug et al., 1991). Also, a new species of protobacteria capable of degradation of azo dyes under aerobic conditions was recently reported (Govindaswami et al., 1993). These observations indicate that bacteria have a higher potential for azo dye degradation than previously thought. We have observed that some *Streptomyces* species decolorize azo dyes at rates comparable to that of *Phanerochaete chrysosporium*, a white-rot fungus known to rapidly degrade many azo dyes. Structural preferences in azo dye aromatic ring substitution were observed in those dyes most susceptible to *Streptomyces* metabolism (Pasti et al., 1992).

We observed that the susceptibility of azo dyes to degradation by *P. chrysosporium* and *Streptomyces* species can be increased by attaching guaiacyl (*o*-methoxyphenol) substituents similar to structures found in softwood lignin (Paszczynski et al., 1991b). We later demonstrated that *P. chrysosporium* was able to degrade azo dyes at concentrations up to 300 parts per million (ppm) (Paszczynski et al., 1991a), whereas only about 50 ppm was degraded by several strains of *Streptomyces* (Pasti et al., 1991). These studies are important because bacteria may have much greater potential than fungi to be industrially used for azo dye degradation. Our general observation was that the increased degradability depended on the presence of an aromatic methoxy or a methyl group, which resembles some naturally occurring substitution patterns on the aromatic rings of lignin. We confirmed the mineralization of sulfonated azo dyes by *P. chrysosporium* using ^{14}C-[U]-ring-labeled compounds, including both dyes and free sulfanilic acid. The latter compound is an important industrial chemical used for the manufacture of dyes and other commercial chemicals. In shaken culture the fungus was able to release up to 15% of the label from sulfanilic acid and 35% of the label from certain dyes as $^{14}CO_2$ after 21 days' incubation, starting from an initial concentration of 200 ppm of substrate. We reaffirmed that very specific changes in the molecular structure of azo dyes enhanced their biodegradability (Pasti-Grigsby et al., 1992; Paszczynski et al., 1991b). Dyes with naphthalene rings were also degraded; those with hydroxyl and azo groups in the 1,2 position were degraded most rapidly by the fungus, while those substituted at the 1,4 position were preferred by streptomycetes. Figure 1 shows that the only difference in the structures of Orange I and Orange II is the position of the hydroxyl group on the naphthalene ring moiety. Orange I was decolorized more rapidly by bacteria, and Orange II was decolorized more rapidly by the fungus. This experiment indicated that the choice of the test organism may lead to quite different results.

Veratryl alcohol, present in the purified ligninase reaction mixture, stimulated the oxidation of Tartrazine and Biebrich Scarlet dyes by lignin peroxidase. Without veratryl alcohol the enzyme was unable to oxidize these dyes. We postulated that lignin peroxidase compound I oxidized the azo dyes and was reduced in the process to compound II. In the absence of veratryl alcohol, the reaction was terminated. When veratryl alcohol was present, the enzyme readily completed its catalytic cycle, and the reaction mixture was decolorized at a rapid rate (Paszczynski and Crawford, 1991). This finding was further investigated by Ollikka et al., (1993). Three major lignin peroxidase isoenzymes (H2, H7, and H8) were used for decolorization experiments. The partly purified isoenzymes decolorized all dyes investigated without veratryl alcohol, but probable contamination of the preparations with endogenous veratryl alcohol could explain these results. With pure isoenzymes, veratryl alcohol was required for decolorization of several dyes. Some of the biodegradable structures of azo dyes and biodegradation procedures have been patented (Paszczynski et al., 1996).

We now have elucidated some of the major chemical steps in the degradation of azo dyes by lignin peroxidases and manganese peroxidases (Goszczynski et al., 1994; Pasti-Grigsby et al., 1994a). Figure 2 shows the aerobic degradation pathway of azo dyes by *P. chrysosporium* and other white-rot fungi (modified from the original report of Goszczynski et al., 1994, and supplemented with data from Chivukula et al., 1995, and Chivukula and Renganathan, 1995). Degradation appears to involve two initial one-electron oxida-

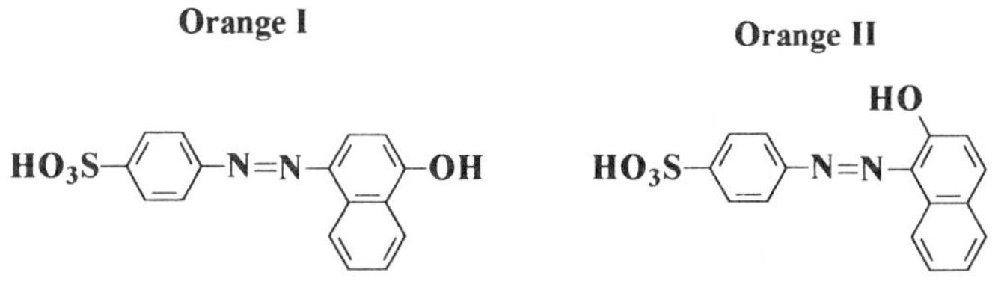

Figure 1. Structures of Orange I [4-(4-hydroxy-1-naphthyl-azo)benzenesulfonic acid] and Orange II [4-(2-hydroxy-1-naphthylazo)benzenesulfonic acid]. Orange I was decolorized more rapidly by bacteria, and Orange II was decolorized more rapidly by the fungus.

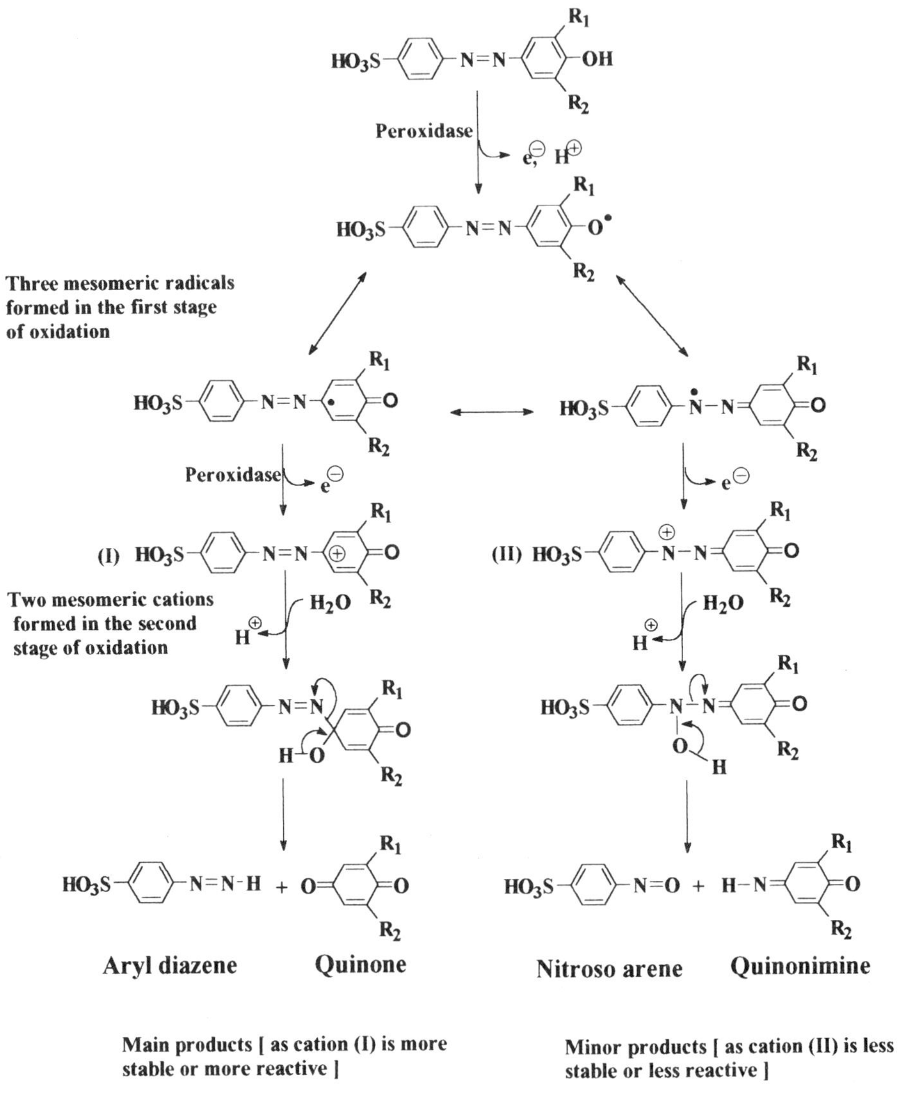

Figure 2. Mechanism of oxidative degradation of sulfonated monoazo dyes by fungi. The pathway involves initial oxidation of the substrate by peroxidase or laccase with formation of a cation that reacts with water to form unstable intermediates. These transient compounds stabilize by splitting as shown.

tion steps catalyzed by ligninase or laccase, and the formation of carbon- or nitrogen-centered cations which undergo nucleophilic attack by water. The unstable transitional intermediates decompose, forming quinone and phenyldiazene or nitrozobenzene derivatives and iminobenzoquinone. The iminobenzoquinone hydrolyzes to 1,4-benzoquinone and ammonia (Figure 3A), and diazonium salt gives rise to the formation of a phenol derivative (Figure 3B).

The formation of aryldiazene and quinone in the first stage of the enzymatic degradation process of azo dyes is an important clue to the mechanism of this process. Two of the possible mesomeric cations formed in the second stage of enzymatic oxidation may form four possible splitting products. Aryl diazene and quinone are the main products, al-

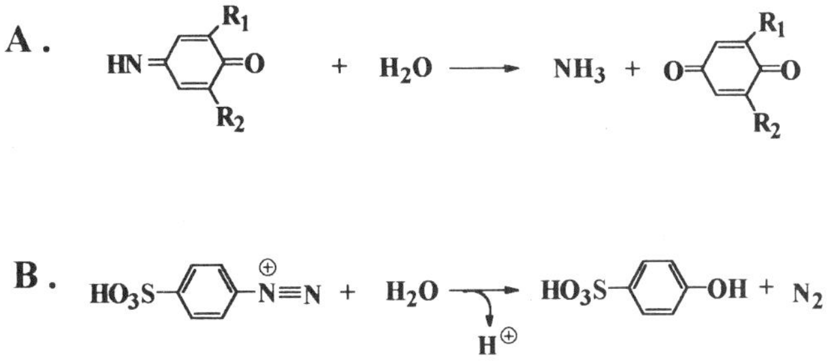

Figure 3. Hydrolysis as a stabilizing process for the primary or secondary splitting products of azo dyes. **A.** Quinonimine giving rise to the formation of ammonia and quinone. **B.** Diazonium cation forming a phenol.

though nitroso and quinonimine may be also formed, but in minor quantities. As molecular orbital calculations have shown, the cation with the charge located on carbon is more stable than the one with the charge on nitrogen. This calculation is consistent with our experimental data.

Aryldiazenes (Ar–N=NH) have long been postulated as unstable intermediates in the course of oxidation of arylhydrazines, reduction of diazonium salts, and fragmentation of certain arylazocarboxylic acids. The formation of aryldiazenes was confirmed directly by UV absorption spectra of several substituted phenyldiazenes (Mannen and Itano, 1973). Aryldiazenes, when used for reduction of some peroxides, give rise to the formation of diazonium salts. Their existence as intermediates was unquestionably demonstrated by a coupling reaction and the Sandmayer reaction (Hoffman and Kumar, 1984). A review of the formation and detection of aryldiazenyl radicals in solution illustrates the increasing significance of these species as intermediates in many transformations (Suehiro, 1988). The interaction of phenyldiazenes with cytochrome P-450 for topological analysis has stimulated interest in this type of compound (Raag et al., 1990; Swanson, et al., 1991). The formation of phenyl-iron complexes by phenyldiazenes has been used lately for evaluation of active site topology in some enzymes (Gerber and Ortiz de Montellano, 1995; Newmyer and Ortiz

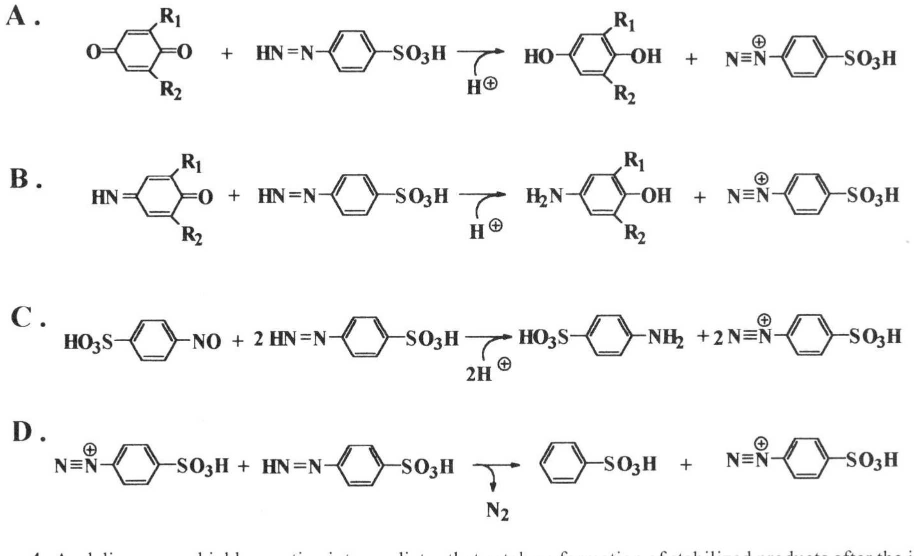

Figure 4. Aryl diazenes as highly reactive intermediates that catalyze formation of stabilized products after the initial splitting of azo dyes. Aryl diazenes are powerful reductive species that readily oxidize to diazonium cations. **A.** Reduction of quinone to hydroquinone. **B.** Reduction of quinonimine to aminophenol. **C.** Reduction of nitroso arene to amine. **D.** Reduction of diazonium salt to arene.

de Montellano, 1995). All possible redox transformations of the primary splitting products by aryl diazenes are shown in Figure 4. The formation of hydroquinones, aminophenols, amines, and aromatic hydrocarbons was confirmed experimentally.

The reduction of 1,4-benzoquinone by an intracellular quinone reductase was proposed by Brock et al., (1995). The further hydroxylation of the aromatic ring and the splitting by a 1,2,4-trihydroxybenzene-1,2-dioxygenase may possibly be the mechanism of the final mineralization (Rieble et al., 1994).

In experiments with ^{14}C-[U]-ring-labeled water insoluble solvent dyes, Spadaro et al., (1992) confirmed our finding that *P. chrysosporium* is capable of mineralizing azo dyes. In further investigations of the degradation mechanism of azo days, sulfophenyl hydroperoxide was postulated as new degradation product (Chivukula et al., 1995) (Figure 5). These data require further confirmation by others, since aromatic hydroperoxides are known to be very unstable (Kropf, 1988).

Under the conditions described by Chivukula et al., (1995), benzenesulfonic acid formed by abstraction of H• from another organic compound, or azobenzene-4,4′-disulfonic acid formed by coupling of diazene and benzene sulfonic acid radicals, would be the expected products (Goszczynski et al., 1994), and probably not the hydroperoxide shown here, which is likely to be very unstable.

Our earlier research identified the azo dye 3,5-dimethyl-4-hydroxyazobenzene-4′-sulfonic acid as the structure most susceptible to degradation of all the azo dyes studied by our group. Six isomers of this dye were synthesized and their degradation studied using *P. chrysosporium* as a model organism (Paszczynski et al., 1995). A whole culture degradation study showed that azo isomers with a hydroxyl in the 2 position were much more resistant to degradation than isomers with the hydroxyl in the 4 position of the aromatic ring. Figure 6 shows the structures of the hydroxy dimethyl isomers of azo dyes used in our pattern recognition degradation study. In reporting the initial degradation of the azo dyes Azure B, Tropaeolin O, Orange II, and Congo Red by *P. chrysosporium*, Cripps et al. (1990) suggested that some decolorization can result from dye binding to fungal hyphae; however, in our experiments binding accounts for a very small percentage of decolorization.

Both the ligninases of *P. chrysosporium* and adsorption to biomass were responsible for decolorizing 8 out of 18 commercially used dyes. From 40 to 73% decolorization was detected in 5-day-old cultures after 72 hours of treatment (Capalash and Shrama, 1992).

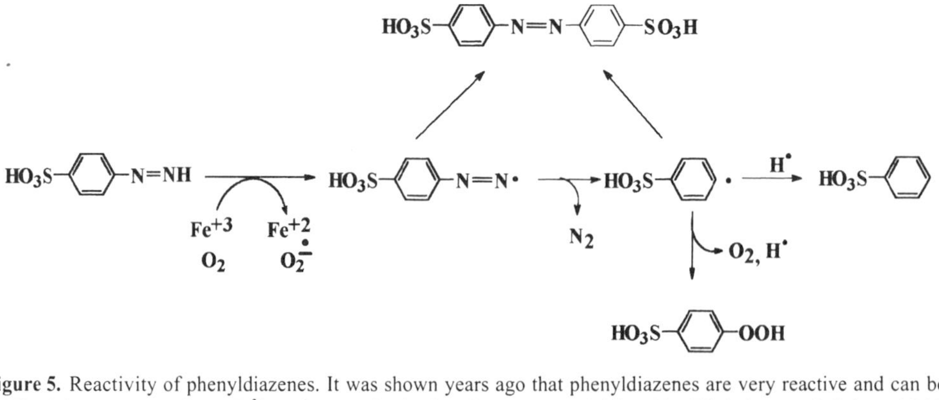

Figure 5. Reactivity of phenyldiazenes. It was shown years ago that phenyldiazenes are very reactive and can be oxidized by agents such as Fe^{3+} to form radicals that decompose, yielding N_2 (Nicholson and Cohen, 1966; $C_6H_5N{=}NH + Fe^{3+} \rightarrow C_6H_5N{=}N\bullet + Fe^{2+} + H^+$; $C_6H_5N{=}N\bullet \rightarrow C_6H_5\bullet + N_2$ or O_2); (Hoffman and Kumar, 1984; $RN{=}H \rightarrow RH + N_2$).

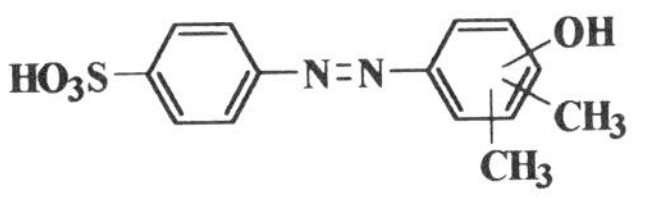

1

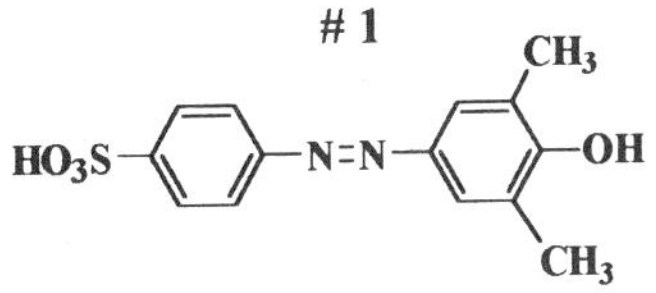

2

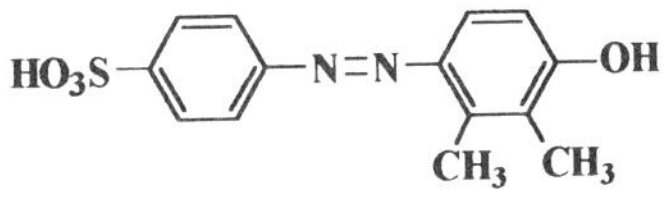

3

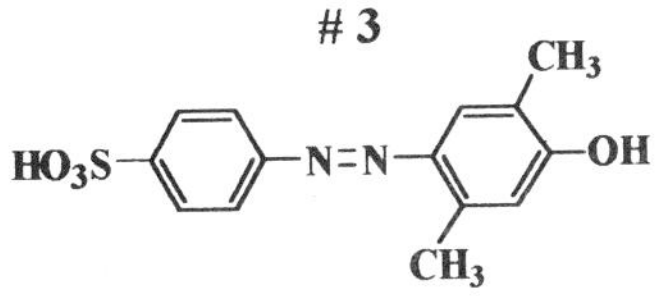

4

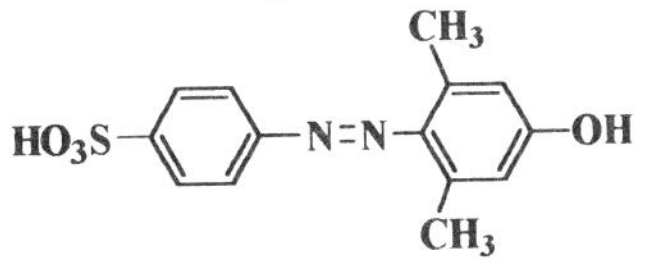

5

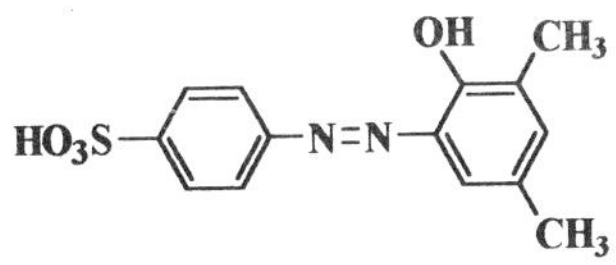

6a,b

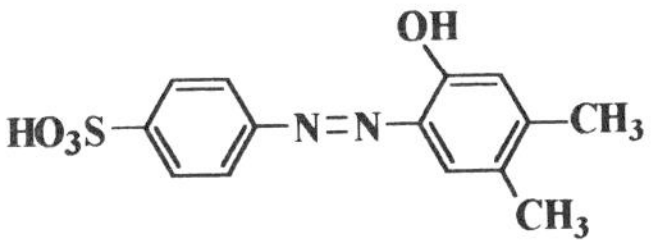

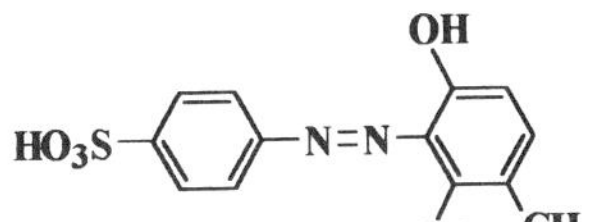

Figure 6. Structures of 4-hydroxy- and 2-hydroxy-dimethyl derivatives of azobenzene-4′-sulfonic acid. Azo dye 1 was the best substrate in the submerged agitated culture of *Phanerochaete chrysosporium*, followed by dyes 4 and 3. Azo dye 4 was bleached more quickly on the solid agar plate culture containing the same medium, followed by dyes 2 and 5. Azo dye 1 was the best substrate for ligninase and Mn peroxidase preparations (Paszczynski et al., 1995).

Later, Muralikrishna and Renganathan (1993) postulated the formation of 2,6-dimethyl-1,4-benzoquinone and sulfide anion during oxidative degradation of this dye. Further research is needed to confirm sulfide formation in these reactions. The same group suggested a similar oxidation pathway of monoazo phenolic dyes by another fungal oxidase, a laccase from *Pyricularia oryze* (Chivukula and Renganathan, 1995).

Azo dyes have been examined for their potential use as substrates for assaying lignin peroxidases and manganese peroxidases of white-rot fungi. The novel dye 3,5-dimethyl-4-hydroxyazobenzene-4′-sulfonic acid or the commercial dye Orange I served well as the substrate for specific assays of Mn(II)-peroxidase, as did 3,5-difluoro-4-hydroxy-4′-sulfonic acid or Orange II for specific assays of lignin peroxidases (Pasti-Grigsby et al.,

1994b). These dyes are therefore potentially useful for both enzyme work and the isolation of specific types of peroxidase mutants.

3. POLYMERIC DYES

White-rot fungi are a unique group among microorganisms, in that they are the only microorganisms known to be capable of complete mineralization of lignocellulosic polymers. Their molecular biology (Gold and Alic, 1993) and physiological and biochemical properties (Tour et al., 1995; Daniel, 1994) have been closely examined by researchers because of the perceived potential of white-rot fungi for use in bioremediation (Reddy, 1995; Paszczynski and Crawford, 1995; Hammel, 1995). Polymeric dyes (Dawson, 1981) such as Poly B-411 (anthraquinone chromophore), Poly R-481 (anthrapyridone chromophore), and Poly Y-606 (nitroaniline chromophore) were used in early studies of *P. chrysosporium* as indicators for lignin-degrading enzymes during fungal growth in liquid medium. The results with Poly B-411, Poly R-481, and Poly Y-606 suggested that polymeric dyes might be useful for selecting mutants and indicating peroxidase activity (Glenn and Gold, 1983). Decolorization of Remazol Brilliant Blue R, a dye whose structure resembles that of Poly-B-411, was also reported (Ulmer et al., 1984) for *P. chrysosporium*. The involvement of an extracellular H_2O_2-dependent ligninolytic activity of another white rot fungus, *Pleurotus ostreatus*, in the decolorization of this compound was reported recently. Manganese peroxidase, manganese-independent peroxidase, and phenol oxidase activities were detected during solid-state fermentation of wheat straw along with Remazol decolorization (Vyas and Molitoris, 1995). The decolorization of the dye Poly B-411 and the correlation of color loss with the ability of fungi to degrade lignin have been investigated (Chet et al., 1985; Platt et al., 1985). After good correlations between the mineralization of ^{14}C-labeled lignin and decolorization of dyes were established, decolorization of Poly R-478 was successfully used for screening 170 strains of white-, brown-, and soft-rot and xylophilous fungi for their lignin peroxidase and oxidase activities (Freitag and Morrell, 1992). Others have also shown good correlations between dye decolorization and production of peroxidase and H_2O_2. The Poly R-478 decolorization activity of 3 of 67 new fungal strains was significantly higher than that of *P. chrysosporium* (de Jong et al., 1992). The use of solid medium containing dyes for screening the ligninolytic activity of large numbers of organisms and their mutants or clones is quickly becoming a standard method in microbiology (Chahal et al., 1995). The method is simple, but the choice of dye substrate is crucial because of the different susceptibilities of dyes to degradation by various groups of microorganisms (see Figure 1).

Our earlier studies with biodegradable dyes involved only changes of substitutions on the terminal benzene ring. Here we report original data from new experiments where the internal structures of dye molecules were changed by substituting dimethyl- or dimethoxy-biphenyl moieties for an internal biphenyl structure. Decolorization results with three such blue dyes are summarized in Figure 7. The dyes were synthesized by coupling diazonium salts of benzidine, *o*-tolidine, and *o*-dianisidine with Chicago Acid (8-amino-1-naphthol-5,7-disulfonic acid; obtained from Pfaltz and Bauer, Inc., Waterbury, Conn.), following the method described by Hartwell and Fieser (1943). Decolorization studies were performed as described earlier (Pasti-Grigsby et al., 1992), using a mineral medium with low nitrogen concentration and 75 or 150 ppm of each dye. In these experiments we were asking how small a change on the biphenyl ring (addition of one or more methyl or methoxyl substituents) might affect the degradability of these compounds by *P. chryso-*

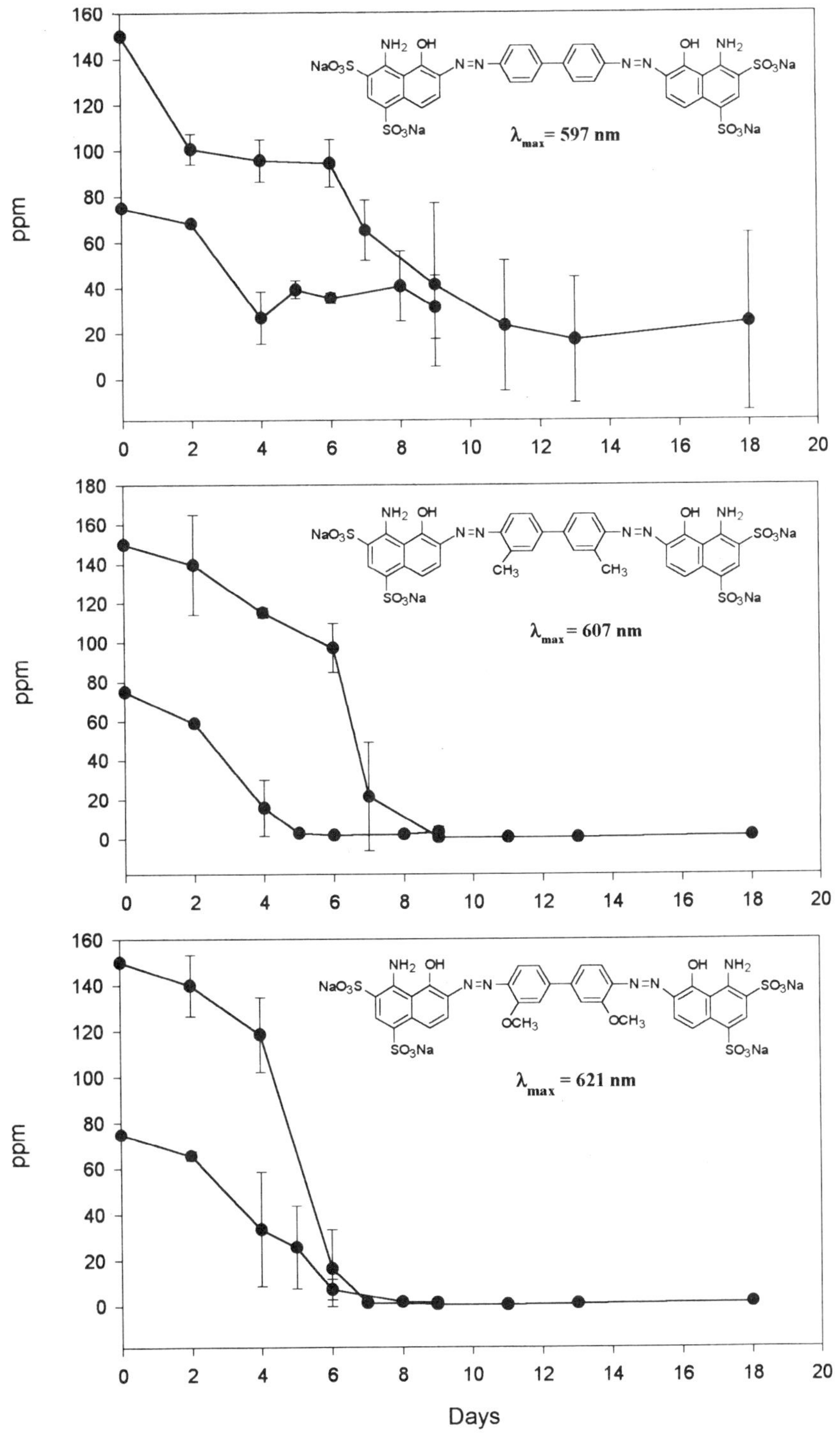

Figure 7. Decolorization by *Phanerochaete chrysosporium* of Chicago Sky Blue, λ_{max} = 621 nanometers (nm), Tolidine 1824, λ_{max} = 607 nm; and Sabine Blue, λ_{max} = 597 nm. Two initial concentrations of each dye were used, 150 and 75 ppm, in mineral medium with low nitrogen content (Bonnarme et al., 1991). The $MnSO_4$ concentration in the medium was 180 micrometers (μm). Adding two methyl groups to the biphenyl rings significantly improved the degradation rate of Tolidine 1824 as compared to Sabine Blue. Culture volume was 25 milliliters; agitation was at 110 rotations per minute. Data points represent mean values, error bar = standard deviation of three cultures.

sporium. Dyes substituted by methyl (Tolidine 1824) and methoxyl (Chicago Sky Blue) were degraded faster by the fungus than dyes with a bare biphenyl (Sabine Blue) ring. Again, we found that these small changes in the dye molecule greatly affected degradability, without affecting the color and textile binding properties of the dyes.

4. DEGRADATION OF COLOR WASTE EFFLUENT

According to a study performed by the Ecological and Toxicological Association of the Dyestuff Manufacturing Industry (ETAD; Basel, Switzerland), under anaerobic conditions in the environment, dyestuffs and azo dyestuffs are likely to undergo substantial primary biodegradation, forming sizable amounts of toxic aromatic amines and their oxidation products. These compounds might slowly desorb from anaerobic muds to the aerobic aqueous environment, where they can be completely degraded (Brown and Laboureur, 1983; Brown and Hamburger, 1987). A recent study showed, however, that fungi and some aerobic bacteria can carry out this type of process in an oxygen-rich environment. A model wastewater containing thickening agents as well as different dyes was treated in a rotating biological contactor, where *Pseudomonas cepacia* BNA was immobilized within a κ-carrageenan gel. The effluent was extensively bleached after passing through three contact reactors (Ogara and Yatome, 1991). Aerobic decolorization of the effluent from a pigment plant by the white-rot fungus *Pycnoporus cinnabarinus* was also reported (Schiephake et al., 1993). *P. cinnabarinus* rapidly decolorized and clarified wastewater samples passed through a packed-bed bioreactor. Lignin peroxidase activity and glucose utilization were monitored during the decolorization of nine textile dyes and an artificial textile effluent. However, since decolorization of the effluent required 7 days, the role of lignin peroxidase was unclear (Kirby et al., 1995). The aerobic bacteria *Pseudomonas alcaligenes*, *Ps. mendocina*, *Ps. putida*, and *Ps. stutzeri* were isolated from samples taken from a factory manufacturing Methyl Violet. These species were able to metabolize Methyl Violet and phenol in liquid culture (Sarnaik and Kanekar, 1995). In a recent report on decolorization of olive mill wastewater by a *P. chrysosporium* culture, a positive role for veratryl alcohol with lignin peroxidase was observed when more than 70% of the color was removed (Sayadi and Ellouz, 1995).

5. TRIPHENYLMETHANE DYES

P. chrysosporium degraded (decolorized) triphenylmethane dyes, including Crystal Violet, Pararosaniline, Cresol Red, Bromophenol Blue, Ethyl Violet, Malachite Green, and Brilliant Green (Bumpus and Brock, 1988). Three identified metabolites of Crystal Violet, N,N,N′,N′,N′′-pentamethylpararosaniline, N,N,N′,N′-tetramethylpararosaniline, and N,N′,N′′-trimethylpararosaniline, were formed by sequential N-demethylations of the parent compound. From this evidence it was suggested that fungal ligninase may catalyze N-demethylation reactions. However, since a nitrogen-rich medium was used, Crystal Violet degradation by *P. chrysosporium* may have involved a non-ligninolytic enzyme system.

6. OTHER DYES

New dye structures that can be degraded by *P. chrysosporium* are continually being found. Recent reports have described the degradation of Methylene Blue, a cationic thiazine dye, by a crude extracellular medium of *P. chrysosporium* (Kling and Neto,

1991). The aerobic biodegradation of the exotic dye Rose Bengal (tetrachloro-tetraiodo-fluorescein) was also described (Gogna et al., 1991). Azure B, the demethylation product of Methylene Blue, was found to be a good replacement for veratryl alcohol for the colorimetric assay of ligninase activity in white-rot fungi (Archibald, 1992).

REFERENCES

Archibald, F. A., 1992. New assay for lignin-type peroxidases employing the dye Azure B, *Appl. Environ. Microbiol.* 58:3110–3116.

Bonnarme, P., Perez, J., and Jeffries, T. W., 1991. Regulation of ligninase production of white-rot fungi, in: *Enzymes in Biomass Conversion* (G. F. Leatham and M. E. Himmel, eds.), American Chemical Society, Washington, D.C., pp. 200–208.

Brock B. J., Rieble, S., and Gold M. H., 1995. Purification and characterization of a 1,4-benzoquinone reductase from basidiomycete *Phanerochaete chrysosporium*, *Appl. Environ. Microbiol* 61:3076–3081.

Brown, D., and Hamburger, B., 1987. The degradation of dyestuff: Part 3, Investigation of their ultimate degradability, *Chemosphere* 16:1539–1553.

Brown, D., and Laboureur, P., 1983. The degradation of dyestuff: Part 1, Primary biodegradation under anaerobic conditions, *Chemosphere* 12:397–408.

Brown, M. A., and DeVito S. C., 1993. Predicting azo dye toxicity, *Crit. Rev. Environ. Sci. Technol.* 23:249–324.

Bumpus, J. A., and Brock, B. J., 1988. Biodegradation of Crystal Violet by the white-rot fungus *Phanerochaete chrysosporium*, *Appl. Environ. Microbiol.* 54:1140–1150.

Capalash, N., and Shrama, P., 1992. Biodegradation of textile azo-dyes by *Phanerochaete chrysosporium*, *World J. Microbiol. Biotechnol.* 8:309–312.

Chahal, D. S., Kluepfel, D., Morosoli, R., Shereck, F., Laplante, S., and Rouleau, D., 1995. Use of dyes in solid medium for screening ligninolytic activity of selective Actinomycetes, *Appl. Biochem. Biotechnol.* 51:137–144.

Chet, I., Trojanowski, J., and Huttermann, A.,1985. Decolorization of the Poly B-411 and its correlation with lignin degradation by fungi, *Microbios. Lett.* 29:37–43.

Chivukula, M., and Renganathan, V., 1995. Phenolic azo dye oxidation by laccase from *Pyricularia oryze*, *Appl. Environ. Microbiol.* 61:4374–4377.

Chivukula, M., Spadaro, J. T., and Renganathan V., 1995. Lignin peroxidase-catalyzed oxidation of sulfonated azo dyes generates novel sulfophenyl hydroperoxides, *Biochemistry*, 34:7765–7765.

Chung K-T., 1983. The significance of azo reduction in the mutagenesis and carcinogenesis of azo dyes, *Mutation Res.* 114:269–281.

Claxton, L. A., Walsh, D. B., Esancy, J. F., and Freeman, H. S., 1990. Structure and activity analysis of azo dyes and related compounds, in: *Mutation and the Environment*, part B, Wiley-Liss, Inc., New York, pp. 11–22.

Cripps, C., Bumpus, J. A., and Aust. S. D., 1990. Biodegradation of azo and heterocyclic dyes by *Phanerochaete chrysosporium*, *Appl. Environ. Microbiol.* 56:1114–1118.

Daniel, G., 1994. Use of electron microscopy for aiding our understanding of wood biodegradation, *FEMS Microbiol. Rev.* 13:199–223.

Dawson, D. J., 1981. Polymeric dyes, *Aldrichimica Acta* 14:23–29.

De Jong, E., De Vries, F. P., Field, J. A., Van Der Zwan, R. P., and De Bont, J. A. M., 1992. Isolation and screening of basidiomycetes with high peroxidase activity, *Mycol. Res.* 96:1098–1104.

Freitag, M., and Morrell, J. J., 1992. Decolorization of the polymeric dye Poly R-487 by wood-inhabiting fungi, *Can. J. Microbiol.* 38:811–822.

Gerber, N. C., and Ortiz de Montellano, P. R., 1995. Neuronal nitric oxide synthase. Expression in *Escherichia coli*, irreversible inhibition by phenyldiazene, and active site topology, *J. Biol. Chem.* 270:17791–17796.

Glenn, J. K. and Gold, M. H., 1983. Decolorization of several polymeric dyes by the lignin degrading basidiomycete *Phanerochaete chrysosporium*, *Appl. Environ. Microbiol.* 45:1741–1747.

Gogna, E., Vohra, R., and Sharma, P., 1991. Biodegradation of Rose Bengal by *Phanerochaete chrysosporium*, *Lett. Appl. Microbiol.* 14:58–60.

Gold, M. H., and Alic, M., 1993. Molecular biology of the lignin-degrading basidiomycete *Phanerochaete chrysosporium*. *Microb. Rev.* 57:605–622.

Goszczynski, S., Paszczynski, A., Pasti-Grigsby, M. B., Crawford, R. L., and Crawford, D. L., 1994. New pathway for degradation of sulfonated azo dyes by microbial peroxidases of *Phanerochaete chrysosporium* and *Streptomyces chromofuscus*, *J. Bacteriol.* 176:1339–1347.

Govindaswami, M., Schmidt, T. M., White, D.C., and Loper, J. C., 1993. Phylogenetic analysis of a bacterial aerobic degrader of azo dyes, *J. Bacteriol.* 175:6062–6066.

Hammel, K., 1995. Organopollutant degradation by ligninolytic fungi, in: *Microbial Transformation and Degradation of Toxic Organic Chemicals.* (L. Y. Young and C. E. Cerniglia, eds.), John Wiley & Sons, Inc., New York, pp. 331–346.

Hartwell, J. L., and Fieser, F. L., 1943. Coupling of *o*-Tolidine and Chicago Acid, in: *Organic Synthesis, Collective Vol 2, A Revised Edition of Annual Volumes 10–19* (A. H. Blatt, ed.), John Wiley & Sons, New York, pp. 145–149.

Haug, W., Schmidt, A., Nortemann, B., Hempel, D., Stolz, A., and Knackmuss, H., 1991. Mineralization of the sulfonated azo dye Mordant Yellow 3 by a 6-aminonaphthalene-2-sulfonate-degrading bacterial consortium, *Appl. Environ. Microbiol.* 57:3144–3149.

Hoffman, R. V. and Kumar, A., 1984. Oxidation of hydrazine derivatives with arylsulfonyl peroxides, *J. Org. Chem.* 49:4014–4017.

Hunger, K. Mischke, P. Rieper, W. and Raue, R., 1985. Azo dyes, in: *Ulmann's Encyclopedia of Industrial Chemistry,* 5th ed. (F. T. Campbell, R. Pfefferkorn, and J. F. Rounsaville, eds.), VCH Publishers, Deerfield Beach, Florida, pp. 245–323.

Joachim, F., Burrell, A., and Anderson, J., 1985. Mutagenicity of azo dyes in the *Salmonella*/microsome assay using *in vitro* and *in vivo* activation, *Mutation Res.* 156:131–138.

Kirby, N., McMullan, G., and Marchant, R., 1995. Decolourisation of an artificial textile effluent by *Phanerochaete chrysosporium*, *Biotechnol. Lett.* 17:761–764.

Kling, S. H., and Neto, J. S. A., 1991. Oxidation of Methylene Blue by crude lignin peroxidase from *P. chrysosporium*, *J. Biotechnol.* 21:295–300.

Kropf, H., Aryl-hydroperoxide, alkyl-aryl-peroxide, diaryl-peroxide, 1988. in: *Methoden der Organischen Chemie (Houben-Weyl)* Volume E13, (H. Kropf, ed.), Georg Thieme Verlag, Stuttgart, pp. 762–763.

Kulla, H. G., 1981. Aerobic bacterial degradation of azo dyes, in: *Microbial Degradation of Xenobiotics and Recalcitrant Compounds* (T. Leisinger, A. M. Cook, R. Hootter, and R. Nuesch, eds.), Academic Press, Inc., Ltd., London, pp. 387–389.

Kulla, H. G., Klausener, F., Meyer, U., Ludeke, B., and Leisinger, T., 1983. Interference of aromatic sulfo groups in microbial degradation of azo dyes Orange I and Orange II, *Arch. Microbiol.* 135:1–7.

Kulla, H. G., Krieg, R., Zimmermann, T., and Leisinger, T., 1984. Biodegradation of xenobiotics. Experimental evaluation of azo dye-degrading bacteria, in: *Current Perspectives in Microbial Ecology* (M. J. Klug and C. A. Reddy, eds.), American Society for Microbiology, Washington DC, pp. 663–667.

Lin, G. H. Y. and Solodar W. E., 1988. Structure activity relationship studies on the mutagenicity of some azo dyes in the *Salmonella*/microsome assay, *Mutagenesis* 3:311–315.

Lyons, C. D., Katz, S., and Bartha, R., 1984. Mechanisms and pathways of aniline and elimination from aquatic environment, *Appl. Environ. Microbiol.* 48:491–496.

Mannen, S., and Itano, H. A., 1973. Stoichiometry of the oxidation of arylhydrazines with ferricyanide. Quantitative measurements of absorption spectra of aryldiazenes, *Tetrahedron* 29:3497–3502.

Meyer, U., 1981. Biodegradation of synthetic organic colorants, in: *Microbial Degradation of Xenobiotics and Recalcitrant Compounds* (T. Leisinger, R. Hutter, A. M. Cook, and J. Nuesch, eds.), Academic Press, London, pp. 387–399.

Muralikrishna, C., and Renganathan, V., 1993. Peroxidase-catalysed desulfonation of 3,5-dimethyl-4-hydroxy and 3,5-dimethyl-4-aminobenzenesulfonic acid, *Biochem. Biophys. Res. Commun.* 197:798–804.

Newmyer S. L. and Ortiz de Montellano, P. R., Horseradish peroxidase His-42 → Ala, His-42 → Valine, and Phe-41 → Ala mutants. Histidine catalysis and control of substrate access to the heme iron, 1995. *J. Biol. Chem.* 270:19430–19438.

Nicholson, J., and Cohen, S. G., 1966. Phenyldiimide. III. Ferric ion catalyzed formation of free radicals in heterolysis of azo compounds, *J. Am. Chem. Soc.* 86:2247–2252.

Ollikka, P., Alhonmaki, K., Leppanen, V.-M., Glumoff, T., Raijola, T., and Suominen, I., 1993. Decolorization of azo, triphenyl methane, heterocyclic, and polymeric dyes by lignin peroxidase isoenzymes from *Phanerochaete chrysosporium*, *Appl. Environ. Microbiol.* 59:4010–4016.

Pasti, M. B., Hagen, S. R., Goszczynski, S., Paszczynski, A., Crawford, R. L., Crawford, D. L., 1991. The influence of guaiacol and syringyl groups in azo dyes on their degradation by lignocellulolytic *Streptomyces* spp., in: *Abstracts of the International Symposium on Applied Biotechnology for Tree Culture, Protection, and Utilization,* Battelle Press, Columbus, Ohio, pp. 119–120.

Pasti, M. B., Paszczynski, A., Goszczynski, S., Crawford, D. L., and Crawford, R. L., 1992. Influence of aromatic substitution patterns on azo dye degradability by *Streptomyces* spp. and *Phanerochaete chrysosporium*, *Appl. Environ. Microbiol.* 58:3605–3613.

Pasti-Grigsby, M. B., Paszczynski, A., Goszczynski, S., Crawford, D. L., Crawford, R. L. 1992. Influence of aromatic substitution patterns on azo dye degradability by *Streptomyces* spp. and *Phanerochaete chrysosporium*, *Appl. Environ. Microbiol.*, *58*, 3605–3613.

Pasti-Grigsby, M. B., Paszczynski, A., Goszczynski, S., Crawford, D. L., and Crawford, R. L., 1994a. Biodegradation of novel azo dyes, in: *Applied Biotechnology for Site Remediation* (R. E. Hinchee, D. B. Anderson, F. B. Metting, Jr., G. D. Sayles, eds.), Lewis Publishers, Boca Raton, Florida, pp. 384–390.

Pasti-Grigsby, M. B., Paszczynski, A., Goszczynski, S., Crawford D. L., and Crawford, R. L. 1994b. Use of dyes in assaying *Phanerochaete chrysosporium* Mn(II)-peroxidase and ligninase, *Proc. Inst. Mol. Agric. Gen. Eng. (IMAGE)* 1, 1–12. http://www.uidaho.edu/~crawford/image.html

Paszczynski, A., and Crawford, R. L, 1991. Degradation of azo compounds by ligninase from *Phanerochaete chrysosporium*: involvement of veratryl alcohol, *Biochem. Biophys. Res. Commun.* 178:1056–1063.

Paszczynski, A., Pasti, M. B., Goszczynski, S., Crawford, D. L., Crawford. R. L., 1991a. Designing biodegradability: lessons from lignin, in: *Abstracts of the International Symposium on Applied Biotechnology for Tree Culture, Protection, and Utilization.* Battelle Press, Columbus, Ohio, pp. 73–78.

Paszczynski, A., Pasti, M. B., Goszczynski, S., Crawford, D. L., Crawford, R. L., 1991b. New approach to improve degradation of recalcitrant azo dyes by *Streptomyces* spp. and *Phanerochaete chrysosporium. Enzyme Microb. Technol.* 13:378–384.

Paszczynski A., and Crawford, R. L., 1995. Potential for bioremediation of xenobiotic compounds by the white-rot fungus *Phanerochaete chrysosporium*, *Biotechnol. Prog.* 11:368–379.

Paszczynski, A., Goszczynski, S., Crawford, R. L., and Crawford, D. L., 1995. Interaction of peroxidases with dyes and plastics, in: *Microbial Processes for Bioremediation* (R. E. Hinchee, F. J. Brockman, and C. M. Vogel, eds.), Battelle Press, Columbus, Ohio, pp. 187–195.

Paszczynski, A., Goszczynski, S., Crawford, R. L., Crawford, D. L., and Pasti, M. B., 1996. Biodegradable azo dyes, United States Patent 5,486,214.

Platt, M. W., Hadar, Y., and Chet, I., 1985. The decolorization of the polymeric dye Poly-blue (polyvinylamine sulfonate-anthroquinone) by lignin degrading fungi, *Appl. Microbiol. Biotechnol.* 21:394–396.

Raag, R., Swanson, B. A., Poulos, T. L., and Ortiz de Montellano, P. R., 1990. Formation, crystal structure, and rearrangement of a cytochrome P—450_{cam} iron-phenyl complex, *Biochemistry* 29:8119–8126.

Reddy, C. A., 1995. The potential for white-rot fungi in the treatment of pollutants, *Curr. Opin. Biotechnol.* 6:320–328.

Reid, T., Morton K. C., Wang C. Y., and King, C. M., 1984. Mutagenicity of azo dyes following metabolism by different reductive/oxidative systems, *Environ. Mutagenesis* 6:705–717.

Rieble, S., Joshi, D. K., and Gold, M. H., 1994. Purification and characterization of a 1,2,4-trihydroxybenzene 1,2-dioxygenase from the basidiomycete *Phanerochaete chrysosporium*, *J. Bacteriol.* 176:4838–4844.

Rothkopf, G. S. and Bartha, R., 1984. Structure-biodegradability correlations among xenobiotics industrial amines, *JAOCS* 61:977–980.

Sarnaik, S. and Kanekar, P., 1995. Bioremediation of colour of Methyl Violet and phenol from a dye-industry waste effluent using *Pseudomonas* spp. isolated from factory soil, *J. Appl. Bacteriol.* 79:459–469.

Sayadi, S. and Ellouz, R., 1995. Role of lignin peroxidase and manganese peroxidase from *Phanerochaete chrysosporium* in decolorization of olive mill wastewater, *Appl. Environ. Microbiol.* 61:1098–1103.

Schiephake, K., Lonergan, G. T., Jones, C. L., and Mainwaring, D. E., 1993. Decolorization of pigment plant effluent by *Pycnoporus cinnabarinus* in packed-bed reactor, *Biotechnol. Lett.* 15:1185–1188.

Shahin, N. M., 1989. Evaluation of the mutagenicity of azo dyes in *Salmonella typhimurium*: a study of structure activity relationship, *Mutagenesis* 4:115–125.

Spadaro, J. T., Gold, M. H., and Renganathan, V., 1992. Degradation of azo dyes by the lignin-degrading fungus *Phanerochaete chrysosporium, Appl. Environ. Microbiol.* 58:2397–2401.

Suehiro, T., 1988. Behavior of aryldiazenyl radicals in solution, *Rev. Chem. Intermed.* 10:101–137.

Swanson, B. A., Dutton, D. R., Lunetta, J. M., Yang, C. S., and Ortiz de Montellano, P. R., 1991. The active sites of cytochromes P450 Ia1, IIB1, IIB2, and IIE1. Topological analysis by in situ rearrangement of phenyl-iron complexes, *J. Biol. Chem.* 266:9258–19264.

Tour, U., Winterhalter, K., and Fiechter, A., 1995. Enzymes of white-rot fungi involved in lignin degradation and ecological determinations for wood decay. *J. Biotechnol.* 41:1–17.

Ulmer, D. C., Leisola, M. S. A., and Fiechter, A., 1984. Possible induction of the ligninolytic system of *Phanerochaete chrysosporium*, *J. Biotechnol.* 1:13–24.

Vyas, B. R. M., and Molitoris, H. P., 1995. Involvement of an extracellular H_2O_2-dependent ligninolytic activity of the white rot fungus *Pleurotus ostreatus* in the decolorization of Remazol Brilliant Blue R, *Appl. Environ. Microbiol.* 61:3919–3927.

5

ENGINEERING ENZYMES AND MICROORGANISMS FOR THE TRANSFORMATION OF SYNTHETIC COMPOUNDS

Joost P. Schanstra, Gerrit J. Poelarends, Tjibbe Bosma, and Dick B. Janssen

Department of Biochemistry
Groningen Biomolecular Sciences and Biotechnology Institute
University of Groningen
9747 AG Groningen, The Netherlands

1. INTRODUCTION

The biotransformation of synthetic chemicals that enter the environment is dependent on the capacity of microbial enzymes to recognize xenobiotic substrates and catalyze reactions with stable structural elements, such as carbon-halogen bonds. The range of compounds that can be enzymatically degraded strongly influences the environmental fate of many potentially harmful compounds. Environmental recalcitrance thus is, in part, a problem of enzyme activity and specificity. This can be well illustrated with the halogenated aliphatics, which are frequently encountered as environmental pollutants due to losses and emissions during their use in industry and agriculture. Halogenated aliphatic compounds are used as blowing agents (methylchloride), cooling liquids (ethylchloride), soil fumigants (1,3-dichloropropylene, methylbromide), insecticides (hexachlorocyclohexane), intermediates in chemical synthesis (1,2-dichloroethane, vinyl chloride, chloroacetates) and as solvents (trichloroethanes, tri- and tetrachloroethene). Several related structures (e.g., chlorinated alkanes, ethers and alcohols) occur as wastes or contaminants of products.

In general, microbial degradation processes can be divided in cometabolic conversions and metabolically productive transformations. The former are usually mediated by organisms that produce aspecific enzymes, and are considered to be the result of the fact that many enzymes are not specific for their physiological substrate. Productive conversions, on the other hand, occur in organisms that use chlorinated compounds as a sole carbon source for growth (Figure 1). Productive, growth-stimulating degradation is the

Biotechnology in the Sustainable Environment, edited by Sayler *et al.*
Plenum Press, New York, 1997

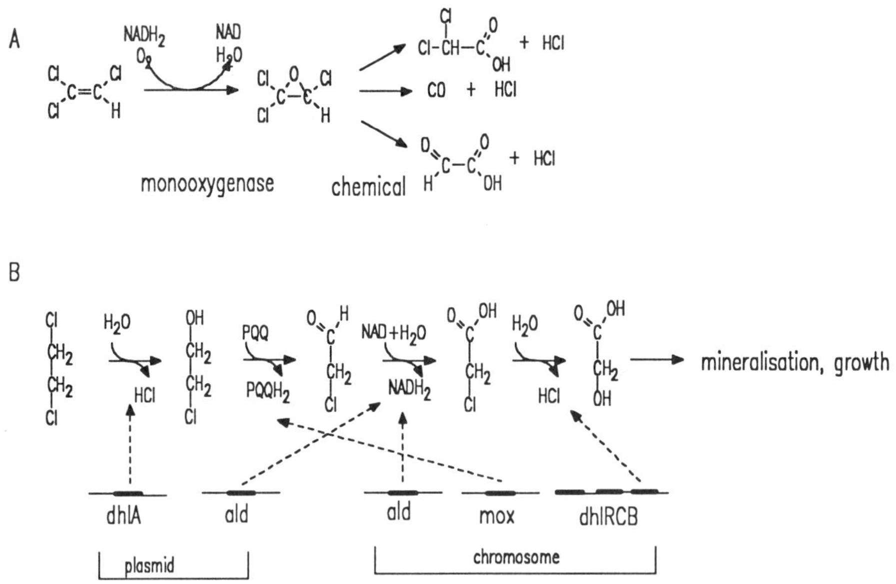

Figure 1. A. Cometabolic conversion of trichloroethylene to an epoxide by soluble methane monooxygenase from *Methylosinus trichosporium* OB3b. The unstable epoxide spontaneously decomposes into a variety of products (Fox et al., 1990). **B.** Catabolic pathway for the productive metabolism of 1,2-dichloroethane in *Xanthobacter autotrophicus* GJ10 (Keuning et al., 1985). The gene coding for haloalkane dehalogenase (*dhlA*) is located on a large (200 kb) plasmid (Tardiff et al., 1991).

preferred process from an application point of view, since the cells benefit and proliferate at the expense of the polluting chemical. Cometabolic transformations do not stimulate growth of the responsible organisms, and an additional C-source is required to induce the necessary cometabolic enzymes. Full scale applications of organisms that grow at the expense of a chlorinated solvent have been developed (Stucki & Thuer, 1995).

General rules in the sense of structural characteristics that determine the possibility that a compound can serve as a carbon source are difficult to establish for haloaliphatics, apart from the obvious thermodynamic condition. The reason for this is that blocks in a degradation pathway may occur at different steps, and that the possibility of carbon-halogen bond cleavage is strongly determined by the specificity of the dehalogenating enzyme. Furthermore, individual enzymes have substrate specificities that are not directly correlated with chemical structure or reactivity. Nevertheless, some rules can be given (Figure 2). Degradability (in the sense of the existence of organisms that can use a compound for growth) certainly decreases with increasing chlorine substitution, and the presence of hydroxyl or oxo functions increases degradability. Significant differences may exist between regioisomers and stereoisomers.

Based on a detailed insight in the structure-function relationship of enzymes that transform chlorinated chemicals, it is becoming possible to modify their specificity by site-directed mutagenesis. Several three-dimensional structures of microbial enzymes that can degrade chlorinated xenobiotics have now been determined. This includes methane monooxygenase (Rosenzweig et al., 1993), cytochrome P450 (Poulos et al., 1987), and haloalkane dehalogenase (Verschueren et al., 1993a). The latter enzyme is involved in the degradation of a whole range of environmentally important chlorinated compounds, and will be discussed in detail here.

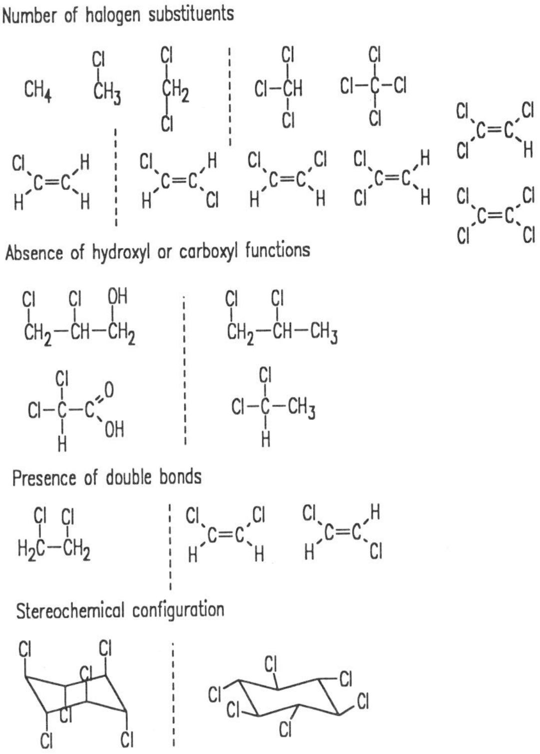

Figure 2. Relationships between structural properties and microbial utilization of halogenated compounds as a growth substrate. On the left side of the broken lines, compounds are listed that are known to support growth of one or more known bacterial cultures. On the right side, recalcitrant analogs are given.

2. ENZYMES CATALYZING DEHALOGENATION OF HALOALIPHATICS

Many organisms that grow on chlorinated aliphatics have been obtained in pure culture. In most cases the enzymes that catalyze the dehalogenation reactions have been identified. These enzymes are likely to act specifically on carbon-halogen bonds, which at the molecular level means that the halogen function is in some way recognized by the enzyme, and that an interaction between enzyme and halogen function is important for catalysis. The synthesis of dehalogenases is often regulated by induction, confirming that the enzymes have a specific physiological function related to utilization of halogenated compounds (Janssen et al., 1994).

Examples of such dehalogenating enzymes are given in Figure 3. A diversity of reaction mechanisms seems to be involved, including glutathione conjugation, hydrolytic dehalogenation, intramolecular substitution, and elimination. The enzymes often have been purified and the gene sequences for glutathione transferases, hydrolytic dehalogenases, lyases, and the enzyme catalyzing elimination of HCl from hexachlorohexane have been published. This has opened the way to obtain mechanistic insight and to study the evolutionary origin of dehalogenating enzymes.

3. HALOALKANE DEHALOGENASES

Haloalkane dehalogenases catalyze the hydrolytic cleavage of carbon-chlorine bonds in a wide range of halogenated alkanes (Keuning et al., 1985). The first example

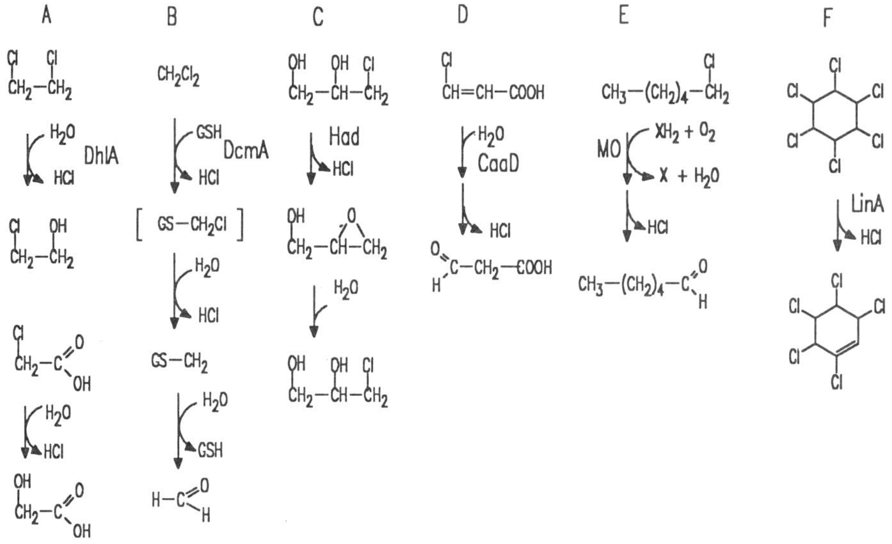

Figure 3. Examples of dehalogenating enzymes. **A.** Hydrolytic cleavage of C-X bonds; **B.** glutathione transferase mediated conversion of dichloromethane; **C.** intramolecular nucleophilic substitution; **D.** hydrolysis of chloroacrylic acid; **E.** oxidative dehalogenation; **F.** elimination of HCl.

came from our studies on 1,2-dichloroethane degradation. The initial step in its utilization by *Xanthobacter autotrophicus* and *Ancylobacter aquaticus* (van den Wijngaard et al., 1992) is catalyzed by a 35 kDa dehalogenase that produces 2-chloroethanol and chloride. The substrate range of the enzyme is very broad and includes also environmental pollutants such as 1,2-dibromoethane and 1,3-dichloropropylene. Haloalkane dehalogenases that catalyze conversion of long-chain chloroalkanes have also been identified (Scholtz et al., 1987; Yokota et al., 1987; Sallis et al., 1989; Curragh et al., 1994).

The three-dimensional structure of the *X. autotrophicus* dehalogenase (DhlA) was solved by X-ray crystallography. The active site of the enzyme is a hydrophobic cavity located between a globular main domain with an α/β-hydrolase fold structure (Ollis et al., 1992) and a separate cap domain (Verschueren et al., 1993a). The catalytic mechanism was studied by X-ray crystallography and isotope incorporation studies (Figure 4, Verschueren et al., 1993b; Pries et al., 1994a). These experiments indicated that DhlA catalyzes cleavage of halogenated compounds by nucleophilic displacement of the halogen by the carboxylate of Asp124. The resulting covalent alkyl-enzyme intermediate is hydrolyzed by attack of water at the carbonyl carbon of the esterified Asp124. His289, together with Asp260, activates this water molecule by subtracting a proton (Pries et al., 1995a). Two tryptophans are involved in binding of the substrate halogen and the halide ion after it is released from the substrate (Verschueren et al., 1993c; Kennes et al., 1995).

Since 1,2-dichloroethane is not known to occur naturally, the dehalogenase must have existed in preindustrial times as an enzyme with a closely related function, or it must have evolved during the last 50 years or so from an enzyme that had a different function. We have proposed, on basis of the sequence of mutant enzymes, that the dehalogenase has recently evolved from a more primitive dehalogenase. The sequence of the dehalogenase gene harbors sequence duplications that are indicative of recent evolutionary changes (Pries et al., 1994b). Furthermore, the constitutive expression of the dehalogenase gene in *Ancylobacter* and *Xanthobacter* strains that degrade 1,2-dichloroethane, the plasmid local-

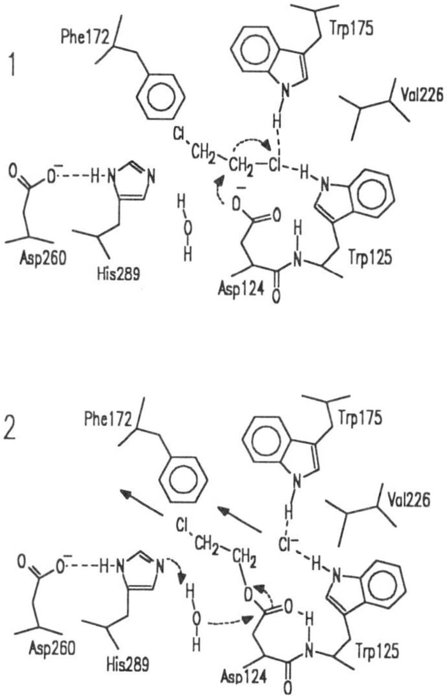

Figure 4. Catalytic mechanism of haloalkane dehalogenase. The reaction is initiated by nucleophilic attack of Asp124 on the Cα of the alkylhalide, leading to formation of a covalent alkyl-enzyme intermediate. In the following step this intermediate is hydrolyzed by an activated water molecule where His289 acts as a general base.

ization of the gene, and the lack of evolutionary divergence between 1,2-dichloroethane dehalogenases isolated from different organisms also indicate that the enzyme is of recent evolutionary origin.

4. ENGINEERING PROSPECTS

4.1. Kinetic Mechanism of Haloalkane Dehalogenase

The recalcitrance of several highly chlorinated alkanes can be attributed to the lack of activity of dehalogenases toward these compounds. Therefore, it would be attractive to expand the specificity of these enzymes by protein engineering, which requires knowledge about the three-dimensional structure, the reaction mechanism, the reaction kinetics, and the relationship between these properties. We have decided to explore the possibilities of this approach with haloalkane dehalogenase. The structure and catalytic mechanism of the enzyme were described above.

Recently, we have used a combination of steady-state and pre-steady-state kinetic experiments to establish the kinetic mechanism and the associated rate constants (Figure 5). Stopped-flow fluorescence quenching experiments are possible since the fluorescence of Trp125 and Trp175 is quenched when halides or halogens are bound in the active site. The results showed that release of the charged halide ion out of the buried active site cavity could limit the overall conversion rate of the enzyme. A complex dependence of the observed rate on halide concentration was observed. It suggested the existence of two parallel routes for halide release at the end of the catalytic sequence. In the most important route, a slow conformational change of the enzyme preceded rapid halide dissociation.

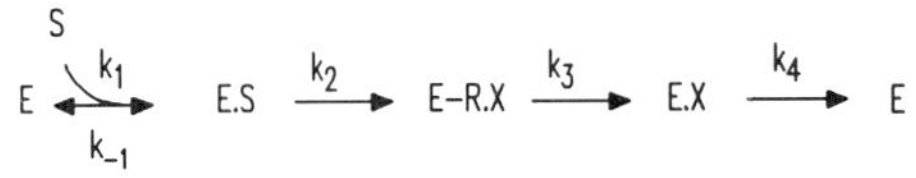

Figure 5. Kinetic scheme of haloalkane dehalogenase. Steps: 1, formation of the Michaelis complex; 2, cleavage of the carbon-halogen bond; 3, cleavage of the covalent intermediate; 4, conformational change and halide dissociation.

This conformational change probably is a motion of a part of the cap domain of the enzyme that is needed to allow water to enter the active site and solvate the halide ion.

The rates of all steps in the dehalogenase reaction were determined for 1,2-dibromoethane and 1,2-dichloroethane conversion (Table 1). The rate of C-Cl bond cleavage was much slower than the rate of C-Br bond cleavage. This caused a much lower K_m for dibromoethane than for dichloroethane. Halide release was indeed found to be the main rate-determining step in the hydrolysis of 1,2-dichloroethane and 1,2-dibromoethane.

Steady-state kinetic analysis of the conversion of a number of other substrates showed that the enzyme has a higher apparent affinity (lower K_m) for brominated compounds than for their chlorinated analogs, which originated from the higher rate of carbon-halogen bond cleavage and a higher second-order association rate constant of enzyme and substrate for brominated compounds. For good substrates at high concentrations, brominated and chlorinated analogs were converted at similar rates since a conformational change required for halide release is rate-limiting. For poor substrates, brominated analogs were converted faster since the rate of carbon-halogen bond cleavage is rate-limiting.

4.2. Construction of Faster Mutants for 1,2-Dibromoethane

A number of mutants were constructed that were expected to show an increased rate of halide release. Kinetic analysis of a Val226Ala mutant dehalogenase with a 2.5-fold higher k_{cat} for 1,2-dibromoethane confirmed the observation that halide release is the main rate-limiting step in the wild-type haloalkane dehalogenase (Table 1). The enzyme isomerization preceding actual halide release was five to seven-fold faster in the mutant than in the wild type. The observation that the increase in k_{cat} was only found with 1,2-dibromoethane and not with 1,2-dichloroethane can be explained by a strongly reduced rate of carbon-halogen bond cleavage. As the rate of carbon-chlorine bond cleavage is already lower than that of carbon-bromine cleavage, carbon-halogen becomes the slowest step for

Table 1. Kinetic constants for a four-step reaction scheme[a] of Val226Ala and wild-type haloalkane dehalogenase

	k_1 ($mM^{-1}.s^{-1}$)	k_{-1} (s^{-1})	k_2 (s^{-1})	k_3 (s^{-1})	k_4 (s^{-1})	k_{cat}[b] (s^{-1})	K_m[b] (mM)
Wild-type enzyme							
1,2-dibromoethane	750	>20	>130	10	4	2.8	0.004
1,2-dichloroethane	9	20	50	14	8	4.6	0.72
Val226Ala mutant							
1,2-dibromoethane	410	45	60	12	43	8.1	0.035
1,2-dichloroethane	4.5	25	14	9	50	4.9	3.1

[a]The reaction scheme used is given in Fig. 5 where E.RX is the Michaelis complex, E-R.X is the alkyl-enzyme intermediate and E.X is halide bound enzyme.
[b]Steady-state kinetic parameters calculated from the derived rate constants.

1,2-dichloroethane conversion in the mutant. Apparently, the rates of carbon-halogen bond cleavage and halide release are inversely correlated, and because these steps are well balanced for 1,2-dichloroethane conversion, it is much more difficult to improve the enzyme for 1,2-dichloroethane conversion than for 1,2-dibromoethane conversion. Thus, the wild-type enzyme is optimized for 1,2-dichloroethane, although 1,2-dibromoethane is a better substrate.

A similar effect on activity was seen in a Phe172Trp mutant of haloalkane dehalogenase (Table 2). Phe172 is one of the hydrophobic cap domain residues that is located in the active site cavity. It is a member of the helix-loop-helix structure covering the active site, and might interact with the R-group of substrate during the reaction. Mutation of Phe172 showed that only a Trp, Tyr, His, Met and Cys at position 172 gave enzyme that retained its catalytic activity with 1,2-dibromoethane. The Phe172Trp enzyme had a 10-fold higher k_{cat}/K_m for 1-chlorohexane (Table 2) and a 2-fold higher k_{cat} for 1,2-dibromoethane than the wild-type enzyme.

The X-ray structure of the Phe172Trp enzyme indicated that the flexibility of a helix-loop-helix structure covering the active site of the dehalogenase was increased compared to the wild-type enzyme (J.P. Schanstra et al., submitted for publication). This can explain the elevated activity and affinity for 1-chlorohexane of the Phe172Trp enzyme, since it allows this large substrate, that does not fit in the active site cavity of the wild type, to bind more easily. Pre-steady-state analysis of 1,2-dibromoethane conversion showed that, like in the Val226Ala mutant, the increase in catalytic rate could be attributed to an increase in the rate of the enzyme isomerization that preceded halide release, and the structural differences indicated that the isomerization may indeed be a conformational change in a part of the cap domain.

4.3. *In Vivo* Selection of New Specificities

Recently, Pries et al., (1994b) investigated the adaptation of the dehalogenase to a new substrate by selecting spontaneous mutants that grow on 1-chlorohexane, a compound hardly hydrolyzed by the wild-type enzyme. The results of these selection experiments indicated that generation of short direct repeats in the N-terminal part of the cap domain plays an important role in the adaptation of haloalkane dehalogenase to new substrates. The N-terminal part of the cap domain of wild-type haloalkane dehalogenase already contains two direct repeats positioned in frame: a 15-bp perfect repeat and a 9 bp repeat carrying a 1 bp substitution. Deletion of the two repeats resulted in loss of 1,2-dichloroethane hydrolysing activity, but not in loss of activity for several brominated dehalogenase substrates. This indicates that the direct repeats in the DNA sequence encoding the N-terminal part of the cap domain could be of recent evolutionary origin and were se-

Table 2. Specificity constants (k_{cat}/K_m values) for 1,2-dichloroethane, 1,2-dibromoethane and 1-chlorohexane conversion for wild type, Phe172Trp (*in vitro* constructed), and Asp170His (*in vivo* selected) haloalkane dehalogenase

	k_{cat}/K_m (mM^{-1}.s^{-1})		
Enzyme	1,2-dichloroethane	1,2-dibromoethane	1-chlorohexane
Wild type	6.2	300	0.063
Phe172Trp	0.56	240	0.67
Asp170His[a]	–	30	0.40

[a]Data from Pries et al., 1994b.

lected during the adaptation of an older dehalogenase to industrially produced dichloroethane.

It also presents another example of improving the catalytic activity of the enzyme for a substrate for which it is not optimized (1-chlorohexane, Table 2). Kinetically, the improved activity seems to be due to an elevated rate of cleavage of the carbon-halogen bond in bound substrate. The strategy used, i.e., selection of a new catalytic activity in a host organism to which the desired activity confers a selective growth advantage, may be exploited more widely for obtaining dehalogenase variants that degrade recalcitrant substrates.

5. NATURAL DIVERSITY AND DISTRIBUTION OF GENES

5.1. α/β-Hydrolase Fold Enzymes in Biodegradation

The structure and mechanism of haloalkane dehalogenase is shared with a group of enzymes commonly classified as α/β-hydrolases (Ollis et al., 1992). In these enzymes, the position of the active site residues that form the catalytic triad is conserved along the sequence and follows the order nucleophile-charge relay carboxylate-histidine, each separated by a variable number of other amino acids. Only the histidine is strictly conserved. The nucleophile may be an aspartate (as in haloalkane dehalogenase), a serine, or a cysteine. An aspartate is used in enzymes that catalyze nucleophilic substitution on a sp^3 hybridized carbon atom.

Various microbial enzymes involved in the degradation of xenobiotics have this α/β-hydrolase fold structure. This includes 3 haloalkane dehalogenases, a fluoroacetate dehalogenase, and an epoxide hydrolase, which all have an Asp as the nucleophile. The reason for the Asp is the requirement for the presence of a carbonyl function in the covalent intermediate (Janssen et al., 1994). In the α/β-hydrolase fold proteolytic enzymes, triacylglyceride lipases, and diene lactone hydrolases (Ollis et al., 1992; Derewenda et al., 1993) the carbonyl function is provided by the substrate, and the enzyme has a serine as the nucleophile. A serine as the nucleophile in a dehalogenase would produce an ether as the covalent intermediate, which is difficult to hydrolyze since there is no possibility to form a tetrahedral intermediate.

5.2. The Haloalkane Dehalogenase Family

Based on sequence homology to haloalkane dehalogenase from *X. autotrophicus*, several other dehalogenases were proposed to belong to the α/β-hydrolase fold group and have the same catalytic mechanism (Figure 6). This includes LinB (Nagata et al., 1993), involved in hexachlorocyclohexane degradation, DehH1 (Kawasaki et al., 1992), involved in fluoroacetate degradation, dehalogenases produced by *Rhodococcus* strains that degrade long chain chloroalkanes (M. Larkin et al., pers. commun.), and a dehalogenase that we have recently identified in a 1,3-dichloropropylene degrading *Pseudomonas cichorii* strain.

The known dehalogenases do not share high sequence homology to each other or to related proteins from which they may have recently evolved. The question thus arises of how these enzymes evolved and were distributed. We are currently searching for natural compounds that might be degraded by dehalogenases, and that could be a source of the genes as they occur in the xenobiotics-degrading strains.

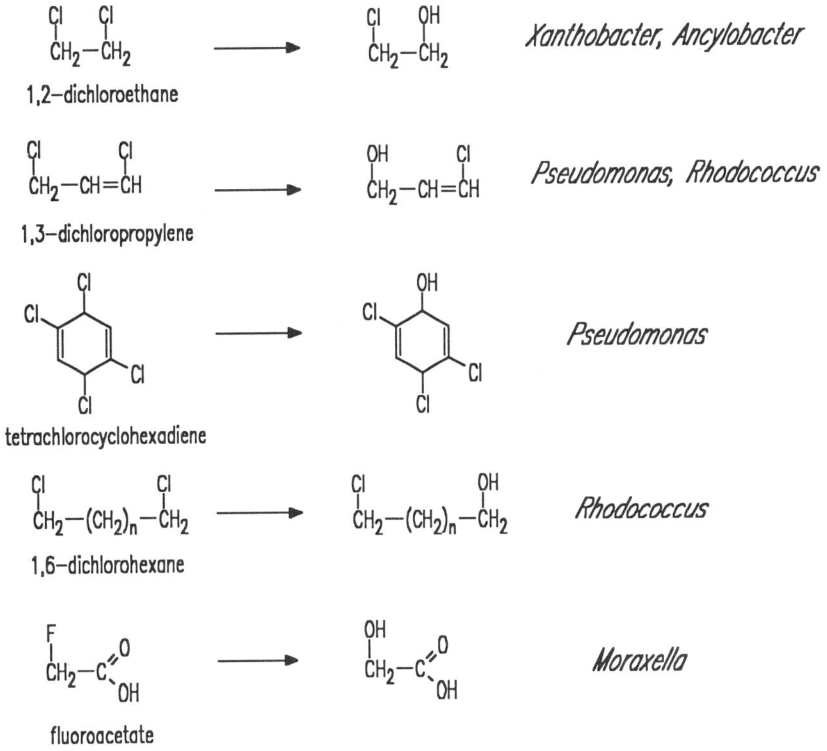

Figure 6. Compounds known to be hydrolyzed by dehalogenases that share the structural topology and overall catalytic mechanism.

The evolution and distribution of these dehalogenases may have been triggered or stimulated by the release of xenobiotics into the environment. A candidate for such a chemical is 1,3-dichloropropylene. This compound is an important nematocide that is applied to agricultural fields in the Netherlands at 3,000 tonnes annually. Bacteria that degrade it have been isolated from agricultural soils that showed accelerated degradation of the nematocide, making it essentially ineffective (Verhagen et al., 1995). The catabolic route for 1,3-dichloropropylene starts with hydrolysis to chloroallylalcohol (Figure 7). Us-

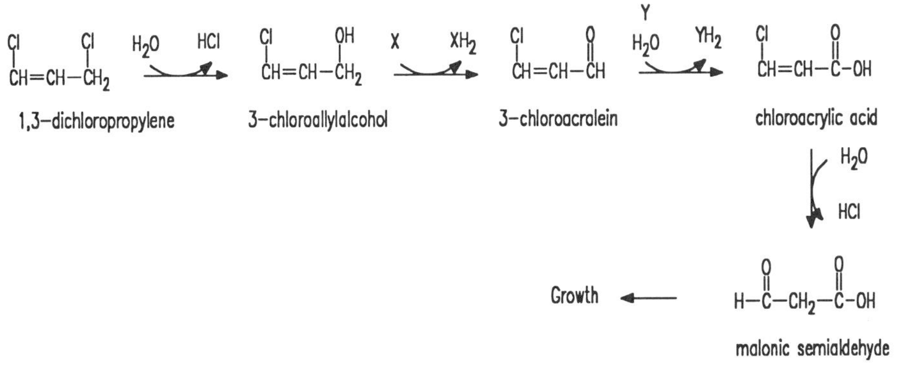

Figure 7. Catabolic route for 1,3-dichloropropene and role of the haloalkane dehalogenase in the first step.

ing polymerase chain reaction (PCR), we have recently amplified a gene from such an organism encoding a protein of which the N-terminus is very similar to the haloalkane dehalogenase from the gram-positive organisms *Rhodococcus rhodochrous* NCIMB 13064 and *Rhodococcus* sp. m15–3 (Curragh et al., 1994; Omori et al., 1991). The occurrence of this gene in adapted agricultural soils indicates that the use of the nematocide may have created an ecological niche in which dehalogenase genes proliferate.

6. BIOCATALYTIC APPLICATIONS

The two most important bottlenecks for obtaining organisms with improved conversion of low molecular weight haloaliphatics are the difficulty of carbon-halogen bond cleavage and toxicity of reactive intermediates. By expressing the dehalogenase genes in suitable host organisms and by constructing recombinant organisms that carry combinations of enzymes that break recalcitrant structures and detoxify reactive intermediates, it may become possible to obtain bacterial cultures for the detoxification of chemicals that remained recalcitrant so far. Wackett et al., (1994) recently showed the successful construction by genetic engineering of a recombinant *Pseudomonas* strain capable of pentachloroethane degradation. If modification of enzyme specificity is important, the best approach probably is to use a combination of semi-random mutagenesis to obtain the desired enzyme specificities and strong selection procedures to enrich the most potent organisms under conditions that allow further mutations and gene transfer.

In addition, microorganisms that can convert synthetic compounds are a wealthy source of enzymes that can be exploited for regio- and stereoselective transformation reactions in synthetic organic chemistry (Kasai et al., 1990, Nakamura et al., 1992). The diversity of activities has hardly been exploited so far, and since these enzymes by definition are capable of recognizing and converting synthetic chemicals, they may often be more suitable for biocatalytic conversions than enzymes evolved for natural compounds. The use of protein engineering for expanding the activities of enzymes that convert xenobiotics may further enhance their applicability for this purpose.

REFERENCES

Curragh, H., Flynn, O., Larkin, M. J., Stafford, T. M., Hamilton, J. T. G., and Harper, D. B., 1994. Haloalkane degradation and assimilation by *Rhodococcus rhodochrous* NCIMB 13064, *Microbiology* 140: 1433–1442.

Derewenda, Z. S., and Sharpe, A. M., 1993. News from the interface: The molecular structure of triacylglyceride lipases, *Trends Biochem. Sci.* 18: 20–25.

Fox, B. G., Borneman, J. G., Wackett, L. P., and Lipscomb, J. D., 1990. Haloalkane oxidation by the soluble methane monooxygenase from *Methylosinus trichosporium* OB3b: mechanistic and environmental implications, *Biochemistry* 29: 6419–6427.

Janssen, D. B., Pries, F., and van der Ploeg, J. R., 1994. Genetics and biochemistry of dehalogenating enzymes, *Ann. Rev. Microbiol.* 48: 163–191.

Kasai, N., Tsujimura, K., Unoura, K., and Suzuki, T., 1990. Degradation of 2,3-dichloro-1-propanol by a *Pseudomonas* sp., *Agric. Biol. Chem.* 54: 3185–3190.

Kawasaki, H., Tsuda, K., Matsushita, I., and Tonomura K., 1992. Lack of homology between two haloacetate dehalogenase genes encoded on a plasmid from *Moraxella* sp. strain B., *J. Gen. Microbiol.* 138: 1317–1323.

Keuning, S., Janssen, D. B., and Witholt, B., 1985. Purification and characterization of hydrolytic haloalkane dehalogenase from *Xanthobacter autotrophicus* GJ10, *J. Bacteriol.* 163: 635–639.

Kennes, C., Pries, F., Krooshof, G. H., Bokma, E., Kingma, J., and Janssen, D. B., 1995. Replacement of tryptophan residues in haloalkane dehalogenase reduces halide binding and catalytic activity, *Eur. J. Biochem.* 228: 403–407.

Nagata, Y., Nariya, T., Ohtomo, R., Fukuda, M., Yano, K., and Takagi, M., 1993. Cloning and sequencing of a dehalogenase gene encoding an enzyme with hydrolase activity involved in the degradation of γ-hexachlorocyclohexane in *Pseudomonas paucimobilis*, *J. Bacteriol.* 175:6403–6410.

Nakamura, T., Nagasawa, T., Yu, F., Watanabe, I., and Yamada, H., 1992. Resolution and some properties of enzymes involved in enantioselective transformation of 1,3-dichloro-2-propanol to (R)-3-chloro-1,2-propanediol by *Corynebacterium* sp. strain N-1074, *J. Bacteriol.* 174: 7613–7619.

Ollis, D. L., Cheah, E., Cygler, M., Dijkstra, B. W., Frolow, F., Franken, S. M., Haral, M., Remington, S. J., Silman, I., Schrag, J., Sussman, J. L., Verschueren, K. H. G., and Goldman, A., 1992. The α/β-hydrolase fold, *Protein Eng.* 5: 197–211.

Omori, T., Kimura, T., Kodama, T., 1991. Cloning of haloalkane dehalogenase genes of *Rhodococcus* sp. m15–3 and expression and sequencing of the genes. pp498–503. In: On-Site Bioreclamation. Processes for Xenobiotic and Hydrocarbon Treatment. R.E. Hinchee & R.F. Olfenbuffel, Eds. Butterworth-Heinemann, Boston.

Pries, F., Kingma, J., Pentenga, M., van Pouderoyen, G., Jeronimus-Stratingh, C. M., Bruins, A. P., and Janssen, D. B., 1994a. Site-directed mutagenesis and oxygen isotope incorporation studies of the nucleophilic aspartate of haloalkane dehalogenase, *Biochemistry* 33: 1242–1247.

Pries, F., van den Wijngaard, A. J., Bos, R., Pentenga, M., and Janssen, D. B., 1994b. The role of spontaneous cap domain mutations in haloalkane dehalogenase specificity and evolution, *J. Biol. Chem.* 269: 17490–17494.

Pries, F., Kingma, J., Krooshof, G. H., Jeronimus-Stratingh, C. M., Bruins, A. P., and Janssen, D. B., 1995. Histidine 289 is essential for hydrolysis of the alkyl-enzyme intermediate of haloalkane dehalogenase, *J. Biol. Chem.* 270: 10405–10411.

Poulos, T. L., Finzel, B. C., and Howard, A., 1987, High resolution crystal structure of P450cam, *J. Mol. Biol.* 195: 687–700.

Rosenzweig, A. C., Frederick, C. A., Lippard, S. J., and Nordlund, P., 1993. Crystal structure of a bacterial non-haem iron hydroxylase that catalyses the biological oxidation of methane, 366: 537–543.

Sallis, P. J., Armfield, S. J., Bull, A. T., and Hardman, D. J., 1990. Isolation and characterization of a haloalkane halidohydrolase from *Rhodococcus erythropolis* Y2, *J. Gen. Microbiol.* 136: 115–120.

Scholtz, R., Leisinger, T., Suter, F., and Cook, A. M., 1987. Characterization of 1-chlorohexane halidehydrolase, a dehalogenase of wide substrate range from an *Arthrobacter* sp., *J. Bacteriol.* 169: 5016–5021.

Stucki, G., and Thüer, M., 1995. Experiences of a large-scale application of 1,2-dichloroethane degrading microorganisms for groundwater treatment, *Environ. Sci. Technol.* 29: 2339–2345.

Tardiff, C., Greer, C. W., Labbe, D., and Lau, P. C. K., 1991, Involvement of a large plasmid in the degradation of 1,2-dichloroethane by *Xanthobacter autotrophicus* GJ10, *Appl. Environ. Microbiol.* 57: 1853–1857.

van den Wijngaard, A. J., van der Kamp, K., van der Ploeg, J., Kazemier, B., Pries, F., and Janssen, D. B. 1992. Degradation of 1,2-dichloroethane by facultative methylotrophic bacteria, *Appl. Env. Microbiol.* 58: 976–983.

Verhagen, C., Smit, E., Janssen, D. B., and van Elsas, J. D., 1995. Bacterial dichloropropene degradation in soil, screening of soils and involvement of plasmids carrying the *dhlA* gene, *Soil Biol. Biochem.* 27: 1547–1557.

Verschueren, K. H. G., Franken, S. M., Rozeboom, H. J., Kalk, K. H., and Dijkstra, B. W., 1993a. Refined X-ray structure of haloalkane dehalogenase at pH6.2 and pH8.2 and implications for the reaction mechanism, *J. Mol. Biol.* 232: 856–872.

Verschueren, K. H. G., Seljée, F., Rozeboom, H. J., Kalk, K. H., and Dijkstra, B. W., 1993b. Crystallographic analysis of the catalytic mechanism of haloalkane dehalogenase, *Nature* 363: 693–698.

Verschueren, K. H. G., Kingma, J., Rozeboom, H. J., Kalk, K. H., Janssen, D. B., and Dijkstra, B. W., 1993c. Crystallographic and fluorescence studies of the interaction of haloalkane dehalogenase with halide ions. Studies with halide compounds reveal a halide binding site in the active site, *Biochemistry* 32: 9031–9037.

Wackett, L. P., Sadowsky, M. J., Newman, L. M., Hur, H-G., and Li, S., 1994. Metabolism of polyhalogenated compounds by a genetically engineered bacterium, *Nature* 368: 627–629.

Yokota, T., Fuse, H., Omori, T., and Minoda, Y., 1986. Microbial dehalogenation of haloalkanes mediated by oxygenase or halidohydrolase, *Agric. Biol. Chem.* 50: 453–460.

6

PHYTOREMEDIATION APPLICATIONS FOR REMOVING HEAVY METAL CONTAMINATION FROM SOIL AND WATER

Burt D. Ensley,[1] Ilya Raskin,[2] and David E. Salt[2]

[1]Phytotech, Inc.
1 Deer Park Drive
Monmouth Junction, New Jersey
[2]AgBiotech Center, Rutgers
The State University of New Jersey
New Brunswick, New Jersey

Over the past several years, scientists have discovered many examples of living plants that can remove heavy metals and other pollutants from soil and water. This potential approach to cleaning the environment, termed phytoremediation, draws on our centuries of experience in cultivating crops and is emerging as a low cost treatment technology. The idea that plants can be used for environmental remediation is not new. Extensive research on using plants and entire ecosystems for treating radionuclide contamination took place in Russia in the early 1960s (Timofeev-Resovsky, et al., 1962). Since then, there have been a number of reports that aquatic plants such as water hyacinth, duckweed and water velvet can accumulate Pb, Cu, Cd, Fe, and Hg from contaminated water (Mo, et al., 1989; Jackson, et al., 1990; Dierberg, et al., 1987). This ability is currently utilized in many constructed wetlands, which can be effective in removing some heavy metals and organics from water (Jain, et al., 1989).

The ability of plants to accumulate metals has usually been considered a detrimental trait. Metal-accumulating plants are directly or indirectly responsible for a proportion of the dietary uptake of toxic heavy metals by humans and other animals (Brown, et al., 1994). While some heavy metals are required for life, their excessive accumulation in living organisms is always toxic, and is aggravated by their almost indefinite persistence in the environment.

Recently the value of metal-accumulating terrestrial plants for environmental remediation has also been recognized (Baker, et al., 1988; Cunningham, et al., 1993; Wenzel, et al., 1993). Crop plants such as Indian mustard have been used to extract heavy metals from soil and sediments and translocate those metals to the harvestable stalks and leaves of the plants (Raskin, et al., 1994). This application, called phytoextraction, may be used in the removal of heavy metals in soils, sludges and sediments.

Most metal accumulating plant species known today were discovered growing on soils containing high levels of heavy metals. These plants are often endemic to these types of soils, suggesting that metal accumulation is associated with heavy metal resistance (Baker and Brooks, 1989). The majority of hyperaccumulating species discovered so far are restricted to tropical areas (Baker and Brooks, 1989; Baker, et al., 1993; Brooks, et al., 1993). Beyond this, we know little about the significance of metal hyperaccumulation. Several current hypotheses include tolerance or disposal of metal from plants, drought resistance, inadvertent uptake and defense against herbivores or pathogens. Some evidence favors hyperaccumulation of metals as a defense against herbivores (Ernst, et al., 1990; Baker, et al., 1989).

The current, expanding list of hyperaccumulating plants includes those that accumulate Ni, Co, Cu, Zn, Mn, Pb, Cd, and Cr. The largest numbers of temperate climate hyperaccumulating species belong to the *Brassicaceae* (Baker and Brooks, 1989), but in the tropics the *Euphorbiaceae* is the best represented group. One of the most striking examples of metal hyperaccumulation is displayed by a New Caledonian tree, *Sebertia acuminata*, which has over 11% of Ni in its latex (Baker, et al., 1988). Many other Ni hyperaccumulators belong to the genera *Alyssum* and *Thlaspi*. *Thlaspi* species also accumulate Pb and Zn (Baker et al., 1988). The search for metal-accumulating plants to date has not been extensive or broad in scope. It is likely that there are many more unidentified metal-accumulating plants growing on metalliferous soils that await identification.

The use of metal-accumulating plants in the phytoextraction of contaminated soils could significantly contribute to improved environmental quality. The optimum plant for phytoextraction would be able to tolerate and accumulate high levels of heavy metals and also grow rapidly and be able to produce a high biomass yield. The first reported field trials of metal accumulators on soils demonstrated the feasibility of phytoextraction (Baker, et al., 1994). This site was contaminated by Ni and Zn through the long term application of heavy metal containing sludges. Although significant metal accumulation was measured during this trial, the practicality of phytoextraction was not conclusively demonstrated. Due to the low biomass production of the plants used in this study, even the best metal accumulator identified in this trial, *Thlaspi caerulescens*, would require 13 to 14 years of continuous cultivation to clean the site.

Several high biomass crop species related to wild mustards have been identified that can accumulate heavy metals in their shoots. Certain cultivars of *Brassica juncea* (Indian mustard) have the ability to accumulate and tolerate Pb, Cd, Cr (VI), Ni, Zn and Cu (Duchenkov, et al., 1995). This work demonstrated that the ability to accumulate heavy metals varies greatly between species and cultivars within a species. Recent work (Brown, et al., 1994) has demonstrated that *T. caerulescens* has a high resistance to the toxic effects of both Cd and Zn, but the more sensitive *B. juncea* produces at least 20 times more biomass than *T. caerulescens* under field conditions, giving it the potential to remove much more metal per cropping, and increased ability to clean a site. Rapidly growing crop plants such as *Brassica juncea* are well suited for remediation activities in soils because of their high biomass accumulation, extensive root systems and ability to translocate metals to the harvestable shoots. Root systems of *B. juncea* often extend well below 50 centimeters (cm) depth in the soil, providing efficient metal removal and preventing leaching of heavy metals into ground water.

During phytoextraction processes several sequential crops of hyperaccumulating plants may be used to reduce soil concentrations of heavy metals to environmentally acceptable levels. Dried, ashed or composted plant residues, highly enriched in heavy metals may be isolated as hazardous waste or recycled as a source of metal ore. While the most heavily contaminated soils do not support plant growth, sites with light to moderate heavy metal contamination would be suitable for growing hyperaccumulating plants. These plants can be grown and harvested relatively economically, leaving the soil or water with a greatly reduced level of heavy metal contamination.

The value of metal-accumulating plants for environmental remediation has only recently been fully realized (Chaney, 1983; Cunningham, 1993). In spite of the growing number of toxic metal-contaminated sites, the most commonly used methods of dealing with heavy metal pollution are either removal and burial or isolation of the contaminated sites. Water treatment facilities also do a relatively poor job of removing certain heavy metals from residential and industrial effluents, contributing to the overall problem (Bubb and Lester, 1991). The advantages of using metal-accumulating plants for removal of metals from contaminated soils and waters are lower costs, generation of a recyclable metal-rich plant residue, applicability to a range of toxic metals and radionuclides, minimal environmental disturbance, elimination of secondary air or water-borne wastes and enhanced regulatory and public acceptance.

Metal-accumulating plants can also be used for the treatment of contaminated water. Ground, surface or process water containing heavy metals can be difficult and expensive to treat. Low cost methods for cleaning this water before discharge are needed to maintain competitive process economics while protecting the environment.

Heavy metals such as chromium, nickel, cadmium, lead, copper and zinc are some of the most toxic inorganic environmental pollutants. Removal of these metals from aqueous streams presents a particularly difficult problem. Heavy metal contaminants in water are usually removed by ion exchange, chemical or biological precipitation or by flocculation, followed by sedimentation and disposal of the resulting sludge. These methods are difficult to implement in many cases and can be prohibitively expensive particularly if large water volume, low metal concentration, and high cleanup standards are involved.

Recently there has been a growing interest in the use of metal-accumulating roots and rhizomes of aquatic or semiaquatic vascular plants for the removal of heavy metals from contaminated aqueous streams. For example, water hyacinth (*Eichhornia crassipes*), pennywort (*Hydrocotyle umbellata*), duckweed (*Lemna minor*) and water velvet (*Azolla pinnata*) take up Pb, Cu, Cd, Fe, and Hg from contaminated solutions (Kay, et al., 1984; Dierberg, et al., 1987; Jain, et al., 1989; Mo, et al., 1989). In a related development, cell suspension cultures of *Datura innoxia* were found to remove a wide variety of metal ions from solutions (Jackson, et al., 1990; Jackson, et al., 1993). Most of the removed metals were tightly chelated by unidentified components of cell walls in a process which did not require metabolic activity.

The observation that hydroponically grown roots of terrestrial plants are extremely effective in removing Pb, Cr, Zn, Cd, Cu, and Ni from water has laid the foundation for the development of rhizofiltration (Duchenkov, et al., 1995). Rhizofiltration is particularly effective and economically compelling when low concentrations of contaminants and large volumes of water are involved. Therefore, rhizofiltration may be applicable to radionuclide contaminated water. Uptake of radionuclides by plants is not well studied, but promising results obtained in our laboratories suggest that many cationic and anionic radionuclide contaminants can be substantially or completely removed from water with selected metal accumulating plants, cultivated in a specially developed and optimized rhizofiltration system. Rhizofiltration of radionuclides may be particularly effective when used in combination with a microorganism-based bioremediation strategy (Macaskie, 1991).

We have evaluated the performance of hydroponically cultivated metal-accumulating plants in treating site water contaminated with uranium, cesium and strontium. These field results demonstrated that rhizofiltration is a practical way to treat radionuclide contamination in site water (Vasudev, et al., 1996). Rapid metal uptake to concentrations below 1 part per billion (ppb) was observed with uranium. Cesium and strontium were removed to concentrations below 10 Baquerals/liter (Bq/L) for Cs-137 and below 200 Bq/L for Sr-90.

The trial with uranium-contaminated water was carried out at U concentrations of 100–400 ppb. Roots of a selected sunflower cultivar (*Helianthus annuus L.*) caused the uranium solution concentration to decline by 95% within the first 24 hours. Following bench-scale studies, a prototype flow-through rhizofiltration system was tested with uranium contaminated site water. The uranium concentration in the water exiting the system never exceeded 5 ppb, and was usually below 1 ppb throughout the 52 day trial. These results met or exceeded our goals for system performance with site water.

A similar approach was used to evaluate the use of plants to treat water contaminated with cesium and strontium in the Ukraine. A natural pond with a surface area of about 75 square meters (m^2) near the village of Yanov was selected as a site for a rhizofiltration field trial. Radionuclide concentrations in this basin were approximately 80 Bq/L for cesium-137 and 1200 Bq/l for strontium-90 as a result of the 1986 release by the nearby Chernobyl nuclear reactor. Radionuclide uptake studies were carried out directly in the contaminated pond. After 4 to 8 weeks of hydroponic growth in the pond, sunflower plants were harvested, dried, and the radionuclide concentrations measured. The plant tissue concentrations of cesium-137 reached 6.4×10^5 Bq per kilogram (dry weight) and 2.5×10^6 Bq of strontium-90 per kilogram dry weight. This quantity of radioactivity was sufficient for the plant material to be held in quarantine in the restricted zone at Chernobyl.

A better understanding of the biochemical processes involved in plant heavy metal uptake, transport, accumulation and resistance will stimulate improvements in phytoremediation using modern genetic approaches. Some successes have been already scored along this path. One strategy for improving the phytoremediation potential of high biomass plant species is the introduction of genes responsible for metal accumulation and resistance from the wild metal accumulators. In the absence of known "phytoremediation" genes this may be accomplished via somatic and sexual hybridization followed by extensive screening and backcrossing of progeny. However, a long term effort should be directed towards developing specific genetic information composed of genes valuable for phytoremediation. Systematic screening of plant species and genotype for metal accumulation and resistance will broaden the spectra of genetic material available for optimization and transfer. Mutagenesis of selected high biomass plant species may also produce improved phytoremediating cultivars.

Optimizing agronomic practices employed during phytoremediation, such as irrigation, fertilization, planting and harvest time and the timing of amendment application, will increase the efficiency of both phytoextraction and rhizofiltration processes. In addition, the problems of design and engineering of phytoextraction and, particularly, rhizofiltration systems must be addressed. Cooperation with federal, state and local environmental agencies is also required to obtain the necessary regulatory approvals and performance standards for the phytoremediation process.

Phytoremediation of heavy metals is designed to concentrate metals in plant tissues, thus minimizing the amount of solid or liquid hazardous waste which needs to be treated and deposited at hazardous waste sites. As an ultimate goal, an economical method of reclaiming metals from plant residue should be developed. This will completely eliminate the need for costly off-site disposal. At present, methods for the further concentration of metals in plant tissues include: sun, heat and air drying, environmentally-safe ashing or incineration, composting, pressing and compacting and acid leaching.

Phytoremediation is a very new field which holds great potential for the future. In order to realize this potential it will be necessary to considerably increase our understanding of the many and varied processes which it involves. This will require a multidisciplinary approach, spanning fields as diverse as plant biology, agricultural engineering, agronomy, soil science, microbiology and genetic engineering.

REFERENCES

Baker, A., Brooks, R., and Reeves, R. 1988. Growing for gold...and copper...and zinc. New Scientist. 10: 44–48.

Baker A., and Brooks R. 1989. Terrestrial Higher Plants which Hyperaccumulate Metallic Elements - A Review of their Distribution, Ecology and Phytochemistry. Biorecovery, 1:81–126.

Baker A.J.M., Proctor J., van Balgooy M., and Reeves, R.D. 1993. Hyperaccumulation of Nickel by Flora of the Ultramafics of Palawan, Republic of the Philippines. In: The Vegetation of Ultramafic (Serpentine) Soils. Proceedings of the First International Conference on Serpentine Ecology. Andover, Hampshire: Intercept Ltd., UK; pp. 291–304.

Baker A.J.M., McGrath S.P., Sidoli C.M.D., and Reeves R.D., 1994. The Possibility of in Situ Heavy Metal Decontamination of Metal-Polluted Soils Using Crops of Metal-Accumulating Plants. Resource, Conservation and Recycling, in press.

Brooks R., Reeves, R.D., and Baker, A.J.M. 1993. The Serpentine Vegetation of Goias State, Brazil. In: The Vegetation of Ultramafic (Serpentine) Soils. Proceedings of the First International Conference on Serpentine Ecology. Andover, Hampshire: Intercept Ltd, UK; 67–81.

Brown, S.L., Chaney, R.L., Angle, J.S., and Baker, A.J.M. 1994. Phytoremediation Potential of *Thlaspi caerulescens* and *Bladder Campion* for Zinc- and Cadmium-Contaminated Soil. J. Environ. Qual. 23:1151–1157.

Bubb, J.M., and Lester, J.N. 1991. The Impact of Heavy Metals on Lowland Rivers and the Implications for Man and the Environment. The Science of the Total Environment, 100:207–233.

Chaney R.L., 1983. Plant Uptake of Inorganic Waste. In Land Treatment of Hazardous Wastes. Edited by Parr J.E., Marsh P.B., Kla J.M. Park Ridge: Noyes Data Corp; pp. 50–76.

Cunningham, S.D., 1993. Plant-Based Environmental Remediation: Progress and Promise. Abstract P-33. In Vitro Cell Develop Biol, 29A:42A.

Cunningham, S.D., and Berti, W.R. 1993. Remediation of contaminated soils with green plants: an overview. In Vitro. Cell. Dev. Biol. 29P: 207–212.

Dierberg, F.E., DeBusk, T.A., and Goulet, Jr. N.A. 1987. Removal of Copper and Lead Using a Thin-Film Technique. In: Aquatic Plants for Water Treatment and Resource Recovery. Edited by Reddy, K.B., and Smith, W.H. Florida: Magnolia Publishing Inc; pp. 497–504.

Ernst, W.H.O., Schat, H., and Verkleij, J.A.C. 1990. Evolutionary Biology of Metal Resistance in *Silene vulgaris*. Evolutionary Trends in Plants, 4:45–51.

Jackson, P.J., Torres, A.P., Delhaize, E., Pack, E., and Bolender, S.L. 1990. The Removal of Barium Ions from Solution using *Datura innoxia* Suspension Culture Cells. J. Env. Quality. 19: 644–648.

Jackson, P.J., DeWitt, J.G., and Kuske, C.R., 1993. Accumulation of Toxic Metal Ions by Components of Plant Suspension Cell Cultures. Abstract P-34. In Vitro Cell Develop Biol, 29A:42A.

Jain, S.K., Vasudevan, P., and Jha, N.K. 1989. Removal of Some Heavy Metals from Polluted Waters by Aquatic Plants: Studies on Duckweed and Water Velvet. Biological Wastes. 28: 115–126.

Kay, S.H., Hailer, W.T., and Garrard, L.A. 1984. Effects of Heavy Metals on Water Hyacinths (*Eichhornia crassipes (Mart.)* Solms). Aquatic Toxicology, 5:117–128.

Macaskie, L.E. 1991. The Application of Biotechnology to the Treatment of Wastes Produced from the Nuclear Fuel Cycle: Biodegradation and Bioaccumulation as a Means of Treating Radionuclide-Containing Streams. Critical Reviews in Biotechnology. 11: 41–112.

Mo, S.C., Choi, D.S., and Robinson, J.W. 1989. Uptake of Mercury from Aqueous Solutions by Duckweed: The Effect of pH, Copper and Humic acid. J. Env. Sci. Health. A24: 135–146.

Raskin, I. Kumar, N., Dushenkov, S., and Salt, D., 1994, Current Opinion in Biotechnology 5: 285–290.

Timofeev-Resovsky, E.A., Agafonov, B.M., and Timofeev-Resovsky N.V. 1962. Fate of radioisotopes in aquatic environments. Proceedings of the Biological Institute of the USSR Academy of Sciences. 22: 49–67.

Vasudev, D., Ledder, T., Dushenkov, S., Epstein, A., Kumar, N., Kapulnik, Y., Ensley, B., Huddleston, G., Cornish J., Raskin, I., Sorochinsky, B., Ruchko, M., Prokhnevsky, A., Mikheev, A., and Grodzinsky D. 1996. Removal of Radionuclide Contamination From Water by Metal-Accumulating Terrestrial Plants. Proceedings AIChE Spring Meeting. In Press.

Wenzel, W.W., Sattler H., and Jockwer, F. 1993. Metal hyperaccumulator plants: a survey on species to be potentially used for soil remediation. In Agronomy Abstracts, p 52.

7

THE ROLE OF MICROBIAL PCB DECHLORINATION IN NATURAL RESTORATION AND BIOREMEDIATION

Donna L. Bedard* and Heidi M. Van Dort

GE Corporate Research and Development
P.O. Box 8, Schenectady, New York 12301

1. INTRODUCTION

From 1929 to 1978 polychlorinated biphenyls (PCBs) were widely used as dielectric fluids, hydraulic fluids, solvent extenders, and plasticizers. Several hundred million pounds of PCBs were released into the environment (Shiu and Mackay, 1986). It has been estimated that 50–80% of the PCBs in the environment have been deposited in the North Atlantic Ocean, and that the major continental sink for PCBs is freshwater sediment (NRC, 1979). PCBs tend to accumulate in biota and have been associated with potential health effects; consequently, their environmental fate is important.

Commercial PCBs, such as Aroclors, were manufactured by catalytic chlorination of biphenyl to produce complex mixtures containing a specified weight percent of chlorine. Aroclors 1242, 1248, 1254, and 1260, which contain, respectively, 42, 48, 54, and 60 weight percent chlorine, were the most commonly used PCB mixtures. These Aroclors each contain 60 to 90 different PCB structures, known as congeners, that differ in the number and position of chlorines.

Microbial dechlorination of PCBs was recently discovered when it became apparent that PCBs in several aquatic sediments had undergone a novel transformation. Relative to the Aroclors that originally contaminated the sites, PCBs in the sediments of the Hudson River (NY) and Silver Lake (Pittsfield, MA) exhibited selective depletion of the more highly chlorinated congeners and elevation of mono- and dichlorinated biphenyls with chlorines in the *ortho* positions (Brown et al., 1987a,b). It was proposed (Brown et al., 1987a,b) and subsequently confirmed (Quensen et al., 1988) that these changes resulted from stepwise *meta* and *para* dechlorination mediated by anaerobic microorganisms. Subsequent research demonstrated that this dechlorination is occurring in situ in many aquatic

* Current address: Chemistry Department, Purdue University, 1393 Brown Chemistry Building, Lafayette, Indiana 47907-1393

Biotechnology in the Sustainable Environment, edited by Sayler *et al.*
Plenum Press, New York, 1997

sediments (Brown and Wagner, 1990; Sokol et al., 1994; Bedard and Quensen,1995; Bedard and May, 1996). This newly discovered environmental fate of PCBs has significant implications for risk assessment and remediation strategies. We will use examples of PCB dechlorination in the Hudson and Housatonic River systems to discuss the implications for detoxication, natural restoration, and bioremediation.

2. PCB DECHLORINATION IN THE HUDSON RIVER

Microbial dechlorination of the Aroclor 1242 that contaminated the upper Hudson River is occurring over a broad area of the river (Brown et al., 1987a,b). Several sediment surveys have found extensive PCB dechlorination in areas receiving a wide range of PCB concentrations (Abramowicz, 1994). In the upper Hudson this dechlorination has converted most of the hotspot PCBs with three or more chlorines to mono- and dichlorinated biphenyls. The PCB homolog distributions of Aroclor 1242 and of its dechlorinated residue in a sediment sample from the upper Hudson River are shown in Figure 1. Nearly 82% of the PCBs in Aroclor 1242 have three or more chlorines, whereas only 25% of the

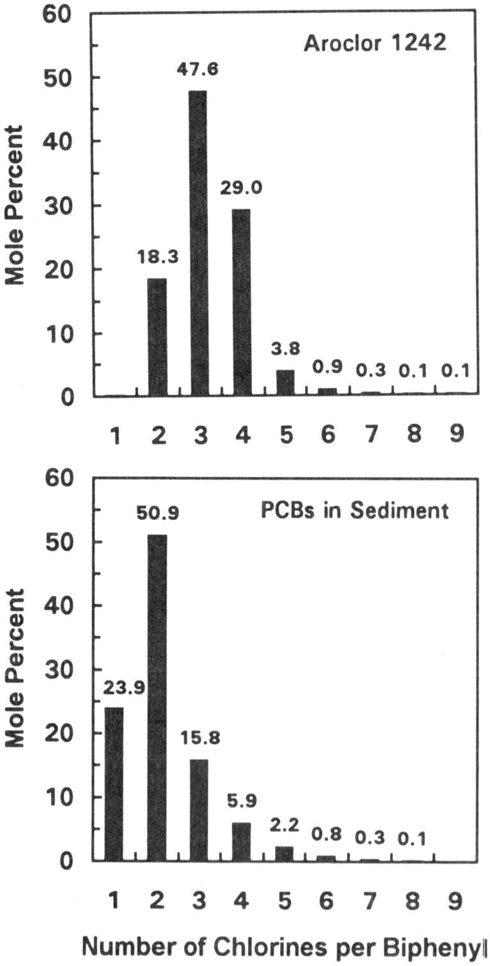

Figure 1. Effect of microbial dechlorination on the PCB homolog distribution of Aroclor 1242. Top panel: Aroclor 1242; Bottom panel: The dechlorinated residue of Aroclor 1242 in Hudson River sediment collected at the H7 site, near the western bank of the river 312 km upstream from New York City (Harkness et al, 1993; Abramowicz, 1994).

PCBs in the dechlorinated Aroclor 1242 residue in the sediment sample have three or more chlorines. This reflects nearly a 70% decrease of the highly chlorinated PCBs. The mono- and dichlorobiphenyls produced by the dechlorination have less tendency to bioaccumulate (Shaw and Connell, 1986; Brown, 1994) and are highly susceptible to microbial oxidation (Bedard et al., 1987; Bedard and Haberl, 1990). Moreover, the discovery of metabolites produced from microbial PCB oxidation in Hudson River sediments provides convincing evidence for in situ biodegradation of the dechlorinated PCBs (Flanagan & May, 1993). Hence, together with physical burial, PCB dechlorination is playing a major role in natural restoration of the upper Hudson.

3. PCB DECHLORINATION IN THE HOUSATONIC RIVER

The sediments of the upper Housatonic River were contaminated with Aroclor 1260, which is primarily composed of hexa- and heptachlorobiphenyls (Figure 2A). Sediments from Woods Pond, an impoundment on the Housatonic River, show evidence of modest in situ dechlorination (Bedard and May, 1996; Figure 2B). In the example shown, in situ dechlorination has decreased the proportion of PCBs with six or more chlorines by about 27%, from ~90% in Aroclor 1260 to ~66% in the dechlorinated residue. This dechlorination also decreased the PCBs that are most persistent in human tissues by about 25% (Bedard and May, 1996), but the dechlorinated PCB residue consists primarily of PCBs with four or more chlorines and hence is not very susceptible to microbial oxidation.

4. PRIMING MICROBIAL PCB DECHLORINATION

We are investigating a way to stimulate or "prime" rapid microbial dechlorination of PCBs in sediments where the PCB dechlorination has not progressed as far as desired. Priming is based on the premise that PCB-dechlorinating microorganisms use PCBs as electron acceptors, and the subsequent hypothesis that PCBs or other halogenated biphenyls might provide a means to selectively enrich PCB-dechlorinating microorganisms in aquatic sediments (Bedard et al., 1993, 1997). We have tested priming in controlled experiments with sediments from Woods Pond using excess amounts of individual tri-, tetra-, or pentachlorobiphenyls. The added congeners primed extensive microbial dechlorination of highly chlorinated PCBs that have persisted in sediment for decades (Bedard et al., 1993, 1996, 1997; Van Dort et al., 1997; Wu et al., in press). Figure 2C shows the effect of priming dechlorination of PCBs in Housatonic River sediment with 2,3,4,5,6-pentachlorobiphenyl (2,3,4,5,6-CB) (final concentration 350 μMol). The 2,3,4,5,6-CB was dechlorinated to 2,4,6-trichlorobiphenyl by sequential loss of two *meta* chlorines. This dechlorination in turn primed *meta* dechlorination of the Aroclor 1260 residue in the sediment, resulting in a further decrease of 42% of the PCBs with six or more chlorines in 20 weeks. Subsequently we were able to drive the dechlorinating even further by successive transfers of actively dechlorinating sediment slurries and 2,3,4,5,6-CB onto autoclaved PCB-contaminated Housatonic River sediment (Figure 2D, Bedard et al., 1997; Van Dort et al., 1997). The volume of active slurry used for each transfer was 10–20% of the volume of the autoclaved sediment slurry, hence if no growth of PCB-dechlorination microorganisms occurred the dechlorination should have become less effective with each transfer. Instead, the dechlorination was more extensive and progressed further with each successive transfer. Hence, our observations are consistent with the hy-

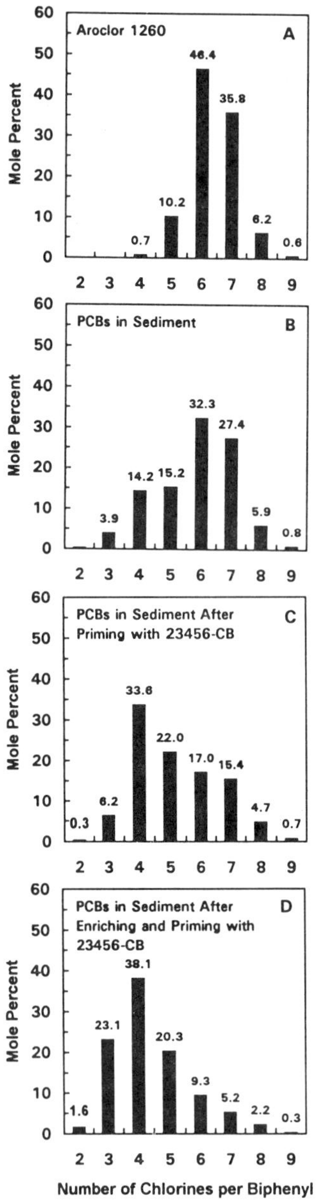

Figure 2. Effect of microbial dechlorination on the PCB homolog distribution of Aroclor 1260 before and after priming and enrichment of PCB-dechlorinating microorganisms. **A.** Aroclor 1260; **B.** The dechlorinated residue of Aroclor 1260 in Housatonic River sediment; **C.** The dechlorinated residue of Aroclor 1260 in Housatonic River sediment 141 days after priming with 2,3,4,5,6-CB (350 μMol); **D.** The dechlorinated residue of Aroclor 1260 in Housatonic River sediment 238 days after priming with 2,3,4,5,6-CB (350 μMol) and inoculating with an actively dechlorinating sediment slurry that had been transferred three times with 2,3,4,5,6-CB to enrich PCB-dechlorinating microorganisms (Bedard et al., 1997; Van Dort et al., 1997).

pothesis that high concentrations of an appropriate PCB congener can selectively enrich PCB-dechlorinating microorganisms. The net result of priming and enriching the indigenous PCB-dechlorinating microorganisms in the sediment was the conversion of hexa- and heptachlorobiphenyls to a mixture of primarily tri- and tetrachlorobiphenyls that are less persistent in human tissue and more susceptible to oxidation by aerobic microorganisms and by higher organisms (Bedard et al., 1997; Van Dort et al., 1997).

Although it is clearly not acceptable to add PCBs to the environment, PCB congeners can be used in laboratory experiments to determine whether it is possible to prime microbial dechlorination of PCBs in a given contaminated sediment. If the results of priming experiments with PCB congeners establish that the indigenous microbial population can be primed to dechlorinate the PCBs, environmentally acceptable compounds that stimulate the same dechlorination can be sought.

Based on our success in priming PCB dechlorination with PCB congeners, we proposed that brominated biphenyls (BB) might also be used to prime microbial dechlorination of PCBs in contaminated sediments. We have tested a variety of mono- through tetrabromobiphenyls and have identified several, including 2,5-BB, 2,6-BB, and 2,5,3′-BB, that prime *meta*-dechlorination of the PCBs in sediments from Woods Pond. The brominated biphenyls are dehalogenated to biphenyl and are considerably more effective than individual PCB congeners for priming dechlorination. For example, priming with 2,6-BB led to a 79% decrease in the PCBs with six to nine chlorines (from 73% to 15% of the total PCBs) in 93 days (Bedard et al., 1993; Bedard et al., in prep.).

5. BENEFITS OF PCB DECHLORINATION

The benefits of microbial PCB dechlorination include reductions in the potential human exposure to and risk from PCBs. Reduced exposure comes from increased degradability, decreased lipophilicity, and decreased persistence of dechlorinated PCBs. Dechlorination generally makes PCBs more susceptible to microbial oxidation and degradation (Bedard and Quensen, 1995) and less persistent in human tissues (Brown, 1994; Bedard and May, 1996; Bedard et al., 1997). In addition, known correlations with octanol/water partitioning coefficients show that each chlorine loss will decrease the log K_{ow} by 0.5 with corresponding reductions in bioaccumulation (Shiu and Mackay, 1986).

PCB dechlorination may reduce risks associated with potential "dioxin-like" toxicity and potential carcinogenity of PCBs. "Dioxin-like" toxicity mediated through the aryl hydrocarbon (Ah) receptor is the form of PCB toxicity for which structure-activity relationships are best understood. The most potent PCB congeners are approximate stereoisomers of 2,3,7,8- tetrachloro-dibenzo-*p*-dioxin (TCDD) (Safe, 1994). These congeners lack *ortho* chlorines and have two *para* chlorines and at least two *meta* chlorines. These "coplanar" PCBs are present in only minute quantities (0.015 to 0.3 weight %, Schwartz et al., 1993) in Aroclors. The mono-*ortho*-derivatives of the coplanar PCBs, such as 2,3,4,3′,4′- and 2,4,5,3′,4′-pentachlorobiphenyls, exhibit similar toxicity but are orders of magnitude less potent (Ahlborg et al., 1994). In all cases that have been described to date, microbial dechlorination specifically targets *meta* and/or *para* chlorines (Bedard and Quensen, 1995). The adjacent *meta* and *para* chlorines that are present in all of the "dioxin-like" congeners make them especially susceptible (Bedard and Quensen, 1995). Hence microbial dechlorination reduces the potential "dioxin-like" toxicity of PCBs by decreasing the concentrations of the "dioxin-like" congeners.

Although there is no conclusive evidence for PCB-related carcinogenicity in humans, even in occupationally exposed groups (ASTDR, 1993), PCBs are considered potential carcinogens because some commercial PCB mixtures have been shown to increase the incidence of hepatic tumors in rats (Kimbrough et al., 1975; Norback and Weltman, 1985; IARC, 1987). Less chlorinated PCB mixtures were not as potent as Aroclor 1260 (ASTDR, 1993) and did not cause statistically significant elevations of liver tumors (IEHR, 1991; Moore et al., 1994). The PCB congeners that constitute Aroclor 1260 are generally more highly chlorinated and more persistent than those of the PCB mixtures that did not cause significant increases in liver cancers. This suggests that PCB congeners that are highly chlorinated and highly persistent may contribute to carcinogenic potential. Other data have described "dioxin-like" PCBs as strong promoters suggesting that this class of congeners may also contribute to carcinogenic potential (USEPA, 1991). As described above, the "dioxin-like" PCBs are especially susceptible to PCB dechlorination. In addition, microbial dechlorination generally targets the more highly chlorinated PCBs (Brown and Wagner, 1990; Bedard and Quensen, 1995) and has been shown to decrease the concentrations of the congeners reported to be most persistent in humans (Bedard and May, 1996; Bedard et al., 1997). Hence PCB dechlorination is expected to reduce the potential carcinogenicity of PCBs in the environment.

6. SUMMARY

The extensive *meta*- and *para*-dechlorination of the PCBs in the Hudson River has reduced their potential toxicity and carcinogenicity by converting the more highly chlorinated toxic and persistent congeners to mono- and dechlorinated PCBs (Brown et al., 1987a, b; Abramowicz, 1994). The dechlorinated PCBs, in turn, can be degraded by aerobic microorganisms in the sediment. Hence PCB dechlorination is playing a significant role in the natural recovery of the Hudson River.

Microbial dechlorination of the Aroclor 1260 in the Housatonic River has significantly reduced the concentrations of the more highly chlorinated PCB congeners and those that are most persistent in human tissue (Bedard and May, 1996) and hence has most likely reduced the potential toxicity and carcinogenity of the PCBs. The dechlorinated PCBs are still too highly chlorinated to be very susceptible to oxidation by aerobic microorganisms, but our success in priming more extensive dechlorination suggests approaches for the development of effective bioremediation technologies.

REFERENCES

Abramowicz, DA, (1994). Aerobic PCB biodegradation and anaerobic PCB dechlorination in the environment, *Research in Microbiology* 145: 42–26.

Ahlborg, U.G., Becking, G.C., Birnbaum, L.S., Brouwer, A., Derks, H.J.G.M., Feeley, M., Golor, G., Hanberg, A., Larsen, J.C., Liem, A.K.D., Safe, S.H., Schlatter, C, Wærn, F., Younes, M. Yrjånheikki, E., 1994. Toxic equivalency factors for dioxin-like PCBs, *Chemosphere* 28:1049–1067.

ATSDR (Agency for Toxic Substances and Disease Registry), 1993. Update: Toxicological profile for selected PCBs (Aroclor-1260, -1254, -1248, -1242, 1232, -1221, and -1016). Atlanta, GA, U.S. Department of Human Health & Human Services, Public Health Service, ATSDR, TP-92/16.

Bedard, D.L., Wagner, R.E., Brennan, M.J., Haberl, M.L., and Brown, J.F., Jr., 1987. Extensive degradation of Aroclors and environmentally transformed polychlorinated biphenyls by *Alcaligenes eutrophus* H850, *Appl. Environ. Microbiol.* 53:1094–1102.

Bedard, D.L. and Haberl, M.L., 1990. Influence of chlorine substitution pattern on the degradation of polychlorinated biphenyls by eight bacterial strains. *Microbial Ecol.* 20:87–102.

Bedard, D.L., Van Dort, H.M., Bunnell, S.C., Principe, J.M., DeWeerd, K.A., May, R.J., and Smullen, L.A., 1993. Stimulation of reductive dechlorination of Aroclor 1260 contaminant in anaerobic slurries of Woods Pond sediment, In *Anaerobic Dehalogenation and Its Environmental Implications,* Abstr. of 1992. Am. Soc. Microbiol. Conf., Athens, GA: Office of Research and Development, U.S. E.P.A., pp. 19–21.

Bedard, D.L. and Quensen, J.F. III, 1995. "Microbial Reductive Dechlorination of Polychlorinated Biphenyls" in Young, L.Y. and Cerniglia, C. (eds) *Microbial Transformation and Degradation of Toxic Organic Compounds*, Wiley-Liss, New York, pp 137–216.

Bedard, D.L. and May, R.J., 1996. Characterization of the polychlorinated biphenyls in the sediments of Woods Pond: Evidence for microbial dechlorination of Aroclor 1260 in situ, *Environ. Sci. Technol.* 30:237–245.

Bedard, D.L., Bunnell, S.B., and Smullen, L.A., 1996. Stimulation of microbial *para* dechlorination of polychlorinated biphenyls that have persisted in Housatonic River sediment for decades, *Environ. Sci. Technol.* 30:687–694.

Bedard, D.L., Van Dort, H.M., and DeWeerd, K.A. Brominated biphenyls prime microbial dechlorination of highly chlorinated PCBs that have persisted in Housatonic River sediment for decades., in preparation.

Bedard, D.L., Van Dort, H.M., May, R.J., Smullen, L.A. 1997. Enrichment of microorganisms that sequentially meta-, *para*dechlorinate the residue of Aroclor 1260 in Housatonic River sediment. *Environ. Sci. Technol.* 31: 3308–3313.

Brown, J.F. Jr., Bedard, D.L., Brennan, M.J., Carnahan, J.C., Feng, H., and Wagner, R.E., 1987a. Polychlorinated biphenyl dechlorination in aquatic sediments, *Science* 236:709–712.

Brown, J.F. Jr., Wagner, R.E., Feng, H., Bedard, D.L., Brennan, M.J., Carnahan, J.C., and May, R.J., 1987b. Environmental dechlorination of PCBs, *Environ. Toxicol. Chem.* 6: 579–593.

Brown, J.F. Jr., and Wagner, R.E., 1990. PCB movement, dechlorination, and detoxication in the Acushnet Estuary, *Environ. Sci. Technol.* 9:1215–1233.

Brown, J.F. Jr., 1994. Determination of PCB metabolic, excretion, and accumulation rates for use as indicators of biological response and relative risk, *Environ. Toxicol. Chem.* 28: 2295–2305.

Erickson, M., 1986. *Analytical Chemistry of PCBs*, Butterworth Publishers, Boston.

Flanagan, W.P. and May, R.J.,1993. Metabolite detection as evidence for naturally occurring aerobic PCB biodegradation in Hudson River sediments, *Environ. Sci. Technol.* 27:2207–2212.

Harkness, M.R., McDermott, J.B., Abramowicz, D.A., Salvo, J.J., Flanagan, W.P., Stephens, M.L., Mondello, F.J., May, R.J., Lobos, J.H., Carroll, K.M., Brennan, M.J., Bracco, A.A., Fish, K.M., Warnner, G.L., Wilson, P.R., Dietrich, D.K., Lin, D.T., Morgan, C.B., and Gately, W.L., 1993. In situ stimulation of aerobic PCB biodegradation in Hudson River sediments, *Science* 259: 503–507.

IARC (International Agency for Research on Cancer), 1987. IARC Monographs on the Evaluation of Carcinogenic Risks to Humans, Suppl. 7, Overall Evaluations of Carcinogenicity: An Updating of IARC Monographs Volumes 1–42, Lyon, France.

IEHR (Institute for Evaluating Health Risks), 1991. Reassessment of liver findings in five PCB studies in rats. Washington, D.C., dated July 1. Report submitted to USEPA.

Kimbrough, R.D., Squire, R.A., Linder, R.E., Strandberg, J.D., Montali, R.J., and Burse, V.W., 1975. Induction of liver tumors in Sherman strain female rats by polychlorinated biphenyl Aroclor 1260, *J. Natl. Cancer Inst.*, 55:1453–1459.

Moore, J.A., Hardisty, J.F., Banas, D.A., Smith, M.A., 1994. A comparison of liver tumor diagnoses from seven PCB studies in rats, *Regul. Toxicol. Pharmacol.* 20:362–370.

Norback, D.H., and Weltman, R.H., 1985, Polychlorinated biphenyl induction of carcinoma in the Sprague-Dawley rat, *Environ. Health Perspective*, 60:97–105.

NRC (National Research Council), 1979. *Polychlorinated Biphenyls*, National Academy of Sciences, Washington, D.C.

Quensen, J.F. III, Tiedje, J.M., Boyd, S.A., 1988. Reductive dechlorination of polychlorinated biphenyls by anaerobic microorganisms from sediments, *Science* 242:752–754.

Safe, S., 1994, Polychlorinated biphenyls (PCBs): environmental impact, biochemical and toxic responses and implications for risk assessment, *Crit. Rev. Toxicol.*, 24:87–149.

Schwartz, T.R., Tillitt, D.E., Feltz, K.P., and Peterman, P.H., 1993. Determination of mono- and non-O-O′-chlorine substituted polychlorinated biphenyls in Aroclors and environmental samples, *Chemosphere*, 26:1443–1460.

Shaw, G.R., and Connell, D.W., 1986. Factors controlling bioaccumulation of PCBs, in Waid, J.S., *PCBs and the Environment*, Vol. 1, CRC Press, Boca Raton, Florida, pp. 121–133.

Shiu, W.Y. and Mackay, D., 1986. A critical review of aqueous solubilities, vapor pressures, Henry's Law constants, and octanol-water partition coefficients of the polychlorinated biphenyls. *J. Phys. Chem. Ref. Data* 15:911–929.

Sokol, R.C., Kwon, O.-S., Bethoney, C.M., Rhee, G.-Y., 1994. Reductive dechlorination of polychlorinated biphenyls in St. Lawrence River sediments and variations in dechlorination characteristics, *Environ. Sci. Technol.* 28:2054–2064.

U.S. EPA, 1991. Workshop report on toxicity, equivalency factors for polychlorinated biphenyl congeners. Risk Assessment Forum, Washington, D.C. EPA/625/3–91/020.

Van Dort, H.M., Smullen, L.S., May, R.J., and Bedard, D.L. 1997. Priming microbial meta-dechlorination of polychlorinated biphenyls that have persisted in Housatonic River samples for decades. *Environ. Sci. Technol.* 31: 3300–3307.

Wu, Q., Bedard, D.L., and Wiegel, J. Temperature determines the pattern of anaerobic microbial dechlorination of Avoclor 1260 primal by 2,3,4,6-tetrachlorobiphenyl inWoods Pond Sediment. *Appl. Environ. Microbiol.*, in press.

8

AN INTEGRATED TREATMENT SYSTEM FOR POLYCHLORINATED BIPHENYLS REMEDIATION

Mary Jim Beck,[1] Alice C. Layton,[2] Curtis A. Lajoie,[2] James P. Easter,[2] Gary S. Sayler,[2] John Barton,[3] and Mark Reeves[3]

[1]Tennessee Valley Authority, CEB 1F
Muscle Shoals, Alabama 35662-1010
[2]Center for Environmental Biotechnology, Department of Microbiology, and the Graduate Program in Ecology
University of Tennessee
10515 Research Drive, Suite 100
Knoxville, Tennessee 37922-2567
[3]Oak Ridge National Laboratory
PO Box 2008
Oak Ridge, Tennessee 37831-6226

1. INTRODUCTION

Bioremediation is an environmental biotechnology with promise for promoting a sustainable environment. Bioremediation makes use of natural processes and applies the metabolic properties of microorganisms for transforming contaminants to forms that are harmless in the environment. The added capability of biotechnology for tailoring microbial processes to specific problems expands the potential of bioremediation for encouraging a sustainable environment. The process for the biotransformation of polychlorinated biphenyls (PCB) described in this paper is a good example of the enhancement of bioremediation through the tools of biotechnology.

PCBs are one of the most notorious classes of man-made chemicals due to their worldwide distribution, persistence, and toxicological impact on humans and the environment. Bioremediation of PCB contaminated sites is a potential alternative technology to the standard PCB remediation technologies of incineration and landfilling. Although anaerobic and aerobic microbial degradation of PCBs has been demonstrated in the laboratory (Abramowicz, 1990; Bedard and Quensen, 1995), a large gap exists between laboratory experiments and commercialization of remediation technologies. To bridge this gap, both regulatory issues and engineering scale up processes which lead to field demonstrations need to be pursued.

The Tennessee Valley Authority (TVA), in collaboration with the Center for Environmental Biotechnology (CEB) at the University of Tennessee, Knoxville (UTK) and Oak Ridge National Laboratory (ORNL), is developing a PCB remediation process which integrates surfactant solubilization of PCBs and aerobic PCB degradation by genetically engineered microorganisms followed by anaerobic bacterial treatment for further dechlorination of the PCBs. The ultimate goal of this project would be the commercialization of this process for use at electric power substation sites contaminated with PCBs. This report documents the progression of this process from laboratory studies to a proposed field trial demonstration and describes site characterization, development of genetically engineered microorganisms, development of process designs and regulatory approval necessary for a field trial.

2. SITE CHARACTERIZATION

PCB contaminated soils at electric power substation sites have unique characteristics that set them apart from other well-known PCB contaminated sites such as lake and river sediments (Bedard and Quensen, 1995). These soils are often subject to high traffic, covered with gravel, subject to harsh chemical and physical treatments such as weed control, and are nutrient poor. The site chosen for our field demonstration has approximately 100–300 parts per million (ppm) of weathered PCBs and was found to have a small natural microbial population that lacked PCB-degrading activity (Layton et al., 1994a). The natural population could not be stimulated to degrade PCBs by the addition of nutrients alone or by the addition of nutrients and biphenyl or chlorobiphenyl (Layton et al., 1994b), thus indicating that these sites are not likely to undergo natural restoration.

Biostimulation is a valid strategy for the degradation of PCBs. Microorganisms can transform PCBs by either growth on specific chlorinated biphenyls, or congeners, as carbon source (Bedard et al., 1987) or by cometabolism (Kohler et al., 1988; Barriault and Sylvestre, 1993). Although improved PCB degradation has been shown to result from nutrient stimulation alone (Fiebig et al., 1993), the addition of biphenyl is a more important factor for PCB degradation (Focht and Brunner, 1985; Brunner et al., 1985). Further improvement in PCB degradation may be achieved through bioaugmentation, the addition of known degrading bacterial strains, in comparison to degradation resulting from stimulation of native populations (Brunner et al., 1985; McDermott et al., 1989; Hickey et al., 1993). The activities of a variety of strains complement each other for the most complete degradation of PCBs (Bedard et al., 1987). Bacterial strains such as *Alcaligenes eutrophus* H850, *Pseudomonas* sp. LB400, and *Cornybacterium* sp. MB1 demonstrate a broad range of PCB congener attack, including a number of highly chlorinated biphenyls. The white rot fungus, *Phanerochaete chrysosporium*, also has potential for the degradation of many PCB congeners (Bumpus et al., 1985; Yadav et al., 1995; Dietrich, 1995).

The potential of PCB bioremediation by bioaugmentation with organisms capable of PCB degradation was tested in a year-long incubation of the soil with various amendments and organisms (Table 1). After periodic additions of the PCB degrading organisms, nutrients, and biphenyl, the PCBs decreased by 57%, indicating that bioaugmentation may be a viable option for this site. However, other alternative strategies were explored because bioaugmentation is a slow process, requires prolonged operations, repeated biphenyl additions and may be limited by bioavailability.

Table 1. Summary of year-long incubations of biotreatment regimes for PCB-contaminated soil

	Treatment regimes[a]						*bph* C positive CFU[b]	
Treatment	Biphenyl	Wheat straw	Nutrients	PCB-degrading fungus[c]	PCB-degrading bacteria[d]	% PCB loss	1 month	1 year
1	+	+				31	BDL	BDL
2	+		+			25	BDL	BDL
3	+	+	+	+		40	BDL	BDL
4	+		+		+	57[e]	1.0×10^5	BDL
5	+	+	+	+	+	57[e]	2.1×10^6	1.8×10^7

[a]All treatments consisted of 1 g soil samples containing 100 ppm PCBs. Applications of biphenyl, nutrients and PCB degrading bacteria were performed approximately every 2 months or as required when depleted.
[b]The number of *bph*C positive colony forming units (CFU) was determined by colony blot DNA hybridization using a gene probe made to the biphenyl operon of *Pseudomonas pseudoalcaligenes* KF707 (Layton et al., 1994b). The detection limit was 0.5 to 1% of the colony forming units on nonselective agar plates.
[c]The PCB degrading fungus used in these treatments was *Phanerochaete chrysosporium* (Farrell et al., 1987).
[d]The PCB degrading bacteria used in these treatments were *P. putida* LB400 (Mondello, 1989), *A. eutrophus* H850 (Bedard et al., 1987), and *Rhodococcus globerulus* P6 (formerly designated as *Corynebacterium* sp. strain MB1 (Asturias and Timmis, 1993)).
[e]Differences between treatments 4 and 5 and treatments 1, 2, and 3 were statistically significant at the 95% confidence level.

3. DEVELOPMENT OF GENETICALLY ENGINEERED STRAINS

The development of the PCB degrading genetically engineered strains was based on field application vectors (FAVs), which are used for expressing non-adaptive genes in a competitive environment (Lajoie et al., 1992, 1993). FAVs consist of a selective substrate, a host, and a cloning vector. For this application, the FAVs are surfactant degrading bacterial strains containing the constitutively expressed PCB-degradative genes (Lajoie et al., 1994) on a transposon. There are several advantages to using surfactant and PCB degrading FAVs over naturally occurring PCB degrading strains. First, the surfactant can be used to remove the PCBs from the soil and enhance the bioavailability of the PCBs. Second, the surfactant serves as a selective growth substrate for the FAVs. Third, PCB co-metabolism is accomplished without the use of biphenyl, which is also a toxic compound (Figure 1).

The parent organisms used in this work were the surfactant degrading bacterial strains *Pseudomonas putida* IPL5 and *Alcaligenes eutrophus* B30P4 (Lajoie et al., 1994; Lajoie et al., 1996). The genes for PCB degradation, under the regulation of a constitutive promotor, were inserted into the chromosome of these strains carried on a mini-Tn5 transposon (Lajoie et al., 1994; Lajoie et al., 1996) (Figure 2). These strains were submitted to U.S. Environmental Protection Agency (USEPA) for Premanufacturing Notification (PMN) of proposed field use in bioremediation of hazardous wastes (Sayler et al., 1996). For this submission, the strains were extensively characterized taxonomically using classic diagnostic criteria, fatty acid analysis, Biolog analysis and partial sequencing of the 16S rDNA (Lajoie et al., 1996; Sayler et al., 1996).

These strains utilize the nonionic surfactant polyoxyetheylene 10 lauryl ether (POL) as a growth substrate by separate metabolic pathways. *Pseudomonas putida* IPL5 degrades the ethoxylate portion of the surfactant resulting in 2- (dodecyloxy)ethanol as a product (Lajoie et al., 1996). *Alcaligenes eutrophus* B30P4 grows on POL without the appearance of degradation products and also degrades 2-(dodecyloxy)ethanol (Lajoie et al., 1996). Surfactant degradation by these two strains is complementary with >90% removal of the surfactant and no accumulation of detectable metabolites (Lajoie et al., 1996).

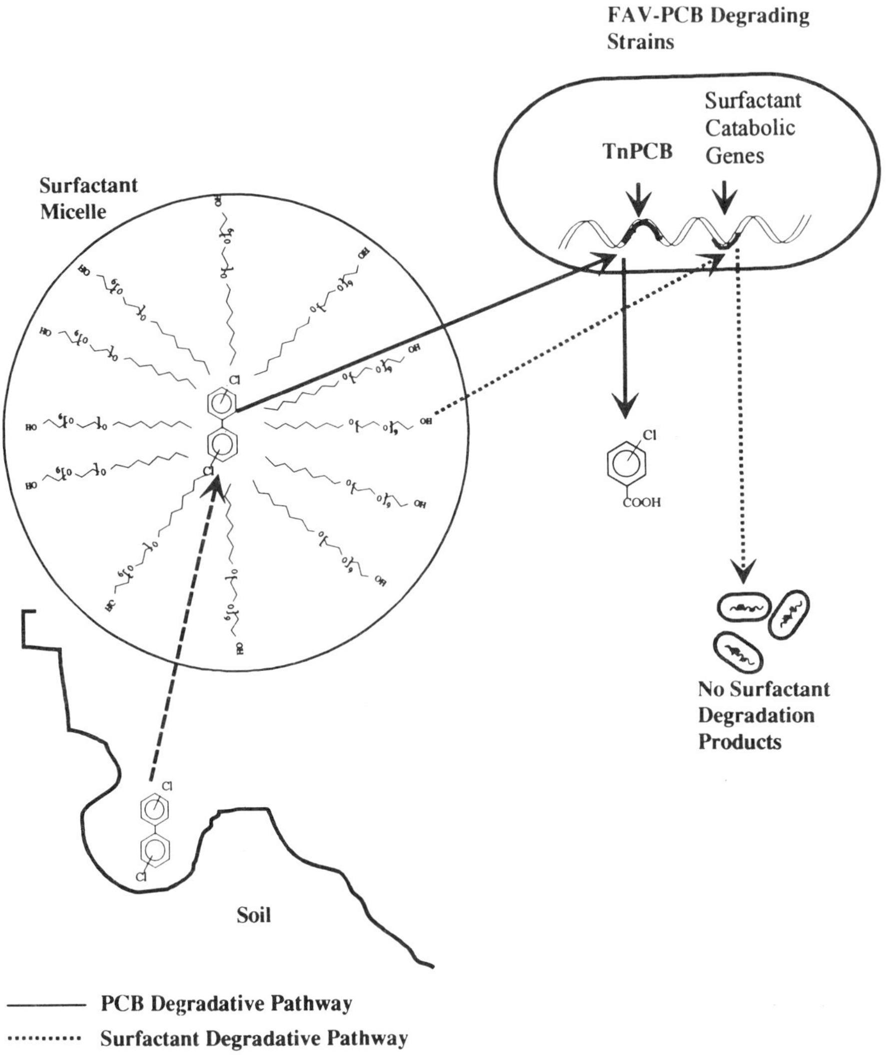

Figure 1. Field application vector concept.

4. TREATMENT PROCESS DESIGN

The design of the integrated soil washing/bioremediation process for the intermediate-scale demonstration consists of four unit operations (Figure 3). These unit operations are soil washing to extract PCBs, aerobic PCB and surfactant degradation, separation of residual solids and anaerobic treatment of the residual PCBs. In the soil washing unit operation, a solution of 10,000 ppm surfactant is pumped onto the soil. The surfactant solution is collected after slowly draining through the soil. This recycling wash is repeated as necessary for PCB recovery. In the aerobic PCB degradation unit operation, the surfactant/PCB wash solution is processed in a bioreactor. Nutrients, diatomaceous earth, and the genetically engineered organisms *Alcaligenes eutrophus* B30P4::TnPCB and *Pseudomonas putida* IPL5::TnPCB are added to the reactor. The bioreactor is aerated and mixed to facilitate PCB and surfactant degradation. As the surfactant is degraded, the undegraded

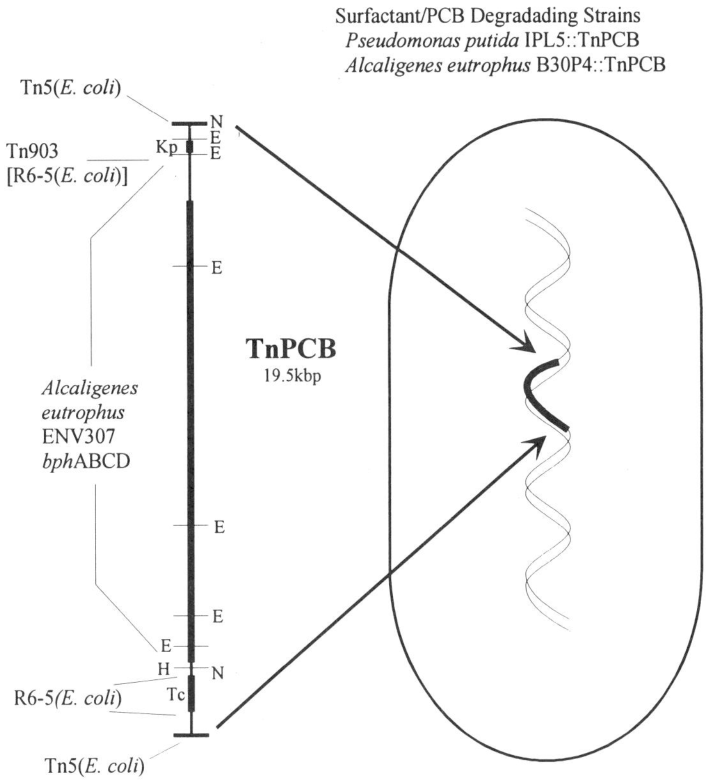

Figure 2. Construction of the Field Application Vectors (FAVs) for surfactant and PCB degradation.

PCBs are desolubilized and deposited onto the diatomaceous earth. In the third unit operation, the residual solids, consisting of cell debris, diatomaceous earth and residual PCBs, are separated from the aqueous phase by centrifugation and the solids are transferred to the anaerobic reactor. In the anaerobic unit operation, the more highly chlorinated PCBs are dechlorinated. Once the PCBs are dechlorinated, they can be returned along with the diatomaceous earth to the aerobic reactor (Figure 3, unit operation 2).

We have proposed a field trial of the treatment process of our research to the USEPA under a research and development (R&D) permit application (Figure 4). Contaminated soil at the site will be placed in the cell indicated in Figure 4. Future operations of the process may be conducted with soil left in place, but partitioned from surrounding clean soil. Liquid and solid streams from the process will be measured and evaluated during the field trial for a materials balance. These same streams will provide inputs to the process in future operations. The liquid stream from the separation phase (unit operation 3) will be used as make-up water in the soil washing phase. The solid stream from the anaerobic phase with its dechlorinated PCB congeners and diatomaceous earth will undergo respective reprocessing and reuse in the aerobic biodegradation operation. The equipment for the field trial is sized to handle about 800 kilograms (kg) of PCB contaminated soil per run. The soil washing and aerobic degradation phase will require about 10 days for their performance and the anaerobic dechlorination phase will require 8 weeks for performance. Since the volume of solids entering the anaerobic phase is a small fraction of the original soil volume, multiple operations of the soil washing and aerobic biodegradation can be conducted while the anaerobic incubation is ongoing.

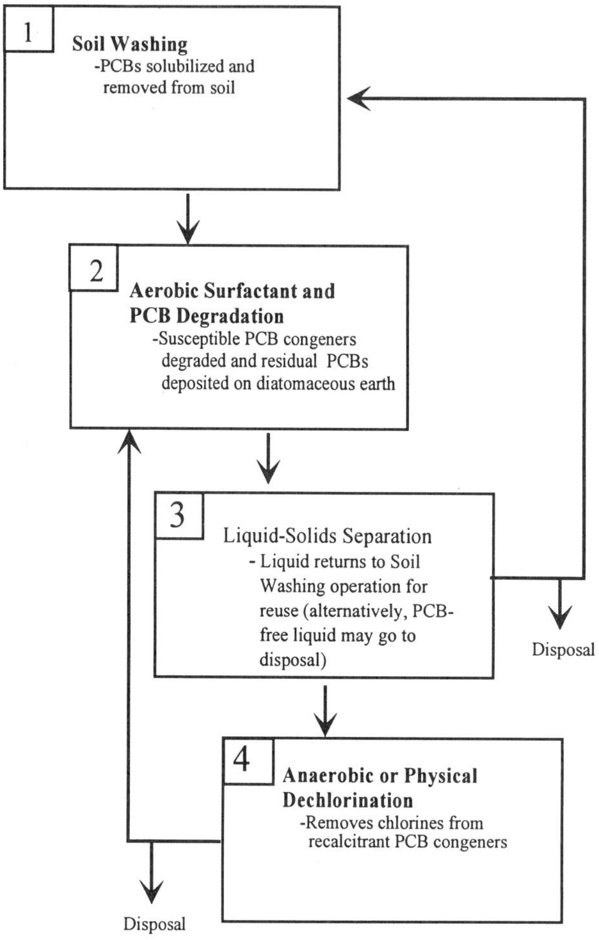

Figure 3. Process concept showing the four unit operations of the integrated treatment system for PCB degradation.

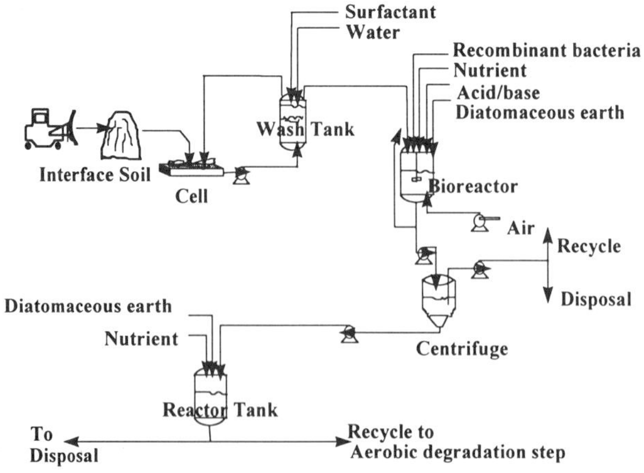

Figure 4. Field process flow diagram for the integrated treatment system for PCB degradation.

4.1. Unit Operation 1. Soil Washing with Surfactant Solution

Surfactants have been shown to enhance the solubility of PCBs and can be used to separate PCBs from soil (Abdul and Gibson, 1991, Abdul et al., 1992, Sun et al., 1995). Surfactant removal of PCBs by soil washing with surfactant concentrations above the critical micelle concentration (CMC) is possible because at supra-CMC dosages, the aqueous micelles show a stronger affinity for the PCBs than do the soil-sorbed surfactants (Sun et al., 1995). In addition, surfactant removal of PCBs from soil does not show congener specificity.

The soil washing efficiency at 10,000 ppm of two polyoxyethylene alcohols, five polyoxyethylene alkylphenols, three polyoxyethylene esters of fatty acids, sodium dodecyl sulfate, and a commercial preparation of mixed surfactants were compared. The best results were obtained with a polyoxyethylene alcohol with removal of >80% of PCBs (data not shown). The surfactant POL (polyoxyetheylene 10 lauryl ether) was chosen for soil washing in these experiments because, in the integrated soil washing/PCB bioremediation process complete degradation was achieved by the FAVs. The CMC for this surfactant is 20 ppm in distilled water and 1,000 ppm when combined with an equal volume soil to water. In a soil wash test with 1 kg soil samples and 1 liter wash volumes of 10,000 ppm POL, approximately 75% of the PCBs were removed from the soil with a 2 day recycling wash (Table 2). Additional PCBs may be removed by a more complete collection of the soil wash solution (approximately 10% of the liquid volume was not recovered) and additional wash cycles.

4.2. Unit Operation 2. Aerobic PCB and Surfactant Degradation

In test tube experiments, the potential for PCB degradation by the FAVs is quite high. In surfactant solutions of 10,000 ppm POL with 25 ppm Aroclor 1242, >70% of the PCBs were degraded by *A. eutrophus* B30P4::TnPCB and *P. putida* IPL5::TnPCB (Lajoie et al., 1996). In 10,000 ppm POL wash solutions containing approximately 100 ppm weathered PCB from the power plant substation soil, 30 to 40% of the PCBs were degraded in 8 days (Layton et al., 1996). Degradation of the PCBs in the soil wash solutions is expected to be less than that observed using Aroclor 1242 due to the more highly chlorinated nature of the weathered PCBs in the soil.

One of the main benefits of the aerobic degradation phase using the FAVs is the desolubilization of the PCBs as the surfactant is degraded. After 8 days of incubation in a 1 liter (L) bioreactor, >90% of the POL was degraded leaving <5 ppm of PCBs in the aqueous phase (Layton et al., 1996).

Table 2. Surfactant solubilization of PCBs from TVA soil with 10,000 ppm of the surfactant polyoxyethylene 10 lauryl ether. Duplicate treatments consisting of 1 kg soil samples were washed with 1 liter surfactant solutions

	PCB concentration (mg PCBs (± SD))	
	Reactor 1	Reactor 2
Initial soil PCBs	107 mg (±18)	107mg (±18)
PCBs in solution after wash	80 mg (±10.0)	78 mg (±4)
Residual PCBs in soil	29 mg (±7)	25 mg (±2)
Wash efficiency	75%	73%
Mass balance	102%	96%

An intermediate scale reactor (50 L) is being used to overcome the engineering obstacles of operating a process with 10,000 ppm surfactant. The agitation and aeration of surfactant at this concentration creates a severe foaming problem. In addition, the aerobic degradation of this surfactant creates a high biological oxygen demand (BOD). These problems have been overcome by addition of foam control with chemical antifoam and the careful regulation of dissolved oxygen with agitation and aeration to maintain a stable oxygen set point.

4.3. Unit Operation 3. Separation of Solids and Recovery of PCBs on Diatomaceous Earth

Centrifugation is the current means of separation of the suspended solids consisting of cell debris, diatomaceous earth and residual PCBs. This process is used to concentrate the PCBs on the solid pellet following surfactant degradation and represents a 65-fold reduction in the mass of PCB contaminated materials (Layton et al., 1996). The effect of surfactant biodegradation and subsequent separation processing steps on soluble PCB concentrations are indicated in Table 3. Results of analyses of the liquid and solids after centrifugation show that after the bioremediation unit operation, the PCBs remain with solids exiting the separation process. If the bacterial inoculum is not included in the process and the surfactant is not degraded, PCBs remain in solution after the separation stage (Table 3).

4.4. Unit Operation 4. Anaerobic Dechlorination

After aerobic degradation and separation of the liquids and solids, the residual PCBs adhering to the diatomaceous earth are the more highly chlorinated PCB congeners that are not readily degraded by aerobic processes. The intent of an anaerobic reactor would be to dechlorinate these PCB congeners and then return the less chlorinated PCBs to the aerobic reactor for ring cleavage. In situ anaerobic dechlorination of PCBs has been demonstrated in river and lake sediments and several different patterns of anaerobic dechlorination have been documented (Bedard and Quensen, 1995).

Table 3. Partioning of PCBs in a soil wash solution with and without aerobic degradation of the PCBs and surfactants by the genetically engineered strains *Pseudomonas putida* IPL5::TnPCB and *Alcaligenes eutrophus* B30P4::TnPCB

	PCB mass balance (%)[b]			
Treatment conditions[a]	Liquid phase	Solid phase	Adsorbed to walls	Loss[c]
Wash solution	90	1	1	8
Wash solution + nutrients	84	2	1	13
Wash solution + nutrients + DE	73	7	2	18
Wash solution + nutrients + GEMS	10	49	1	34
Wash solution + nutrients + GEMS + DE	15	72	1	12

[a]Starting PCBs were obtained from TVA soil samples after surfactant washing with 10,000 ppm POL. All treatment conditions consisted of 100 ml soil wash solution containing 130 ppm PCBs and approximately 10,000 ppm POL. Treatments were incubated for four days in fabricated stainless steel vessels with stirring, followed by separation of the solid and liquid phases by centrifugation (16,000 xg). Nutrients - PAS minimal salts media from Bedard et al., 1986. D.E. - Diatomaceous earth added to a concentration of 5 g/liter.GEMs - Genetically engineered microorganisms *Pseudomonas putida* IPL5::TnPCB and *Alcaligenes eutrophus* B30P4::TnPCB.

[b]The percentage of PCBs recovered was determined based on the starting concentration of PCBs in the soil wash solution.

[c]The PCB loss = 100 – (liquid + solids + walls). Excess PCB loss in treatment 3 represents PCB degradation by the GEMs.

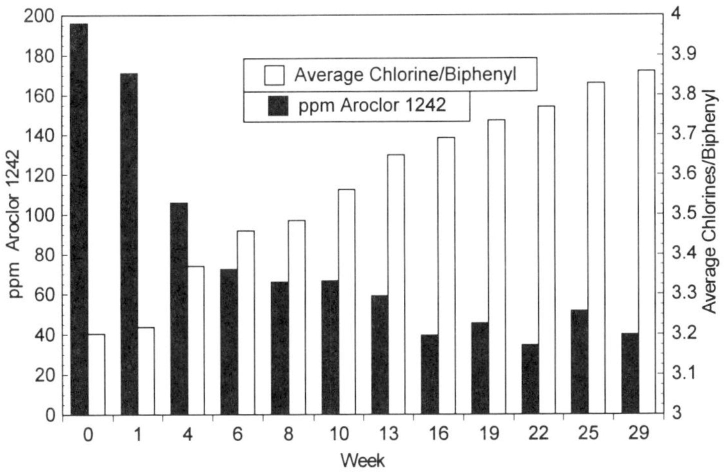

Figure 5. Change in PCB concentration and chlorine/biphenyl ratio during anaerobic incubation of Aroclor 1242 on diatomaceous earth (200 milligrams per kilogram) in a solution of 200 milligrams per liter polyoxyethylene 10 lauryl ether inoculated with a preincubated Hudson River sediment.

In laboratory experiments using anaerobic conditions and anaerobic bacteria transferred from Hudson river sediments (Quensen et al., 1990), the more highly chlorinated PCBs in TVA soils were dechlorinated following typical dechlorination patterns (Evans et al., 1996). In relation to the integrated PCB bioremediation process, experiments were performed to determine the effect of residual surfactants and diatomaceous earth on PCB dechlorination. Unexpectedly, the anaerobic dechlorination pattern of PCBs on diatomaceous earth differed from that of PCBs in soil slurries, with a disappearance of less chlorinated PCBs at a faster rate than the more highly chlorinated PCBs. The anaerobic bacterial degradation of lower-chlorinated PCBs has also been reported by other researchers (Chen et al., 1988; Rhee et al., 1989). In an experiment using 200 ppm Aroclor 1242 on diatomaceous earth and a surfactant concentration of 200 ppm incubated with an inoculum prepared by preincubation of actively-dechlorinating river sediment, a 75% reduction of PCBs was observed after 25 weeks of incubation (Figure 5). The best efficiency achieved with solids recovered from the separation step following aerobic biodegradation was a 68% reduction in PCBs after 8 weeks of incubation.

5. SUMMARY

The integrated process consists of four unit operations: surfactant solubilization of PCBs, aerobic biodegradation of PCBs and surfactant, liquid-solid separation, and anaerobic dechlorination of PCBs. The efficiency of the process is a summation of both the physical/chemical and biological operations included in the overall process (Table 4). The efficiency of the soil washing operation is governed by the removal and containment of the PCBs from the contaminated site. Removal of the PCBs from the soil is accomplished by surfactant solubilization of the PCBs. Containment of the PCBs is accomplished by desolubilization and deposition of the PCBs in the solid phase following the aerobic degradation of PCBs. Our bench scale tests indicate that we can remove large amounts of PCBs (75%) from the soil in a relatively short period of time (2 days). The target effi-

Table 4. Current process efficiencies for individual process operations based on laboratory evaluations of a single pass through and target efficiencies for the integrated soil washing/PCB bioremediation process

Unit operation	Fraction of PCB removed/step	Fraction of starting PCB degraded	Target efficiency
Soil washing	0.75	na	0.99
Aerobic degradation	0.35	0.75 × 0.35 = 0.26	0.55
Separation of solid and liquid phases	0.90	na	0.99
Anaerobic degradation	0.68	(0.75 – 0.26) × 0.9 × 0.68 = 0.30	0.85

ciency for the surfactant soil washing unit operation is 99% removal of PCBs from the soil. This target may be attainable through the use of an improved surfactant solubilization regiment. The containment of PCBs is accomplished during the liquid-solids separation unit operation. The resulting effluent from the operation contains <5 ppm PCBs. Greater than 90% of the PCBs are removed from with the solids. The resulting solids represent a 50-fold reduction in PCB-contaminated material from the original soil sample. A target efficiency of 99% in the separation unit operation may be attainable by improvements in separation.

PCBs are degraded in both the aerobic and anaerobic unit operations of the process. In bench scale studies, 35% of the PCBs were degraded in 8 weeks in the anaerobic phase. The combination of the chemical/physical and biological unit operations in a single pass results in approximately 56% degradation of the original PCBs (Table 4). Improvements in the aerobic and anaerobic degradative potential as well as PCB removal and containment are ongoing.

The combination of each of these unit operations is more efficient for PCB remediation than any one of these operations alone and may allow for the near complete removal and destruction of PCBs from contaminated sites. Additional ongoing work seeks to improve the efficiency of each of these steps and the development of an effective integrated process for PCB bioremediation.

REFERENCES

Abdul, A.S. and T.L. Gibson. 1991. Laboratory studies for surfactant-enhanced washing of polychlorinated biphenyl from sandy material. Environ. Sci. Technol. 25:665–671.

Abdul, A.S., T.L. Gibson, C.C. Ang, J.C. Smith, and R.E. Sobezynski. 1992. In situ surfactant washing of polychlorinated biphenyls and oils from a contaminated site. Ground Water 30: 219–231.

Abramowicz, D.A. 1990. Aerobic and anaerobic biodegradation of PCBs: A review. Critical Reviews in Biotechnology 10: 241–251.

Asturias, J.A. and K.N. Timmis. 1993. Three different 2,3-dihydroxybiphenyl-1,2-dioxyygenase genes in the gram-positive polychlorobiphenyl-degrading bacterium *Rhodococcus globerulus* P6. J. Bacteriol. 175: 4631–4640.

Barriault, D. and M. Sylvestre. 1993. Factors affecting PCB degradation by an implanted bacterial strain in soil microcosms. Can. J. Microbiol. 39: 594–602.

Bedard, D.L., R.E. Wagner, M.J. Brennan, M.L. Haberl and J.F. Brown, Jr. 1987. Extensive degradation of Aroclors and environmentally transformed polychlorinated biphenyls by *Alcaligenes eutrophus* H850. Appl. Environ. Microbiol. 53: 1094–1102.

Bedard, D.L. and J.F. Quensen III. 1995. "Microbial reductive dechlorination of polychlorinated biphenyls." In: Microbial transformation and degradation of toxic organic chemicals. L. Young and C. Cerniglia, eds. pp.127–216.

Brunner, W., F.H. Sutherland and D.D. Focht. 1985. Enhanced biodegradation of polychlorinated biphenyls in soil by analog enrichment and bacterial inoculation. J. Environ. Qual. 14:324–328.

Bumpus, J.A., M. Tiem, D. Wright and S.D. Aust. 1985. Oxidation of persistent environmental pollutants by a white rot fungus. Science. 228: 1431–1436.

Chen, M., C.S. Hong, B. Bush, and G.-Y. Rhee. 1988. Anaerobic biodegradation of polychlorinated biphenyls by bacteria from Hudson River sediments. Ecotoxicology and Environmental Safety. 16: 95–105.

Dietrich, D, W.J. Hickey, and R. Lamar. 1995. Degradation of 4,4′-dichlorobiphenyl, 3,3′,4,4′-tetrachlorobiphenyl and 2,2′,4,4′,5,5′-hexachlorobiphenyl by the white rot fungus *Phanerochaete chrysoporium*. Appl. Environ. Microbiol. 61:3904–3909.

Evans, B.S., C.A. Dudley, and K. T. Klasson. 1996. Sequential anaerobic-aerobic biodegradation of PCBs in soil slurry microcosms. Appl. Biochem. and Biotech. 57/58: 885–894.

Farrell, R.L., T.K. Kirk, and M. Tien. 1987. Novel enzymes which catalyze the degradation and modification of lignin. U.S. Patent 4,687,741.

Fiebig, R., D. Schulze, P. Erlemann, M. Slawinski, and H. Dellweg. 1993. Microbial degradation of polychlorinated biphenyls in contaminated soil. Biotechnol. Lett. 15:93–98.

Focht, D. D. and W. Brunner. 1985. Kinetics of biphenyl and polychlorinated biphenyl metabolism in soil. Appl. Environ. Microbiol. 50:1058–1063.

Hickey, W.J., D. B. Searles, and D.D. Focht. 1993. Enhanced mineralization of polychlorinated biphenyls in soil inoculated with chlorobenzoate-degrading bacteria. Appl. Environ. Microbiol. 59: 1194–1200.

Kohler, H.-P.E., D. Kohler-Staub, and D.D. Focht. 1988. Cometabolism of polychorinated biphenyls: Enhanced transformation of Aroclor 1254 by growing bacterial cells. Appl. Environ. Microbiol. 54:1940–1945.

Lajoie, C.A., S.-Y. Chen, K.C. Oh, and P. F. Strom. 1992. Development and use of field application vectors to express non-adaptive foreign genes in competitive environments. Appl. Environ. Microbiol. 58:655–663.

Lajoie, C.A., G.J. Zylstra, M.F. DeFlaun and P.F. Strom. 1993. Development of field application vectors for bioremediation of soils contaminated with polychlorinated biphenyls. Appl. Environ. Microbiol. 59:1735–1741.

Lajoie, C.A., A.C. Layton, and G.S. Sayler. 1994. Cometabolic oxidation of polychlorinated biphenyls in soil with a surfactant-based field application vector. Appl. Environ. Microbiol. 60: 2826–2833.

Lajoie, C.A., A.C. Layton, J.P. Easter, F.-M. Menn and G.S. Sayler. 1996. Degradation of surfactants and polychlorinated biphenyls by recombinant field application vectors. Submitted for publication.

Layton, A.C., C.A. Lajoie, J.P. Easter, R. Jernigan, M.J. Beck, and G.S. Sayler. 1994a. Molecular diagnostics for polychlorinated biphenyl degradation in contaminated soils. The New York Academy of Science 721:407–422.

Layton , A.C. C.A. Lajoie, J.P. Easter, R. Jernigan, J. Sanseverino, and G.S. Sayler. 1994b. Molecular diagnostics and chemical analysis for assessing biodegradation of polychlorinated biphenyls in contaminated soils. J. Indust. Microbiol. 13: 392–401.

Layton, A.C., C.A. Lajoie, J.P. Easter, and G.S. Sayler. 1996. Integration of surfactant solubilization of PCBs and PCB biodegradation using surfactant/PCB degrading FAVs. Manuscript in preparation.

McDermott, J.B., R. Unterman, M. J. Brennan, R.E. Brooks, D. P. Mobley, C.C. Schwartz, and D. K. Dietrich. 1989. Two strategies for PCB soil remediation: biodegradation and surfactant extraction. Environ. Progress. 8:46–51.

Mondello, F.J., 1989. Cloning and expression in *Escherichia coli* of *Pseudomonas* strain LB400 genes encoding polychlorinated biphenyl degradation. J. Bacteriol. 171: 1725–1732.

Quensen, J.F. III, S.A. Boyd, and J. M. Tiedje. 1990. Dechlorination of four commercial polychlorinated biphenyl mixtures (Aroclors) by anaerobic microorganisms from sediments. Appl. Environ. Microbiol. 56: 2360–2369.

Rhee, G.-Y., B. Bush, M.P. Brown, M. Kane, and L. Shane. 1989. Anaerobic biodegradation of polychlorinated biphenyls in Hudson river sediments and dredged sediments in clay encapsulation. Water Research 23: 957–964.

Sayler, G.S., U. Matrubutham, C. Steward, A. Layton, C. Lajoie, J. Easter, and B. Applegate. 1996. Towards field release of engineered strains for bioremediation. Proceedings from the 7th Symposium on Environmental Release of Biotechnology Products: Risk assessment methods and research progress. In review.

Sun, S. W.P. Inskeep, and S.A. Boyd. 1995. Sorption of nonionic organic compounds in soil-water systems containing micelle-forming surfactant. Environ. Sci. Tech. 29:903.

Yadav, J.S., J.F. Quensen III, J.M. Tiedje, and C.A. Reddy. 1995. Degradation of polychlorinated biphenyl mixtures (Aroclor 1242, 1254, and 1260) by the white rot fungus *Phanerochaete chrysosporium* as evidenced by congener-specific analysis. Appl. Environ. Microbiol. 61:2560–2565.

9

TEN YEARS OF RESEARCH IN GROUNDWATER TRANSPORT STUDIES AT COLUMBUS AIR FORCE BASE, MISSISSIPPI

Thomas B. Stauffer,[1] J. Mark Boggs,[2] and William G. MacIntyre[3]

[1]Armstrong Lab, AL/EQC
139 Barnes Drive, Suite 2
Tyndall AFB, Florida 32404
[2]Tennessee Valley Authority Engineering Lab
P.O. Drawer E
Norris, Tennessee 37828
[3]College of William and Mary (VIMS)
Gloucester Point, Virginia 23062

1. INTRODUCTION

Groundwater research studies at Columbus Air Force Base (AFB), Mississippi (MS) are currently underway and have been under active pursuit for over ten years. This chapter is an overview of some of the key research findings from the groundwater test site. The individual sections are in chronological order and provide a quick summary of the site activities associated with various research projects.

2. MACRODISPERSION EXPERIMENT (MADE-1)

Groundwater transport research at the Columbus AFB site began in 1983 with MADE-1, a large-scale natural-gradient tracer experiment sponsored by the Electric Power Research Institute (EPRI). The study was part of a major EPRI research initiative known as the "Solid Waste Environmental Studies" program which was intended to develop improved models for predicting the transport and fate of utility-related waste contaminants. The principal goals of MADE-1 were to (1) develop a data base for validating the advection-dispersion component of solute transport models, and (2) develop practical methods of estimating hydraulic conductivity variability from which aquifer macrodispersivity can be predicted (e.g., using the methods of Gelhar and Axness, 1983).

The hydrogeology of the experimental site was characterized in detail using a variety of methods during the first two years of the study. The emphasis of these investigations was on the estimation of the three-dimensional spatial structure of hydraulic conductivity of the alluvial aquifer at the test site. The heterogeneity of the alluvial aquifer was evident at several scales. Borehole flowmeter measurements at 15-centimeter (cm) vertical increments in fully-screened wells commonly showed variations in hydraulic conductivity ranging over two-to four-orders of magnitude at each test well (Figure 1). The variance of log conductivity based on 2187 flowmeter measurements was estimated to be approximately 4.5 (Rehfeldt et al., 1992). By comparison, the variances of log hydraulic conductivity for the aquifers at the Borden and Cape Cod research sites are 0.29 (Sudicky, 1986) and 0.26 (Hess, 1989), respectively. In addition, the mean hydraulic conductivity along the tracer plume pathway was shown to vary from approximately 10^{-3} centimeters/second (cm/s) in the near-field region of the test site to 10^{-1} cm/s portions of the far-field. These large-scale conductivity variations produced horizontally converging groundwater flow in the region of the tracer test and a strong vertical flow component in the near-field region of the site. The increase in magnitude of the mean conductivity along the plume travel path and the locally converging flow field strongly affected plume migration (Boggs et al., 1992).

The MADE 1 tracer experiment was initiated in October 1986 with a pulse injection of 10 cubic meters (m^3) of groundwater containing bromide and three organic tracers (pentafluorobenzoic acid, o-trifluoromethylbenzoic acid, and 2,6-difluorobenzoic acid). Over a 20-month period, seven snapshots of the tracer plume were performed at approximately 1- to 4-month intervals using an extensive three-dimensional sampling well network (Figure 2). The hydraulic head field in the aquifer was also monitored during the natural-gradient tracer test. The dominant feature of the tracer plume that evolved during the study was the highly asymmetric concentration distribution in the longitudinal direction. At the conclusion of the experiment, the more concentrated region of the plume remained within approxi-

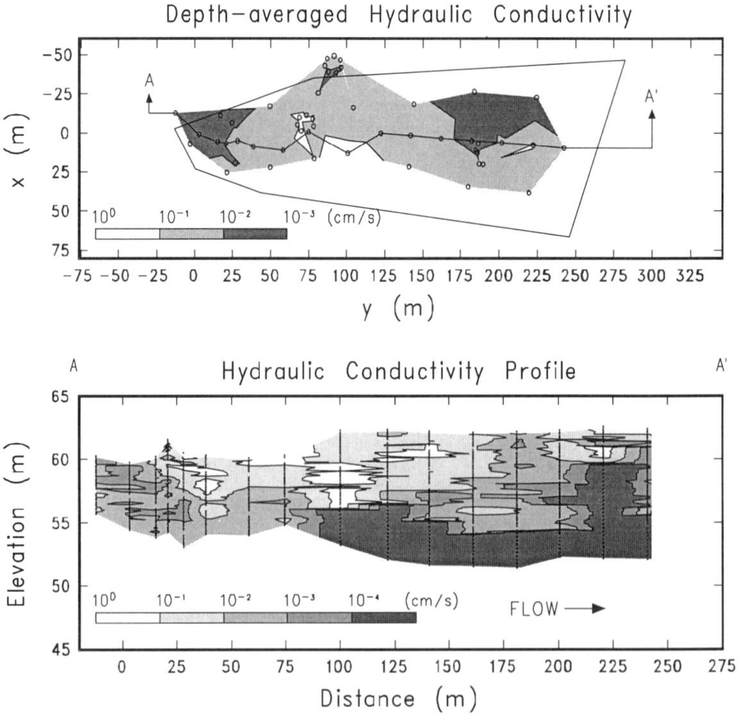

Figure 1. Map of vertically averaged hydraulic conductivity (upper diagram) and hydraulic conductivity profile along section A-A′ (lower diagram).

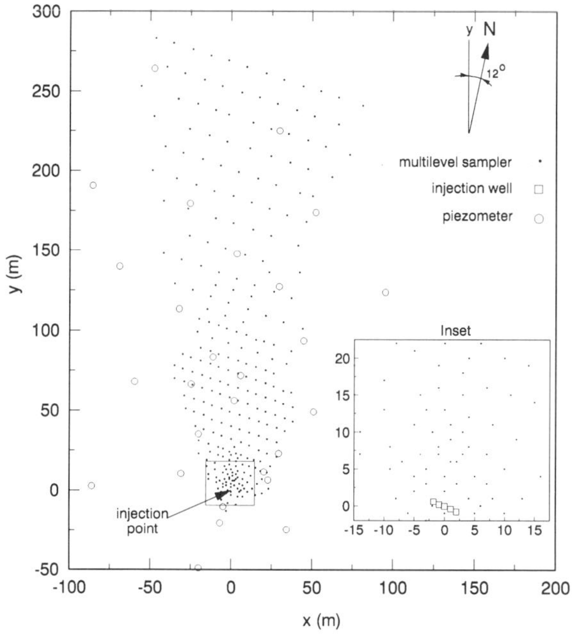

Figure 2. Sampling well network showing injection wells, MSL, and piezometers.

mately 20 m of the tracer injection point, while the advancing side of the plume extended down grading a distance of more than 260 m. The extreme skewness of the plume was caused by an increasing mean groundwater velocity along the plume travel path (Figure 3).

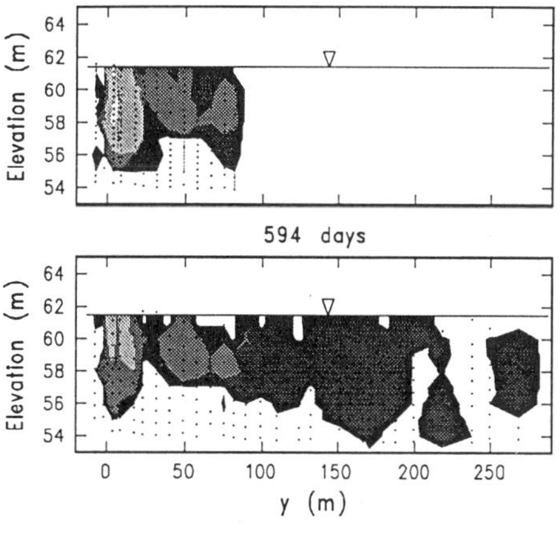

Figure 3. Vertical cross sections along the longitudinal axis of bromide plume 49 days, 279 days, and 594 days since injection. The approximate water table position is indicated by the horizontal line with the triangle in each diagram.

Adams and Gelhar (1992) provide a spatial moments analysis of the bromide data for the MADE-1 field experiment. Their analysis indicated that the bromide mass recovery decreased to approximately 50 percent by the end of the study due to the combined effects of plume truncation, sampling bias, and sorption. However, they showed that the relationship of longitudinal variance and mean plume displacement was relatively insensitive to the bromide mass loss during the experiment. Application of a non-uniform advection-dispersion model to the bromide plume moments leads to an estimate of longitudinal dispersivity of 5 to 10 meters (m) for the alluvial aquifer (Adams and Gelhar, 1992).

A series of field and laboratory investigations designed to determine the contributions of sorption and sampling bias to the anomalous bromide mass balance trend during MADE-1 are described in Boggs and Adams (1992). Their laboratory column experiments showed that up to 20 percent of the bromide may have adsorbed during the field study. Comparisons between bromide measurements in the multilevel samplers (MLS) and in water samples extracted from adjacent soil cores showed a systematic difference. A matrix diffusion type process in conjunction with a natural tendency for preferential sampling from permeable regions in the heterogeneous aquifer was proposed to explain the apparent bias in the MLS bromide data. Although sampling bias could not be reliably quantified, this process was shown to be qualitatively consistent with the anomalous bromide mass balance.

An important achievement of MADE-1 was the development and demonstration of the borehole flowmeter as a practical technique for obtaining extensive measurements of hydraulic conductivity in three dimensions. Rehfeldt et al., (1989a) evaluated the accuracy and reliability of the technique through a series of field and laboratory tests, and provided theoretical and practical guidelines for its implementation. Application of the flowmeter and other methods for estimating hydraulic conductivity variability and aquifer dispersivity at the test site is discussed by Rehfeldt et al., (1989b, 1992).

3. STUDY OF AQUIFER CHARACTERIZATION METHODS FOR ENHANCED *IN SITU* BIOREMEDIATION

A series of field experiments were conducted during 1987–89 to develop an aquifer characterization methodology to support optimal design of well networks or infiltration galleries associated with enhanced in-situ bioremediation systems. The research was conducted by the Tennessee Valley Authority (TVA) for the USAF Armstrong Laboratory. The study involved performing and analyzing a series of aquifer tests and recirculating tracer tests at a site located approximately 200 m northeast of the MADE-1 sampling well field. The test site was approximately one hectare in size and included 37 fully-screened test wells.

The focus of the study was on development of methods for characterizing the hydraulic conductivity field of contaminated aquifers, since the spatial distribution of hydraulic conductivity largely controls the migration rates and pathways of artificially introduced nutrients or dissolved oxygen. The electromagnetic borehole flowmeter was the primary technique used to measure hydraulic conductivity at the test site. All test wells were logged at 0.3 m intervals to develop detailed hydraulic conductivity profiles at each test well site. Single- and multiple-well aquifer tests were also conducted to establish large-scale trends in aquifer hydraulic conductivity and transmissivity. The borehole flowmeter results were evaluated by comparisons with 14 small-scale and one large-scale forced-gradient tracer tests. Small-scale tests involved doublet pumping-injection wells spaced from 4 to 7 m apart, and test duration of 8 to 70 hr. The large-scale test was performed over a 168 hour period using a five-spot well arrangement with wells up to 31 m apart. Investigations showed that detailed layer hydraulic conductivity data derived from the borehole flowmeter tests compared favorably with the estimates based on tracer test breakthrough data. In cases where inter-well spacing was less than 7 m, conductivity estimates were usually within a factor of two. Likewise, the geometric mean hydraulic conductivity estimated from the flowmeter data was consistent with the results of the large-scale aquifer tests and tracer tests in terms of both magnitude and large-scale spatial trends.

The results of the field studies lead to the development of a general protocol for characterizing contaminated sites under consideration for in situ bioremediation. The principal elements of the protocol included, (1) an initial site investigation involving installation and short-term aquifer testing of up to 13 fully-screened test wells, (2) a preliminary assessment of the feasibility of enhanced in-situ bioremediation based on the degree of aquifer heterogeneity as determined from initial investigations (i.e., bioremediation may not be feasible if the site is too heterogeneous), (3) delineation of three-dimensional hydraulic conductivity distribution of aquifer using the borehole flowmeter method, (4) spatial interpolation of hydraulic conductivity data, and (5) design of injection/withdrawal well nutrient application system using three-dimensional groundwater flow and transport models.

4. MACRODISPERSION EXPERIMENT (MADE-2)

The purpose of this experiment was to perform a controlled field experiment, involving the injection of several aromatic hydrocarbons and a nonreactive tracer into an uncontaminated aquifer. By monitoring the plume development of these solutes, and by measuring a number of physical and chemical characteristics of the aquifer, this study was designed to provide data on those properties which significantly control the propagation of dissolved contaminants in groundwater systems. Data of this type are rare, and it is extremely useful for validating numerical fate and transport models which include complex processes, such as solute advection-dispersion, geochemical attenuation, and microbial degradation. A secondary objective was to measure the in situ degradation rates of the selected organic compounds, in order to investigate the possibility of the natural attenuation processes as an alternative to active remediation techniques.

Groundwater contamination is an important environmental concern for both military and civilian industrial operations. The organic constituents of fuels such as JP-4, gasoline, and diesel (e.g., benzene, toluene, naphthalene, xylene, etc.) and cleaning solvents (e.g., trichloroethylene) have resulted in serious contamination problems at many US Department of Defense facilities. Many of these chemical components of fuels and solvents are regulated by the US Environmental Protection Agency (USEPA) as hazardous substances, and have stringent concentration limits in drinking water (Geraghty and Miller, Inc., 1991). Their presence in groundwater at concentrations exceeding the USEPA's maximum contaminant levels generally requires the use of costly water purification methods. The lack of an adequate quantitative understanding of the processes that govern the environmental fate of these contaminants has constrained the development of cost-effective groundwater remediation techniques.

A great number of research efforts have studied various physical, chemical and biological mechanisms that influence the transport of pollutants in the environment. Most have been performed within laboratories, but some relatively recent experiments have been conducted at field sites to determine the effects of natural conditions and heterogeneities on these processes (Sudicky, 1986; Boggs et al., 1992; Harvey et al., 1993). Additionally, the accidental discharge of contaminants into groundwater systems has allowed some limited studies of the in situ transport of pollutants. However, the lack of specific information about the initial conditions of contaminant mass and location hinders the ability to draw meaningful conclusions from such research.

Previously, a few field experiments have observed the attenuation of test solutes, and claimed that the results were possibly due to microbial degradation processes (e.g., Sutton and Barker, 1985). While laboratory studies have proven that soil organisms are capable of utilizing a great variety of chemical compounds as food sources, similar work showing natural degradation in the field has not previously been possible, for a number of reasons. Although it is generally accepted that the in situ biodegradation of chemical contaminants does occur, conclusive experimental evidence is lacking in the current literature.

In order to effectively monitor the development of the plume of hydrocarbons and the tritium tracer, a well field consisting of 328 multilevel and 56 BarCad® samplers was used by the researchers. Due to time and cost constraints, five complete, three-dimensional sample sets were taken at about 100-day intervals during the study. Additional, smaller sampling events were used to observe plume activity on a shorter time scale. Since it is neither physically nor economically possible to sample an entire aquifer, a three-dimensional network of wells must be used to determine the location and concentration of groundwater contaminants. In order to obtain a complete "picture" of this plume, geostatistical techniques (e.g., kriging) were employed to mathematically fill in the spatial gaps between sampling wells. This technique is currently the most sophisticated and accurate method available for use in generating stochastic models and computer graphics of subsurface contaminant transport.

Although the measurement of degradation rates was a principal objective of this study, no attempts were made to classify or quantify the microbial populations. While a detailed analysis of enzymatic activity before, during and after the movement of the plume through a given area would greatly support observations of biodegradation, such work was beyond the scope of this study and is left for future research efforts. In order to distinguish solute degradation from losses due to sorption, dilution or evaporation, and to permit the detection of degradation products, ^{14}C-labeled *p*-xylene was included in the injection solution.

Previous work at this site characterized many of the important physical properties of this aquifer (Boggs et al., 1992). During this study soil samples were taken at various locations and depths, and then analyzed for particle sizes, organic carbon content, surface area, and partition coefficient (Kd). Groundwater samples were also tested for a number of elemental components and chemical characteristics. Finally, a set of piezometers was used to measure temporal fluctuations in the hydraulic gradient of the aquifer, which were due to a combination of daily rain events, evapotranspiration and seasonal factors.

The natural gradient field experiment was conducted to investigate the transport and degradation of four dissolved organic compounds (benzene, naphthalene, *p*-xylene, and *o*-dichlorobenzene) and one non-retained tracer in an aquifer. The study was done at Columbus AFB, Mississippi, in a shallow, unconfined alluvial aquifer having considerable heterogeneity in physical and chemical properties. A two-day pulse of 9.7 m^3 of tritiated water and the hydrocarbons in a dilute aqueous solution was injected into the aquifer. The initial concentrations of the organic test compounds in solution ranged from about 7 to 70 milligrams per liter (mg/L), which was prepared and stored on-site using ambient groundwater from an upgradient local well. The injection was made into five wells, spaced 1 meter apart and aligned transverse to the direction of groundwater flow. Each injection well was screened over a 0.6 m interval, which was placed within the saturated zone of the aquifer.

The solute concentration distributions were monitored at one to three month intervals over a 15-month period, using a three-dimensional network consisting of over 6,000 sampling points. Selected points within the network were sampled more frequently to produce time-series sets of solute concentrations. Spatial moments calculated from the three-dimensional network concentration data and from the time-series concentration data, were used to determine the fate and transport behavior of each solute.

Stable organic compounds in groundwater samples were analyzed by gas chromatography (GC). Deuterated toluene and 4-bromofluorobenzene internal standards were added to 20 milliliter (mL) water samples, which were then extracted with 2 mL of n-pentane. These extracts were then analyzed for benzene, *p*-xylene, naphthalene and *o*-dichlorobenzene, also by GC. Sensitivity of the method was 50 mg/L for benzene and 4 mg/L for the other compounds. All samples were extracted and analyzed within 21 days of collection. Quality control for the organic analyses was ensured by using several techniques, including the use of sample duplicates, certified standard control solutions, internal standards for all analyses, and independent verification of our analytical results by another laboratory.

Tritium and ^{14}C in samples from the wells were analyzed by liquid scintillation counting in dual isotope mode, by the Water Resources Research Center at Mississippi State University. Samples were counted for 20 minutes or to a one-percent error at the 95 percent probability level, whichever was attained first. Background levels of tritium and ^{14}C in ambient groundwater at the test site dictated the analytical sensitivity for these measurements, and were approximately 2 and 3 picocuries per milliliter, respectively. The statistical errors associated with 20-minute measurements at these levels are about ±20 percent at the 95 percent probability level. A set of tritium and ^{14}C standards was counted every 108 samples to check analytical accuracy and verify instrument performance. Every twenty-fifth sample was counted twice to estimate measurement precision and repeatability. As an overall check on the radiological measurements, duplicate field samples amounting to roughly five percent of the total field samples were analyzed independently by the TVA Western Area Radiological Laboratory. Round robin known samples were prepared by Environmental Quality Control and distributed to each analytical group as a final quality control method. Dissolved oxygen (DO) measurements were performed at two to three month intervals at selected locations during the experiment. DO data were intended to demonstrate whether aerobic conditions for possible degradation of the organic compounds were maintained, and were collected from 16 multilevel samplers (MLS) located along the longitudinal axis of the tracer plume, and six piezometers. Samples were collected in 60 mL BOD bottles after purging approximately 100 mL from the MLS tubes, and measurements were made with a calibrated DO probe immediately after sample collection. The DO measurements at the 6 piezometers were made in conjunction with pH, conductivity, and oxidation-reduction potential measurements.

The tritium plume was characterized by extreme skewness in the longitudinal direction. The evolution of the plume was consistent with the measured hydraulic head and conductivity fields, and was primarily due to the increasing mean hydraulic conductivity with distance from the injection site. Measurements of hydraulic conductivity produced a mean Darcy's Law groundwater velocity of 5 meters per year (m/yr) near the injection wells, whereas local velocities in the far field were found to exceed 400 m/yr. The dispersivity of the alluvial aquifer was estimated by fitting the tritium data to general spatial moments equations for two dimensional advective-dispersive transport in a non-uniform flow field. A longitudinal dispersivity of approximately 10 m provided the best representation of the first and second longitudinal moments of the tritium plume.

The aromatic hydrocarbons degraded significantly during the experiment. Analysis for the degradation products of ^{14}C-*p*-xylene revealed that about 80 to 90 percent of the ^{14}C present was associated with dissolved $^{14}CO_2$ and intermediate products, indicating aerobic degradation of the *p*-xylene. Dissolved oxygen in the pulse plume maintained aerobic conditions throughout the experiment, and was always greater than 2.6 mg/L. Degradation kinetics calculated from the complete field data set were approximately first-order with the following rate constants: benzene, 0.0070 day^{-1} (d^{-1}); *p*-xylene, 0.0107 d^{-1}; naphthalene, 0.0064 d^{-1}; *o*-dichlorobenzene, 0.0046 d^{-1}. Similar reaction rates were also obtained from a near field subset of the data, using a model based on the hydrologic characteristics of the aquifer. The shapes of the degradation rate curves were consistent with microbial degradation processes. Maximum degradation rates obtained are presumed to be characteristic of the microbial population's metabolism. A least-squares inversion method of fitting time-series data to an analytical one dimensional transport model, containing first-order decay and linear equilibrium sorption terms, yielded the following mean first-order rate constants: benzene, 0.010 d^{-1}; naphthalene, 0.013 d^{-1}; *p*-xylene, 0.016 d^{-1} and *o*-dichlorobenzene, 0.005 d^{-1}.

Sorption had some effect on the transport of the organic compounds, but it was a relatively minor process compared with the degradation. Because of the dominant influence of the degradative processes, field-average retardation factors could not be estimated from a comparison of the mean displacement rates of the aromatic compound and tritium plumes. However, analysis of time-series data at each of 17 sample points located within a distance of approximately 20 m of the injection point produced mean retardation factors for benzene, naphthalene, *p*-xylene, and *o*-dichlorobenzene of 1.20, 1.45, 1.16, and 1.33, respectively. Agreement between these results and retardation factors calculated from sorption coefficients obtained in laboratory batch and column experiments, suggests that laboratory retardation measurements may be reliable indicators of sorption behavior in the field.

The disappearance and transformation of the organic solutes during this experiment demonstrated that natural degradation processes were able to effectively reduce these levels of dissolved organic contaminants in a reasonable time frame. This result suggests that, for similar organic compounds, it might be best to restrict aquifer remediation activities to the contaminant source region. By reducing the source, the resulting plume of organic solutes could possibly maintain a steady-state limit, given the correct physical, chemical and biological conditions. In such an aquifer, the spatial boundary of this plume would be determined primarily by the hydrology and geochemistry of the site, solute sorption, biodegradation by indigenous microbes, redox capacity, and oxygen and nutrient supply.

This observation represents the most important result of this field study, and could have important implications for the restoration of aquifers contaminated by organic chemicals. Active remediation would not be needed in situations where natural degradation rates were sufficient to reduce contaminant concentrations to safe levels before they were transported to regions where they might be dangerous to human health or the environment. Remediation by natural degradation processes is likely to be applicable to residual organic contaminants (i.e., non-aqueous phase liquid organic chemicals, or NAPL), immobilized by capillary forces that cannot be effectively removed from the aquifer by pumping. The monetary and environmental cost savings of allowing natural biological restoration of the residual contaminants are potentially enormous. After a review of the current scientific literature, it appears that this study is the first field experiment to prove conclusively that hydrocarbon solute losses were due to chemical degradation rather than physical losses, such as sorption and vaporization.

5. NATURAL ATTENUATION STUDY (NATS)

Further research is needed to build upon results of MADE 2, to further support the observation of natural attenuation and to attempt to quantify the environmental requirements for this process. The degradation rates determined for the four aromatic compounds in the Columbus aquifer have been used in the design and modeling stages of a new field test now in progress which uses an emplaced NAPL source. This test at the Columbus site will be a demonstration of the practicality of natural degradation (natural attenuation) as a remediation action for a steadily leaching source in an aquifer. This experiment more closely emulates the effects of a fuel spill. Assuming that a steady-state contaminant concentration situation is attained in this test, natural degradation will be a verified groundwater contaminant treatment option.

The NATS experiment has several research objectives that will provide quantitative information to support implementation of natural attenuation as a remediation action at groundwater contamination sites. The need for better understanding of natural attenuation processes in aquifers so that natural attenuation can be developed as a remediation action is made apparent from conclusions of the Groundwater Cleanup Alternatives Committee report (National Research Council, June 22, 1994). This report states the unfeasibility of conventional active groundwater cleanup processes for most contamination situations. It thus supports the investigators' conclusion, based on qualitative thermodynamic analysis of contaminated aquifers, that existing active remediation actions are not economically effective or economically feasible at most sites.

Natural attenuation of the hydrocarbons leaching from the source region will be demonstrated by observation of spatial and temporal distributions of (at least) hydrocarbon concentration, dissolved oxygen concentration, dissolved ferrous iron concentration, dissolved carbonate species concentration, and carbon stable isotope abundance throughout the approximately four year duration of the experiment. Quantification of natural attenuation of the emplaced NAPL requires sampling of the source and downgradient regions to obtain information for a mass balance of carbon. The field data obtained during the study will be used to test predictive models of the course of natural attenuation, and to determine the steady state condition of the plume which is assumed to be reached during the experiment. Since this is the first field experiment to investigate natural attenuation of a large NAPL in an aquifer, development of tools for prediction of the steady state plume extent is an objective of the study. Therefore, the investigators cannot calculate the steady state plume boundary position from the specified initial conditions until the test is begun and some monitoring data has been obtained during the early evolution of the plume. Accordingly, the initial conditions for the experiment were chosen with the intent of attaining a steady state plume within the test site boundaries. These choices constitute subjective best estimates (educated guesses) by the investigators, and it is emphasized that there can be no prior certainty that the plume will reach a steady state in the experiment region.

The NATS test will be used to test the hypothesis that ferric iron oxide minerals commonly occurring in oxygenated aquifers oxidize (degrade) hydrocarbon contaminants after oxygen has been locally depleted by aerobic degradation. This information is needed to determine whether the aquifer has an oxidizing capacity considerably in excess of that provided by the dissolved oxygen in ground water. In sites where oxidation by iron oxides is significant, existing fate and transport prediction models for hydrocarbons in groundwater (such as USEPA's BIOPLUME II) that treat only oxidation by dissolved oxygen are inadequate. If hydrocarbon oxidation by iron oxides is appreciable in the NATS study, the resulting data set can be used to develop a more descriptive fate and transport model. The resultant model would then be applied to other site situations where iron oxides are present. In support of this work on iron oxides, research will be done on activities of iron bacteria, on the aquifer material mineralogy and chemical composition, and on oxidation capacities of the aquifer material.

The MADE-2 test results showed that hydrocarbons degraded at definite rates reported by MacIntyre et al., (1993). Biodegradation has been assumed to be the primary route of aerobic degradation of hydrocarbons in aquifers, but that assumption has not yet been verified. Biochemical techniques will be used in the NATS test to investigate the degradation process at a molecular level in order to demonstrate that the observed hydrocarbon degradation is a result of biodegradation. In one method the induction of an enzyme specific for biodegradation of a particular hydrocarbon component of the source will be measured. If this enzyme induction is found to occur within, but not outside the region of the aquifer affected by the source, "proof" of biodegradation in the aquifer will have been demonstrated. This work will also involve identification and enumeration of aerobic bacterial species in the aquifer.

The NATS started in the fall of 1995. At that time sheet piles were driven into the same source area as the MADE 2 experiment to form a rectangle 15 m and 1 m in width. Sheet piles were driven to a depth of approximately 57.6 m or about 1 meter below the intended bottom of the source materials. After the sheet piles were in place, a dewatering operation for the rectangular source area began. The water table was lowered over several days to "dewater" the source emplacement area. After the water table had been sufficiently lowered, the soil and aquifer material within the sheet piled area was excavated.

Aquifer material was coated with the mixture of hydrocarbons shown in Table 1 to a residual saturation of about 16%, i.e., 16% of the pore space was filled with hydrocarbon mixture. This coating procedure was done by placing approximately 4 m^3 of aquifer material in a pre-cleaned cement mixer and adding the required amount of pre-mixed hydrocarbons. Two hundred gallons of an aqueous solution containing KBr was also added to the hydrocarbon loaded aquifer material. The bromide solution concentration was 128 mg/L in bromide. After about 10 minutes of mixing, the coated aquifer material was placed in the rectangular source area. Subsamples of the hydrocarbon mix showed little loss from volatilization during the mixing process. Eight batches of the material were

Table 1. Source composition and dissolution concentrations calculations

Compound	Mol wt. (g)	Mass (kg)	Mass fraction	Mole fraction	Water sol. (ppm)	Mixture (ppm)
Decane	144.23	1200	0.7495	0.7365	0.08	0.08
Benzene	78.12	0.8	0.0005	0.0007	1760	1.5
Toluene	92.14	100	0.0625	0.0761	515	49.0
Ethylbenz.	106.17	100	0.0625	0.0660	152	12.0
p-Xyl.	106.17	100	0.0625	0.0660	198	15.7
Naphth.	128.19	100	0.0625	0.0547	100	10.42
Total		1600.8	1.00	1.00		89.1

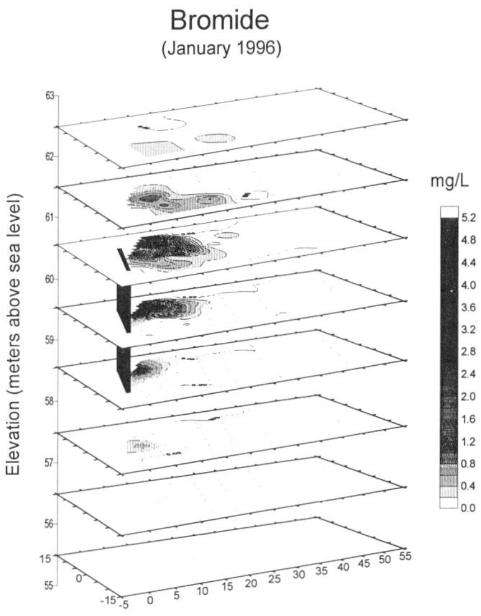

Figure 4. One meter vertically averaged bromide concentration contours from NATS, January 15, 1996.

mixed and placed in the excavated trench forming a rectangular hydrocarbon source of approximately 30 m^3. This procedure produced a known residual phase reactive tracer and known conservative tracer.

The source material was immediately covered with a layer of excavated aquifer material to prevent hydrocarbon volatilization. The trench was subsequently completely back filled with the excavated aquifer material and the dewatering pumps stopped. The water table was allowed to rebound for several days and the sheet piling was removed to start the study on November 28, 1996.

Field samples have been collected once to determine the location of the conservative tracer and the hydrocarbon aqueous plume. Preliminary results show that the hydrocarbon plume is moving more slowly than the Br^- plume. Figure 4 shows the distribution of the Br- as of January 15, 1996. Quarterly sampling of groundwater and aquifer materials is planned for the next several years. Resulting data showing changes in groundwater chemistry, aquifer mineralogy, and aquifer microbiological activity will be used to support adoption of natural attenuation as a scientifically valid remediation method, if it can be demonstrated that the plume is no longer advancing and the source hydrocarbons are being converted to methane and carbon dioxide. Research studies conducted at the Columbus site have been invaluable in learning about contaminant transport in heterogeneous settings. Data derived from extensive sampling of the solid aquifer material has led to increased knowledge of the distribution of both chemical and physical properties within heterogeneous aquifers. The opportunity to introduce known quantities of conservative and reactive tracers has produced a new understanding of the fate and transport of hydrocarbons in natural aquifer settings. The data sets derived during the various studies have provided the most extensive information currently available for testing mathematical contaminant fate and transport models. The availability of other natural gradient aquifers for scientific research would greatly enhance the body of groundwater contaminant knowledge, and therefore, enable the development of more cost-effective cleanup methods.

ACKNOWLEDGMENTS

Funding for this research work was provided from several sources. Those providing funding for different aspects of the work were, Electrical Power Research Institute, Tennessee Valley Authority, US Environmental Protection Agency, and the US Air Force Armstrong Laboratory.

REFERENCES

Adams, E.E. and Gelhar, L.W., 1992, Field study of dispersion in a heterogeneous aquifer, 2, Spatial moments analysis, *Water Resour. Res.*, 28(12), 3293–3307.

Boggs, J. M.,. Young S.C, Beard, L.M, Gelhar, L. W., Rehfeldt, K.R. and Adams, E.E., 1992, Field study of dispersion in a heterogeneous aquifer, 1, Overview and site description, Water Resour. Res., 28(12), 3281–3291.

Boggs, J. M. and Adams, E.E.,1992, Field study of dispersion in a heterogeneous aquifer, 4, Investigation of tracer adsorption and sampling bias, Water Resour. Res., 28(12), 3325–3336.

Gelhar, L.W. and Axness, C.L., 1983, Three-dimensional stochastic analysis of macrodispersion in aquifers, Water Resour. Res., 19(1), 161–180.

Geraghty and Miller, Inc., 1991, Existing and proposed USEPA maximum contaminant levels in drinking water as of August 1991, compiled by R. A. Saar, Geraghty and Miller, Inc., Albuquerque, NM.

Harvey, R. W.,Kinner, N.E., MacDonald, D, Metge, D.W., and Bunn, A., 1993, Role of physical heterogeneity in the interpretation of small-scale laboratory and field observations of bacterial, microbial-sized microsphere, and bromide transport through aquifer sediments, Water Resour. Res. 29(8), 2713–2721.

Hess, K. M., 1989, Use of borehole flowmeter to determine spatial heterogeneity of hydraulic conductivity and macrodispersivity in a sand and gravel aquifer, Cape Cod, Massachusetts, in Proceedings of NWWA Conference on New Field Techniques for Quantifying the Physical and Chemical Properties of Heterogeneous Aquifers, Houston, TX.

MacIntyre, W. G., Antworth, C. P., Stauffer, T. B., Boggs, J. M., 1993, Degradation kinetics of aromatic organic solutes introduced into a heterogeneous aquifer, Water Resour. Res. 29,(12), 4045–4051.

Rehfeldt, K. R., Gelhar, L.W., Southard, J.B., and Dasinger, A.M., 1989a, Estimates of macrodispersivity based on analyses of hydraulic conductivity variability at the MADE site, EPRI Topical Rept. EN-6405, Electric Power Res. Inst., Palo Alto, CA.

Rehfeldt, K. R., Hufschmied, P., Gelhar, L.W.,and Schaefer, M. E., 1989b, Measuring hydraulic conductivity with the borehole flowmeter, EPRI Topical Rept. EN- 6511, Electric Power Res. Inst., Palo Alto, CA.

Rehfeldt, K.R., Boggs, J.M., and Gelhar, L.W., 1992, Field study of dispersion in a heterogeneous aquifer, 3, Geostatistical analysis of hydraulic conductivity, Water Resour. Res., 28(12), 3309–3324.

Sudicky, E. A., 1986, A natural gradient experiment on solute transport in a sand aquifer: spatial variability of hydraulic conductivity and its role in the dispersion process, Water Resour. Res., 22(13), 2069–2082.

Sutton, P.A. and Barker, J.F., 1985, Migration and attenuation of selected organics in a sandy aquifer - a natural gradient experiment, Groundwater, 23(1).

10

BIOAUGMENTATION OF TCE-CONTAMINATED SOIL WITH INDUCER-FREE MICROBES

Takeshi Imamura, Shinya Kozaki, Akira Kuriyama, Masahiro Kawaguchi, Yoshiyuki Touge, Tetsuya Yano, Etsuko Sugawa, and Yuji Kawabata

Canon Research Center, Canon Inc.
Atsugi-shi, Kanagawa 243-01, Japan

1. INTRODUCTION

Environmental pollution is a world-wide problem since these pollutants cause serious disease not only for human beings but also for other plants and animals. Trichloroethylene (TCE) is a typical pollutant in soil and groundwater, and its concentration is strictly regulated since many people are using groundwater as drinking water. TCE is very stable in the environment, and soil vapor extraction (SVE) is frequently used for the removal of TCE. However, the removal efficiency in SVE decreases with low concentrations of TCE. Charcoal adsorption and photocatalytic degradation are other approaches to detoxify soil and groundwater, but in situ treatment is needed for cost-effective remediation. Bioremediation for TCE-contaminated soil is an attractive approach for in situ treatment, and many microbes which can decompose TCE have been reported.

Aerobic degradation of TCE is very rapid in comparison with anaerobic degradation. Thus many aerobic microbes have been isolated, and their degradation activities have been evaluated for bioremediation use. Such microbes are characterized as either aromatic compounds-oxidizers[1,2] or methanotrophs,[3,4] since TCE is co-oxidized by oxygenase enzymes that these microbes contain. However, phenol-oxidizers can degrade TCE only during the production of phenol hydroxylase, and phenol is always required as an inducer for TCE degradation. The same problem is encountered for biodegradation of TCE by other types of microbes used.

In order to enhance the degradation activity, some aromatic compounds or methane should be added into the soil and groundwater. But these inducers are exogenous chemicals, and environmental hazards caused by such chemicals may create other problems. Furthermore, the oxidizing enzyme is competitively used for oxidization of the inducer, thus the inducer is a competitive inhibitor in TCE degradation. Inducer-free microbes for TCE degradation are necessary to overcome these problems.

In this study, inducer-free microbes are obtained by a chemical mutation of phenol-degrading microbes. Degradation of TCE by these microbes are evaluated in various con-

Biotechnology in the Sustainable Environment, edited by Sayler *et al.*
Plenum Press, New York, 1997

ditions such as liquid and soil media, and a basic characterization of the microbes is achieved for bioaugmentation of TCE-contaminated soil.

2. EXPERIMENTAL

2.1. Mutation of Strain J1

First, several phenol-oxidizing microbes for TCE degradation were isolated from a variety of natural soils. Then, their degradation activities in soil were compared. Since Strain J1 had a high degradation activity, it was selected as a parent microbe.

Wild-type strain J1 was incubated at 30°C in the M9 medium containing 0.1% of yeast extract, 200 parts per million (ppm) of phenol, and trace amounts of mineral components (such as magnesium). This M9 medium contained 6.2 grams (g) of disodium hydrophosphate, 3.0 g of potassium dihydrophosphate, 0.5 g of sodium chloride, and 1.0 g of ammonium chloride in 1 liter (l) of deionized water. After incubation for 14 hours (h), the late-log-phase microbes were harvested and resuspended in the M9 medium containing 200 ppm of phenol. Then, nitrosoguanidine was added to the medium as a mutagen, and the concentration of nitrosoguanidine was adjusted to 30 micrograms per milliliter (μg/ml) culture. After shaking for 1 to 2 h, this culture was inoculated on the agar plate containing the M9 medium, 0.1% of yeast extract, and 200 ppm of indole. Indole was well known to be transformed to indoxyl with oxygenase, and indoxyl was rapidly oxidized to indigo. After transformation from indole to indigo, the color turned dark blue in the agar plate. Thus, the presence of oxygenase was ascertained by the color change of indole.

When the oxygenase was constitutively produced by the microbe, the color of this colony was immediately changed. On the other hand, these was some time-lag between the colony growth and the color change when production of oxygenase was induced by phenol. Therefore, the colonies of the wild-type strain J1 and the inducer-free strain JM1 could be distinguished by the time course of the color change.

The agar plate was incubated for 3 days at 30°C, and the dark blue colony was obtained as the inducer-free strain JM1. However, the color of the wild-type J1 was also changed after incubation for 4 days since indole was an inducer for the production of oxygenase by J1.

2.2. Constitutive Degradation in Liquid Experiments

The colony of strain JM1 was inoculated into the medium of 200 ml containing 2.0% sodium malate and the M9 medium, and the concentration of JM1 was roughly adjusted to 10^6 colony forming units per milliliter (cfu/ml). After incubation for 3 days at 15°C, strain JM1 grew to 3–5 × 10^8 cfu/ml. Another medium containing 1.0% sodium malate and M9 was prepared, and this medium (10 ml) was poured into the glass vial (68 ml). The culture of JM1 (0.1 ml) was added to the vial, and the vial was sealed with a butyl rubber septum and an aluminum crimp. It was incubated for appropriate time until the culture reached the late-log-phase. The aqueous solution of TCE (saturated, approximately 1000 ppm) was prepared in a closed glass bottle, and gaseous TCE was sampled by a syringe. Then, the gaseous TCE of 0.1 ml was injected into the vial containing JM1 to observe degradation of TCE. The headspace gas (0.1 ml) in the vial was sampled, and the TCE concentration was determined by gas chromatography.

2.3. Constitutive Degradation in Soil Experiments

The experimental procedure for the measurement of TCE degradation in soil is shown in Figure 1. The preculture of strain JM1 was obtained by the same manner as in the liquid experiments. This culture was diluted 100 times with a medium containing 1.0% sodium citrate and the M9 medium. Sodium citrate was used in soil experiments instead of sodium malate since a high degradation activity was obtained using sodium citrate in soil as described in the Section 3.1. An appropriate amount of unsterilized sandy soil was put into the glass vial (68 ml). The moisture of the soil was approximately 10%. Then, a diluted culture (10 weight % against the amount of soil) was added to the soil. The moisture of the soil was increased to 20% after the addition of the diluted culture. The initial concentration of the microbe was 10^5 cfu/g wet soil. It was incubated for appropriate time without sealing. When the growth of the microbe reached to the late-log-phase, the concentration of JM1 was 10^8–10^9 cfu/g wet soil. Then, the glass vial was sealed with a butyl rubber septum and an aluminum crimp, and the gaseous TCE was injected to degrade TCE by the same manner in the Section 2.2. The headspace gas (0.1 ml) was sampled by a syringe, and analyzed by gas chromatography. TCE is well known to adsorb to the organic carbon in the soil. But the sandy soil used in this study had little organic carbon, and adsorption of TCE was negligible.

2.4. Degradation of TCE in High Concentration

Strain JM1 was grown at 25°C by 2.0% sodium citrate in the soil as described in the previous section, and the concentration of TCE was adjusted to 12, 25, 50, 75, and 180 ppm when TCE was concentrated in soil water. Then, the TCE degradation was observed by the gas chromatography.

2.5. Degradation Period by JM1

First, JM1 was added to the aqueous solution containing 2.0% sodium citrate and M9 medium, and it was added to unsterilized soil in several glass vials by a similar manner as described in the Section 2.3. After incubation for 1 to 6 days, TCE was added to the vial and the concentration of TCE was adjusted to 40 ppm. The TCE concentration in the

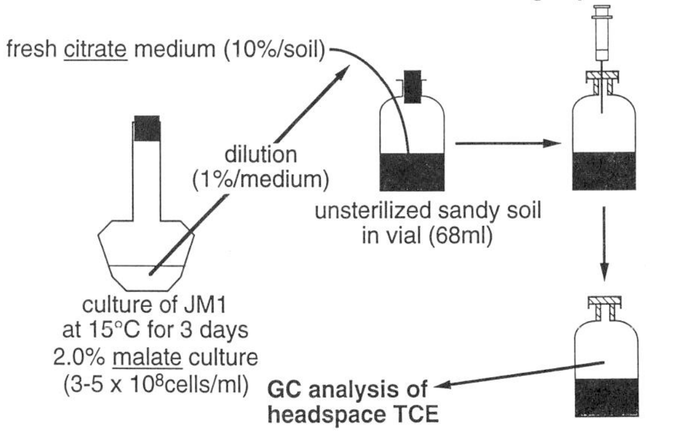

Figure 1. Experimental procedure for TCE degradation in soil.

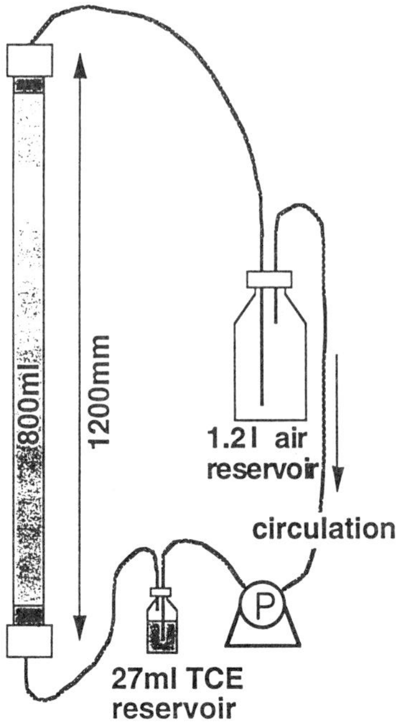

Figure 2. Schematic diagram of soil column experiments.

headspace was measured by gas chromatography and the degradation rate was calculated from the amount of the degraded TCE for 3 h after the addition of TCE.

2.6. TCE Degradation in Soil Column

The schematic diagram of the experiments is shown in Figure 2. The glass column had a diameter of 30 millimeters (mm) and a length of 1200 mm. The sandy soil was sterilized, and it was mixed with 2.0% sodium citrate diluted with the M9 medium (172 ml). Secondly, the soil was mixed with the solution of JM1, and the concentration of JM1 was adjusted to 10^6 cfu/g wet soil. The water content in the soil was approximately 14%/g wet soil. Then the glass column was filled with this soil. Fresh air was passed through the column for 20 h at 23°C, and JM1 was grown to 10^8–10^9 cfu/g wet soil. TCE-saturated water (15 ml) was poured in the glass reservoir (27 ml), and the vapor phase was circulated through the column by an air reservoir (1.2 l), and a pump. The flow rate was 30 milliliters per minute. Oxygen was consumed during the growth of JM1 and TCE degradation, and it was supplied by the air reservoir. The headspace gas (0.1 ml) of the TCE reservoir was sampled, and the concentration was measured by gas chromatography.

3. RESULTS AND DISCUSSION

3.1. Suitable Nutrient for TCE Degradation

For biological experiments, yeast extract was frequently used as a nutrient. When indigenous microbes were separated from the natural soil and mixed with JM1 in the liquid culture, the number of indigenous microbes rapidly increased by the addition of yeast extract (0.1%). Thus JM1 could not grow up selectively in the liquid culture. Then, we evaluated several organic acids as a nutrient, and found that sodium malate (2.0%) was suitable for the selective growth of JM1. Furthermore, the growth of JM1 was relatively rapid and the degradation activity was high at 15°C. Therefore, the liquid culture experiments were achieved at 15°C with sodium malate as a nutrient.

For the soil experiments, the growth of JM1 was very rapid when sodium malate was used as the nutrient. Some soil components might be suitable for the growth of JM1 because of the physical and chemical interactions between JM1 and the soil. However, as described later, the degradation activity was significantly decreased when the growth was rapid. Thus we evaluated several organic acids again as a nutrient, and found that sodium citrate (2.0%) was suitable for gradual and selective growth of JM1 in the natural soil. Therefore, sodium citrate was used through the soil experiments.

3.2. TCE Degradation by Strain J1 and JM1

J1 shows a high activity in TCE degradation when yeast extract and phenol are used and J1 is only incubated in the liquid culture. But, it still requires phenol as an inducer for TCE degradation, since this strain is a wild-type phenol-oxidizer. For strains J1 and JM1 incubated with yeast extract (0.1%) at 30°C, TCE degradation and the growth of J1 and JM1 were determined as shown in Figure 3. The growth curve of JM1 was similar to that of J1, and the final concentrations were approximately 10^8 cfu/ml. The concentration of TCE decreased with time by JM1, resulting in 70% of the TCE being degraded. On the other hand, strain J1 could not degrade TCE since phenol was absent. Slight decreases in the TCE concentration with J1 was caused by leakage of TCE from the vial. These data effectively serve as a negative analytical control for comparison with TCE degradation by JM1. These results show that JM1 can degrade TCE without an inducer, and thus the TCE-

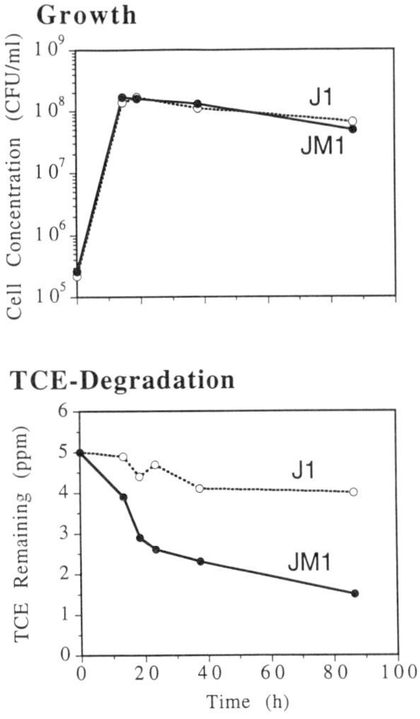

Figure 3. Degradation and growth of J1 and JM1.

degradation by JM1 is constitutive. The degradation rate by JM1 is similar to that by J1 when J1 is incubated with phenol, and no loss in the degradation activity is observed by the mutation of J1.

The strain JM1 may be putatively classified as *Acinetobacter* due to its shape and mobility, the homology of its 16S rRNA to reference strains, the Gram stain, and the G/C content of its DNA. However, identification of these strains is not complete at present. As far as our investigation for JM1, this strain has the same biological characteristics to that of J1, except for the TCE degradation activity.

3.3. Constitutive Degradation in Liquid Experiments

Strain JM1 can degrade TCE without inducers as shown above. However, it has been reported that TCE degradation by some microbes is induced by TCE. In order to clarify constitutive degradation by JM1, TCE degradation is observed after incubation without phenol since phenol causes production of phenol hydroxylase. The results are shown in Figure 3. The concentration of TCE remaining in the vial was immediately decreased in the JM1 vial. On the other hand, the control vial without JM1 shows roughly the same TCE concentration. Thus it is apparent that JM1 can degrade TCE. If the TCE degradation by JM1 is induced by TCE, the degradation would begin later after the addition of TCE, since the induction time is necessary to produce the oxidizing enzyme[5]. From Figure 4, TCE is degraded just after the addition of TCE, and thus the degradation by JM1 is considered to be constitutive. The time between the addition of TCE and the first measurement of TCE is about 14 minutes, due to the time necessary to prepare the sample for gas chromatography. Furthermore, TCE is not degraded by J1 without phenol as shown in Figure 3, and it implies that TCE does not act as an inducer even for J1.

3.4. Constitutive Degradation in Soil Experiments

The growth and the degradation activity of JM1 is quite different between the liquid and soil experiments as described in section 3.1. There are two dominant differences related to the degradation activity in the liquid and soil experiments. One is the difference in the concentration of JM1. The concentration of JM1 reaches 10^9 cfu/g wet soil, corresponding to 10^{10} cfu/ml since the water content in soil is 10%. On the other hand, the concentration of JM1 is about 10^8 cfu/ml in the liquid medium. This concentration difference is due to the physical and chemical interaction between the microbes and soil. Another

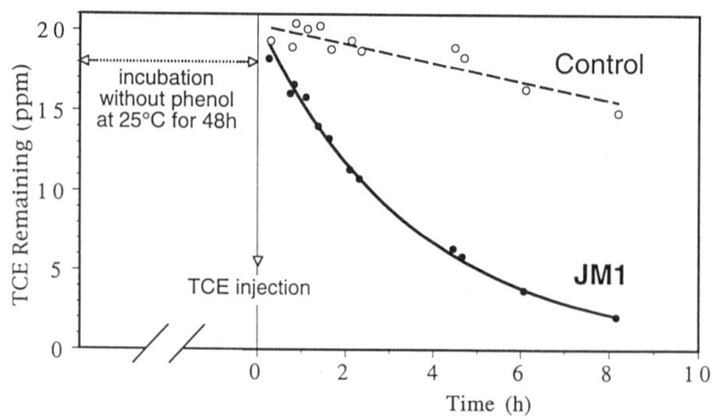

Figure 4. Constitutive degradation in soil culture.

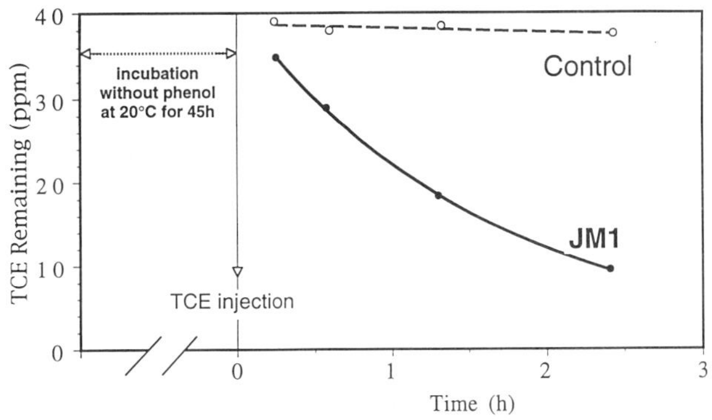

Figure 5. Constitutive degradation in soil culture.

reason relates to the difference in the transport of TCE. Since soil is porous, gaseous TCE is completely exposed to the microbes and diffusion of TCE is higher than that in the liquid medium. The degradation activity is strongly related to the concentration of JM1 and diffusion of TCE, and the constitutive degradation is also ascertained in the soil experiments. The results are shown in Figure 5. Similar to the liquid experiments, the TCE concentration decreases immediately after the addition of TCE. This means that the degradation by JM1 is constitutive in the soil as well as in the liquid. The time of 14 minutes is also required for the preparation of the sample for the determination of TCE concentration.

3.5. Degradation of TCE in High Concentration

Many microbes which can degrade TCE with inducers have been reported; however their activity significantly decreases at a high concentration of TCE[6]. This is due to the toxicity of TCE to the living cell. Thus 10 ppm of TCE is usually difficult to degrade by the microbes.

The TCE degradation by JM1 is observed at a high TCE concentration, and the results are shown in Figure 6. When the TCE concentration is below 50 ppm, JM1 can degrade TCE completely. It is noted that JM1 can degrade TCE even for the initial concentration of 75 and 180 ppm. However, TCE is still remaining after the decrease in the degradation activity. This might be due to the lack of nutrients and oxygen. A toxic

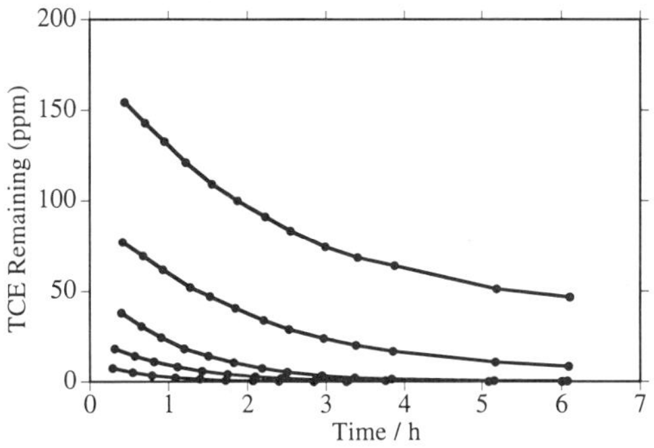

Figure 6. TCE degradation at various concentrations.

degradation intermediate such as TCE epoxide might also be produced, causing a decrease in the degradation activity.

3.6. TCE Degradation at Various Temperatures

Since the standard temperature of the soil in Japan is around 15°C, this is the temperature at which degradation activity was examined in these experiments. However, microbes which are able to degrade TCE in a wide temperature range are desired for practical application. Thus the degradation dependence on temperature was evaluated. The results are shown in Figure 7.

The culture of JM1 was added to the soil with sodium citrate at various temperatures. JM1 was grown up and the degradation of TCE was observed simultaneously. The cell concentration increased from 10^6 to 10^9 cell/ml water in the soil in all cases after incubation for 2–3 h. The degradation rate was rapid at 25°C, and decreased at lower temperatures. Although the final cell concentration was similar, the degradation rate was different since the growth rate is strongly related to temperature. The growth rate was relatively rapid at 25°C, resulting in a fast decrease in the TCE concentration. On the other hand, the slow rate at 10°C caused slow degradation of TCE. But 38 ppm TCE is almost degraded completely at 10–25°C. Further experiments were conducted at 7–35°C, and JM1 shows degradation activity through this temperature range.

3.7. Nutrient Dependence in Degradation

As described above, sodium malate is a good nutrient for TCE degradation in liquid culture experiments at 15 and 30°C. For the soil experiments, the degradation activity at 15°C is comparable to that in the liquid culture experiments, however it is significantly decreased at 30°C even though the growth of the microbe is rapid. This suggests that TCE degradation is affected by the medium and temperature. In order to obtain high degradation activity in the soil at 15 and 30°C, several organic acids were used as nutrients and the degradation activities were evaluated. Sodium citrate was found to be one of the suitable nutrients for TCE degradation in soil. JM1 is an aerobic microbe, and it consumes oxygen in growth. Thus the oxygen consumption during TCE degradation was observed using sodium malate and sodium citrate at 15 and 30°C, and the results are shown in Figure 8. It is noted that TCE degradation increases at low oxygen consumption, and this im-

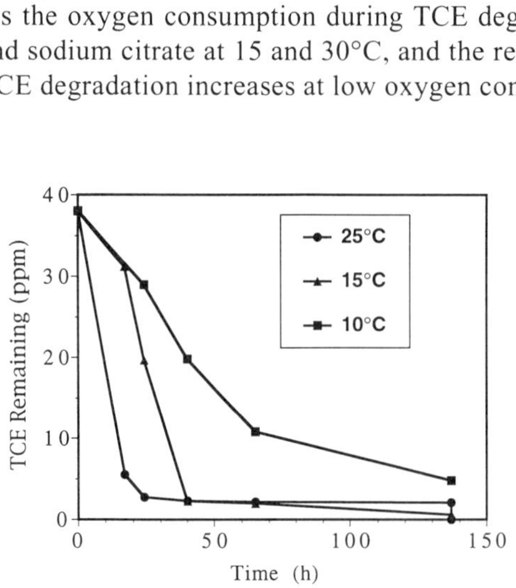

Figure 7. Temperature dependence in degradation.

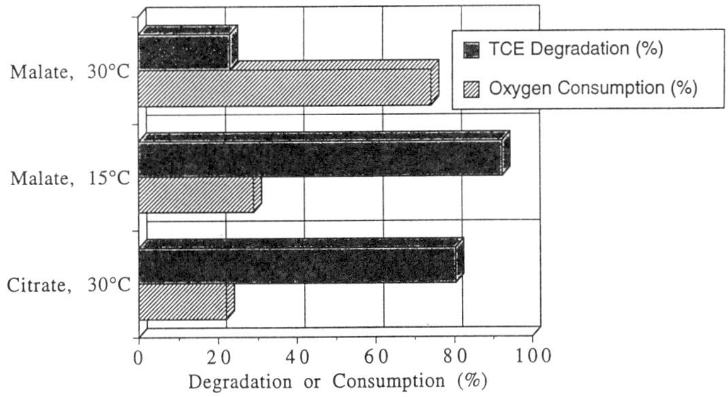

Figure 8. TCE degradation and oxygen consumption by JM1.

plies that the high degradation activity of JM1 is obtained during a relatively low growth rate. Furthermore, TCE degradation by JM1 is evaluated under various conditions by changing the kind of nutrients and temperature. These results suggest that JM1 shows low degradation when oxygen is rapidly consumed and the growth rate is high. The reason of the growth rate dependence on the degradation activity is not clear at present, but it might be due to catabolite repression[7]. We have started a detailed evaluation of TCE degradation related to the growth rate and the growth phase of the microbe.

3.8. Degradation Period by JM1

JM1 can degrade TCE constitutively, however the degradation activity decreases in a few days after incubation of JM1. Thus the degradation activity of JM1 was measured for several days, and the results are shown in Figure 9. JM1 shows a high degradation rate from 1 to 2 days since the concentration of JM1 is 10^8–10^9 cfu/g wet soil and the enzymatic activity is high due to the growth of JM1. But the degradation rate significantly decreases after 3 days. The concentration of JM1 decreased after 3 days since the indigenous microbes were grown by the addition of sodium citrate. Thus the decrease in the degradation activity may be due to the decrease of the enzyme concentration, the lack of the energy source, and the lack of oxygen or toxicity. In this experiment, TCE degradation could

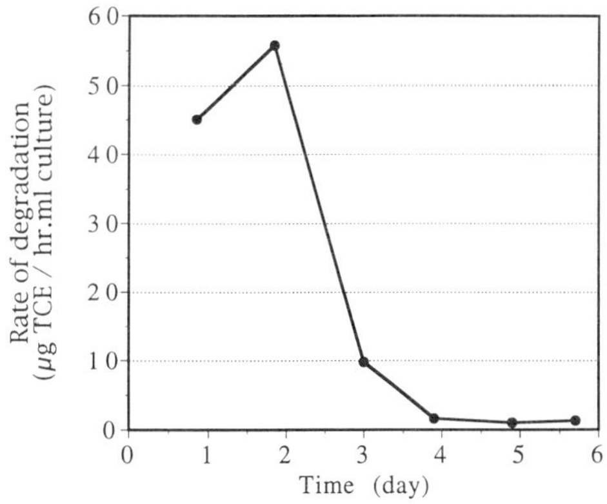

Figure 9. Degradation activity during incubation.

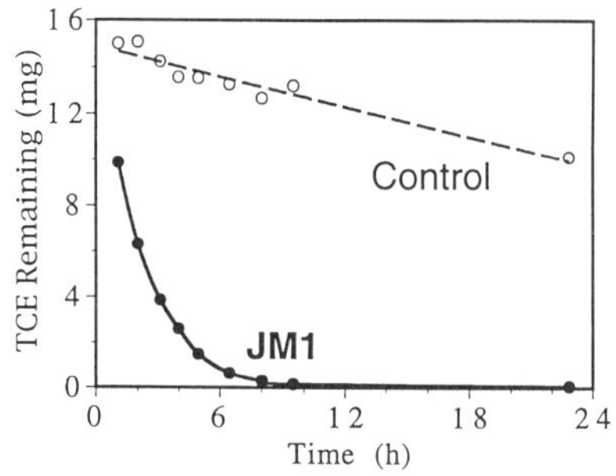

Figure 10. TCE degradation in soil column.

not be restored by the addition of nutrients and oxygen since indigenous microbes were already grown in the glass vial. However, TCE degradation can be restored when sterilized soil is used in the degradation experiments.

3.9. Soil Column Experiments

The degradation experiments using the glass vial provide various basic results in TCE degradation by JM1. However, the actual remediation of soil and groundwater have further problems such as transport or mass transfer of microbes within the contaminated zone. Thus the feasibility of TCE degradation was evaluated using a soil column for the large scale experiments. The results are shown in Figure 10. The initial amount of TCE circulated through the column was 15 mg, and it was almost degraded after 10 h. The soil column without JM1 was also evaluated in the same manner, and it showed a slight decrease in the TCE concentration due to the leakage from the system. It is apparent that JM1 can degrade TCE even in the soil column. In future experiments, TCE degradation will be further evaluated by looking at the effect of water content, temperature, and flow rate.

4. CONCLUSION

Strain JM1 of an inducer-free TCE degrader was developed by the chemical mutation technique. JM1 can degrade 20 ppm of TCE for 10 h in liquid and 50 ppm for 5 h in soil without any inducers. Its degradation activity might depend on the growth rate which also relates to environmental conditions such as the concentration of nutrients and temperature. Furthermore, JM1 shows a high activity in TCE degradation in soil column experiments, degrading 15 mg of TCE in 10 h.

REFERENCES

1. M. J. K. Nelson, S. O. 1987. Montogomery, W. R. Mahaffey, and P. H. Pritchard, Appl. Environ. Microbiol., *53*, 949–954.
2. L. P. Wackett and D. T. Gibson, 1988. Appl. Environ. Microbiol., *54*, 1703–1708.
3. H. Tsien, G. A. Brusseau, R. S. Hanson, and L. P. Wackett, 1989. Appl. Environ. Microbiol., *55*, 3155–161.
4. H. Uchiyama, T. Nakajima, O. Yagi, and T. Tabuchi, 1989. Agric. Biol. Chem., *53*, 2903–2907.
5. K. Mcclay, S. H. Streger, and R. J. Steffan, 1995. Appl. Environ. Microbiol., *61*, 3479–3481.
6. L. P. Wackett and S. R. Householder, 1989. Appl. Environ. Microbiol., *55*, 2723–2725.
7. W. A. Duetz, S. Marques, B. Wind, J. L. Ramos, and J. G. van Andel, 1995. Appl. Environ. Microbiol., *62*,601–60.

11

IS BIOREMEDIATION A VIABLE OPTION FOR CONTAMINATED SITE TREATMENT?

Integrated Risk Management — a Scientific Approach to a Practical Question

A. Heitzer,[1] R. W. Scholz,[1] B. Stäubli,[2] and J. Stünzi[3]

[1]Chair of Environmental Sciences, Natural and Social Sciences Interface
Swiss Federal Institute of Technology
ETH-Zentrum VOD, CH-8092 Zürich
Switzerland
[2]Cantonal Institute for Water Protection and Hydraulic Engineering
Walchetor, CH-8090 Zürich
Switzerland
[3]GOE Professor Schulz & Partner, Wiesenstrasse II
CH-8700 Küsnacht
Switzerland

1. INTRODUCTION

During the last decade systematic investigations on the contaminated site situation were conducted in many European countries. As a result federal and state registers of sites suspected for contamination were established. These registers include industrial sites, locations of former accidents with pollutant spills as well as landfills. In Table 1 an overview of suspected contaminated sites for some selected European countries and states are provided. The figures illustrate the dimensions of the problem and the large number of sites requiring prioritization, further detailed investigations and potential remediation. These contaminated sites present a hazard and potential risk not only to human health but also to ecosystems and natural resources such as ground water, soil and land area.

Since land area is an increasingly scarce resource, particularly in highly populated industrialized areas, used land should be recycled in order to reduce the pressure to use new undeveloped land. In addition, excavated contaminated soil material should be treated such that it becomes suitable again for unrestricted reuse as filling or recultivation material. Similarly, ground water should be protected as a drinking water source.

The potential for reuse of remedied contaminated sites and soil materials depends largely on the efficacy of a treatment technology to reduce the pollutants to the cleanup

Biotechnology in the Sustainable Environment, edited by Sayler *et al.*
Plenum Press, New York, 1997

Table 1. Suspected contaminated sites for selected countries and states in Europe

Country, region, state	Population [106]	Area [kilometer^2]	No. of suspected contaminated sites[a]	Reference
Germany	77.62	357,050	143,252	RSU, 1995
"Old states"	61.02	248,717	73,559	RSU, 1995
"New states"	16.6	108,333	69,693	RSU, 1995
Nordrhein-Westfalen	16.68	34,068	19,505	RSU, 1995
Netherlands	14.9	41,864	>100,000	Verheul et al., 1993
Norway	4.22	323,895	2,696	Folkestad, 1993
Switzerland	6.87	41,293	40,000	BUWAL, 1994
Zürich	1.16	1,729	12,200	AGW, 1993
Berne	0.96	6,050	2,452	GSA, 1995

[a]The assessment procedures used for site identification may be different for the countries and states reported.

levels required and on their overall environmental performance. It is obvious that the goal should be the maximum quality that can be efficiently achieved.

During the past few years numerous innovative remediation technologies have been developed and are currently awaiting full scale implementation (USEPA, 1993). Because of the resulting uncertainties associated with novel treatment technologies, it is necessary to develop strategies to characterize their risks to achieve the cleanup requirements. For this purpose it is necessary to integrate existing knowledge on treatment efficacy and efficiency into the planning process for the management of contaminated sites. Thereby it is not only important to consider the selection of a treatment technology from a local site-specific perspective, but also at the regional scale, taking all the contaminated sites into account. The latter is important to ensure an overall efficiency of contaminated site management.

From an ecological but also economical point of view, bioremediation belongs to the most promising technologies. This is largely due to its characteristics to potentially detoxify or mineralize many organic pollutants, thereby providing a sustainable solution to hazard and risk reduction.

To date probably most experience with full scale in situ and ex situ applications of bioremediation (Baker and Herson, 1993) has been acquired for the biodegradation of petroleum hydrocarbons (PHC). These include refined oil-products such as gasoline, the middle distillates heating oil, diesel and jet fuel, as well as the wide cut gas oils (Neff, 1990). Their chemical composition ranges from a variety of straight and branched chain, saturated, unsaturated and cyclic aliphatics to mono-, di- and polyaromatic hydrocarbons (Riser-Roberts, 1992). The microbiology of petroleum hydrocarbons is a well studied process and has been extensively reviewed (Atlas, 1984; Leahy and Colwell, 1990).

The relative number of full scale biological applications for contaminated site treatment are still rather limited. In the German state Nordrhein-Westfalen only 17 biological applications were recorded from a total of 207 remediation treatments (RSU, 1995). For the Netherlands, it was noted that use of bioremediation was still small due to limited availability on the commercial market (Verheul et al., 1993). Various factors might be responsible for this situation such as limited experience and demonstration data, limited acceptance of the technology, but also failures to achieve the cleanup levels required due to unqualified applications or because of intrinsic process limitations. In addition, there is still a need to demonstrate the specific biological and general environmental efficacy of these technologies (Heitzer and Sayler, 1993; Shannon and Untermann, 1993). Technology demonstration programs such as, for example, the SITE program initiated in 1986 in the US (USEPA, 1993) or the demonstration program of Baden-Württemberg in Germany (LfU, 1993a) provide important contributions to overcome some of these problems.

For the application of novel innovative treatment technologies it is therefore important that experience data are efficiently used in order to identify opportunities and niches according to their specific technology risks.

In order to address the question "is bioremediation a viable option", a conceptual and methodical approach is presented for the comparative evaluation of two ex situ treatment technologies, i.e., biological land treatment and physico-chemical soil washing, with respect to their general performance risks to achieve given cleanup standards for PHC contaminated soil materials.

2. REMEDIAL ACTION OBJECTIVES AND CLEANUP STANDARDS

2.1. General Remedial Action Objectives

The general strategies most widely pursued for the cleanup of contaminated sites include as a minimum action requirement the reduction of human health risks and ecological risks associated with the pollution. Some European countries, including Switzerland, have developed cleanup strategies that go beyond this minimum requirement (BUWAL, 1994). In the Netherlands, for example, the concept of multifunctionality has been introduced more than ten years ago (Verheul et al., 1993). According to this concept the cleanup of a contaminated site should result in the establishment of a site quality that allows for any type of land use including residential use and horticulture.

In the state of Zürich in Switzerland the general strategy for remedial action objectives for contaminated sites was specified in the waste management act from September 24, 1994. In this act, general remedial action objectives were defined which are organized in a priority cascade (Suter, 1995). The first priority level involves the reestablishment of conditions at the site similar to a natural background level, typical for the site location. In practice, it should be evaluated if treatment technologies are available to effectively and efficiently achieve this cleanup goal. If the achievement of this first priority goal is unrealistic based on best available technologies and economic considerations, a multifunctional land use should be considered as a second priority cleanup goal. Should this goal again turn out to be impossible to achieve, the treatment should result in a site quality which allows for the reestablishment of the actual use level or the establishment of the planned use. If it was demonstrated that even this third priority goal cannot be achieved, then use restrictions may be considered. In order to find an effective and efficient technology it is a prerequisite that several potentially feasible treatment alternatives are evaluated simultaneously. In addition to the priority cascade, further constraints were defined in the act which require that a treatment results in products which can be reused. It should be also demonstrated, that the environmental impact of the treatment is significantly smaller than the contamination problem which is eliminated. The deposition of contaminated materials in a landfill is only possible if it has been demonstrated that no other treatment is possible and, finally, it is required that treated clean materials should be reused or left at the site, if logistically possible.

The two approaches presented above implement many of the ideas for a sustainable development (World Resources Institute, 1992). The concepts of multifunctionality and background levels contribute to an economic use of limited environmental resources such as land area, soil and groundwater, thereby providing an environmental quality where future generations will have all possible options for land use according to their needs. In addition, the requirement to look at the general environmental impact forces to look at the overall treatment efficacy, including pre- and post-treatment actions. Thus, spacial and temporal shifts of the actual problem can be systematically detected and avoided which is considered a fundamental requirement for sustainability (Minsch, 1993).

2.2. Cleanup Standards

The establishment of specific remedial action objectives for the cleanup of a contaminated site is a complex integration and optimization process within a set of numerous constraints. These include legal and political, socio-economic, ecological and technical aspects reflecting the often conflicting goals of the parties involved in the site management process.

One of the most difficult and politically sensitive aspects in defining remedial action objectives is the development of quantitative cleanup requirements. Two general approaches can be identified: the site-specific risk-based establishment of cleanup levels and the use of generally developed cleanup standards. Irrespective of the approach it is most important that pollutant levels are derived according to toxicological and scientific criteria for relevant exposure scenarios (Lühr et al., 1991; Von der Trenck and Fuhrmann, 1991; Paustenbach et al., 1993).

From a practical point of view cleanup standards have several advantages: They facilitate the decision making process and provide an equal legal basis for everybody. In addition, decisions of regulators become transparent and can be anticipated. However, there are also several disadvantages, including the fact that cleanup standards are usually derived from standardized exposure scenarios and often involve safety factors. It is obvious that under these circumstances local site conditions affecting the fate of pollutants might not be represented adequately, which could negatively affect the decision making process. Further, cleanup standards result, if strictly applied, in rigid yes- or no- decisions that can be inappropriate if the actual contaminant situation involves only small deviations from a given cleanup standard.

In order to avoid some of these problems, cleanup standards with different functions were established. A well known example is the "Dutch list" first published in 1983. The updated version includes two S- and I-values, for both soil and groundwater pollutant concentrations (Vegter, 1992). The S-value (formerly A-value) represents a target pollutant concentration for multifunctional use of the soil or water. If the contaminant concentration exceeds $0.5(S + I)$ (function of the former B-value), then further investigations on the contaminant situation are required. The I-value (formerly C-value) is a trigger value for treatment intervention. If the contaminant concentration exceeds the I-value, cleanup is considered necessary. The degree of urgency is then determined based on the actual exposure risk. Similar conceptual approaches based on trigger values exist also in the UK (Barry, 1995) and in the German state Baden-Württemberg (Innenministerium, 1993). In the state of Zürich in Switzerland target cleanup standards were defined for pollutant concentrations in excavated soil materials for multifunctional reuse and for recultivation at the source site (AGW, 1994). However, even though several countries and states have defined cleanup standards, in practice very often compromises have to be found between ecologically desirable and economically feasible cleanup solutions which means that in many instances the concept of multifunctionality cannot be met and land use specific cleanup levels have to be derived.

Table 2. Cleanup standards for petroleum hydrocarbons and related products in soil for some countries and states in Europe

Compound	Value (ppm)	Type of standard	Function	Country or state	Reference
Mineral oil[a]	50	target value	S-(former A) value of the Dutch list. Clean up target and reference value for multifunctional use.	Netherlands	Vegter, 1992
	5000	trigger value	I-(former C) value of the Dutch list. Intervention value when cleanup is considered necessary.	Netherlands	Vegter, 1992
Mineral oil[a]	50	target and trigger value	H-B value. Background value for pollutant levels in soil. If significantly exceeded further investigations are necessary. Target value for clean up if economically possible.	State of Baden-Württemberg (Germany)	Innen-ministerium, 1993
	100	target and trigger value	See previous above. Applies if top soil humics content >1%.	State of Baden-Württemberg (Germany)	Innen-ministerium, 1993
	400	trigger value	P-P value. For soil and plant growth protection. If value is exceeded, agricultural and horticultural land use may not be safe anymore and further investigations are necessary.	State of Baden-Württemberg (Germany)	Innen-ministerium, 1993
Mineral oil[a]	50	target value	Target value for excavated soil materials for multifunctional reuse without restrictions and for recultivation at the source site.	State of Zürich (Switzerland)	AGW, 1994
Gasoline	10	target value	See above.	State of Zürich (Switzerland)	AGW, 1994

[a]The term "mineral oil" used in the standard lists is analogous to the term petroleum hydrocarbon.

Table 2 provides an overview of cleanup target and trigger values for PHC related pollutants in soil for some countries and states in Europe. Because of the toxicological characteristics of PHC, in all of the guidelines mentioned in Table 2, cleanup standards are also included for the monoaromatics benzene, ethylbenzene, toluene and xylene as well as for polyaromatic hydrocarbons and must be considered in addition to the sum parameter PHC. For the US a detailed review of state cleanup levels for hydrocarbon contaminated soils was provided by Bell et al., (1991) where remediation goals for total petroleum hydrocarbon content were found to range from background levels to 10,000 parts per million (ppm) depending on site and compound specific factors.

In order to characterize the utility of a treatment technology it is necessary to evaluate its potential to effectively and efficiently achieve these cleanup levels for PHC.

3. THE GENERAL INTEGRATED RISK MANAGEMENT MODEL

The general integrated risk management model provides a conceptional framework for a systematic risk-based evaluation and selection of action alternatives. Its structure has been described in detail by Scholz et al., (1996) and is outlined in Figure 1. The model consists of five distinct phases which are subsequently explained.

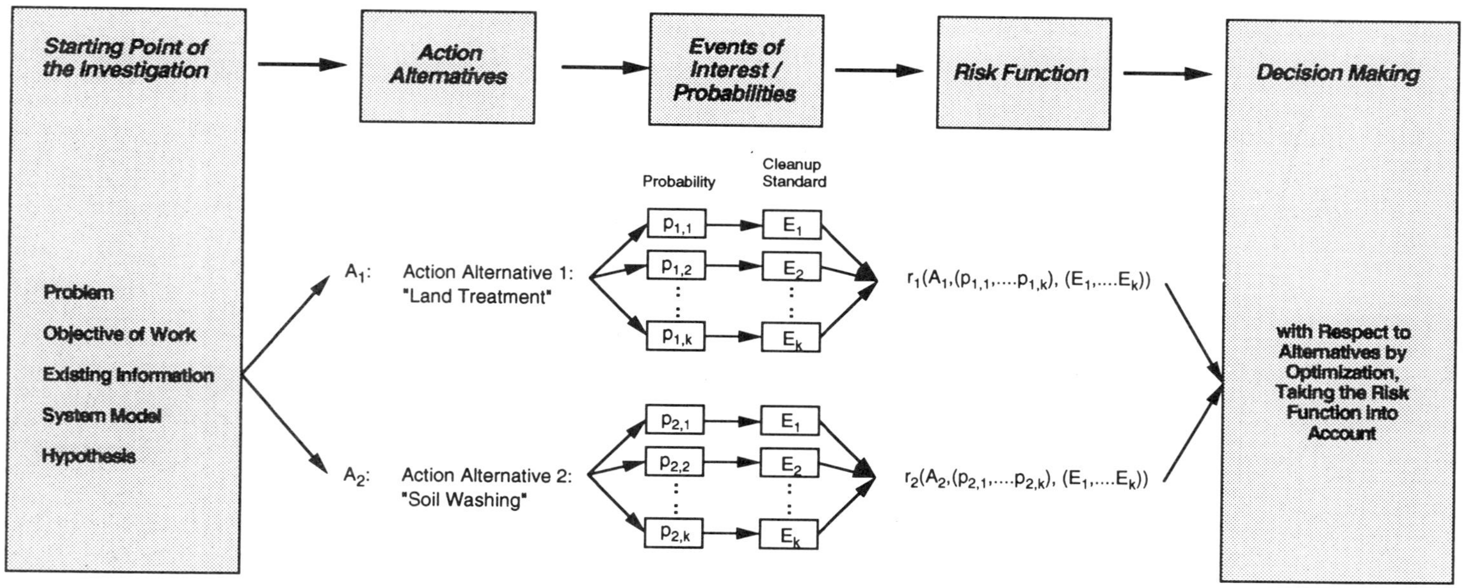

Figure 1. The general risk management model.

The first phase entitled "starting point of the investigation" involves the definition of the scope of an investigation. This includes the characterization of the problem and the relevant environmental system. Based on existing information several hypotheses might be formulated. In the analysis of the treatment technologies subsequently presented we may hypothesize that technology A is better than technology B. Alternatively, we may simply ask: How can the two technologies be applied most efficiently for a given problem? In the second phase the action alternatives need to be specified. For our considerations these are biological land treatment and physico-chemical soil washing for the remediation of PHC contaminated soils. The third phase, "events and probabilities", requires as a first step the identification of the events of interest. For the example subsequently presented, the events are the cleanup standards which have to be met. The second step involves the identification and quantification of performance indicators. These are used to derive the probabilities to achieve the events of interest for each action alternatives. In phase four a risk function is being developed which integrates events, probabilities and treatment alternatives. The fifth "decision making" phase then seeks to achieve an overall integration of the previous four phases which usually involves the identification of the most effective and efficient action alternative.

3.1. The Treatment Alternatives Considered

3.1.1. Land Treatment. Land treatment has been one of the first methods for bioremediation of soils and sludges, and has been widely applied in the petroleum industry (Anderson, 1995). The process can be conducted in situ or ex situ. Ex situ applications involve the distribution of contaminated excavated soil materials as a 30 to 50 centimeter (cm) layer into a treatment bed surrounded by a berm. The treatment bed usually consists of a graded area covered with a high density polyethylene membrane to prevent uncontrolled migration of leachate into the groundwater. The leachate can be either returned to the soil or treated in a treatment plant. For protection of the liner and to improve drainage, a layer of sand is usually placed onto the liner before the contaminated soils are added. Treatment units may be also fully covered to avoid uncontrolled air emissions (Baker and Herson, 1993). In order to stimulate and increase the biodegradation of the contaminants, oxygen transfer is improved by tilling the soil on a regular basis. The soil moisture and pH can be controlled by an irrigation system, which may be also used for the application of nutrients, microorganisms and other additives. However, the degradation process may be significantly affected by temperature fluctuations.

Land treatment is a technically simple process which can be very cost effective for the remediation of PHC contaminated soils, provided that sufficient space and time are available. Treatment costs range between 35 and 100 US dollars per ton (Anderson, 1995). Critical factors affecting treatment efficacy include the type, concentration, toxicity, biodegradability and bioavailability of the contaminant. Further, the potential formation of toxic biotransformation products in complex pollutant mixtures should be considered. In addition, pollutant emissions to air and water have to be controlled, minimized and quantified to ensure and demonstrate sound overall environmental performance.

3.1.2. Soil Washing Process. Soil washing is an ex situ treatment process applicable to a broad range of organic, inorganic and radioactive contaminants in soil (Anderson, 1993; LfU, 1993b). The process is based on chemical and physical extraction and separation techniques to remove the pollutants by separation and volume reduction. The process can be conducted on site or off site using mobile or stationary treatment technologies, respectively. Soil washing requires the excavation and transportation of the contaminated soil to the treatment plant. The treatment consists of distinct processes such as pretreatment to remove oversize particles, separation processes to divide the soil into coarse and fine grained fractions, the washing of these fractions using various physical but also chemical treatments including addition of surfactants and chelating agents. Finally, a clean product fraction representing between 70 and 90% of the initial soil amount can be recycled. The residual fraction which represents 10 to 30% of the initial contaminated materials contains the concentrated pollutants and has to be managed either by further treatment, incineration or deposition in landfills (Anderson, 1993). The process water is treated and can be recycled depending on the process technology. Treatment cost including sludge disposal and all additional known cost components range between 150 and 250 US dollars per ton (Anderson, 1993).

Critical factors limiting the performance of the technology are complex mixtures of contaminants which are difficult to treat with a single washing fluid. Soils containing high concentrations of mineralized metals or hydrophobic organics may be difficult to clean to levels acceptable for reuse of the materials. Treatment of soils with high clay and silt contents (30–40% of the soil with particle sizes < 63 micrometers) or highly porous materials is not cost effective. High humics content in soil makes separation very difficult. The main risk of using this technology is inaccurate site characterization such that the treatment soils might be different from the tested materials.

3.2. Treatment Performance Indicators

In order to characterize the remediation technologies outlined above with respect to their associated risks to achieve a given cleanup level, treatment performance indicators have to be defined and subsequently quantified based on experience data. It is important that such performance indicators are derived and interpreted in relation to the purpose and scale of their application. If risk-based general treatment evaluations are to be conducted for the planning at a regional scale, data from a wide range of successful applications may be pooled in order to obtain an overall performance indicator. In contrast, for site-specific analysis a more detailed resolution of the data is required with respect to site and contamination relevant factors affecting the treatment performance.

The performance of a treatment technology can be evaluated statically or dynamically. A useful static performance indicator for both land treatment and soil washing is given by the relative remaining contaminant fraction, F, which is defined by equation (1),

$$F = c/c_o \tag{1}$$

where c_o and c are the contaminant concentrations in milligrams per kilogram (mg/kg) in the soil before and after remediation, respectively. It should be noted that, F, reflects the sum of all transport and reaction processes contributing to the disappearance of the pollutant. However, quantitative data on the relative contribution of these processes at the field scale are not available from the literature.

The treatment performance can also be characterized dynamically. For land treatment the biodegradation of PHC can be modeled using the first order kinetic equation (2), (Troy et al., 1994).

$$dc/dt = -kc \tag{2}$$

The first order degradation rate, k, with the dimension (day^{-1}) can be derived by linear regression from time course data using the integrated form of equation (2) which is provided by equation (3),

$$\ln c = \ln c_0 - kt \tag{3}$$

where c_0 is the initial contaminant concentration. Often, data on the biodegradation characteristics are reported as half life of the contaminant, $t_{1/2}$, which is related to the first order degradation rate constant, k, by equation (4).

$$k = (\ln 2)t_{1/2} \tag{4}$$

In contrast to land treatment, the time factor is less critical for the soil washing, where between 10 and 80 tons of soil can be processed per hour, depending on the process type and on the characteristics of the contaminated soil (LfU, 1993b).

In order to obtain some overall estimates for the performance indicators F_r and k at the field scale level, a literature survey was conducted for land treatment and soil washing processes. For land treatment, data were obtained either directly or by using equations (1) to (4) from the following literature sources: Reynolds et al., (1994); Hogg et al., (1994); Ellis (1994); Fogel (1994); Jensen and Miller (1994); Ying et al., (1992); Troy et al., (1992), and Poetzsch (1993). The data used comprise treatments at temperatures ranging from 15 to 25°C, where average temperatures were around 20°C. For the performance indicators F and k for land treatment, 25 and 41 observations were used, respectively. For the soil washing process data 23 observations for PHC contaminated soils were used (LfU, 1993b; Eberhard, 1993).

It was assumed that the data reported on land treatment and soil washing resulted all from processes that were principally found feasible based on treatability and/or pilot investigations.

The frequency distributions for the performance indicators are shown as box plots for F and k in Figure 2A and 2B, respectively. The corresponding 10, 25, 50, 75 and 90th percentile of F for land treatment were 0.02, 0.04, 0.06, 0.13 and 0.17, respectively. For the soil washing process the 10, 25, 50, 75 and 90th percentile of F were 0.01, 0.022, 0.042, 0.073

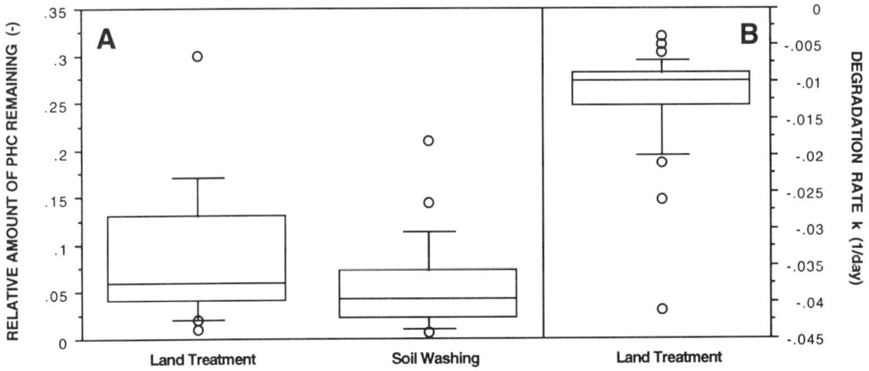

Figure 2. A. Box plots for petroleum hydrocarbon reduction, expressed as remaining fraction of initial pollutant concentration, for biological land treatment and soil washing processes at the field scale level for 25 and 23 observations, respectively. **B.** Box plot for petroleum hydrocarbon first order degradation rate constants reported for biological land treatment processes form 41 observations.

and 0.113, respectively. For the biodegradation rate constant, k, the 10, 25, 50, 75 and 90th percentile for land treatment were −0.02, −0.013, −0.01, −0.009 and −0.0072 day^{-1}, respectively. For all the performance indicators, the frequency distributions were significantly skewed. The relative pollutant fraction, F, in Figure 2A exhibited a much wider frequency distribution for the land treatment process than for the soil washing process.

3.3. The Risk Function

The general treatment-based risk can be expressed in terms of the probability to achieve a cleanup standard from a given initial PHC concentration. These probabilities and corresponding scores for the relative remaining pollutant fraction, F, were derived from cumulative frequency distribution data according to equation (5) (Runyon and Haber, 1980),

$$F_P = F_{P,l} + (i(\text{cum } f_P - \text{cum } f_{P,l}))/ f_i \tag{5}$$

where F_P is the relative remaining pollutant fraction at a given percentile rank, P, of the source data, cum f_P is the cumulative frequency of the score at the percentile rank, cum $f_{P,l}$ the cumulative frequency at the lower real limit of the interval containing cum f_P, $F_{P,l}$ is the relative pollutant fraction at the lower real limit of the interval containing cum f_P, f_i is the number of cases within the interval containing cum f_P, and i is the width of the interval. The cumulative frequency cum f_P at a given percentile is defined by equation (6),

$$\text{cum } f_P = PN/100 \tag{6}$$

where P is the percentile rank and N the total number of observations. The results for F_P are then inserted into equation (7),

$$c_{o,s,P} = c_s/F_P \tag{7}$$

where $c_{o,s,P}$ is the initial contaminant concentration from which a given cleanup standard, c_s, is achieved with a probability, P, expressed in percent. The results for the land treatment and soil washing process are presented in Figure 3A and 3B, respectively. The curves indicate the relatively smaller probabilities to achieve a cleanup standard from a given initial PHC concentration for the land treatment process as compared to soil washing.

In Figure 4 the land treatment process is considered from a dynamic perspective. Figure 4A presents the time course of PHC concentrations for the degradation rate constant, k, at the 10, 25, 50 , 75 and 90th percentile of the source data frequency distribution. Data were calculated according to equation (8),

$$c = (1-r)c_o e^{-kt} + rc_o \tag{8}$$

where r represents the recalcitrant residuals fraction of the total initial PHC contamination c_o. For simplicity, the factor r is considered constant. The curves for the PHC concentration approach rc_o asymptotically. For heavy oils, these residuals can make up to 5–10% of the total initial amount of PHC (Ying et al., 1992). For the calculations in Figure 4A an initial PHC concentration of 5,000 mg/kg was used for c_o and r was assumed to be 0.005, corresponding to 0.5% of the initial PHC mass. Figure 4A illustrates that according to the survey data, the fastest degradation is the least likely one.

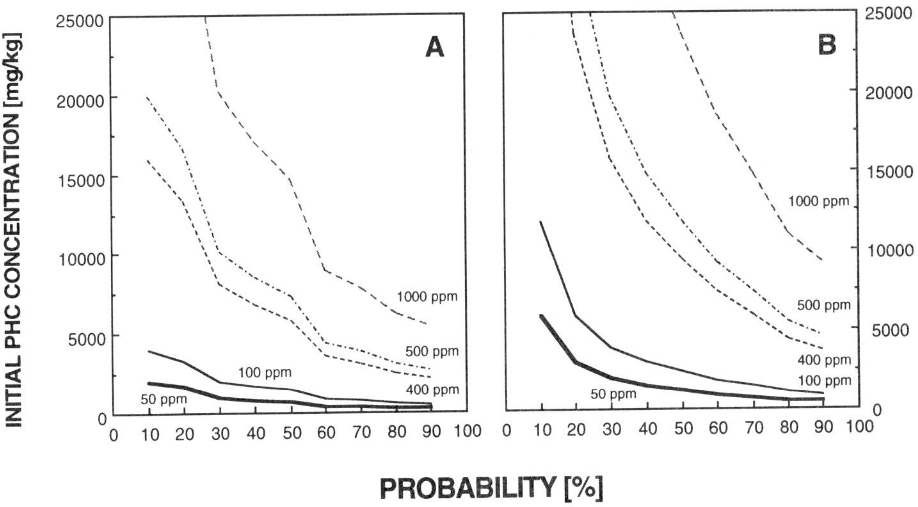

Figure 3. Probability to achieve given cleanup standards based on literature data analysis. Initial pollutant concentration and corresponding probabilities to achieve cleanup standards of 50, 100, 400, 500 and 1000 ppm for petroleum hydrocarbons. **A.** Biological land treatment. **B.** Soil washing process.

In Figure 4B, the relation between cleanup time and probability to achieve a given cleanup standard is presented. The calculations for the degradation rate constant, k, at different percentiles were conducted, analogous to the procedure outlined above, for the relative pollutant fraction F_P. The treatment time to achieve a given cleanup standard, c_s, was then calculated according to equation (8), for the conditions for c_0 and r as outlined above. The curves in Figure 4B demonstrates that the more stringent the cleanup standard, the

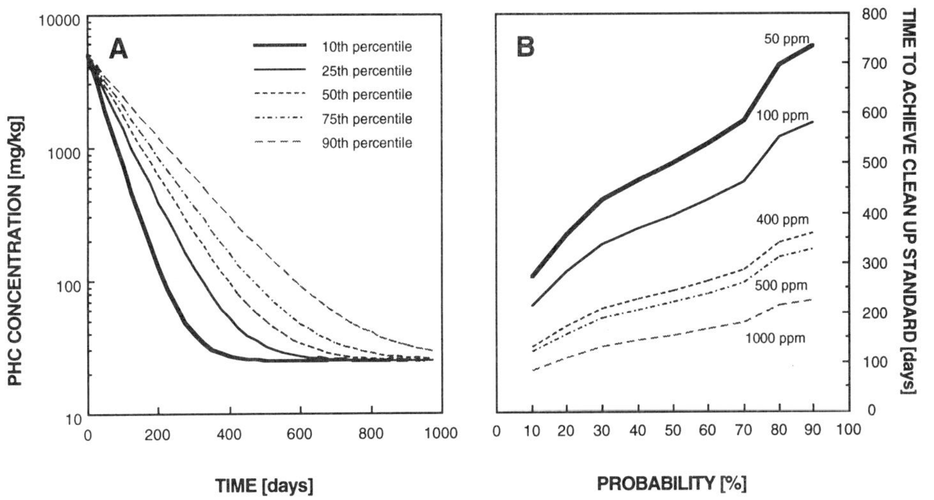

Figure 4. Probability to achieve cleanup standards within a given time frame for biological land treatment. **A.** Petroleum hydrocarbon concentration as function of time for different first order degradation rate constants corresponding to the 10th, 25th, 50th, 75th and 90th percentile of the analyzed literature data set. **B.** Available time and corresponding probabilities to achieve given cleanup standards for petroleum hydrocarbons of 50, 100, 400, 500 and 1000 ppm at an initial pollutant concentration of 5000 mg/kg.

longer the time required to achieve this cleanup level at a given probability. In other words it might be risky using land treatment for stringent cleanup levels if only limited time is available.

Based on the risk functions for the two remediation technologies to achieve given cleanup standards it becomes possible to identify application ranges of the technologies for contaminated site scenarios within specified constraints. A methodical approach how this can be done is subsequently presented.

3.4. The Optimization of Treatment Selection

With the establishment of site registers and subsequent detailed site investigations on the contaminant situation, databases are being developed that may support regional planning and management of remediation activities in the future. Such information could help identifying application ranges for remediation technologies based on their risk within a set of constraints as defined by the contaminated site situation and by the goals of the parties involved in the cleanup process. Using the method of linear optimization (Ossenbruggen, 1994; Nash and Sofer, 1996), all these aspects can be integrated into the decision making process.

This approach is illustrated on a hypothetical example, outlined as network diagram in Figure 5: In a region a total of at least 35,000 tons of PHC contaminated soil is expected as a result of construction work at several different contaminated sites. Of these soils a maximum of 5,000, 20,000 and 10,000 tons are contaminated with PHC at concentrations of 10,000, 5,000 and 2,500 mg/kg, respectively. The two remediation technologies available to treat this type of excavated soils are land treatment and soil washing. In order to implement a sustainable resource management of contaminated soils, the most efficient use of the two technologies should be found such that treatment results in the maximum possible amount of soil that meets the practical cleanup standards of 50, 100, 500 and 1000 ppm, within a financial budget of 12 million US dollars. It is assumed that treated soil not meeting the cleanup standard has to be deposited in a landfill. The process cost and relevant probabilities to characterize the treatment risks to achieve given cleanup standards are summarized in Table 3.

The procedure to solve the optimization problem includes the definition of control variables, the formulation of the objective function and of the relevant restrictions (Ossen-

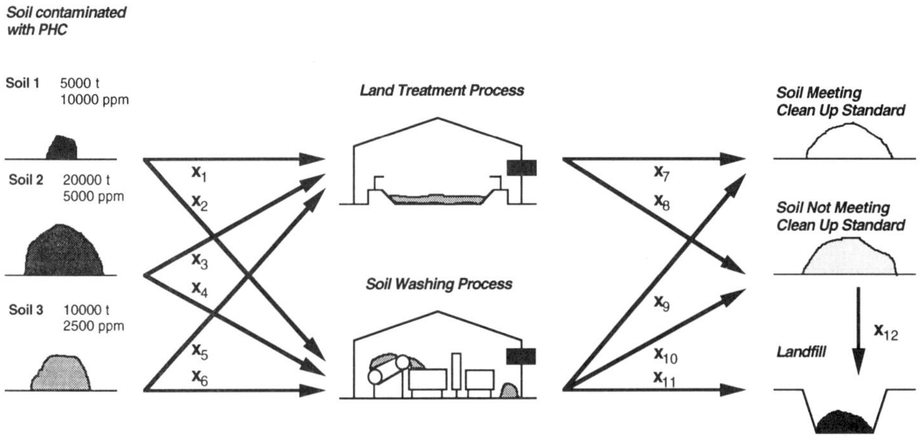

Figure 5. Network diagram for the optimization problem.

Table 3. Process cost and probabilities to achieve given cleanup standards for different initial PHC concentrations

Remediation technology	Cost range (average) [$/t]	Soil [no.]	Initial PHC concentration [mg/kg]	Probabilities, pL (land treatment) and pS (soil washing) to achieve cleanup standard				Reference
				50 ppm	100 ppm	500 ppm	1000 ppm	
Land treatment	35–100 (67.5)							Anderson, 1995
		S1	10000	0	0	0.31	0.52	Equ. 7
		S2	5000	0	0.07	0.52	0.92	Equ. 7
		S3	2500	0.07	0.20	0.92	1	Equ. 7
Soil washing	118–224[a] (171)							Anderson, 1993
		S1	10000	0.6	0.12	0.57	0.85	Equ. 7
		S2	5000	0.12	0.24	0.85	0.96	Equ. 7
		S3	2500	0.24	0.47	0.96	1	Equ. 7
Landfill	200–300 (250)							Anderson, 1995

[a]The cost data for soil washing do not contain disposal of the residual fraction (Anderson, 1993).

bruggen, 1994). The control variables are listed and explained in Table 4. The following objective function (9) has to be maximized for this problem,

$$y = x_7 + x_9 \tag{9}$$

where x_7 and x_9 can be substituted by equations 9a and 9c from Table 4, respectively, resulting in equation (10).

$$y = p_{L,S1}x_1 + p_{L,S2}x_3 + p_{L,S3}x_5 + 0.9[p_{S,S1}x_2 + p_{S,S2}x_4 + p_{S,S3}x_6] \tag{10}$$

Table 4. Description of the control variables

Control variable	Description	Equation	
x_1	Amount of soil 1, S1, processed with land treatment		
x_2	Amount of soil 1, S1, processed with soil washing		
x_3	Amount of soil 2, S2, processed with land treatment		
x_4	Amount of soil 2, S2, processed with soil washing		
x_5	Amount of soil 3, S3, processed with land treatment		
x_6	Amount of soil 3, S3, processed with soil washing		
x_7	Amount of soil processed with land treatment meeting cleanup standards	$x_7 = p_{L,S1}x_1 + p_{L,S2}x_3 + p_{L,S3}x_5$	(9a)
x_8	Amount of soil processed with land treatment not meeting cleanup standards	$x_8 = (1-p_{L,S1})x_1 + (1-p_{L,S2})x_3 + (1-p_{L,S3})x_5$	(9b)
x_9	Amount of soil processed with soil washing meeting cleanup standards (90% of initial amount)	$x_9 = 0.9[p_{S,S1}x_2 + p_{S,S2}x_4 + p_{S,S3}x_6]$	(9c)
x_{10}	Amount of soil processed with soil washing not meeting cleanup standards (90% of initial amount)	$x_{10} = 0.9[(1-p_{S,S1})x_2 + (1-p_{S,S2})x_4 + (1-p_{S,S3})x_6]$	(9d)
x_{11}	Amount of soil residuals to be deposited on landfill after soil washing (10% of initial amount)	$x_{11} = 0.1(x_2 + x_4 + x_6)$	(9e)
x_{12}	Amount of soil not meeting cleanup standard to be deposited on landfill	$x_{12} = x_8 + x_{10}$	(9f)

The corresponding probabilities for the different cleanup levels are provided in Table 3.

The following restrictions apply to the problem. Equation (11) specifies the budget limit. The specific treatment cost per ton of soil, c_l for land treatment (Anderson, 1995), c_s for soil washing (Anderson, 1993) and c_w for deposition on a hazardous waste landfill (Anderson, 1995) are provided in Table 3 and can be substituted into equation (11). The variables x_{11} and x_{12} are substituted by the equations given in Table 4.

$$12{,}000{,}000 \geq c_l(x_1 + x_3 + x_5) + c_s(x_2 + x_4 + x_6) + c_w(x_{11} + x_{12}) \tag{11}$$

The total amount of soil subjected to remediation is at least 35,000 tons which results in restriction equation (12).

$$35{,}000 \leq x_1 + x_2 + x_3 + x_4 + x_5 + x_6 \tag{12}$$

The maximum amounts of 5,000, 20,000 and 10,000 tons of the soils which need remediation are specified with the restrictions (13), (14) and (15), respectively:

$$5{,}000 \geq x_1 + x_2 \tag{13}$$

$$20{,}000 \geq x_3 + x_4 \tag{14}$$

$$10{,}000 \geq x_5 + x_6 \tag{15}$$

$$x_1, x_2, x_3, x_4, x_5, x_6 \geq 0 \tag{16}$$

Equation (16) represents the non-negativity condition for the control variables.

The maximization problem in equation (10) was solved using the linear programing routine ConstrainedMax as implemented in the Software package Mathematica (Wolfram, 1991). The optimum solutions for y in equation (10) and for the control variables x_1 to x_6 are listed in Table 5 for the cleanup standards 50, 100, 500 and 1000 ppm. The results demonstrate that the maximum amount of soil meeting cleanup requirements increases with increasing cleanup standards. This response reflects the increasing probabilities of the two remediation technologies to achieve the less stringent cleanup standards.

A very important tool to analyze the results of optimization problems is the use of sensitivity analysis. With this method, the effect of variations in a given parameter on the results of the optimization are investigated, while holding all other conditions constant. As an example, the effect of variations in the available financial budget on the optimum application of land treatment and soil washing for the three soils is presented in Figure 6.

Table 5. Solutions of the optimization problem for different cleanup standards[a]

Cleanup standard [ppm]	Control variables—amount of soil treated [t]						Amount of soil meeting cleanup standard[t]
	x_1	x_2	x_3	x_4	x_5	x_6	y
50	5000	0	14869	5131	0	10000	2714
100	5000	0	1194	18806	0	10000	8376
500	0	5000	0	20000	10000	0	27065
1000	0	5000	20000	0	10000	0	32225

[a]The restrictions are specified in the text.

The relative amounts of each soil treated with the technologies are specified for different cleanup standards of 50, 100, 500 and 1000 ppm in Figures 6A, 6B, 6C and 6D, respectively. The figures can be divided into three distinct ranges with respect to budget variation: at too low available budget no solutions can be found within the constraints defined earlier, such that all the soil can be treated. At a narrow intermediate range a transitional complex response pattern is observed, where the budget limits the response. The third region is characterized by the fact that budget increases do not result in any further changes in the optimum use pattern of the two remediation technologies. This latter region coincides also with the maximum possible amount of soil meeting cleanup standards that can be achieved (data not shown). Based on these figures general application ranges can be identified for the two technologies as function of cleanup standard and initial PHC concentration: (1) for optimal solutions at standards of 50 and 100 ppm all the soils should be subjected to soil washing only (Figures 6A and 6B), (2) at 500 ppm the least contaminated

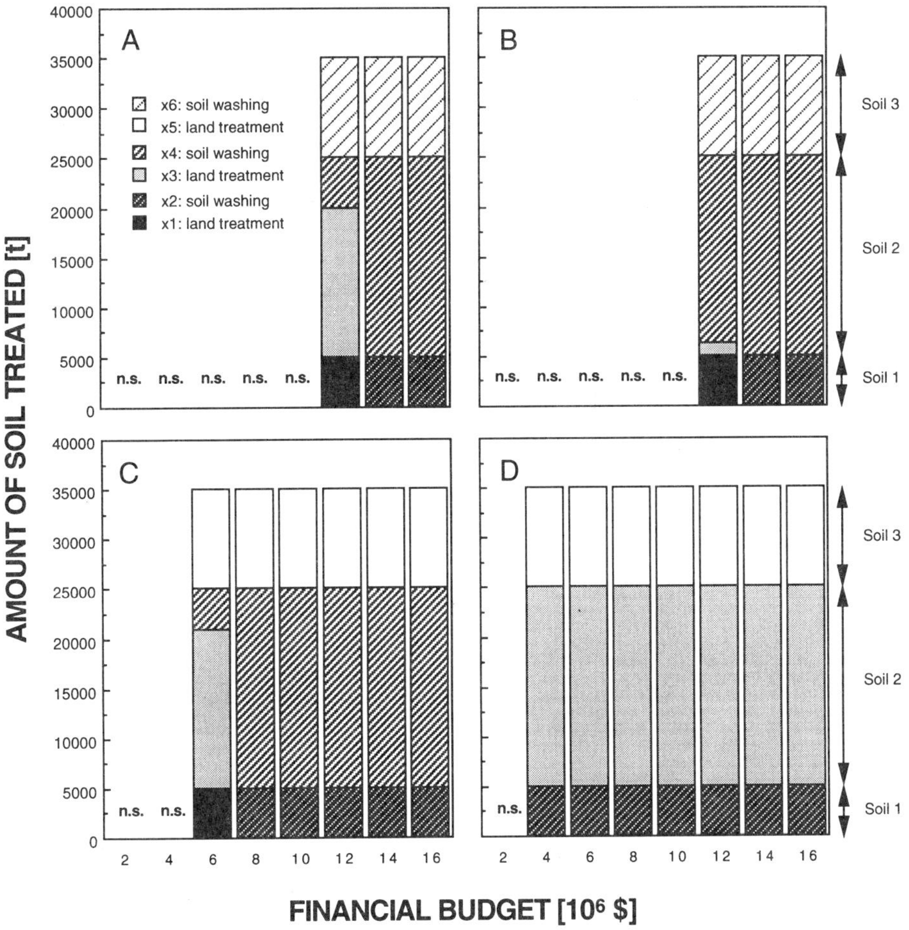

Figure 6. Sensitivity diagrams illustrating the effect of variations in the financial budget on the control variables x_1 to x_6 for optimum solutions. Bars indicate the relative amounts of soil 1 (black), soil 2 (grey) and soil 3 (white) to be treated with biological land treatment (plain colored bars) and soil washing (hatched bars) at different cleanup standards: (A) 50 ppm; (B) 100 ppm; (C) 500 ppm; (D) 1000 ppm. Areas where no solutions were found are indicated with n.s.

soil 3 should be treated biologically and (3) at 1000 ppm the soils 2 and 3 should be bioremediated. These risk-based application ranges identified for the biological land treatment are in good accordance with practical experience data reported for land treatment in the Netherlands (Verheul et al., 1993). While cleanup levels of 1000 ppm were reported to be achievable, the lower target cleanup S-levels of 50 ppm may be difficult to achieve within reasonable time.

In Figure 7A the maximum achievable amounts of soil meeting the cleanup requirements are presented for different standards, including the relative contributions of land treatment and soil washing processes. The amount of clean soil increases with increasing cleanup standards, while the amount of soil to be deposited on a landfill decreases simultaneously. The corresponding minimum cost at which the optimum solutions in Figure 7A can be achieved were found to decrease with increasing cleanup standard as shown in Figure 7B. These findings reflect the significant contribution of landfill cost for treated soils not meeting cleanup standards, which is generated at stringent cleanup levels.

4. DISCUSSION AND CONCLUSION

The procedure outlined provides an approach to integrate information and knowledge on specific remediation technologies based on their risk to achieve cleanup standards. The results indicate that the use of land treatment technology is in general a viable option for the bioremediation of PHC contaminated soils, if cleanup standards are in the range of several hundred ppm and initial PHC concentrations are in the range of a few thousand mg/kg. However, several additional aspects concerning the overall efficacy of the processes should also be taken into account for a general technology evaluation. These include the relative contribution of individual pollutant fate processes to the pollutant re-

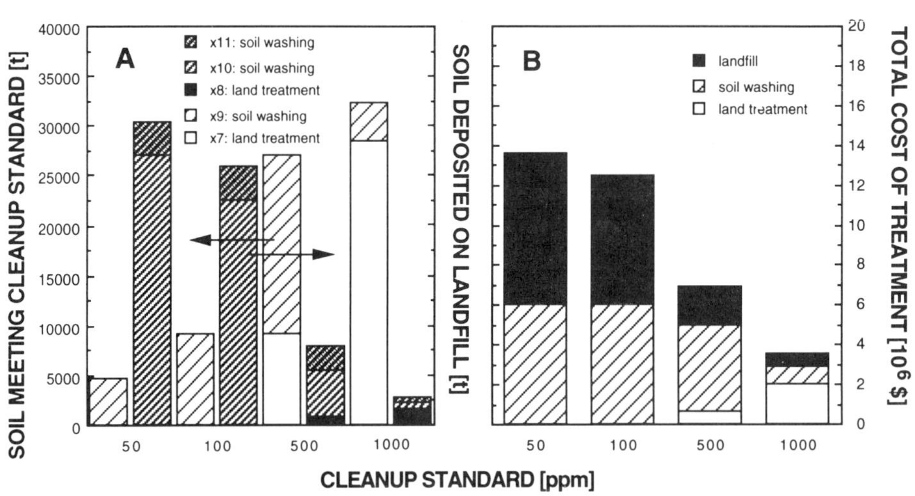

Figure 7. Optimum solutions at different cleanup standards for petroleum hydrocarbons. **A.** Maximum achievable amounts of soil products meeting cleanup standards (light bars) and corresponding amount of soil products to be deposited on landfill (dark bars) and corresponding amount of soil products to be deposited on landfill (dark bars) for biological land treatment (open) and soil washing (hatched). **B.** Corresponding total cost and relative contributions of biological land treatment (white), soil washing (hatched) and landfill (black) for the optimum solutions shown in **A.**

duction, the quality and potential for reuse of the clean product, the amount of wastes generated, as well as the use of additional resources such as energy, water and chemical additives. More recent developments in land treatment technology such as the use of completely covered fields and leachate control (Baker and Herson, 1993) will facilitate the quantification of non-biological emissions, and combined with effective sampling (Scholz et al., 1994) and monitoring methods, (Madsen, 1991; Heitzer and Sayler, 1993) biodegradation may also be quantified.

Another important issue concerns the general cleanup standard policy. Based on the optimization results shown in Figure 7 some fundamental questions may be asked: Should remediation go towards small amounts of high quality products at high cost and at the potential expense of landfill volume or would it be more efficient to produce a product of lesser quality at reduced cost and landfill volume? Although the use of soil washing seems favorable at low cleanup standards according to the optimization analysis, it has to be emphasized that the clean product is not an intact soil anymore due to size fractionation of the initial soil during the washing process (Anderson, 1993). In contrast, the product of the bioremediation process does not exhibit an altered overall soil composition. Therefore, rather than only considering the pollutant concentration in relation to fixed cleanup standards, a risk-based evaluation of the treated soil with respect to the bioavailability, toxicity and general mobility of the remaining pollutant fraction needs to be conducted. Promising diagnostic approaches to analytically address the question of pollutant bioavailability have been developed and investigated during the past few years (King et al., 1990; Heitzer et al., 1994). If it can be demonstrated that the treated material is not toxic and no relevant emissions are observed, the soils should be considered for reuse under appropriate conditions rather than being deposited in a landfill. Such practice helps to avoid the exclusion of effective and efficient treatment technologies due to fixed stringent cleanup standards that do not consider site-specific risk situations.

The general integrated risk management model (Scholz et al., 1996) provided a systematic and logical conceptual framework for the risk-based evaluation of treatment alternatives. It was demonstrated that risk indicators can be developed using experience data from successful treatment applications. However, it should be emphasized that the reliability of these indicators is determined not only by the quality and number of experience data available but also by the scale of the problem for which they are used. The treatment risk indicators presented in this contribution provide information on the general performance of land treatment and soil washing processes to reduce petroleum hydrocarbons. Therefore, they should be used for general planning considerations at a regional scale rather than for the site-specific evaluation of treatment alternatives. For this purpose, data on the indicators have to be further resolved with respect to pollutant type and other site relevant factors. In a recent comparative analysis on bioremediation and conventional technologies for PHC, cost data were analyzed with respect to some site-specific factors such as pollutant concentration and size of the site (Davis et al., 1995). Along with the increasing number of field scale applications more data will become available which should be organized and continuously updated in a generally accessible database. Although the initial effort will be significant, the ultimate benefit of such information for regulators and planners should not be neglected.

Using the method of linear optimization, efficient application ranges for specific treatment technologies could be identified based on their risks to achieve the required cleanup standards. This method allows for the simultaneous integration of various aspects such as risk-based characteristics of treatment technologies, cleanup standards and more general remedial action objectives into the optimization and evaluation process of treatment technologies, e.g., for a given regional contaminated site situation. Recent applications of this method for the optimized use of emission control equipment in industrial plants demonstrates the practical utility of this approach in environmental planning and management (Lou et al., 1995). The good accordance of the application ranges found for the land treatment technology with practical experience data are very promising and indicate the potential utility of this approach as a useful tool to support planning and decision making in the management process of contaminated sites.

ACKNOWLEDGMENT

This work was partially supported by a grant from the Cantonal Institute for Water Protection and Hydraulic Engineering in Zürich.

REFERENCES

AGW. (1994). "Wegleitung für die Klassierung von Bauabfällen." Amt für Gewässerschutz und Wasserbau des Kantons Zürich.

AGW. (1993). "Altlastenbearbeitung: Einführung in die Altlastenpraxis des Kantons Zürich.," Amt für Gewässerschutz und Wasserbau des Kantons Zürich.

Anderson, W. C. (1993). *Soil washing/Soil flushing*, American Academy of Environmental Engineers, Annapolis, MD.

Anderson, W. C. (1995). *Bioremediation*, American Academy of Environmental Engineers, Annapolis, MD.

Atlas, R. M. (1984). *Petroleum Microbiology*, Macmillan, New York. Innenministerium Baden-Württemberg (1993). "Gemeinsame Verwaltungsvorschrift des Umweltministeriums und des Sozialministeriums über Orientierungswerte für die Bearbeitung von Altlasten und Schadensfällen." *GABl*, 41(33), 1115–1123.

Baker, K. H., and Herson, D. S. (1993). *Bioremediation*, McGraw Hill, New York.

Barry, D.L. (1995). Recycling derelict sites: UK experiences. Brachfluachen und Flächenrecycling, D.D. Genske, H.-P. Noll, eds., Ernst & Sohn, Berlin, 281–289.

Bell, C. E., Kostecki, P. T., and Calabrese, E. J. (1991). "Review of state cleanup levels for hydrocarbon contaminated soils." Hydrocarbon contaminated soils and groundwater, P. T. Kostecki, E. J. Calabrese, and C. E. Bell, eds., Lewis Publishers, Chelsea, 77–89.

BUWAL. (1994). "Altlasten-Konzept für die Schweiz: Ziele und Massnahmen." *220*, BUWAL.

Davis, K. L., Smith, H. R., and Day, S. M. (1995). "A comparative cost analysis: Conventional treatment technologies and bioremediation." Bioremediation of pollutants in soil and water, B. S. Schepart, ed., ASTM, Philadelphia, 5–17.

Eberhard. (1993). "Bodenwaschanlage Eberhard Recycling AG." .

Ellis, B. (1994). "Reclaiming contaminated land: In situ/ex situ remediation of creosote- and petroleum hydrocarbon-contaminated sites." Bioremediation: Field experience, P. E. Flathman, D. E. Jerger, and J. H. Exner, eds., Lewis Publishers, Boca Raton, 107–141.

Fogel, S. (1994). "Full scale bioremediation of no.6 fuel oil-contaminated soil: 6 months of active and 3 years of passive treatment." Bioremediation: Field experience, P. E. Flathman, D. E. Jerger, and J. H. Exner, eds., Lewis Publishers, Boca Raton, 161–175.

Folkestad, B. (1993). "Approaches to cleanup standards in Norway." *Developing cleanup standards for contaminated soil, sediment and groundwater: How clean is clean?*, Washington, 187–193.

GSA (1995). Altlasten- und Verdachtsflächenkataster des Kantons Bern. Schlussbericht. Amt für Gewässerschutz und Abfallwirtschaft des Kantons Bern.

Heitzer, A., and Sayler, G. S. (1993). "Monitoring the efficacy of bioremediation." *Trends in Biotechnology*, 11, 334–343.

Heitzer, A., Malachowsky, K., Thonnard, J.E., Bienkowski, P.R., White, D.C., and Sayler, G.S. (1994). Optical Biosensor for environmental on-line monitoring of naphthalene and salicylate bioavailability with an immobilized bioluminescent catabolic reporter bacterium. Appl. Environ. Microbiol., 60:1487–1494.

Hogg, D. S., Piotrowski, M. R., Masterson, R. P., Jorgensen, M. R., and Frey, C. (1994). "Bioremediation of hydrocarbon-contaminated soil." Hydrocarbon Bioremediation, R. E. Hinchee, B. C. Alleman, R. E. Hoeppel, and R. N. Miller, eds., Lewis Publishers, Boca Raton, 398–404.

Jensen, R. E., and Miller, J. A. (1994). "Field demonstrations of bioremediation and low-temperature thermal treatment technologies for petroleum contaminated soil." Hydrocarbon contaminated soils and groundwater, E. J. Calabrese and P. T. Kostecki, eds., The Association for the environmental health of soils, Amherst, 227–261.

King, J.M.H., DiGrazia, P.M., Applegate, B., Burlage, R., Sanseverino, J., Dunbar, P., Larimer, F. and Sayler, G.S. (1990). Rapid, sensitive bioluminescent reporter technology for naphthalene exposure and biodegradation. *Science*, 249:778–781.

Leahy, J. G., and Colwell, R. R. (1990). "Microbial degradation of hydrocarbons in the environment." *Microbiol. Rev.*, 54(3), 305–315.

LfU. (1993a). "Das Modellstandortprogramm des Landes Baden-Württemberg." *Band 12*, Landesanstalt für Umweltschutz Baden-Württemberg.

LfU. (1993b). "Handbuch Bodenwäsche." *11*, Landesanstalt für Umweltschutz Baden-Württemberg.

Lou, J. C., Chung, K. F., and Chen, K. S. (1995). "An optimization study of total emission controls for particulate pollutants." *J. Environ. Management*, 43, 17–28.

Lühr, H.P., Hefer, B. and Scholz, R.W. (1991). Das Donator-Akzeptor-Modell (DAM). In: Ableitung von Sanierungswerten für kontaminierte Böden, Inst. für wassergefährdende Stoffe (ed.) Erich Schmidt Verlag, Berlin.

Madsen, E. L. (1991). "Determining in situ biodegradation: facts and challenges." *Environ. Sci. Technol.*, 25, 1663–1673.

Minsch, J. (1993). "Nachhaltige Entwicklung, Idee - Kernpostulate: Ein ökologisch-ökonomisches Referenzsystem für eine Politik des ökologischen Strukturwandels in der Schweiz." *Nr. 14*, Institut für Wirtschaft und Ökologie, Hochschule St, Gallen.

Nash, S. G., and Sofer, A. (1996). *Linear and nonlinear programming*, McGraw Hill Comp. Inc., New York.

Neff, J. M. (1990). "Composition and fate of petroleum and spill treating agents in the marine environment." Sea mammals and oil: Confronting the risks, J. R. Geraci and D. J. St. Aubin, eds., Academic Press, San Diego, 1–33.

Ossenbruggen, P. J. (1994). *Fundamental principles of systems analysis and decision making*, John Wiley & Sons, Inc., New York.

Paustenbach, D. J., Kalmes, R., Jernigan, J. D., Bass, R., and Scott, P. (1993). "A proposed approach to regulating contaminated soil: Identify safe concentrations for seven of the most frequently encountered exposure scenarios." Principles and practices for petroleum contaminated soils, E. J. Calabrese and P. T. Kostecki, eds., Lewis Publishers, Boca Raton, 511–552.

Poetzsch, E. (1993). "Biologische Bodensanierung auf dem Gelände eines Tanklagers in Saalfelden, Österreich." Bewertung und Sanierung mineralöl-kontaminierter Böden, G. Kreysa and J. Wiesner, eds., DECHEMA, Frankfurt, 471–481.

RSU, Rat der Sachverständigen für Umweltfragen. (1995). *Altlasten II: Sondergutachten Altlasten, Februar 1995*, Metzler-Poeschel, Stuttgart.

Reynolds, C. M., Travis, M. D., Braley, W. A., and Scholze, R. J. (1994). "Applying field expedient bioreactors and landfarming in Alaskan climates." Hydrocarbon Bioremediation, R. E. Hinchee, B. C. Alleman, R. E. Hoeppel, and R. N. Miller, eds., Lewis Publishers, Boca Raton, 100–106.

Riser-Roberts, E. (1992). *Bioremediation of petroleum contaminated sites*, C.K. Smoley, Boca Raton.

Runyon, R. P., and Haber, A. (1980). *Fundamentals of behavioral statistics*, Addison-Wesley, Reading.

Scholz, R.W., Nothbaum, N., and May, W. (1994). Fixed and hypothesis-guided soil sampling methods - principles, strategies, and examples. Environmental sampling for trace analysis, B. Markert, ed., VCH, Weinheim, 335–345.

Scholz, R. W., Heitzer, A., May, T. W., Nothbaum, N., Stünzi, J., and Tietje, O. (1996). "Datenqualität und Risikoabschätzung - Das Risikohandlungsmodell zur Altlastenbearbeitung." Altlasten - Gefährdungsabschätzung: Datenanalyse und Gefahrenbewertung, S. Schulte-Hostede and S. U., eds., Ecomed-Verlag, Neuherberg, (in press).

Shannon, M. J. R., and Unterman, R. (1993). "Evaluating bioremediation: Distinguishing fact from fiction." *Annu. Rev. Microbiol.*, 47, 715–738.

Suter, J. (1995). "Sanierungsziele bei Altlasten." *Schweizer Ingenieur und Architekt*, 13, 321–324.

Troy, M.A., McGinn, S. R., Greenwald, B. P., Jerger, D. E., and Allen, B. (1992). "Bioremediation of diesel fuel contaminated soil at a former railroad fueling yard." Contaminated soils: Diesel fuel contamination, P. T. Kostecki and E. J. Calabrese, eds., Lewis Publishers, Boca Raton, 63–80.

Troy, M. A., Berry, S. W., and Jerger, D. E. (1994). "Biological Land treatment of diesel fuel contaminated soil: Emergency response through closure." Bioremediation: Field experience, P. E. Flathman, D. E. Jerger, and J. H. Exner, eds., Lewis Publishers, Boca Raton, 145–160.

USEPA. (1993). "Superfund Innovative Technology Evaluation Program: Technology Profiles." *EPA/540/R-93/526*, US Environmental Protection Agency.

Vegter, J. J. (1992). "Development of soil and groundwater cleanup standards in the Netherlands." *Developing cleanup standards for contaminated soil, sediment and groundwater*, Washington, 81–92.

Verheul, J., de Bruijn, P., and Herbert, S. (1993). "Biological remediation: A European perspective." Remedial processes for contaminated land, M. Pratt, ed., Institution of Chemical Engineers, Rugby, 53–85.

von der Trenck, K. T., and Fuhrmann, P. (1991). "Environmental cleanup objectives standard procedure." *Chemosphere*, 23(8–10), 1323–1335.

Wolfram, S. (1991) Mathematica: A system for doing mathematics by computer. Addison-Wesley Publishing Company inc., Redwood City.

World Resources Institute (1992). "World resources 1992–93: Toward sustainable development. Al. Hammond, M.E. Paden, R.T. Livernash, eds., Oxford University Press, New York.

Ying, A. C., Balba, M. T., Shepherd, G., and Wright, D. (1992). "Enhanced Biodegradation of heavy engine oil in soil from railroad maintenance yard: Phase II field demonstration." Contaminated soils: Diesel fuel contamination, P. T. Kostecki and E. J. Calabrese, eds., Lewis Publishers, Boca Raton, 47–61.

12

MONITORING THE POPULATION DYNAMICS OF BIODEGRADABLE CONSORTIA DURING BIOREMEDIATION

Karen Budwill,[1] Mark Roberts,[1] David B. Knaebel,[2] and Don L. Crawford[1]

[1]Department of Microbiology, Molecular Biology, and Biochemistry
University of Idaho
Moscow, Idaho 83844-3052
[2]Department of Biology
Clarkson University
Postdam, New York 13699-5805

1. INTRODUCTION

The use of bioremediation technology has gained widespread popularity as an efficient, cost-effective means of cleaning up contaminated sites (Bouwer, 1992; Forsyth et al., 1995). Bioremediation relies on indigenous microorganisms present at a site to transform the polluting compound to a non-toxic form. It is, therefore, imperative during the designing and testing of a bioremediation process to distinguish between abiotic chemical reactions and microbial processes. Only by demonstrating that there is a potential for microbial transformation of the contaminating chemical, as well as that biodegradation is occurring in situ, can one have confidence that bioremediation is occurring.

Successful bioremediation relies on microbial diversity. Diversity results from microorganisms evolving to occupy new environmental niches so that, in general, microorganisms have become relatively specific for particular substrates and for particular environmental conditions. Microorganisms have evolved mechanisms to degrade structurally similar compounds, or to use similar modes of attack on different compounds (Boyle, 1992). Microbial adaptation is important for most bioremediation processes since a contaminating compound that has an analogous structure found in nature may be readily, or eventually, degraded by the microorganisms that normally would use the naturally occurring substrate (Tiedje, 1993). Diversity might also ensure that there will be species present in the contaminated environment that are capable of transforming the contaminating compound at different concentration levels and/or under different environmental conditions. Often, diversity among the degrading microorganisms will lead to the emergence of the competent, most fit organisms for the desired biodegradation.

Indigenous microorganisms often are incapable of transforming contaminants, particularly synthetic chemicals, upon initial exposure to them (Lappin et al., 1985; Spain and

Van Veld, 1983; Wiggens et al., 1987). However, upon prolonged exposure, the microorganisms adapt and either acquire and/or develop the capability to degrade the contaminant. This adaptation can result from a number of factors including enzyme induction, genetic or physiological changes within the individual bacterium, such as gene exchange or recruitment, among the microbes present, and/or enrichment followed by an increase in the number of specific biodegrading bacteria (Boyle, 1992; Karns et al., 1984; Tiedje, 1993). Adaptation is not limited to single microbial communities, but can also occur among distinct microbial communities that have developed cooperative relationships for the degradation of contaminants. For example, one community may start the degradative reaction and transform or partially degrade the compound, while a second community completes the transformation. Some compounds may require consortia, and both anaerobic and aerobic processes for complete degradation (Funk et al., 1993).

It is often crucial to identify any rate-limiting factors in the bioremediation process under study, and subsequently to modify them so that the contaminant is transformed into a non-hazardous end product or completely mineralized in as short a time as possible. Factors to investigate are the nature and recalcitrance of the contaminating compound(s), the diversity, activity and numbers of biodegrading microorganisms and the state of the environment itself. Often, the simple additions of nutrients and/or the adjustment of pH can alter the physiological state of the microorganisms to positively affect the rate of biodegradation.

There are numerous microbiological techniques to detect, enumerate, and measure activity of bacteria involved in a bioremediation process. This paper will review some of those techniques and how they can be applied to monitoring the extent and rate of biodegradation. The emerging, powerful, molecular biology tools that offer new insights into microbial diversity and interactions will be presented as well. These new techniques can provide vast information on the bioremediation process.

2. MICROBIOLOGICAL TECHNIQUES FOR MONITORING MICROBIAL NUMBERS AND ACTIVITY

2.1. Enumeration

One indication that active bioremediation is occurring is by establishing a correlation between removal of contaminant and an increase in the number of degrading bacteria above the background level. In order to enumerate the degrading bacteria, they must be culturable and a knowledge of their physiology is required. There are several established methods for counting bacteria, but for meaningful data to be obtained, it is crucial that the sampling techniques used will not introduce any biases, and that a representative subsample of the environment under study is obtained (Atlas and Bartha, 1987).

Direct microscope counting is a fast method for obtaining an estimation of microbial numbers, but the major drawbacks are that living microbes generally cannot be distinguished from dead microbes and no information on cell types or metabolic activity is obtained. Viable count procedures such as plate counts can provide considerable information on the metabolic capability of bacteria since a range of growth substrates and nutrients can be used to prepare the media. However, this technique can underestimate the number and diversity of bacteria in the environmental sample since significant growth must occur for visible colonies to form and only a small portion of a microbial community is culturable (Ward et al., 1990). Another viable counting technique is the most probable number

(MPN) technique. Enumeration is done using statistical methods. A drawback to the MPN method, like the plate count method, is that the bacterium or bacteria to be counted must be culturable and significant growth must occur.

2.2. Measuring Rates of Bacterial Activity

It is not enough to simply demonstrate bioremediation is occurring as measured by an increase in bacterial numbers when a contaminant is introduced, or when optimal conditions are provided within an already contaminated environment. Demonstration that the bacteria are actively degrading the contaminant at significant rates provides further evidence of successful bioremediation. By measuring rates of bacterial activity, an indication, on the one hand, of the potential of the indigenous bacteria to degrade contaminating compounds is revealed. On the other hand, any limitations to bacterial activity may be discovered and then remedied.

Most techniques for measuring bacterial activity make use of radiolabeled substrates. The advantage of using radiolabeled tracers is that only small amounts of tracer need to be added to the experiment. In this way the tracer will not alter natural substrate levels thus giving a more representative picture of the true metabolic activity present. The microbial heterotrophic potential for the utilization of a given substrate is determined by measuring the rate of uptake of tracer levels of the radiolabeled substrate under study (Atlas and Bartha, 1987). This method allows one to study the effect of a pollutant on microbial activity. The mineralization rate of a specific substrate is often determined by measuring $^{14}CO_2$ release from labeled organic substrates (Fedorak et al., 1982).

The use of radiolabeled thymidine and leucine is gaining usefulness as a tool to estimate bacterial productivity in an ecosystem (Kaplan et al., 1992, and Tibbles and Harris, 1996). Radiolabeled [methyl-^{3}H]thymidine incorporation is used to measure DNA synthesis, whereas radiolabeled [^{14}C]leucine incorporation is used to measure protein synthesis.

3. MOLECULAR BIOLOGY TECHNIQUES FOR MONITORING MICROBIAL POPULATION DYNAMICS

3.1. Gene Probing

The use of molecular biology techniques in the study of microbial ecology can provide more precise and faster means of identifying and enumerating bacteria in environmental samples and characterizing microbial communities than can classical microbiological techniques (Ogram and Sayler, 1988; Pace et al., 1986). Oligonucleotide probing, for example, is a powerful technique for identifying which types of bacteria are present in a given environmental sample (Holben et al., 1988). Most of the molecular biology techniques make use of the sequence variation in ribosomal (r) RNA genes. By comparing rDNA sequences, phylogenetic relationships between bacterial strains and species can be deduced (Giovannoni et al., 1988 and Lane et al., 1985). This method can also be used to detect viable but uncultured microorganisms in the environment (Ward et al., 1990), and bacteria present in low numbers that would otherwise be undetected (Tsai and Olsen, 1992).

The polymerase chain reaction (PCR) has also been used to directly detect and enumerate soil microorganisms (Picard et al., 1992). A direct lysis of soil bacteria was performed and at least 90% of the soil DNA was extracted and recovered. The MPN technique was used to enumerate bacteria by correlating the number of amplifiable target DNA sequences to the number of bacterial cells. Specific primers for the plasmid-borne *vir* genes for *Agrobacterium tumefaciens* and for the variable regions of the 16S rDNA for *Frankia* spp. were used. A good correlation was found to occur between the MPN-PCR estimation and the initial inoculum quantity.

Probing can be performed using function-specific probes for genes of a particular biodegradation pathway (Sayler et al., 1985). In this way information whether the community as a whole has the potential to degrade a certain chemical or chemicals can be obtained. Primers for PCR can be designed so that they encode degrading genes (Neilson et al., 1992). Such primers can reveal information about similar and also different pollutant-degrading enzyme encoding genes found among closely or distantly related microorganisms in the environment (Bej and Mahbubani, 1994). It is possible to monitor the degrading microorganisms during bioremediation by using such function-specific primers or probes. The expression of degrading genes can be followed over the bioremediation period by reverse-transcriptase PCR (Tsai et al., 1991).

Careful attention must be paid during the extraction and purification of DNA for PCR to remove substances, such as humic material, that may otherwise inhibit the PCR reaction (Farrelly et al., 1995). It is also important to rule out any possible false positives arising from the formation of chimeric PCR products (Liesack et al., 1991).

3.2. Random Amplified Polymorphic DNA (RAPD)

When it is necessary to detect the presence of a bacterial strain in a mixed sample, and there is no prior knowledge of the strain's sequence, the technique of random amplified polymorphic DNA (RAPD) may be very useful (Hadrys et al., 1992). RAPD uses a short, single synthetic oligonucleotide of random sequence to amplify total genomic DNA present in nanogram amounts (Williams et al., 1990). The technique relies on the presence of several priming sites close to one another in inverted orientation in a genome so that DNA segments of variable length between these small inverted repeats are amplified. When resolved electrophoretically, the segments give a characteristic banding pattern, or fingerprint, for the particular genome under study (Hadrys et al., 1992). By using different primers, more than one fingerprint can be generated for a microbial community.

The general strategy taken during a RAPD experiment is to determine RAPD markers that are diagnostic for a particular strain (Fani et al., 1993). The degree of specificity can be achieved by changing the type of primer used and the conditions of the PCR reaction. The markers are cloned and sequenced, and all or part of the sequences can be used as probes to hybridize to DNA from unknown isolates that had been amplified with the same random primers.

The RAPD technique has been applied in our laboratory to generate both genus- and species-specific probes for *Streptomyces* strains. Some streptomycetes are known to produce antifungal agents and thus are being examined for use as biocontrol agents in agricultural settings (Yuan and Crawford, 1995). In particular, we are interested in being able to track the movement of *Streptomyces lydicus* WYEC108, a biocontrol agent isolated in our laboratory, upon application to soil. Figure 1 shows a RAPD profile of PCR-amplified total DNA using a single primer (primer 70–40). DNA from various *Streptomyces* strains and other actinomycetes were used, as well, DNA from non-actinomycetes (outgroups) were used as controls. A 0.9 Kb DNA fragment from *S. lydicus* WYEC108 (Figure 1) was isolated and used as a probe in Southern blot hybridizations. The probe hybridized with only *S. lydicus* WYEC108 DNA (Figure 2), and thus it is a species-specific RAPD-derived probe.

Xia et al., (1995) reported the use of RAPD fingerprinting in the analysis of three soil microbial community responses to the presence of 2,4-dichlorophenoxyacetic acid

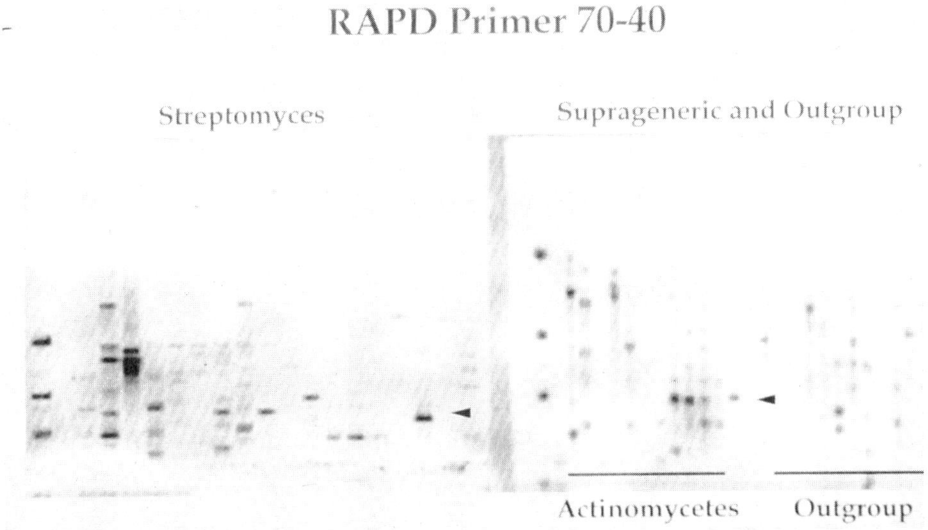

Figure 1. An example of a RAPD profile using primer 70–40 for PCR amplification. The 0.9 Kb DNA fragment from *Streptomyces lydicus* WYEC108 was the fragment isolated and used as a probe in Southern blot hybridizations. The left photograph is an ethidium bromide-stained agarose gel profile for strains of *Streptomyces*, while the right photo is the profile for other actinomycetes and eubacterial outgroup strains.

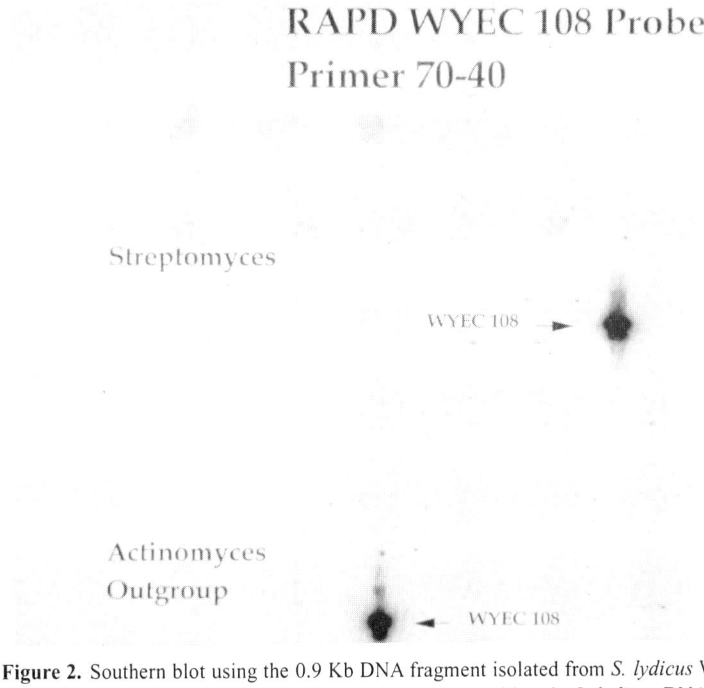

Figure 2. Southern blot using the 0.9 Kb DNA fragment isolated from *S. lydicus* WYEC108 as the hybridization probe. These results show that this fragment hybridizes with only *S. lydicus* DNA; thus, it is species (or, strain) specific RAPD-derived probe.

(2,4-D). The RAPD fingerprints revealed no differences between the soil community before and after treatment with 2,4-D. However, a significant change in the culturable degrading microorganism numbers was measured when the MPN technique was used. This indicates that perhaps, even though the number of 2,4-D degrading microorganisms increased dramatically after the application of 2,4-D, the majority of the non-2,4-D degrading microorganisms did not and were unaffected by the addition of 2,4-D. The RAPD technique may be useful in monitoring significant changes in community structure due to the presence of stressors.

3.3. Denaturing Gradient Gel Electrophoresis (DGGE)

Denaturing gradient gel electrophoresis (DGGE) is another method that can be used to gain an overall profile of microbial community diversity. DGGE was first developed to detect single base substitutions, deletions and insertions in cloned and genomic DNA sequences (Fischer and Lerman, 1979; Lerman et al., 1984). A DNA sequence has high- and low-temperature melting domains, and when a specific sequence encounters a temperature or denaturant gradient, the low-temperature melting domain will become single-stranded at a specific temperature or denaturant concentration (Fischer and Lerman, 1983). This is very sequence dependent; and a single base-pair substitution in the low temperature melting domain will often alter the melting temperature at which the DNA will become single-stranded.

Polyacrylamide gels are used consisting of a linear gradient of urea and formamide. In parallel DGGE, DNA molecules migrate into increasing concentrations of urea and formamide in a polyacrylamide gel. At the concentration of urea and formamide that causes the lowest melting domain to melt, the DNA molecule will change from helical to a partially melted form (Myers et al., 1987). The mobility of this partially melted DNA molecule will be greatly retarded to that of the helical molecule, thus the molecules will form discrete bands at different positions in the gel according to their sequences.

Muyzer et al., (1993) were the first to report the application of the DGGE method to profile the genetic diversity of a microbial population. The authors used the fact that 16S rRNA genes are present in all eubacteria with analogues present in the archaebacteria. Fragments of similar size of amplified regions from the V3 region of 16S rRNA gene from genomic DNA that had been extracted from uncharacterized mixtures of microorganisms, were resolved by DGGE. The observed banding patterns provided a profile of the populations under study. The authors speculated that the relative intensity of each band and its position most likely represented the relative abundance of a particular species in the population. The great advantage of the DGGE technique is that it gave the authors an immediate display of the constituents of a population in both a qualitative and a semiquantitative way.

Separation patterns can be transferred to hybridization membranes and probed with species- or group-specific oligonucleotides. In this way one can determine whether a particular species or a number of different species belonging to a certain group is present (Muyzer et al., 1993). Alternatively, sequences can be obtained and compared by excising the DNA fragments from the DGGE gel, reamplifying and sequencing directly, without the PCR product being cloned and thus eliminating cloning biases (Ferris et al., 1996). DGGE can also be used to determine phylogenetic relationships of species in communities (Muyzer et al., 1995). Wawer and Muyzer (1995) used the expression of the [NiFe] hydrogenase gene to determine the distribution and metabolic activities of *Desulfovibrio* species in environmental samples. They applied reverse transcription to convert mRNA into DNA which was then followed by PCR. In a similar procedure, Teske et al., (1996) examined the distribution of sulfate-reducing bacteria in a marine environment by DGGE analysis of

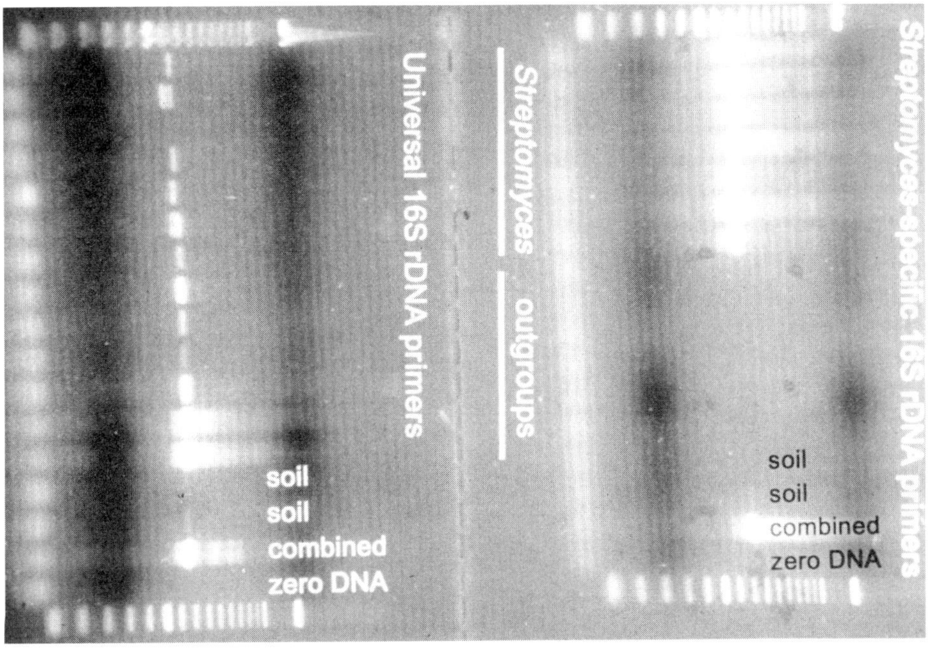

Figure 3. Agarose gel electrophoresis analyses of 16S rDNA fragments obtained after PCR amplification with *Streptomyces*-specific primers, giving an 800 bp product (top half of figure), and with non-specific (universal) primers (bottom half of figure) of different *Streptomyces* strains and eubacteria outgroups.

PCR-amplified 16S rRNA genes from total sample DNA. Simultaneously, the activity of sulfate-reducing bacteria in the environmental sample was determined by DGGE analysis of PCR-amplified total rRNA.

In our laboratory, *Streptomyces*-specific primers for DGGE were developed by comparing 16S rDNA sequences from various *Streptomyces* species and determining conserved regions. Two sequences were thus tested for their specificity and for their ability to serve as PCR primers. Figure 3 shows that a PCR product of 800 bp was only produced from *Streptomyces* DNA, and not from outgroup DNA.

4. 2,4,6-TRINITROTOLUENE BIOREMEDIATION

Soils contaminated with munition compounds such as 2,4,6-trinitrotoluene (TNT) have traditionally been incinerated, though this is a very costly procedure (Roberts et al., 1993). Microbial remediation is a feasible alternative for the cleaning of nitroaromatic-contaminated sites (Funk et al., 1993; Funk et al., 1995). It has been established that nitroaromatic compounds are transformed by anaerobic bacteria. Gorontzy et al., (1993) observed the transformation of several mono- and dinitroaromatic compounds by pure cultures of methanogenic bacteria, sulfate-reducing bacteria and clostridia. While nitro-group reduction strongly correlated with the growth phase of a sulfate-reducing bacterium and with *Clostridium* species, transformation of nitroaromatic compounds by methanogenic bacteria only occurred while the cells lysed. The bacteria also seemed to be sensitive to the addition of the nitroaromatic compound when in the lag phase.

A bench-top methanogenic bioreactor has been maintained in our laboratory for several years under strictly anaerobic conditions with a mixture of munition compounds, primarily TNT and RDX, extracted from contaminated soil. Strains of *Clostridium bifermentans* have repeatedly been isolated as one of the main TNT-degrading bacteria in the bioreactor (Regan and Crawford, 1994; Shin and Crawford, 1995). *Clostridium bifermentans* strain CYS-1 reductively transforms TNT in a co-metabolic process (Shin and Crawford, 1995). The initial reductive reactions of degradation of TNT by this *Clostridium* have recently been elucidated (Shin and Crawford 1996, submitted for publication). The appearance of *Clostridium bifermentans* strains as the dominant degrading species in the bioreactor points to the acclimation of these strains to the presence of nitroaromatic compounds under strong selective pressure over a long period of time.

Sulfate-reducing bacteria have been isolated from the bioreactor. Methanogens are also present in the bioreactor; however, methanogenesis was not detected when the bioreactor was first set up. This indicates that perhaps the microbial populations within the bioreactor have fluctuated in response to the presence of contaminating nitroaromatic compounds. We are interested, therefore, in monitoring the population dynamics within the bioreactor during bioremediation.

The DGGE technique has been used to profile the microbial community from the munitions-fed bioreactor. PCR primers from conserved regions of the 16S rRNA gene were used to amplify 16S rDNA fragments from total bioreactor DNA and from various isolated and culture collection strains. Figure 4 shows a DGGE separation pattern of 600

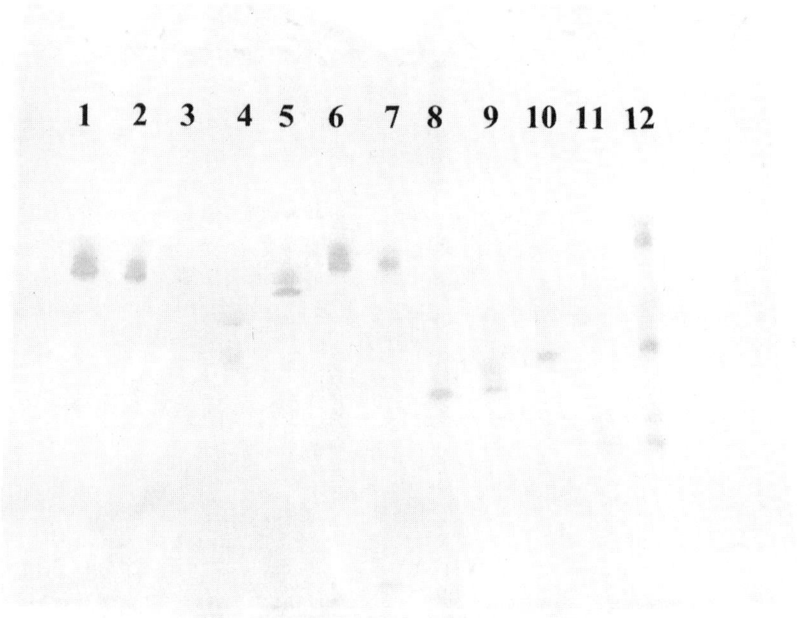

Figure 4. Negative image of a DGGE profile of a mixture of approximately 600 bp 16S rDNA fragments from various *Clostridium* strains separated on a 3 to 7 M urea gradient, polyacrylamide gel. Lane 1, *Clostridium bifermentans* strain KMR; lane 2, *C. bifermentans* strain SBF; lane 3, *C. bifermentans* ATCC 638; lane 4, *C. sordelli* ATCC 9714; lane 5, *C. sporogenes* ATCC 11437; lane 6, *C. acetobutylicum* NCIB 8052; lane 7, *C.* species (bioreactor isolate); lane 8, bioreactor DNA; lane 9, *Bacillus subtilus* BR151 ATCC 33677; lane 10, *Lactobacillus casei* ATCC 393; lane 11, *Escherichia coli* BW545; lane 12, mixture of *C. bifermentans* ATCC 638, *C. sordelli* ATCC 9714, *C. acetobutylicum* NCIB 8052, *E. coli* BW545, *L. casei* ATCC 393.

bp (approximate size) 16S rDNA fragments from various *Clostridium* strains. The fragments were separated on a polyacrylamide gel containing a urea gradient from 3 to 7 M after electrophoresis for 3.5 hr. Two *Clostridium bifermentans* strains isolated from the bioreactor could be resolved from the culture collection strains *C. sordelli* ATCC 9714, *C. sporogenes* ATCC 11437, and *C. acetobutylicum* NCIB 8052, but not from *C. bifermentans* ATCC 638 (Figure 4). A narrower urea gradient may improve the resolution and indicate sequence differences in 16S rDNA between the *C. bifermentans* bioreactor strains and the *C. bifermentans* culture collection strain. As observed in the DGGE profile (Figure 4), the *Clostridium* strains could be separated from other Gram positive bacteria (*Bacillus subtilus* BR151 ATCC 33677 and *Lactobacillus casei* ATCC 393) and from *Escherichia coli* BW545. A mixture of 16S rDNA fragments applied to one lane on the polyacrylamide gel could also be resolved.

Sulfate-reducing bacteria from culture collections were also analyzed by DGGE (Figure 5), by separating 600 bp (approximate size) 16S rDNA fragments on a polyacrylamide gel containing a urea gradient from 3 to 7 M. The various sulfate-reducing bacteria could be resolved from each other, with the exception of *Desulfovibrio gigas* NCIB 9332 and *Desulfotomaculum orientis* NCIB 8445 and NCIB 8382, and from the *Clostridium* strains. None of the fragments from the SRB culture collection strains, however, matched with any of the fragments appearing in the SRB enrichment culture samples or from total

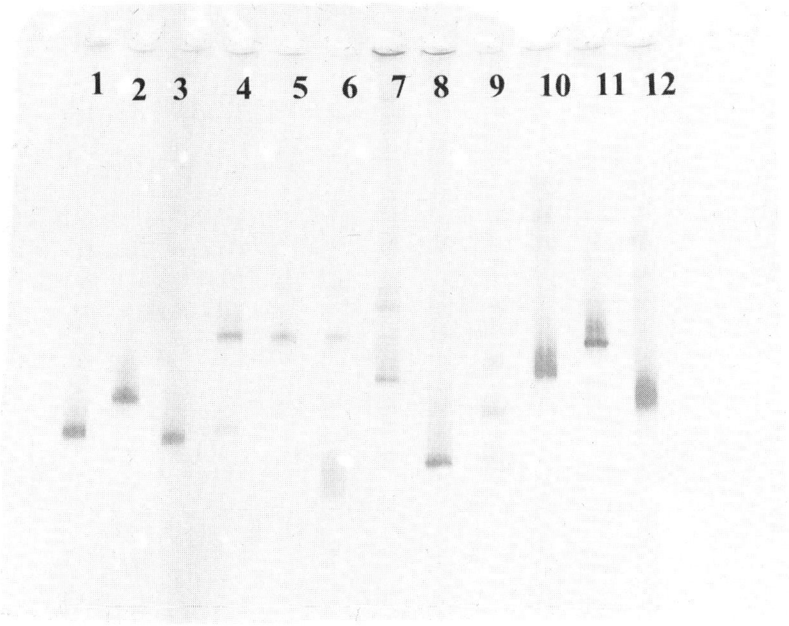

Figure 5. Negative image of a DGGE profile of a mixture of approximately 400 bp 16S rDNA fragments from sulfate-reducing bacteria (SRB) separated on a 3 to 7 M urea gradient, polyacrylamide gel. Lane 1, *Desulfovibrio desulfuricans* Beijerinck ATCC 7757; lane 2 *D. desulfuricans* MB ATCC 27774; lane 3, *D. vulgaris* subsp. *vulgaris* ATCC 29579; lane 4, *D. gigas* NCIB 9332; lane 5, *Desulfotomaculum orientis* NCIB 8445; lane 6, *Desulfotomaculum orientis* NCIB 8382; lane 7, SRB enrichment culture from bioreactor; lane 8, SRB enrichment culture from bioreactor (using defined medium); lane 9, bioreactor DNA; lane 10, *Clostridium bifermentans* strain KMR; lane 11, *C. sporogenes* ATCC 11437; lane 12, *C. sordelli* ATCC 9714.

TNT-bioreactor DNA. DNA from a sulfate-reducing bacterium isolated from the bioreactor did not migrate to the same position in the gel with any of the culture collection strains (data not shown).

5. CONCLUSIONS

The monitoring of bacterial numbers, activities, and population dynamics is an important part of understanding and optimizing any bioremediation process. The continued use of classical microbiological, enzymological, and biochemical techniques for enumeration and measuring microbial activity will an remain important part of monitoring bioremediations, as it provides excellent, needed quantitative data. However, the added use of molecular biology techniques for the study of the microbial populations as they develop during bioremediation is now providing much enhanced qualitative data. The RAPD and DGGE methods can, for example, give an overview of the microbial community structure under study at different genus and species levels, and they can do so in a relatively short period of time with a high degree of accuracy and sensitivity. Of course, all of the techniques described in this paper have inherent limitations, and one technique should not be relied on solely to monitor microbial numbers, population dynamics, and/or activities. Instead, by using of a mixture of classical microbiology techniques and molecular biology techniques, weaknesses of a given technique will be compensated for, and a much better overall picture of the bioremediation process will be obtained.

ACKNOWLEDGMENTS

This research was supported in part by USEPA grant number R-819772–01–0 and by the Idaho Agricultural Experiment Station.

REFERENCES

Atlas, R.M., and Bartha, R., 1987. *Microbial Ecology: Fundamentals and Applications*, 2nd ed., Benjamin/Cummings Publishing Company, Inc., Menlo Park, pp. 195–232.

Bej, A.K., and Mahbubani, M.H., 1994. Applications of the polymerase chain reaction (PCR) in vitro DNA-amplification method in environmental microbiology, in: *PCR Technology: Current Innovations* (H.G. Griffin, and A.M. Griffin, eds.), CRC Press, Boca Raton, pp.327–339.

Bouwer, E.J., 1992. Bioremediation of organic contaminants in the subsurface, in: *Environmental Microbiology* (R. Mitchell, ed.), Wiley-Liss Inc., New York, pp.287–318.

Boyle, M., 1992. The importance of genetic exchange in degradation of xenobiotic chemicals, in: *Environmental Microbiology* (R. Mitchell, ed.), Wiley-Liss Inc., New York, pp. 319-333.

Fani, R., Damiani, G., Di Serio, C., Gallori, E., Grifoni, A, and Bazzicalupo, M., 1993. Use of random amplified polymorphic DNA probes for microorganisms, *Molecular Ecology* 2:243–250.

Farrelly, V., Rainey, F.A., and Stackebrandt, E., 1995. Effect of genome size and *rrn* gene copy number on PCR amplification of 16S rRNA genes from a mixture of bacterial species, *Appl. Environ. Microbiol.* 61:2798–2801.

Fedorak, P.M., Foght, J.M., and Westlake, W.S., 1982. A method for monitoring mineralization of ^{14}C-labeled compounds in aqueous samples, *Water Res.* 16:1285–1290.

Ferris, M.J., Muyzer, G., and Ward, D.M., 1996. Denaturing gradient gel electrophoresis profiles of 16S rRNA-defined populations inhabiting a hot spring microbial mat community, *Appl. Environ. Microbiol.* 62:340–346.

Fischer, S.G., and Lerman, L.S., 1979. Length-independent separation of DNA restriction fragments in two-dimensional gel electrophoresis, *Cell.* 16:191–200.

Fischer, S.G., and Lerman, L.S., 1983. DNA fragments differing by single base-pair substitutions are separated in denaturing gradient gels: Correspondence with melting theory, *Proc. Natl. Acad. Sci. USA*, 80:1579–1583.

Forsyth, J.V., Tsao, Y.M., and Bleam, R.D., 1995. Bioremediation: When is augmentation needed?, in: *Bioaugmentation for Site Remediation* (R.E. Hinchee, J. Fredrickson and B.C. Alleman, eds.), Battelle Press, Columbus, pp. 1–14.

Funk, S.B., Crawford, D.L., and Crawford, R.L., 1995. Bioremediation of nitroaromatic explosives contaminated soils and waters, in: *Bioremediation: Principles and Applications* (R.L. Crawford and D.L. Crawford, eds.), in press.

Funk, S.B., Roberts, D.J., Crawford, D.L., and Crawford, R.L., 1993. Initial-phase optimization for bioremediation of munition compound-contaminated soils, *Appl. Environ. Microbiol.* 59:2173–2177.

Giovannoni, S.J., DeLong, E.F., Olsen, G.J., and Pace, N.R., 1988. Phylogenetic group-specific oligonucleotide probes for identification of single microbial cells, *J. Bacteriol.* 170:720-726.

Gorontzy, T., Kuver, J., and Blotevogel, K-H., 1993. Microbial transformation of nitroaromatic compounds under anaerobic conditions, *J. Gen. Microbiol.* 139:1331–1336.

Hadrys, H., Balick, M., and Schierwater, B., 1992. Applications of random amplified polymorphic DNA (RAPD) in molecular ecology, *Molecular Ecology* 1:55–63.

Holben, L.E., Jansson, J.K., Chelm, B.K., and Tiedje, J.M., 1988. DNA probe method for the detection of specific microorganisms in the soil bacterial community, *Appl. Environ. Microbiol.* 54:703–711.

Kaplan, L.A., Bott, T.L., and Bielick, J.K., 1992. Assessment of [^{3}H]thymidine incorporation into DNA as a method to determine bacterial productivity in stream bed sediments, *Appl. Environ. Microbiol.* 58:3614–3621.

Karns, J.S., Kilbane, J.J., Catterjee, D.K., and Chakrabarty, A.M., 1984. Microbial biodegradation of 2,4,5-trichlorophenoxyacetic acid and chlorophenols, in: *Genetic Control of Environmental Pollutants. Basic Life Sciences*, Volume 28 (G.S. Omenn, and A. Hollaender, eds.), Plenum Press, New York, pp. 3–21.

Lane, D.J., Pace, B., Olsen, G.J., Stahl, D.A., Sogin, M.L., and Pace, N.R., 1985. Rapid determination of 16S ribosomal RNA sequences for phylogenetic analyses, *Proc. Natl. Acad. Sci. USA* 82:6955–6959.

Lappin, H.M., Greaves, M.P., and Slater, J.H, 1985. Degradation of the herbicide Mecoprop [2-(2-methyl-4-chlorophenoxy)propionic acid] by a synergistic microbial community, *Appl. Environ. Microbiol.* 49:429–433.

Lerman, L.S., Fischer, S.G., Hurley, I., Silverstein, K., and Lumelsky, N., 1984. Sequence-determined DNA separations, *Ann. Rev. Biophys. Bioeng.* 13:399–423.

Liesack, W., Weyland, H., and Stackebrandt, E., 1991. Potential risks of gene amplification by PCR as determined by 16S rDNA analysis of a mixed-culture of strict barophilic bacteria, *Microb. Ecol.* 21:191–198.

Muyzer, G., Teske, A., Wirsen, C.O., and Jannasch, H.W., 1995. Phylogenetic relationships of *Thiomicrospira* species and their identification in deep-sea hydrothermal vent samples by denaturing gradient gel electrophoresis of 16S rDNA fragments, *Arch. Microbiol.* 164:165–172.

Muyzer, M., De Waal, E., and Uitterlinde, A.G., 1993. Profiling of complex microbial populations by denaturing gradient gel electrophoresis of polymerase chain reaction-amplified genes coding for 16S rRNA, *Appl. Environ. Microbiol.* 59:695–700.

Myers, R.M., Maniatis, T., and Lerman, L.S., 1987. Detection and localization of single base changes by denaturing gradient gel electrophoresis, *Methods Enzymol.* 155:501–527.

Neilson, J.W., Josephson, K.L., Pillai, S.D., and Pepper, I.L., 1992. Polymerase chain reaction and gene probe detection of the 2,4-dichlorophenoxyacetic acid degradation plasmid, pJP4, *Appl. Environ. Microbiol.* 58:1271–1275.

Ogram, A.V., and Sayler, G.S., 1988. The use of gene probes in the rapid analysis of natural microbial communities, *J. Indust. Microbiol.* 3:281–292.

Pace, N.R., Stahl, D.A., Lane, D.J., and Olsen, G.J., 1986. The analysis of natural microbial populations by ribosomal RNA sequences, *Adv. Micro. Ecol.* 9:1–55.

Picard, C., Ponsonnet, C., Paget, E., Nesme, X., and Simonet, P., 1992. Detection and enumeration of bacteria in soil by direct DNA extraction and polymerase chain reaction, *Appl. Environ. Microbiol.* 58:2717–2722.

Regan, K.M., and Crawford, R.L., 1994. Characterization of *Clostridium bifermentans* and its biotransformation of 2,4,6-trinitrotoluene (TNT) and 1,3,4-triaza-1,3,5-trinitrocyclohexane (RDX), *Biotech. Letters* 16:1081–1086.

Roberts, D.J., Kaake, R.H., Funk, S.B., Crawford, D.L., and Crawford, R.L., 1993. Field-scale anaerobic bioremediation of dinoseb-contaminated soils, in: *Biotreatment of industrial and hazardous wastes* (Levin, M.A., and Gealt, M.A., eds.), McGraw-Hill, New York, pp. 219–244.

Sayler, G.S., Shields, M.S., Tedford, E.T., Breen, A., Hooper, S.W., Sirotkin, K.M., and Davis, J.W., 1985. Application of DNA:DNA colony hybridization to the detection of catabolic genotypes in environmental samples, *Appl. Environ. Microbiol.* 49:1295

Shin, C.Y., and Crawford, D.L., 1995. Biodegradation of trinitrotoluene (TNT) by a strain of *Clostridium bifermentans*, in: *Bioaugmentation for Site Remediation* (R.E. Hinchee, J. Fredrickson and B.C. Alleman, eds.), Battelle Press, Columbus, pp. 57–69.

Spain, J.C., and Van Veld, P.A., 1983. Adaptation of natural communities to degradation of xenobiotic compounds: effect of concentration, exposure time and inoculum and chemical structure, *Appl. Environ. Microbiol.* 45:428–435.

Teske, A., Wawer, C., Muyzer, G., and Ramsing, N.B., 1996. Distribution of sulfate-reducing bacteria in a stratified fjord (Mariager Fjord, Denmark) as evaluated by most-probable number counts and denaturing gradient gel electrophoresis of PCR-amplified ribosomal DNA fragments, *Appl. Environ. Microbiol.* 62:1405–1415.

Tibbles, B.J., and Harris, J.M., 1996. Use of radiolabelled thymidine and leucine to estimate bacterial production in soils from continental Antarctica, *Appl. Environ. Microbiol.* 62:694–701.

Tiedje, J.M., 1993. Bioremediation from an ecological perspective, in: *In situ Bioremediation: When does it Work?* National Research Council Committee on In situ Bioremediation, Water Science and Technology Board, Commission on Engineering and Technical Systems, National Academy Press, Washington, D.C., pp. 110–120.

Tsai, Y., and Olsen, B.H., 1992. Detection of low numbers of bacterial cells in soils and sediments by polymerase chain reaction, *Appl. Environ. Microbiol.* 58:754–757.

Tsai, Y., Park, M.J., and Olsen, B.H., 1991. Rapid method for direct extraction of mRNA from seeded soils, *Appl. Environ. Microbiol.* 57:765–768.

Ward, D.M, Weller, R., and Bateson, M.M., 1990. 16S rRNA sequences reveal numerous uncultured microorganisms in a natural community, *Nature* 343:63–65.

Wawer, C., and Muyzer, G., 1995. Genetic diversity of *Desulfovibrio* spp. in environmental samples analyzed by denaturing gradient gel electrophoresis of [NiFe] hydrogenase gene fragments, *Appl. Environ. Microbiol.* 61:2203–2210.

Williams, J.G.K., Kubelik, A.R., Livak, K.J., Rafalski, J.A., and Tingey, S.V., 1990. DNA polymorphisms amplified by arbitrary primers are useful as genetic markers, *Nucleic Acids Res.* 18:6531–6535.

Xia, X., Bollinger, J., and Ogram, A, 1995, Molecular genetic analysis of the response of three soil microbial communities to the application of 2,4-D, *Molecular Ecology*, 4:17–28.

Yuan, W.M., and Crawford, D.L., 1995, Characterization of *Streptomyces lydicus* WYEC108 as a potential biocontrol agent against fungal root and seed rots, *Appl. Environ. Microbiol.* 61:3119–3128.

13

BIOLOGICAL TREATMENT OF AIR POLLUTANTS

Hinrich L. Bohn

Bohn Biofilter Corp., POB 44345
Tucson, Arizona 85733-4235

1. INTRODUCTION

Reducing pollutant production at the source is probably the best method of pollution control. If this is incomplete, gases must be treated at the end-of-pipe to prevent release to the atmosphere. The treatment is usually oxidation; volatile organic compounds (VOCs) are oxidized to CO_2, H_2S and SO_2 to sulfate, and NO_x to nitrate. These reactions are spontaneous; the pollutant states are chemically unstable in air. The reaction rates are too slow in the atmosphere, however, to prevent environmental and health hazards and nuisances.

Thermal, conventional catalytic, and adsorption (physico-chemical) methods of air pollution control (APC) are wasteful and create secondary pollution. Pollutant concentrations in industrial emissions, for example, are of the order of 100 parts per million by volume (ppmv). To burn (oxidize) these gases in an incinerator, we have to add at least 50,000 ppmv of methane, i.e., add 500+ parts of methane to burn 1 part VOC. Sometimes the combustion energy is utilized, but usually the energy is released to the atmosphere along with the secondary pollutants—CO, NO_x, and any partially oxidized VOCs—created in the flame. If a gas burns in air or is chemically oxidizable, it will also oxidize biologically. The reactions are the same:

$$VOC + O_2 \longrightarrow CO_2 + H_2O \tag{1}$$

Such gases include almost all air pollutants except tri- and higher-halogenated hydrocarbons which oxidize too slowly biochemically to be feasible under most conditions. An incinerator requires high temperatures to carry out the reaction; a scrubber requires strong oxidants (hypochlorite, permanganate); a bioreactor requires microbes, water, and the other nutrients that sustain life. The bioreactions require no fuel or chemicals, create no secondary pollutants, and the reaction rates increase with pollutant emission rates. The microbes carry out reaction presented by Equation 1 for their metabolic energy.

Although the reaction shown by Equation 1 occurs on leaf, soil and water surfaces in nature and controls the natural concentrations of carbon, nitrogen, and sulfur gases in the

Biotechnology in the Sustainable Environment, edited by Sayler *et al.*
Plenum Press, New York, 1997

atmosphere, this reaction is too slow to rapidly remove the higher and localized concentrations of anthropogenic air emissions. The reaction is catalyzed by microbial enzymes.

One way to increase these natural VOC oxidation rates is to increase the surface of contact, by growing far more plants and blowing soil into the atmosphere. Far better would be to carry out the reactions before the contaminated air is released to the atmosphere.

Treating air pollutants before release is possible and feasible in soils, and is called biofiltration. The reaction shown in Equation 1 requires that VOCs interact closely with degradative microbes, oxygen, water, and microbial nutrients. The countless tiny pores of soils and soil-like media provide this interaction for organic gases. Additional catalysis by Fe and Mn oxide, aluminosilicate surfaces and acid neutralization oxidizes inorganic gases.

Soil treatment of waste gases before release is a very old but poorly recognized technology. Every home beyond a municipal sewer line treats the odorous VOCs associated with waste water by a small biofilter. More than 25 million such biofilters operate continuously in North America alone. The VOC treatment is so effective that it is taken for granted.

Those familiar with biological waste water treatment will recognize its similarity to biofiltration. The reaction in Equation 1 occurs when organic compounds, microbes, water, oxygen, and nutrients are present whether a pollutant is air- or water-borne. Biofilter rates tend to be faster and more complete than waste water treatment because biofiltration is oxygen-rich, mixing is complete, and the organic loading rate is low. VOC concentrations in air are characteristically a thousand-fold lower than water pollutant concentrations, on a mass and volume basis, and reactions are faster because the pollutant is molecularly-dispersed. As a rule of thumb, a reaction that requires days in the atmosphere and hours in water treatment requires minutes in biofilters.

The reaction, however, must happen between 1–55°C (33–130°F), the optimum is 37°C (97°F). Few microbes are active above 55°C and microbial activity stops when water freezes. This temperature range is low compared to the 700–800°C range of incinerators. So biological reaction rates, despite the catalysis of microbial enzymes, are minutes to hours as opposed to fractions of a second in incinerators. The volume of a bioreactor therefore must be much larger than that of an incinerator in order to remove/destroy pollutant gases.

Second, bio-oxidation rates vary with chemical composition because enzyme catalysis is compound-specific (Table 1). Generally speaking, organic gases containing O, N, and S ions oxidize rapidly; hydrocarbons oxidize more slowly; halogenated hydrocarbons biodegrade much more slowly if at all. The reactions do not interact or interfere except at high (>1000 ppmv) concentrations (Deschusses, 1994). The presence of hydrocarbons or halocarbons, for example, does not repress biofiltration of alcohols, ketones, or aldehydes. The microbes consume what they like first and let the rest pass by.

Table 1. Biodegradability and chemical reactivity of gases in biofilters

- Rapidly biodegradable
  Alcohols, glycols, ketones, aldehydes, ethers, organic acids, esters, organo-nitrogen, -sulfur, and -phosphate compounds.
- Rapidly chemically-reactive
  H_2S, SO_2, NO_X, PH_3, P oxides, silanes, HCl, HF.
- Slowly biodegradable (in approx. decreasing order)
  Phenols, acrylates, terpenes, aromatic hydrocarbons, CO, aliphatic hydrocarbons, vinyl chloride, methylene chloride.
- Very slowly biodegradable (unfeasible for biofiltration)
  Tri- and quadri-halogenated hydrocarbons, polyaromatic hydrocarbons.

Third, microbes require moisture for activity so the input air should be at 100% relative humidity to prevent the bed from drying out. The 100% relative humidity in air corresponds to soil moisture at −1/3 bar hydraulic head—the moisture content of wet and freely-draining soil. Microbial activity slows as the soil dries but recovers rapidly when moisture is added.

Fourth, biological APC control has been troubled by psychology, ignorance, inertia, and marketing. Physicochemical techniques are chemically simple, mechanically complex and can be, indeed must be, controlled and monitored carefully. Their piping and controls are dazzling and control is egocentrically appealing.

Biological methods have little or no glamour. Biology has many variables, many of which are difficult to measure and in some cases unknown, so design must allow for contingencies. Control of biological methods is less precise but that does not mean that biological reactions are out of control. The complexity of biological methods is in their biochemistry, but this is of little appeal to chemical engineers who have been given the responsibility of air pollution control. Biological reactions are mostly positive feedback, but employing them requires a trust in nature that many engineers lack.

Although well accepted for waste water, using dirt and bugs to clean air seems counterintuitive to many engineers. For example, the exhaust air from computer chip manufacture is readily treatable by biological means. But after going to enormous trouble to create clean rooms to make computer chips, using dirt to clean the waste air seems absurd to the industry.

The large size of biological reactors is counter to the current notion that space optimization is important, even more important than cost optimization and resource utilization. Layering and other configurations reduce biofilter size, but at the expense of installation cost and maintenance requirements.

The physicochemical APC industry is well entrenched. Introducing and marketing an apparently new technology that is low cost and has a low profit margin is difficult and slow.

2. PHYSICOCHEMICAL METHODS OF AIR POLLUTION CONTROL

Understanding the role of biological APC requires some knowledge of physicochemical APC. Incineration (thermal oxidation) uses high temperatures and O_2 to oxidize organic gases to CO_2 and water. Incineration consumes a lot of fuel to burn a little air pollution. The lowest methane concentration that will burn is 50,000 ppmv (20,000 ppmv for propane) so considerable fuel must be added. Fuel is a major expense, about $10/cfm-yr (cfm/yr is cubic feet per minute per year) of air to be treated. To minimize costs the fuel content has to be as "lean" as possible, but lean flames tend to blow out and require frequent maintenance. Catalysts, recuperative and regenerative thermal oxidation, cogeneration, and other incinerator modifications reduce the fuel requirement but increase capital and maintenance costs.

Incineration can destroy 99.9+% of the VOCs in the air, but it creates NO_x, carbon monoxide, and sometimes smoke, odors, dioxins, and other undesirable byproducts. These bad associations have made incineration a dirty word. The industry prefers thermal oxidation.

If incineration includes heat recovery, it comes closer to a sustainable system. If contaminated air were the feed air of a community heating plant, power station, dryer, or cement kiln, the pollutants would usefully contribute their energy and the added fuel would not be wasted.

Scrubbing washes pollutants out of the air with water sprays and sometimes oxidizes them to CO_2 and water with the strong oxidants chlorine, ozone, and permanganate. Scrubbing sometimes only transfers the pollutants to water, generates large amounts of waste water, and the unreacted oxidants are hazardous. If chlorine and ozone are added in excess or if the reaction time is insufficient, they are released as air pollutants. Spent permanganate is a hazardous waste. A related process removes NO_x in stack gases by injecting ammonia to produce N_2. Any excess ammonia is released as an air pollutant.

Adsorbents such as activated carbon and zeolite concentrate pollutants for later recovery, disposal, or destruction. Recovering the pollutant for reuse and regenerating the sorbent is involved, is a fire hazard with activated carbon, the pollutants require purification before reuse, and is worthwhile only if the pollutant is valuable. The sorption capacity decreases with each regeneration. Instead the spent adsorbent is usually sent to a landfill or is incinerated. Water vapor in the air stream and high temperatures interferes with gas adsorption and gas removal decreases as the sorbent nears saturation. One modification is to link the adsorbent to a regeneration-incineration unit to intermittently remove and destroy the pollutant and regenerate the adsorbent.

Condensation removes gases by cooling the gases so that the pollutants condense out as liquids. The technique requires considerable energy, removes water as well as the VOCs, and allows considerable pollution to pass through untouched.

High stacks are less favored now than heretofore, when "the solution to pollution is dilution," was acceptable. High stacks can dilute gases to below harmful concentrations but, because oxidation is usually first order kinetically, dilution slows degradation. Air currents do not always cooperate so the exhaust is sometimes inadequately diluted before it creates harm.

Masking and counteractants are the industrial equivalent of perfumes and room fresheners. They add chemicals to cover the odors of other chemicals. Since the net effect is to increase the organic loading in the air, these methods are discouraged by regulatory agencies. Counteractants are claimed to form droplets or aerosols that adsorb and react with air pollutants. Their sorption capacities are low. The descriptions of the reaction mechanisms seem deliberately misleading and hide the large amounts that must be added to remove a little air pollutant.

Ozone injection can oxidize air pollutants as O_3 reverts to O_2. Maintaining the ozone generation equipment and matching ozone production to the VOC requirement have caused difficulties.

In summary, physicochemical APC methods hit air pollutants hard and fast, they have stainless steel and computers, they afford many opportunities for tinkering, they are shiny and brightly painted, and they can be attached easily to a waste air stream or stack. They have glamour and pizzazz.

3. BIOLOGICAL METHODS OF AIR POLLUTION CONTROL

The biological methods—biofilters, bioscrubbers, and bio-trickle filters—are compared schematically in Figure 1.

3.1. Biofilters

Biofiltration is the oxidation of air pollutants as contaminated air flows slowly through porous media—soils, compost, peat, activated carbon, plastic packing material, ceramic packing material—that support a degradative microbial population attached to the

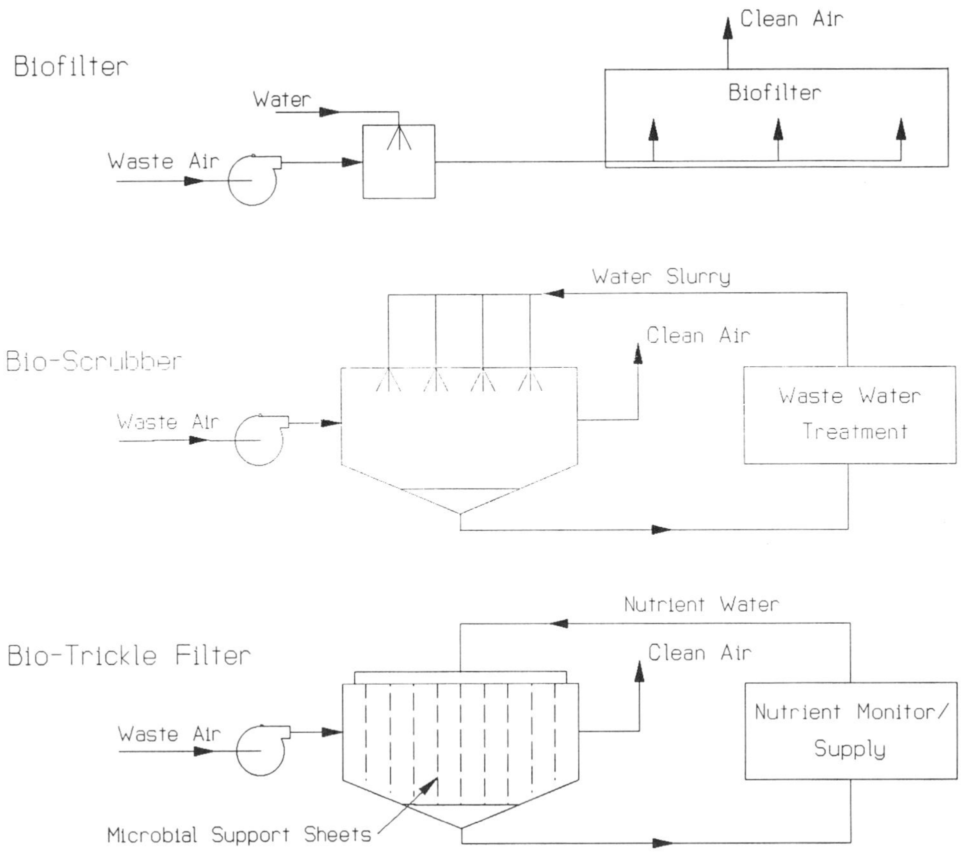

Figure 1. Schematic drawings of a biofilter, bioscrubber, and bio-trickle filter.

walls of the pores. Air pollutants adsorb on the pore walls and flow more slowly than the air. Given adequate reaction time, microbes oxidize the VOCs to CO_2 and water without any additional fuel or chemicals. The high oxygen concentration and the intimate mixing of oxygen with the VOCs ensure complete oxidation, without the noisome byproducts associated with biological waste water treatment.

Biofiltration does not pollute the soil because soils can not accumulate gases or their metabolic products to any significant extent. Organic gases are instead converted to CO_2 and water. Inorganic gases are converted to their calcium oxy salts, e.g.,SO_2 and $H_2S \rightarrow CaSO_4$, $NO_x \rightarrow Ca(NO_3)_2$.

The porosity of soils is 40–50% and is higher for the other materials. Air permeability depends on pore size rather than porosity because flow increases exponentially with pore size. Since soil pores tend to be small, the 1 6 inches water head (2 15 centimeters or 0–150 Pa) backpressure of soil beds tends to be higher than for the other media.

The various media have advantages and disadvantages. Soil is inexpensive, neutralizes acids well, is hydrophilic so moisture content is easier to control, has high bearing strength so deep beds are possible, and is permanent. Soil requires the most area and is heavy.

Compost is inexpensive, lighter than soil, and requires less area. Compost neutralizes acids poorly, is hydrophobic when dry, is prone to compaction, and has to be replaced every 2–5 years.

The inorganic media—plastic packing materials, ceramics, and activated carbon—offer promise of smaller size because uniform air flow is easier to attain. The synthetics, however, are only in the pilot stage of development, are many times more expensive, the microbial populations have to be introduced, sloughing of the microbes and pore plugging have occurred. Activated carbon is hydrophilic and has to be replaced every 4–5 years. Biofilters are damp so activated carbon should not be a fire hazard.

At usual VOC loading rates, the reaction shown in Equation 1 in biofilters is first order kinetics:

$$\frac{C_{output}}{C_{input}} = DRE(\%) = 100\left(1 - e^{-kt}\right) \quad (2)$$

where C is VOC concentration, DRE is the destruction removal efficiency, k is the reaction rate coefficient, and t is the residence time of the air in the biofilter. Over a wide range of loading rates, the oxidation rate varies with the loading rate but DRE remains constant. Approximate reaction rate coefficients are known for many VOCs, (Table 1), and can often be inferred for others based on the their chemical composition and molecular structure. Any degree of destruction removal efficiency can be attained by adjusting the residence time.

Overcoming the size disadvantage of biofilters means increasing the VOC loading rate. This can change the Equation 1 reaction to approach zero order kinetics where the oxidation rate is constant, regardless of input rate. This is the maximum rate of VOC oxidation, but the biofilter is inflexible. The rate does not respond to input changes, DRE varies, and VOC degradation rates may interact.

Biofiltration requires only a fan and some means of maintaining a high moisture content in the biofilter bed. This mechanical simplicity has been a handicap. It has encouraged many to design biofilters without understanding that the complexity of biofiltration is in providing long term microbial requirements and that the air and water flow patterns in porous natural media may change with time. Many biofilters are only marginally or temporarily effective. Others are expensive because the design incorporates extra bells and whistles.

Dyer and Mulholland (1994) concluded that the most cost-effective way to reduce APC cost is to reduce air flow and to employ biofiltration. They reached this conclusion despite using cost data of expensive, modular biofilters. Because of the modules, these biofilters are costly for air flows less than 1,000 cfm (cubic feet per minute), or 1,600 cubic meters per hour. Had they investigated non-modular designs, they would have found biofilters cost-effective down to very low flow rates.

Biofiltration can be very inexpensive compared to physicochemical APC methods. Some biofilter designs are more expensive than others. "You get what you pay for," makes many suspicious of low cost methods. Comparing biofiltration to physicochemical methods that require continuous fuel and chemical input, however, is comparing apples to oranges. Biofilter designs that optimize space and use self-degrading media are more expensive than others.

3.2. Bioscrubbers

Bioscrubbers are modified water scrubbers. The spray into the gas stream is a sludge of suspended microbes which increases the aqueous solubility of the VOCs and removes VOCs more effectively than water alone. The sludge then flows to a water treatment plant to biodegrade the dissolved VOCs and generate a new sludge.

Bioscrubbers appear best suited for large air flows because of their low backpressure and small size. Their disadvantage is operating a waste water treatment plant in addition to one that destroys the VOCs, rather than just stripping VOCs from the water, generating a nonodorous sludge.

One bioscrubber was operating successfully in Stuttgart, Germany, ten years ago at an automobile painting facility. It required weekly additions of nitrogen to maintain the microbial population.

3.3. Bio-Trickle Filters

Bio-trickle filters are sheets of a plastic or other microbial support medium hung in the contaminated air stream. The sheets are bathed continuously by a recirculating stream of water containing the nutrients and vitamins required by the microbes.

Bio-trickle filters hold promise where space utilization is paramount. Bio-oxidation rates per unit volume are high so bio-trickle filters can be as small as physicochemical units. Their disadvantages are that at high loading rates, biological reactions approach zero order and are incapable of responding to "incidents" (e.g., peak loads, accidental discharges, explosions). In addition, their nutrition requires continuous monitoring of a property which is incompletely known. Bio-trickle filters would seem to be susceptible to microbially-produced bactericides and to invasion of less-desirable microbial populations. We think that the microbes are there to remove VOCs, but their interest is survival of the fittest.

4. DISCUSSION

The input air to a biological treatment system should be 10–55°C temperature, 100% relative humidity, particulate-free, and contain a reactive pollutant gas.

Biological APC is related to waste water treatment, an old technology, but is only slowly being accepted on an industrial scale for several reasons. Waste water is usually a mixture which contains a suitable mixture of nitrogen, phosphate, and the other chemical microbial growth factors. Industrial waste air, in contrast, might contain only a single carbon-hydrogen molecule. The biofilter medium must supply the other growth factors including water in order for microbes to metabolize the "empty calories" in contaminated air.

Biological waste water treatment grew out of very small scale facilities around homes and villages where cleanup by natural means was obvious. The cleanup rates were acceptable because the pollutant inflow rates were small relative to biodegradation rates. Sanitary engineers speeded these processes to treat large municipal and industrial waste water flows, in part because no alternative was available. Biological treatment has become well accepted in waste water treatment.

Biological APC, on the other hand, has few obvious small scale examples from which to expand to industrial scale. Gaseous removal rates are slow in nature compared to emission and dispersion rates to avoid odor and visual problems, even on home or farm scales. The natural removal/destruction reactions did not enter our collective consciousness in relation to industrial air emissions. Only the rapid physicochemical techniques seemed fast enough to feasibly combat industrial air pollution.

In addition, APC was taken on largely by chemical engineers who are generally unfamiliar with natural and microbial processes but are comfortable with incineration, chemical scrubbing, and adsorption. Yet we all possess an underlying awareness that soils can reduce air pollution. Removing smell by covering an odorous source with soil is instinctive for cats and humans. The source continues to emit VOCs but they are destroyed by the soil before they reach the air. Transferring this knowledge to an industrial scale has been slow.

Biological and physicochemical treatment have been advancing in opposite directions for water and air treatment. Now the ratio of biological/physicochemical treatment for water is greater than 1000:1. As physicochemical techniques improve, the ratio may decrease to as much as 100:1.

The current ratio of biological/physicochemical APC treatment of industrial gases is less than 1:1000. I believe that biological treatment of contaminated air from chemical, petroleum, food, and waste processing, and other sources could increase over the next decades to a 1:1 ratio of biological to physicochemical treatment.

For those who like buzz words, the biological methods are very-low-temperature catalytic oxidation. The catalysts are microbial enzymes which are free and self-regenerating. Biofilters, bioscrubbers, and bio-trickle filters are not panacea but have the potential to treat a large fraction of industrial air emissions effectively, economically, and sustainably.

REFERENCES

Deschusses, M.A. 1994. Ph.D. Dissertation. Swiss Federal Institute of Technology. Zürich Switzerland.

Dyer, J., and Mulholland, K., 1994. Toxic air emissions: What is the full cost to your business?,*Chem. Engr. Environ. Suppl.*, Feb., pp. 4–8.

14

ENVIRONMENTAL BIOTECHNOLOGY ISSUES IN THE FEDERAL GOVERNMENT

D. Jay Grimes

Office of Energy Research, ER-74
U.S. Department of Energy
19901 Germantown Road
Germantown, Maryland 20874-1290

1. INTRODUCTION

In 1995, the National Science and Technology Council (NSTC), a cabinet-level council chaired by the President of the United States, released two major reports that focussed on environmental technologies. One report prepared by the Environmental Technology Working Group (ETWG) described those environmental technologies that had been identified by the NSTC as the bridge to a sustainable future (ETWG, 1995). The Biotechnology Research Subcommittee (BRS) addressed biotechnology research in the second report, outlining opportunities for federal investments in four areas: (i) agricultural biotechnology, (ii) environmental biotechnology, (iii) manufacturing and bioprocessing, including energy research, and (iv) marine biotechnology and aquaculture (BRS, 1995). The environmental technologies discussed in both reports were cited for their potential to create high-quality, high-paying jobs, while improving and sustaining the environment in the United States (BRS, 1995; ETWG, 1995).

Biotechnology sales in the U.S. are expected to reach $50 billion in the next decade (BRS, 1995), and the industry is presently responsible for approximately 100,000 high-skill jobs generated by 1,300 U.S. biotechnology firms (Lee and Burrill, 1995). Clearly, biotechnology has captured a significant portion of the U.S. marketplace, and all projections point to a sustained honeymoon for this "high tech" industry. Market projections for environmental biotechnology are less definitive, but still point to expansion. Current estimates for the worldwide environmental market are $200 billion annually. Recently, the Organisation for Economic Co-Operation and Development (OECD) estimated that the world market potential for all environmental biotechnologies will reach $75 billion by the year 2000 (OECD, 1994). The United States is presently expanding its research and development (R&D) activities in the area of environmental biotechnology, and this review will discuss the issues and opportunities that face the federal agencies that support research and development in environmental biotechnology.

Biotechnology in the Sustainable Environment, edited by Sayler *et al.*
Plenum Press, New York, 1997

Table 1. Federal bioremediation research priorities

1. Develop an understanding of the structure of microbial communities and their dynamics in response to normal environmental variation and novel anthropogenic stresses.
2. Determine the biochemical mechanisms, including enzymatic pathways, involved in aerobic and particularly anaerobic degradation of pollutants.
3. Expand understanding of microbial genetics as a basis for enhancing the capabilities of micro-organisms to degrade pollutants.
4. As a standard practice, conduct microcosm/mesocosm studies of new bioremediation techniques to determine in a cost-effective manner whether they are likely to work in the field, and establish dedicated sites where long-term field research on bioremediation technologies can be conducted.
5. Develop, test, and evaluate innovative biotechnologies, such as biosensors, for monitoring bioremediation *in situ*; models for the biological processes at work in bioremediation; and reliable, uniform methods for assessing the efficacy of bioremediation technologies.

Source: BRS (1995).

Environmental biotechnology research and development is supported by several federal agencies, and all aspects of this R&D area are supported, including applications for understanding, monitoring, and management of the environment. Agencies are investing most of their environmental biotechnology resources in bioremediation research, because of the need to develop more cost-effective and environmentally benign approaches to environmental cleanup. Table 1 lists the research priorities that were identified by the participatory agencies.

Bioremediation has been projected by the National Research Council (NRC) to become a major environmental biotechnology industry, with annual U.S. sales exceeding $500 million by the year 2000 (NRC, 1993). Unfortunately, federal environmental biotechnology investments in the U.S. have not been commensurate with the potential payoffs reported by the NRC. In fiscal year (FY) 1994, the last year for which detailed biotechnology budget numbers were collected by the Office of Management and Budget (BRS, 1993), the U.S. government invested $90 million in environmental biotechnology R&D. This amount represented 2% of the total FY 1994 $4.3 billion federal investment in biotechnology research and infrastructure (BRS, 1993). Table 2 lists the agencies involved and their level of support in FY 1994 (BRS, 1993). In FY 1996, support for environmental biotechnology, especially bioremediation, R&D has become more prominent in federal agency biotechnology portfolios, and several new programs are discussed in this paper.

Table 2. Federal agency distribution of environmental biotechnology research budgets for fiscal year 1994 ($ Million)

Agency	Budget
Agency for International Development (AID)	0.2
Department of Agriculture (USDA)	2.5
Department of Commerce (DOC)	0.6
Department of Defense (DOD)	18.5
Department of Energy (DOE)	22.3
Department of Health and Human Services (DHHS)	0.5
Department of the Interior (DOI)	4.8
Environmental Protection Agency (USEPA)	20.3
National Aeronautics and Space Administration (NASA)	1.5
National Science Foundation (NSF)	19.1
Total	90.2

Source: BRS (1993).

2. FEDERAL AGENCY PROGRAMS IN BIOREMEDIATION RESEARCH AND DEVELOPMENT (BRS, 1995)

2.1. Agency for International Development

The Agency for International Development (AID), the foreign assistance arm of the U.S. government, has an active and long-standing record of work in more than 50 countries. The AID supports small individual research projects involving bioremediation of petroleum spills and heavy metal contamination of soil and water. It also assists U.S. companies in the transfer of environmental technologies, including biotechnologies, to other nations.

2.2. US Department of Agriculture

The mission of the US Department of Agriculture (USDA) biotechnology research program is to solve problems associated with agriculture through the use of the new tools of molecular biology. The goal of the USDA's environmental biotechnology research is to acquire knowledge that will improve the management and protection of natural resources. Programs address all aspects of agriculture, including forestry and aquaculture, with the aim of improving operational efficiency while minimizing negative environmental impacts. USDA research ranges from molecular characterization of basic processes such as photosynthesis, to improving reproductive and productive efficiency of plants and animals, to understanding the structure and function of both natural and managed ecosystems.

2.3. US Department of Commerce

The US Department of Commerce supports bioremediation R&D through two of its agencies, the National Institute of Standards and Technology (NIST) and the National Oceanic and Atmospheric Administration (NOAA). NIST biotechnology research emphasizes DNA technologies, biocatalysis, bioseparations, biosensors, and protein engineering. Environmental projects at NIST include sensor development, new instruments for detecting environmental damage to DNA, and the use of cytochromes for dehalogenation of hazardous compounds in water. NOAA supports biotechnology research through the National Marine Fisheries Service and the National Sea Grant College Program. All R&D activities at NOAA dealing with the environment were categorized under Marine Biotechnology and Aquaculture. Environmental projects include pollution control through bioprocessing and bioremediation of contaminants in coastal areas.

2.4. US Department of Defense

The US Department of Defense (DOD) biotechnology program seeks to develop products and processes that might be of military and commercial use, especially in areas dealing with maintenance and synthetic processing costs and with preventing the deleterious effects of chemical, biological, and physical agents of warfare. Through the Office of Naval Research, the Army Research Office, and the Air Force Office of Scientific Research, DOD supports bioremediation of petroleum hydrocarbons, explosives, propellants, chlorinated solvents, and organophosphates. Biodegradative processes are being developed that will exploit the metabolism of natural communities of aerobic and anaerobic microorganisms to clean up contaminated soils and sediments at military installations.

2.5. US Department of Energy

Biotechnology plays an important role in the US Department of Energy (DOE) mission. The DOE supports bioremediation R&D, primarily through its Offices of Energy Research (ER) and Environmental Management (EM). From 1984 to 1996, ER's Subsurface Science Program (SSP) developed fundamental biological, chemical, and physical data on the mechanisms that control the reactivity, mobilization, and transport of chemical mixtures in the subsurface. In 1996, ER launched a new program in bioremediation research—the Natural and Accelerated Bioremediation Research (NABIR) Program. NABIR builds on the data and techniques developed by the SSP and focuses on developing the scientific foundation needed to bioremediate the complex mixtures of wastes at DOE facilities. ER's Energy Biosciences Division has supported research on energy-related biotechnologies (photosynthesis, extremophiles, metabolism of renewable resources), and is also partnering with EM in the support of phytoremediation research. ER's Microbial Genome Program is supporting the complete genomic sequencing of extremophiles, a radioresistant bacterium, and a biodegradative bacterium, all of which will provide useful information for bioremediation, bioprocessing, biocatalysis, and bioenergy production. EM, through its Office of Science and Technology, conducts applied R&D in soil/groundwater waste minimization and processing of hazardous and mixed wastes. It is also partnering with ER in a new basic research program, the Environmental Management Science Program. Finally, the DOE Office of Energy Efficiency (EE) also supports some bioremediation research, with a focus on conversion of municipal wastes to alternative energy sources and chemical feedstocks.

2.6. US Department of Health and Human Services

The US Department of Health and Human Services (DHHS) is the federal government's agency for promoting the health of Americans and providing essential human services. To this end, three DHHS components (the National Institutes of Health, the Food and Drug Administration, and the Centers for Disease Control and Prevention) conduct and support biotechnology R&D. The NIH supports basic and applied research in environmental biotechnologies within a framework of health-related R&D. Areas of interest include the mechanisms of degradation and transformation of toxic chemicals.

2.7. US Department of the Interior

Projects in the US Department of Interior (DOI), the nation's principle conservation agency, are focussed on diseases of plants and animals, structure and function of microbial communities and ecosystems in response to environmental changes, and conversion of toxic compounds to innocuous forms. Most of the work that relates to bioremediation is supported by contracts with the US Geological Survey.

2.8. US Environmental Protection Agency

The US Environmental Protection Agency (USEPA) is responsible for anticipating, assessing, reviewing, and implementing policies related to the environmental consequences of using or developing technologies. Biotechnology at the USEPA seeks to advance the understanding, development, and safe application of bioremediation solutions to hazardous waste problems. The program is designed to balance basic research on biological degradation processes with engineering activities leading to practical techniques for cleaning up the environment. Research is conducted to explore factors that limit biodegradation in the field, define the impact of waste characterization on biotreatment, define the mechanisms controlling rate and extent of biodegradation, translate biodegradation process concepts into well-engineered biosystems, develop in situ biostimulation and bioaugmentation delivery systems, and develop tools to assess performance and ecological and human health effects.

2.9. National Aeronautics and Space Administration

Although The National Aeronautics and Space Administration (NASA) currently reports all of its biotechnology R&D under the area of Manufacturing and Bioprocessing (BRS, 1995), it contributes to environmental understanding through its research on life support systems. NASA is interested in life support systems that provide for plant growth, food processing, waste processing, and all the biogeochemical cycles necessary to recreate the cycles of nature in a closed system.

2.10. National Science Foundation

The National Science Foundation (NSF) supports comprehensive fundamental research that underpins advances in biotechnology, including basic research on ecosystem analysis and environmental restoration, maintenance, and remediation. The NSF also supports basic research on the genetic, physiological, taxonomic, and ecological traits of organisms and on how organisms relate to one another and to their physical environment. Other studies address the application of engineering principles to reduce adverse effects of discharges that impair resource value, and innovative biological processes used to restore usefulness of polluted environments.

3. JOINT DOE/EPA/NSF/ONR BIOREMEDIATION RESEARCH ANNOUNCEMENT

In 1995, the Bioremediation Working Group of the BRS Committee on Fundamental Science, NSTC began planning an interagency effort in bioremediation research support, based on the research priorities identified by the BRS (BRS, 1995). The result was an interagency memorandum of understanding and an announcement of interest in supporting field-based research on the bioavailability of waste mixtures in 1996, and it was the first example of a request for applications and proposals resulting from a NSTC planning process. Over 100 applications were received and peer reviewed by a panel of experts jointly selected by the four participating agencies—DOE, USEPA, NSF, and the US Office of Naval Research (ONR). The four agencies selected 12 applications for funding, representing the bioavailability of metals, chlorinated solvents, polyaromatic hydrocarbons, polychlorinated biphenyls, and petroleum hydrocarbons in soil, sediment, and groundwater. It is the intent of the DOE, USEPA, NSF, and ONR to continue this collaboration, and a planning process is now underway for the second year of the program.

4. FUTURE ACTIVITY

It is clear that several federal agencies have strong interest in environmental biotechnology, including applications for understanding, monitoring, and managing the environment. That interest has, to date, focussed on bioremediation R&D, because of the need to develop more economical, environment-friendly approaches to the cleanup of wastes associated with energy production, mining, agriculture, defense, chemical (including drug) production, and other industrial activities. Over the past decade, there has been a discernable movement from waste cleanup to waste prevention, in terms of R&D support. Now, as the United States moves towards sustainable development, biotechnology will become key to understanding and maintaining a sustainable environment. It will be necessary for federal agencies to begin a collective planning process for programs that will develop the biotechnologies needed to achieve a sustainable environment.

5. DISCLAIMER

"The views and opinions expressed in this chapter are those of the author and do not necessarily reflect those of the United States Government or any agency thereof. Neither the United States Government nor any agency thereof, nor any of their employees, makes any warranty, express or implied, or assumes any legal liability or responsibility for the accuracy, completeness, or usefulness of any information, apparatus, product or process disclosed, or represents that its use would not infringe privately owned rights. Reference herein to any specific product, process or service does not necessarily imply its endorsement, recommendation, or favoring by the United States Government or any agency thereof."

REFERENCES

BRS, 1993, Biotechnology for the 21st Century: Realizing the Promise, Biotechnology Research Subcommittee, Committee on Life Sciences and Health, Federal Coordinating Council for Science, Technology and Engineering, Washington, DC.

BRS, 1995, Biotechnology for the 21st Century: New Horizons, Biotechnology Research Subcommittee, Committee on Fundamental Science Research, National Science and Technology Council, Washington, DC.

ETWG, 1995, Bridge to a Sustainable Future, Environmental Technology Working Group, National Science and Technology Council, Washington, DC.

Lee, K.B. and G.S. Burrill, 1995, Biotech '95: Reform, Restructure, Renewal. Industry Annual Report, Ernst & Young, Palo Alto, CA.

NRC, 1993, In Situ Bioremediation: When Does it Work? National Research Council, National Academy Press, Washington, DC.

OECD, 1994, Biotechnology for a Clean Environment: Prevention, Detection, Remediation, Organisation for Economic Co-Operation and Development, Paris, France.

15

ENVIRONMENTAL BIOTECHNOLOGY ISSUES IN RUSSIA

Alexander M. Boronin, Nickolai P. Kuzmin, Ivan I. Starovoytov, Irina A. Kosheleva, Andrei E. Filonov, Renat R. Gaiazov, Alexander V. Karpov, and Sergei L. Sokolov

Institute of Biochemistry and Physiology of Microorganisms of the Russian Academy of Sciences
Pushchino, Moscow region, 142292, Russia

1. INTRODUCTION

Environmental pollution in Russia is growing at an alarming rate and is becoming an ever more immediate danger. To a great extent Russia has inherited its environmental problems from the former Soviet Union, where they arose due to:

- extensive development of heavy industry and large-scale mining;
- a highly militarized economy;
- use of inefficient technologies and equipment for environmental protection at the final stages of technological processes;
- a system for evaluating economic benefits which did not adequately take into consideration ecological losses; and
- the absence in the country of a well-organized system for ecological education.

As a result, more than 500 billion tons of solid wastes have accumulated, occupying about 250 thousand hectares of land.

Moreover, 130 million tons of solid wastes are annually deposited in open dumps, landfills and test sites. Approximately 4 billion curies (Ci) of radioactive wastes are amassed in storage tanks, basins and water reservoirs.

Stocks of chemical weapons amount to about 40 thousand tons by the weight of toxic chemicals. Among them, organophosphorus compounds represent 32.3 thousand tons and compounds of dermatovesical action (Mustard, lewisite and their mixtures) represent 7.7 thousand tons. In Paris, in January 1993, Russia, together with the United States and other countries, signed the Convention on Prohibition of the Development, Production, Stockpiling and Use of Chemical Weapons and on their Destruction. In accordance with this Convention, Russia must destroy all stockpiles of chemical weapons within ten years. Experts have estimated the cost of destroying the stockpile of chemical weapons in the United States at 13 billion US dollars (USD). The activity on destruction of chemical

Biotechnology in the Sustainable Environment, edited by Sayler *et al.*
Plenum Press, New York, 1997

weapons in Russia evidently should be of the same scope, which will pose significant problems for achieving these goals in our country. Providing ecological safety is a particularly urgent question while carrying out the destruction of chemical weapons, given the high toxicity of the compounds to be destroyed.

Nevertheless, the opinion that all environmental problems in Russia have been inherited from the past regime is mistaken. During the transition to a market economy in Russia, old legislation is no longer in effect and federal agencies responsible for the implementation of regulations have ceased to function, while new legislation and structures which coordinate the activity of enterprises and prevent environmental pollution are only beginning to take shape. When industries are privatized, new proprietors often continue to use existing equipment, lacking the stimulus or opportunity to invest in more modern technologies. Despite the tendency towards the lowering of oil production and transportation, the ecological situation in oil-producing regions and those through which pipelines pass is worsening and in a number of regions has become critical. More than 200 thousand hectares of land contaminated by oil up to 10 centimeters in depth are found only in sites in Western Siberia where oil deposits are being exploited. This year in Western Siberia alone up to 35 thousand cases of depressurization of industrial oil pipelines occur, resulting in spills of more than one million tons of oil. This situation is likely to continue for some time, since at least 4 thousand kilometers of pipelines require repair and no time for their repair has been proposed.

These examples of the extent of existing pollution and outlook for environmental contamination in the future clearly demonstrate the necessity to take positive measures to clean up existing and anticipated unpreventable pollution by stimulating the development and application of innovative technologies for treatment of polluted soils and waters.

2. LEGISLATION AND REGULATIONS ON ENVIRONMENTAL PROTECTION AND BIOSAFETY

On July 13, 1993 the President of Russian Federation issued a Decree on organization of the Interagency Commission on Ecological Safety of the Security Council of the Russian Federation. The goals of this Commission are to develop strategic conceptions and co-ordinate all governmental activity in the area of ecological safety in country (Ecological Safety in Russia, 1993). Federal agencies dealing directly with such issues and represented on the Commission include:

- Ministry of Environmental Protection and Natural Resources of the Russian Federation,
- State Committee on States of Emergency,
- State Committee for Sanitary and Epidemiological Inspection of the Russian Federation,
- Federal Inspectorate on Nuclear and Radiation Security of Russia,
- Federal Mining and Industrial Inspectorate of Russia,
- State Committee on the Problems of Chemical and Biological Weapons, and
- other specialized federal agencies.

On the basis of a Decree of the President of the Russian Federation dated December 16, 1993 and the Law of the Russian Federation on Environmental Protection, the Ministry of Environmental Protection and Natural Resources of the Russian Federation has given the mandate of preventing negative impact of industrial toxicants including oil and oil products on the environment.

The goals of the Ministry are:

- to monitor environmental conditions and environmental changes due to industrial and other impact;
- to oversee the implementation of scheduled activities on environmental protection, rational use of natural resources and remediation of the environment;
- to oversee the fulfilment of requirements and standards on proper use of the environment and environmental quality; and
- the implementation of environmental legislation and regulations.

Ecological monitoring is now carried out through a three-level system headed by the Ministry of the Environmental Protection and Natural Resources of the Russian Federation, consisting of a federal level, an oblast and republic level, and a district level.

The Russian Federation inherited most of its biosafety regulations from the USSR. The scope and developmental history of these regulations have been covered in a review by Rimmington (1994), which deals with biotechnology and industrial microbiology, and emphasizes also the issue of genetically engineered organisms.

More recently the effort to bring biosafety regulations into correspondence with those operating in Europe and the United States has led to the establishment of a multidisciplinary commission charged with the task of preparing drafts of new regulations relevant to use and translocation (including release) of genetically modified organisms (GMO). The commission includes representatives from the Ministries of Science and Technological Policy, Environmental Protection and Public Health as well as the Russian Academy of Sciences and the Academy of Agricultural Sciences. The Commission relied extensively upon recommendations provided by several foreign experts as well as organizations such as the Organisation for Economic Co-Operation and Development (OECD). After being circulated through main regions of the country and among the independent experts for comments, the draft law ("On state policy in genetic engineering") was submitted to the Duma (the lower house of Russian Parliament). The Duma passed the draft law in the middle of 1995, but it was not passed by the President on the grounds that it required additional work. This work is in progress at the time of writing. One significant aspect of the draft law was the declaration of supremacy on the Russian territory of international agreements related to biosafety over internal regulations. Thus, if this law is passed in its present form, any document, such as the International Technical Guidelines for Safety in Biotechnology, which assumes a binding nature under the Convention on Biodiversity (to which Russia is a signatory) would influence internal biosafety regulations.

At the present time Russia is establishing legislation in the field of biotechnology. To receive permission to use biopreparations in the environment applicants must get the following forms of approval:

- The Research Center of Toxicology and Hygienic Regulation of Biopreparations of the Ministry of Health and Medical Industry of the Russian Federation conducts a toxicological assessment to show that biopreparations are non-toxic for warm-blooded animals, birds, fish, earthworms, insects (bees), soil microorganisms, agricultural plants.
- The State Committee for Sanitary and Epidemiological Inspection of the Russian Federation issues certification on biological safety.
- The Central Board on the Use of Agrochemicals and the State Committee on Agrochemicals of the Ministry of Agriculture and Food of the Russian Federation grant registration of biopreparations and permanent or temporary permission to use biopreparations.

3. GOVERNMENTAL SCIENTIFIC PROGRAM ON ENVIRONMENTAL BIOTECHNOLOGY

In 1993 by the decision of the Ministry of Science and Technology Policy of the Russian Federation the governmental scientific and technology program "Environmental Biotechnology" was established. The goals of this program are to develop fundamental research studying living organisms, first of all, microorganisms able to degrade and detoxify various kind of pollutants; to carry out applied research focused on the investigation of the possibility of their use as the basis for biotechnology of environmental protection; and in collaboration with other interested agencies and companies to introduce biotechnology and environmental protection in practice.

The complexity of the problem is associated with the fact that in the environment of Earth there are already about 60,000 xenobiotics and their number increases by about 1,000 compounds per year. The range of these compounds and the intensity of their release into the environment outrun the rate of microbial evolution and therefore, the natural associates formed in waste water treatment plants are frequently inefficient in the degradation of xenobiotics which occur in effluents. The lagging of microbial evolution behind the intensity of pollution by xenobiotics predetermines the need for highly efficient microbial strains to be developed and introduced into natural consortia of microorganisms.

One of the major approaches in pollution control is to intensify the microbial activities in the detoxification and total degradation of inorganic and organic compounds, many of which are highly toxic. In fact, the idea is to develop various biotechnologies to provide for the use of microorganisms, both for preventing the pollution of the environment and for its cleanup. It should be stressed that introduction of mutant or recombinant microbial strains is yet a complex problem. Further experiments are required to investigate the behavior of these strains under natural (or close to natural) conditions, as well as the possibility of transferring genetic information from them into other microorganisms and their effect on the natural ecological processes. Table 1 outlines the framework of this program.

The program is made up of projects selected on a competitive basis. At present this program includes 39 projects. Among them, half (19 projects) are carried out by researchers of the Institute of Biochemistry and Physiology of Microorganisms of the Russian Academy of Sciences in Pushchino and 9 projects are implemented by scientists of the Institute of Microbiology of the Russian Academy of Sciences in Moscow (see Table 1).

Some results of the work on two projects of this program implemented by our laboratory, which studies polyaromatic hydrocarbons degradation and biotechnology of degradation of chemical weapons, are given below.

Table 1. Scientific institutions participating in the Russian scientific program "Environmental Biotechnology"

- Institute of Biochemistry and Physiology of Microorganisms, Russian Academy of Sciences, Pushchino
- Institute of Microbiology, Russian Academy of Sciences, Moscow
- Moscow State University, Moscow
- Bakh Institute of Biochemistry, Russian Academy of Sciences, Moscow
- Research Institute of Genetics, Moscow
- Saint Petersburg University, Saint Petersburg
- Institute of Biochemistry of Plants and Microorganisms, Russian Academy of Sciences, Saratov Komarov Botanical Institute, Saint Petersburg

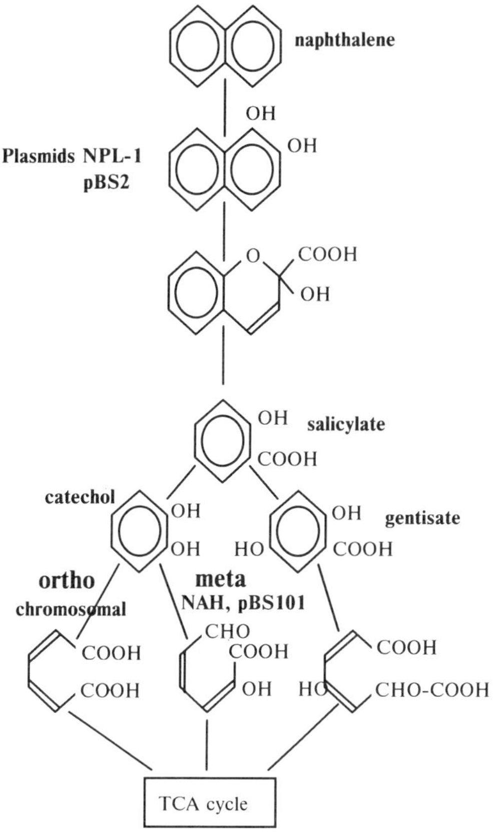

Figure 1. Plasmid-encoding naphthalene degradation pathways in *Pseudomonas putida*.

4. BIODEGRADATION OF POLYCYCLIC AROMATIC HYDROCARBONS

4.1. Naphthalene Catabolic Plasmids

Recent studies show that microbiological degradation of polycyclic aromatic hydrocarbons (PAHs) is the major process that results in the decontamination of sediments and surface soil. Bacterial degradation of naphthalene has been investigated extensively as a model for bioremediation of PAH-contaminated environments. Naphthalene degradation is mostly determined by plasmid-harboring microorganisms. Strain PpG7 contains the archetypal NAH7 plasmid that encodes the genes for the degradative pathway of naphthalene (Figure 1). In plasmid NAH7 naphthalene catabolism genes are organized into two operons, *nah* and *sal*. The two operons are controlled by a positive regulator gene, *nahR*, that is located upstream of the *nahG* gene. Induction of both operons is controlled by the naphthalene metabolite, salicylate (Yen, 1988).

Plasmids controlling the degradation of naphthalene exhibit great diversity of their properties such as molecular size, conjugal transfer and regulation of their expression. Naphthalene catabolic plasmids were found to be in either incompatibility group P7 or P9 (Table 2). Available data suggest that all of the naphthalene catabolic plasmids encode a

Table 2. Naphthalene degradative plasmids

Plasmid	Incompatibility group	Molecular size (kb)	Pathway of catechol degradation of plasmid-bearing strains
NAH7	Inc p-9	85	META
pBS3	Inc P-7	160	
pBS8909	ND	85	
pBS216	Inc P-9	100	ORTHO + META*
pBS1141	ND	110	
NPL-1	Inc P-9	100	ORTHO + "silent" META
pBS2	Inc P-7	130	
pBS4	Inc P-7	170	via GENTISIC ACID

*Low level activity of catechol-2,3-dioxygenase before intraplasmid rearrangements
ND = not determined.

single upper pathway. These observations suggest that all naphthalene catabolic plasmids may be related.

Plasmids controlling degradation of aromatic compounds normally provide catechol *meta*-oxidation. But it has been shown that some naturally occurring *Pseudomonas* plasmid-bearing strains, for instance *P. putida* BS202 (NPL-1) and *P. putida* BS238 (pBS2), degrade naphthalene via catechol *ortho*-cleavage. The growth of these strains on the methylated substrates (2-methylnaphthalene, 5-methylsalicylate) results in the isolation of mutants BS575 and BS359 respectively with the functional expression of the *meta*-pathway (Fig. 2). The presence of the *meta*-pathway genes in these systems expands the range of potentially utilizable aromatic compounds, but oxidation of some halogen-substituted aromatic compounds *via meta*-pathway may be harmful to the cell. This can explain the fact that in many cases *meta*-pathway genes are silent.

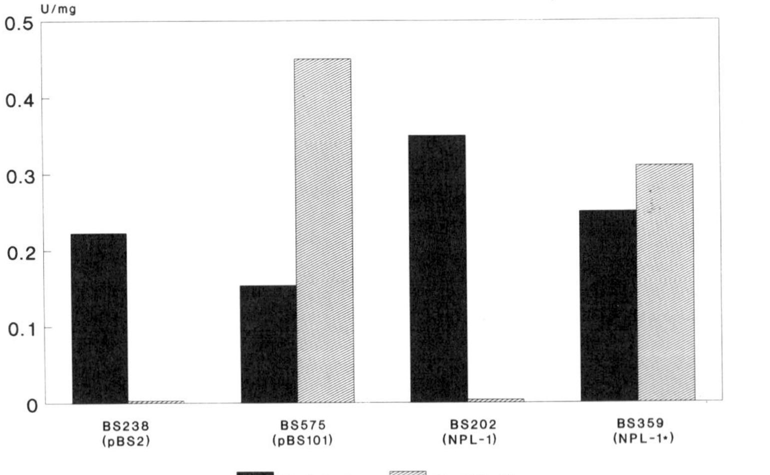

C-1,2-O C-2,3-O

P.putida BS575(BS101), BS359(NPL-1*) are the derivatives with functioning meta-pathway of catechol degradation of strain BS238(pBS2), BS202(NPL-1), respect.

Figure 2. Enzyme activities of *P. putida* strains with silent and functioning genes of the catechol degradation *meta*-pathway.

In some cases plasmids reveal low constitutive levels of catechol-2,3-dioxygenase, the first enzyme for the *meta*-pathway, but these levels may be increased after some genetic rearrangements as it was shown for plasmids pBS216 and pBS1141. An example of such rearrangement is the deletion of DNA fragment about 3.0 kb from the adjacent region of structural genes for naphthalene biodegradation in case of NPL-1* (derived from NPL-1). Similar deletion was shown for plasmids pBS216 and pBS1141 when the *meta*-pathway was activated.

In strains with "switching-on" *meta*-pathway genes the activity of *ortho*-pathway genes is retained but the ratio of enzyme activities of the *ortho*- and *meta*-pathway changed depending on the growth substrate. Thus, when microorganisms were grown on methylated naphthalene the activity of *ortho*-pathway genes was almost absent.

As a rule, plasmids carry all the genetic information necessary for the conversion of naphthalene and its intermediate salicylate to TCA cycle (NAH7, pBS8909, pBS3). However, *P. putida* BS202 plasmid NPL-1 controls only primary stages of naphthalene degradation to salicylate. Salicylate oxidation to catechol with further *ortho*-cleavage of the latter is determined by chromosomal genes. This strain contains two non-homologous salicylate hydroxylase genes, plasmid and chromosomal, and the plasmid salicylate hydroxylase gene is silent (Figure 3). Intraplasmid rearrangements of NPL-1 could result in expression of plasmid *salG* gene. The expression of NPL-1 *nah*-operon is controlled either by the chromosomal *nahR1* gene or by the plasmid *nahR* gene. The plasmid *nahR* gene in the parental NPL-1 does not function; however, it may be "switched-on" simultaneously with *salG*.

We have cloned 6.3 kb fragment of NPL-1 into the broad host range vector pUCP22. The recombinant plasmid was designated pBS1150. This plasmid consists of the gene encoding the salicylate hydroxylase and probably the plasmid regulator gene. *Pseudomonas* strains carrying pBS1150 returned to *Sal*⁺ phenotype.

Genes involved in naphthalene and salicylate degradation may be located in trans as was shown for BS202 (NPL-1). There is a similar situation in the case of *P. putida* BS3701 (pBS1141, pBS1142) carrying *nah*- and *sal*- operons in two different plasmids.

4.2. Physiological Evaluation of Naphthalene Degraders

The transfer of the same NAH plasmid into several *Pseudomonas putida* strains results in constructing strains significantly differing from each other in growth characteristics with naphthalene or salicylate as a sole carbon source. A similar effect is produced by using the same *P. putida* strain as a recipient of different NAH plasmids (Boronin, 1992).

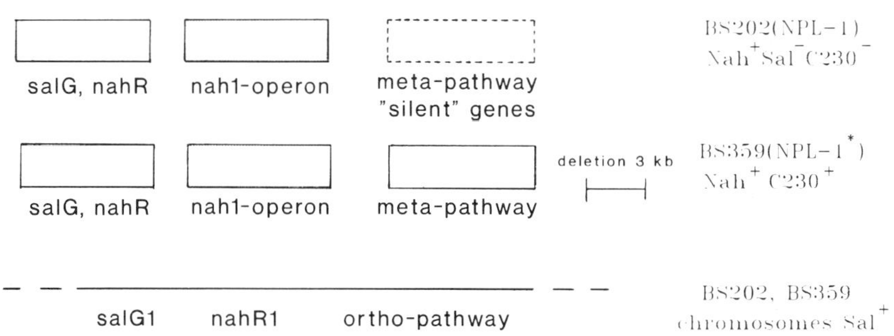

Figure 3. Organization of Nah-system of the strain *P. putida* BS202 and its derivate with functioning *meta*- pathway.

This poses the question: which of the pathways of naphthalene degradation is the most efficient under particular conditions? To clarify this, it is necessary to determine which of the plasmids controlling a certain pathway is the most effective, which strain is the best host of naphthalene degradative plasmid, and which plasmid-bacterium combination would be optimal. The role of such combinations is also of interest from the viewpoint of environmental biotechnology, in particular for selecting the most promising degraders.

Dependence of growth rate (μ) on substrate concentration (S) can be described by the Monod equation (Pirt, 1975):

$$(S) = \mu_{max} S / (S + K_S) \tag{1}$$

where K_S is the saturation constant and μ_{max} is the maximal specific growth rate. There may be two extreme cases: $\mu(S) \approx \mu_{max}$ for $S >> K_S$ in batch culture in excess of substrate and $\mu(S) \approx \mu_{max} S/K_S$ for $S << K_S$ in a chemostat with substrate limitation.

To compare the functioning of various pathways of naphthalene degradation in batch culture we determined the specific growth rate of isogenic plasmid-bearing strains BS202 (NPL-1), BS374(pBS105), BS359(NPL-1*) and BS594 (pBS217), BS573(pBS103) and BS238 (pBS2), BS575(pBS101) with silent and functioning *meta*-pathway genes (Boronin et al., 1993). The data obtained shows that during the growth on naphthalene as the sole source of carbon, the highest specific growth rate was ensured by the plasmids controlling naphthalene metabolism via the *meta*-pathway (Figures 4, 5). Thus, the *meta*-pathway of catechol oxidation allows bacteria to grow faster than the *ortho*-pathway in batch culture with excess naphthalene. In nature, due to low solubility and absorption to soil particles, concentration of hydrophobic substances like PAHs in particular cases may be very low, even in soil samples from areas that are heavily contaminated with PAHs. In these cases the ability to utilize PAHs at low concentrations may be the important determinant for the survival of the PAH degrading strains.

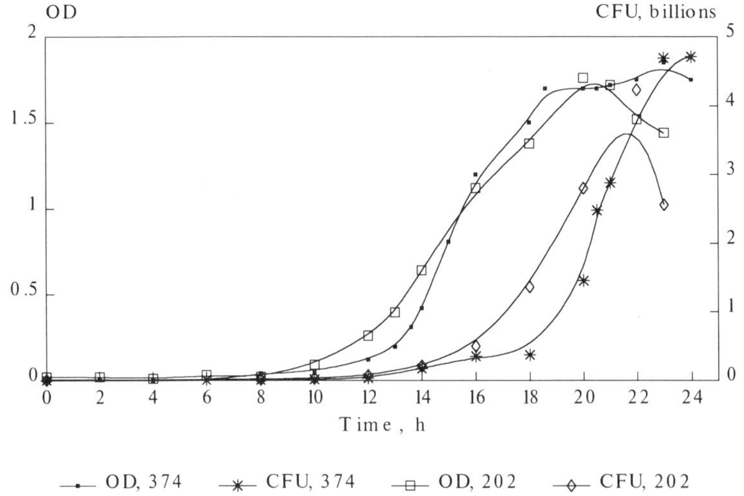

Figure 4. Growth of *P. putida* strains in batch culture on naphthalene. The strain BS374(pBS105) with functioning *meta*-pathway ($\mu^{OD}_{max} = 0.75\ h^{-1}$, $\mu^{CFU}_{max} = 0.65\ h^{-1}$) grew faster than BS202 (NPL-1) with *ortho*-pathway ($\mu^{OD}_{max} = 0.35\ h^{-1}$, $\mu^{CFU}_{max} = 0.37\ h^{-1}$).

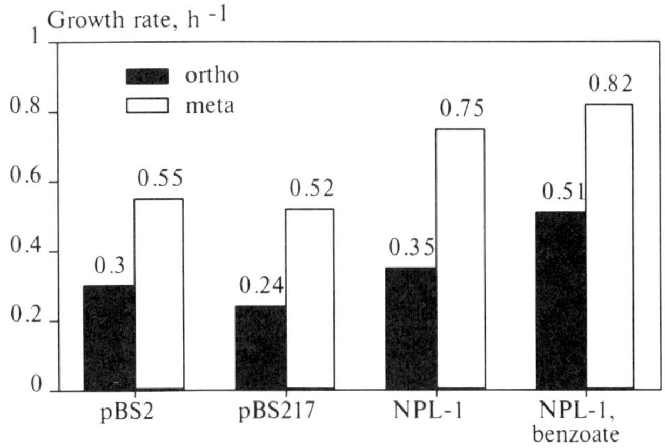

Figure 5. Comparison of the specific growth rates of the strains carrying plasmids pBS2, pBS217, NPL-1 (the *ortho*-pathway) and their derivatives harboring mutant plasmids pBS101, pBS103, pBS105 (the *meta*-pathway) in batch culture on naphthalene and benzoate.

Under laboratory conditions, substrate-limited growth can be accomplished in a chemostat. We assume that the competitive behavior of different plasmid-bacterium combinations in a chemostat may represent one of the important factors determining survival in a polluted environment.

Known approaches (Duetz et al., 1991a,b), allow to estimate the ratio of growth rates for any pair of strains in chemostat culture. Relative growth rates with reference to the most competitive strain BS203(pBS2) were estimated by comparing different pairs of strains (Figure 6). The plasmid-bearing strains degrading naphthalene via the *ortho*-pathway of catechol oxidation were the most selectively advantageous under naphthalene limitation. Less competitive were the strains with plasmids which control naphthalene catabolism via the *meta*-pathway. The least competitive under these conditions were the strain BS291 carrying plasmid pBS4 which encodes for naphthalene catabolism via gentisate and the strain WD25(pBS2) degrading naphthalene only to salicylate.

Dissimilarity from results obtained in batch culture may be explained using the Monod equation. While in batch culture growth rate represents μ_{max}, in chemostat culture

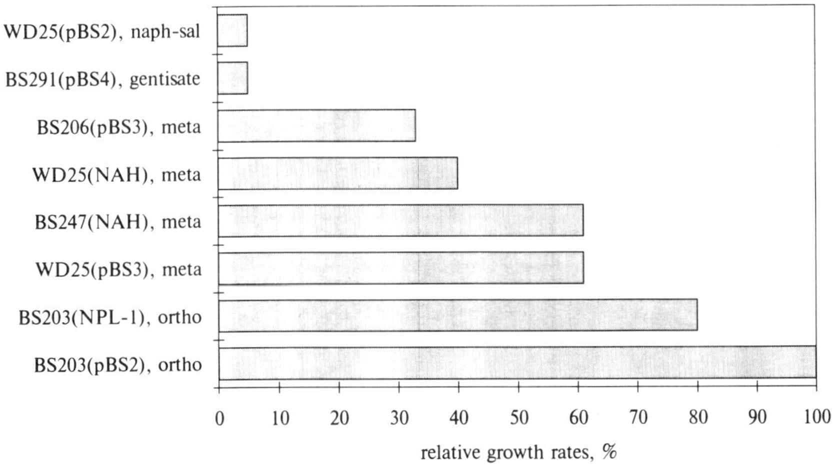

Figure 6. Comparison of relative growth rates in chemostat competition experiments under naphthalene limitation (dilution rate $D = 0.05\ h^{-1}$). The growth rate of the most competitive strain BS203(pBS2) is taken to be 100%.

under substrate limitation not only μ_{max} but also K_S values determine the growth rate. In the latter case the most competitive strains catabolizing naphthalene via the *ortho*-pathway appear to have lower K_S as compared to strains using the *meta*-pathway. With a lower K_S value, a lower residual xenobiotic concentration can be reached after introduction of degrader strains into a polluted environment.

Since the pathway of naphthalene degradation to salicylic acid is the same for all known *Pseudomonas* strains, we assumed that the use of salicylate as growth substrate would permit to establish more precisely the relationship between the growth parameters and catechol oxidation cleavage, as well as to estimate the efficiency of salicylate degradation.

Figure 7 represents experimental data on growth of plasmid-encoded naphthalene-degrading *P. putida* strains in batch cultures on salicylate in the example of BS202(NPL-1), BS359(NPL-1*) and BS374 (pBS105).

To evaluate the growth of microorganisms and substrate consumption on salicylate in batch culture the following simple initial value problem has been used (K_S is presumed to be about zero):

$$ds / dt = -q\, x\, \theta(s),\ s\,(0) = s_0, \\ dx / dt = \mu\, x\, \theta(s),\ x\,(0) = x_0; \qquad (2)$$

where $\theta(s) = 1$ for $s > 0$ and $\theta(s) = 0$, otherwise; s = substrate concentration, x = microorganism concentration measured in CFU/ml, μ = specific growth rate constant, and q = metabolic coefficient. Parameter values of this model can be obtained by fitting the exponential growth phase.

The specific growth rate μ is the basic parameter characterizing growth rate in the exponential phase rather than efficiency of xenobiotic degradation. In environmental biotechnology degradative abilities of microorganism can be characterized by metabolic and economic coefficients. The metabolic coefficient q represents the specific rate of substrate consumption. The economic coefficient (biomass yield) $Y = \mu/q$ reflects yield of microorganisms per unit of substrate. Strains with a lower metabolic coefficient and higher economic coefficient are usually preferred in the microbiological industry. But problems

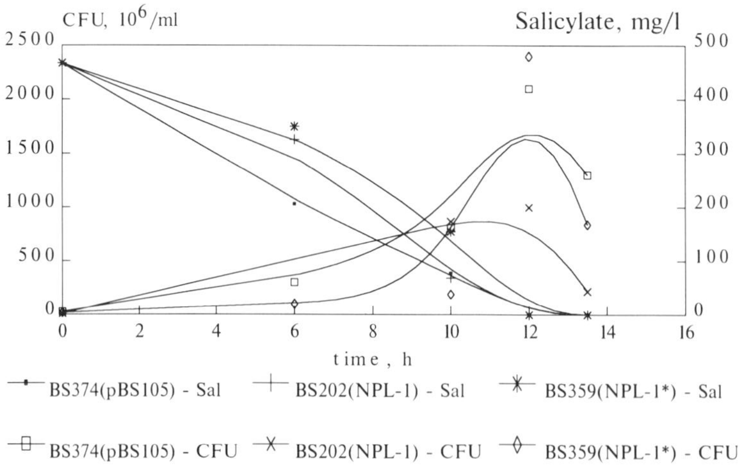

Figure 7. Growth of plasmid-encoded naphthalene degradative *P. putida* strains in batch cultures on salicylate.

of environmental protection suggest using degraders with (1) a higher metabolic coefficient for rapid xenobiotic destruction and (2) a lower economic coefficient to limit concentration of microorganisms introduced in the open environment.

Figure 8 presents growth parameters of different *P. putida* strains. The original strain *P. putida* BS238(pBS2) with an *ortho*-pathway of catechol oxidation has a lower specific growth rate and metabolic coefficient in comparison with its mutant derivative BS238(pBS101) with a functioning *meta*-pathway. Conjugal transfer of pBS2 plasmid into a different host background, *P. putida* BS394, leads to further increases in both the specific growth rate and the metabolic coefficient. In contrast, transfer of pBS101 into *P. putida* BS394 revealed decreases of both the specific growth rate and metabolic coefficient even compared to the original strain *P. putida* BS238(pBS2). These data show the significance of constructing and combining various strains and plasmids for effective xenobiotic degradation.

4.3. Phenanthrene Degrading Strains

It has been shown that NAH7-like plasmids can mediate metabolism of phenanthrene, pyrene and anthracene as well as naphthalene. Generally the rate of PAH degradation is inversely proportional to the number of rings in the PAH molecule. Thus, the lower weight PAHs are biodegraded more rapidly than the higher weight compounds. Bacterial oxidation of phenanthrene (3-fused benzene rings) may be another model for elucidation of principles in the microbial degradation of PAHs.

In our experiments, twenty strains growing on naphthalene as a sole carbon and energy source were also capable to transform phenanthrene. But only strains *Pseudomonas* sp. 8803F, *P. putida* BS3701 and *P. syringae* BS3702 were able to grow on phenanthrene as a sole carbon and energy source. Figure 9 presents data on specific growth rates of these strains and the well known strain *P. putida* PpG7(NAH7) on naphthalene and phenanthrene in batch culture. It is interesting to note the absence of a clear correlation between specific growth rates on these substrates. For example, *P. syringae* BS3702 has the highest specific growth rate on phenanthrene but the lowest one on naphthalene. In all cases specific growth rates on phenanthrene were much lower than that on naphthalene. This fact can be explained by lower bioavailability of phenanthrene.

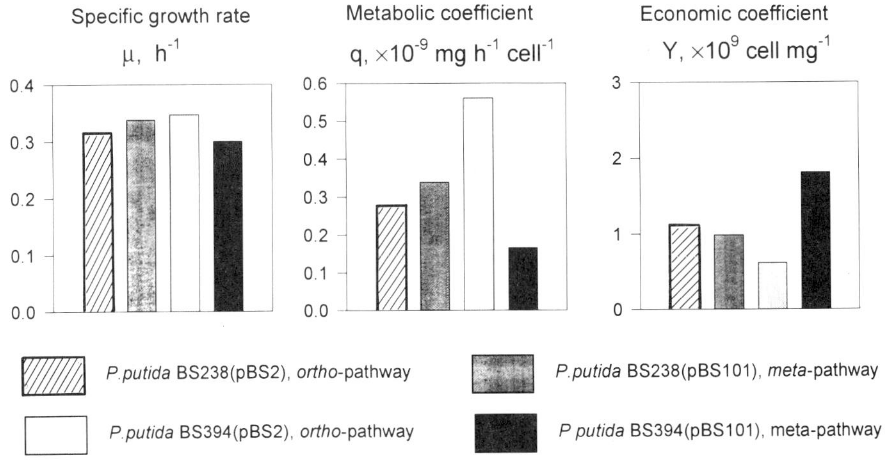

Figure 8. Growth parameters of different *P. putida* strains in batch culture on salicylate.

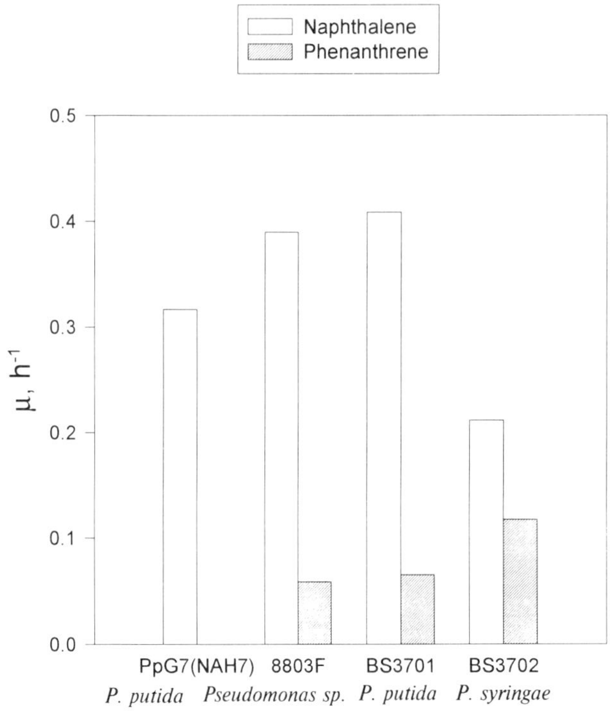

Figure 9. Specific growth rates of *Pseudomonas* strains under batch cultivation on naphthalene or phenanthrene as a sole carbon source.

Pseudomonas strains degrade phenanthrene to 1-hydroxy-2-naphthoic acid by using a single system of enzymes that has broad substrate specificity for the upper catabolic pathways from naphthalene to salicylic acid. The ability of microorganisms to grow on phenanthrene as a sole source of carbon and energy depends on the ability to metabolize 1-hydroxy-2-naphthoic acid. The 1-hydroxy-2-naphthoic acid is consequently decarboxylated to 1,2-dihydroxynaphthalene (the product of naphthalene degradation) which is further metabolized through salicylate. *P. putida* strain BS3701 isolated in the laboratory of plasmid biology IBPhM RAS reveals this pathway of phenanthrene oxidation (unpublished data).

In other cases, the 1-hydroxy-2-naphthoic acid may be oxidized through o-phthalate and protocatechuate (Kiyohara, 1978).

5. DESTRUCTION OF CW

In line with the International Chemical Weapons (CW) Convention (1993), the development of environmentally safe technology for destruction of Mustard gas has become a pressing problem. Up to a certain period, the term "destruction" as applied to Mustard meant only its detoxification. Contemporary ecological norms necessitated the development of the Mustard destruction technology to obtain ecologically safe end products. Thus, there are two aspects of Mustard destruction problem: detoxification of this agent by chemical neutralization and conversion of detoxification products into ecologically safe compounds.

To eliminate incineration of a waste stream resulting from the neutralization step, we have developed a two-step method: electrochemical pre-treatment of Mustard neutraliza-

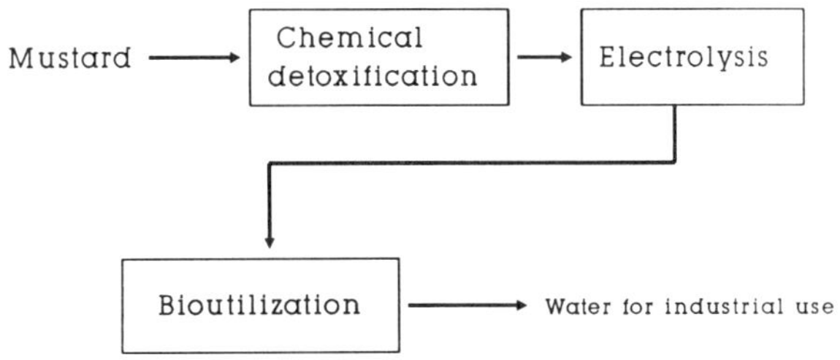

Figure 10. Functional diagram of the proposed combined technology for Mustard destruction.

tion products followed by biological oxidation of the waste stream (Boronin et al., 1996a) (Figure 10). The reason of using electrochemical pre-treatment of Mustard neutralization products is the presence of other products in the industrial grade Mustard. In particular, besides the main product, bis(2-chloroethyl) sulphide, the industrial grade Mustard may contain 1,4-dithiane, bis(2-chloropropyl) sulphide, 2-chloroethyl-2′-chlorobutyl sulphide, bis(2-chloroethyl) disulphide, bis(2-chloroethyl) trisulphide, 1.2-bis(2-chloroethylthiol) ethane, dichloroethane, sulphur, hydrogen chloride and various additives that increase its chemical stability. Such a "bouquet" of chemicals poses a problem for using biological methods for treatment of Mustard neutralization products directly after the neutralization step. Besides, compounds recalcitrant to biodegradation, such as ethanolamine, sulphanol and ethylene glycol, are used for Mustard neutralization. To make possible the use of the biotechnology at the final step of waste stream treatment, we applied electrochemical oxidation of Mustard neutralization products.

After Mustard detoxification by the alkali, sulphanol and ethanolamine method, the solution contained the following constituents in grams per liter (g/l): sodium carbonate, 50; sodium chloride, 72; ethanolamine, 50; sulphanol, 1; and thiodiglycol, 77. Laboratory studies indicated that the chemical composition of the solution obtained after Mustard chemical detoxification was suitable for electrochemical destruction of its organic component to biodegradable products, even with no additives.

Electrolysis (4–6 hours) of thiodiglycol (3 g/l), the product of bis(2-chloroethyl) sulphide neutralization, yielded the following products (in milligrams per liter): bis(2-hydroxyethyl) sulphoxide, 580; bis(2-hydroxyethyl) sulphone, 60; 2-hydroxyethylsulphonate, 1390; and acetate, 100.

The curve of thiodiglycol electrochemical destruction (Figure 11) suggests that the process takes 4 hours. At the beginning of the process bis(2-hydroxyethyl) sulphoxide, a product of thiodiglycol oxidation, was accumulated, but later on its concentration decreased. Bis(2-hydroxyethyl) sulphone was formed by sulphoxide oxidation, and 2-hydroxyethylsulphonate was apparently a product of sulphone destruction. The products of electrolysis were biologically oxidized in a biosorber.

The laboratory model of the biosorber (Figure 12) was a plexiglas column (50 millimeters in diameter and 700 to 900 millimeters in height). The lower part of the column was cone-shaped and had a socket for transit and circulation of waste water; the upper part on the column was a cylinder of large diameter preventing occasional extrusions of the loaded material and had a socket for purified water discharge to an airlift.

The initial water was fed into the lower part of the biosorber; while ascending through granulated matter at a rate of 35–45 meters per hour, the water caused pseudoliquefaction of the layer. The treated water left the biosorber together with the circulatory flow through the upper socket for a water-elevating tube of the airlift. Along with providing circulation in the

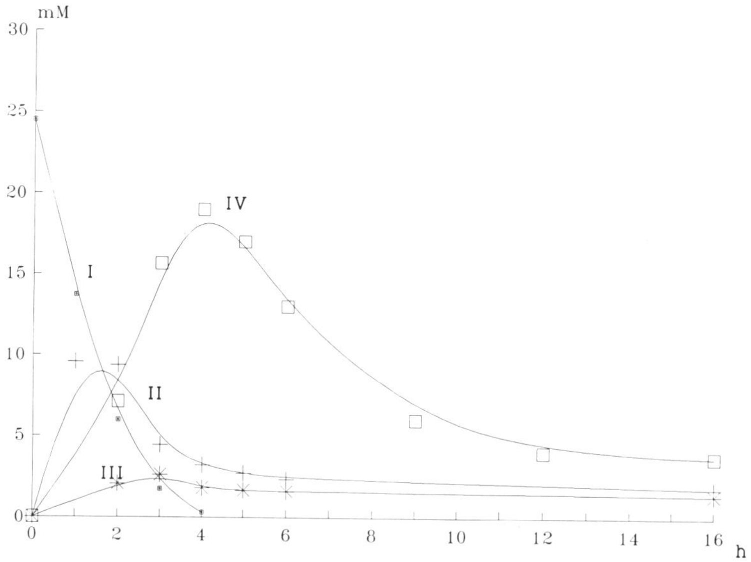

Figure 11. Profile of accumulated products of thiodiglycol electrochemical destruction (thiodiglycol 3 g/l; NaCl 1%; initial pH 14). I - bis(2-hydroxyethyl) sulphide, II - bis(2-hydroxyethyl) sulphoxide, III - bis(2-hydroxyethyl) sulphone, IV - 2-hydroxyethanesulphonate.

biosorber, the airlift acted as an aerator; the air was fed into the airlift by a microcompressor. Duration of water treatment in the biosorber was about 1–2 hours with the total content of thiodiglycol electrochemical destruction products of 3 g/l.

The biologically degradable components of waste water organic contaminants are oxidized by the biofilm formed by microorganisms adsorbed on the packed material particles in a way similar to that taking place on biological filters. Adsorption characteristics of the carrier material exert practically no effect on oxidation processes. Degradation of more

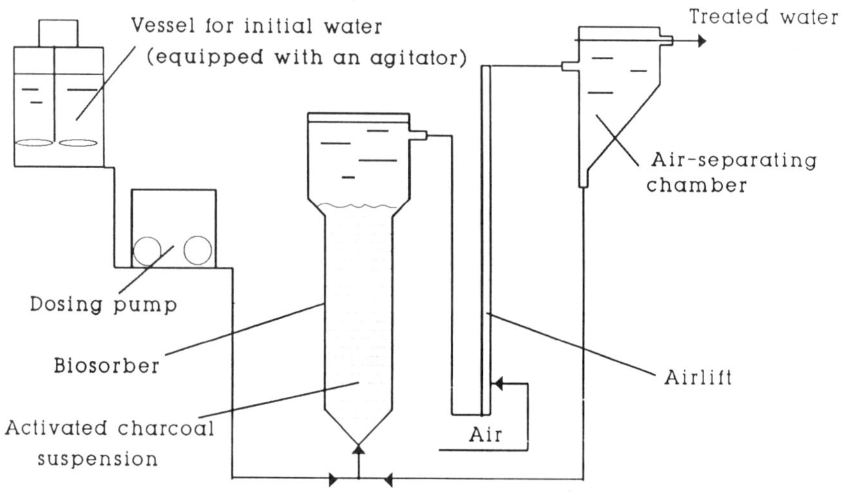

Figure 12. The scheme of laboratory biosorber.

Table 3. Chemical composition of water at the biosorber input and output after 20 days of operation

COD, mg/l		Ammonium nitrogen, mg/l		Nitrites, mg/l		Phosphorus, mg/l	
Input	Output	Input	Output	Input	Output	Input	Output
1404–2908	230–345	56–93	20–42	0	0.02–0.12	9.4–16.0	0.27–7.12

persistent organic contaminants occurs only if activated charcoal is used as the carrier material, since no other packing material was shown to display similar qualities. Continuous biological recovery of the sorbent is achieved.

The mean data on microbial biofilm formation in the biosorber during processing of thiodiglycol electrochemical pre-treatment products over the course of 20 days are indicated in Table 3. The initial ten-day period of biosorber operation could be considered the step of biofilm biocenosis formation. The efficiency of decontamination during this period was 40–50%. Decreases in chemical oxygen demand (by 84–88%), ammonium nitrogen, and phosphorus indicated an efficient removal of organic contaminants.

Comparison of HNMR spectra of input and output water samples from the biosorber with the formed biocenosis confirmed that thiodiglycol electrochemical destruction products were bioutilized. The bioutilization was most efficient under anaerobic conditions (Table 4).

The ecological safety of electrochemical destruction of the organic compounds should be noted particularly. Mass spectrometric profiles of the main components of accumulated gaseous phase showed that the hydrogen-to-oxygen volume ratio was 7:30 after 4 hours of destruction. The gaseous phase contained no toxic agents (chlorine, hydrogen chloride, formaldehyde, or volatile sulphur compounds). The content of organochlorine compounds in it was low (no more than 0.001% of the oxygen content according to mass spectrometry); therefore, both open-type and closed-type reactors with introduced air could be used for electrochemical pre-treatment of the process solution.

According to preliminary estimates, the volume of the detoxyfier can be up to 5 cubic meters (m^3); of the electrolyzer, 5 m^3; and of the biosorber, 10 m^3. These parameters of the main technological units will permit the construction of mobile plants that can be used for Mustard destruction at the sites of its storage. The estimated capacity of a combined plant is 100–150 tons of Mustard per year.

To summarize, our studies have enabled us to work out the principles of a combined, ecologically safe Mustard destruction technology comprising chemical Mustard detoxifi-

Table 4. Bioutilization of bis(2-hydroxyethyl) sulphide electrochemical destruction products

	Concentration of products, mg/l			
	II	III	IV	Acetate
Time, days	Anaerobic process			
0	580	60	1390	100
10	264	14	386	<1
14	200	8	398	<1
17	126	5	181	<1

II - bis(2-hydroxyethyl) sulphoxide
III - bis(2-hydroxyethyl) sulphone
IV - 2-hydroxyethanesulphonate

cation, electrochemical pre-treatment of the detoxification products, and bioutilization of pre-treatment products in a biosorber.

A similar approach was also developed by us for Lewisite destruction (Boronin et al., 1996b).

6. CONCLUSIONS

Russia faces significant environmental problems mostly created over the last few decades. Due to the shift in Russia from a centrally-planned economy to a market-based economy and the resulting chaotic situation in the country with regards to legislation, regulations and their implementation, increased pressure is placed on natural resources and environmental quality. All this means that attention has to be paid to both development of biotechnology to prevent pollution and bioremediation techniques in order to deal with already existing pollution.

REFERENCES

Boronin A.M. (1992). Diversity of *Pseudomonas* plasmids: To what extent? FEMS Microbiol. Lett., 100:461–468.

Boronin A.M., Filonov A.E., Gayazov R.R., Kulakova A.N., Mshensky Y.N. (1993). Growth and plasmid-encoded naphthalene catabolism of *Pseudomonas putida* in batch culture. FEMS Microbiol. Lett., 113:303–308.

Boronin A.M., V.G. Sakharovsky, I.I. Starovoitov, K.I. Kaschparov, A.M. Zyakun, V.N. Schvetsov, K.M. Morosova, I.A. Nechaev, V.I. Tugoschov, N.P. Kuzmin, A.I. Kochergin. (1996a). Principles of a Complex, Ecologically Safe Technology for Mustard Gas Destruction. Applied Biochemistry and Microbiology, 32: 61–68. (In Russian)

Boronin A.M., V.G. Sakharovsky, K.I. Kaschparov, I.I. Starovoitov, EV Kaschparova, V.N. Schvezov, K.M. Morosova, I.A. Nechaev, V.I. Tugoschov, P.A. Schpilkov, N.P. Kuzmin, A.I. Kochergin. (1996b). A complex approach to utilization of the Lewisite. Applied Biochemistry and Microbiology, 32: 211–218. (In Russian)

Duetz W.A. and van Andel J.G. (1991a). Stability of TOL plasmid pWWO in *Pseudomonas putida* mt-2 under non-selective conditions in continuous culture. J. Gen. Microbiol. 137:1369–1374.

Duetz W. A., Winston M., van Andel J. G. and Williams, P. A. (1991b). Mathematical analysis of catabolic function loss in a population of *Pseudomonas putida* mt-2 during non-limited growth on benzoate. J. Gen. Microbiol. 137:1363–1368.

Ecological safety in Russia. First Issue. Proceedings of Interagency Commission on Ecological Safety (October 1993 - July 1994), Moscow, Juridicheskaya literatura, 1994, 224 p. (In Russian)

Kiyohara H., Nagao K. (1978). The catabolism of phenanthrene and naphthalene by bacteria. J.Gen. Microbiol. 105: 69–75.

Pirt S. J. (1975). Principles of microbe and cell cultivation. Blackwell Scientific Publications, Oxford.

Rimmington A. Biotechnology and industrial microbiology regulations in Russia and the former Soviet republics. In: Biosafety in Industrial Biotechnology, Ed. by P. Hambleton, I. Melling, T.T. Salusbury, Blackie Academic and Professional, London, 1994, ch.5, p. 67–89.

Yen K.-M., Serdar C.M. (1988). Genetics of naphthalene catabolism in Pseudomonas. CRC Crit.Rev.Microbiol. 15:247–268.

16

SUSTAINABLE DEVELOPMENT AND RESPONDING TO THE CHALLENGES OF THE EVOLUTION OF ENVIRONMENTAL BIOTECHNOLOGY IN CANADA

The First Fifteen Years (1981–1996)

Terry McIntyre

Biotechnology Advancement Program
Clean Technologies Advancement Division
Environmental Protection Service
Environment Canada, 18th Floor P.V.M.
351 St. Joseph Blvd.
Hull, Quebec, Canada

1. BACKGROUND: PLANTING THE SEED — THE EARLY AWARENESS OF BIOTECHNOLOGY POTENTIAL AND SUSTAINABLE DEVELOPMENT IN CANADA

The prosperity and growth of several industrialized countries including Canada in the late 1960s was less a result of innovation, new technology, export initiative and general entrepreneurship, than a function of a rapidly expanding labor force and what was then viewed as an inexhaustible supply of natural resources. At the same time other countries were making a conscious effort to expand their innovative capacity. As a result, Canada, among others, was faced with a serious decline in its industrial technological capability relative to its international competitors (Ministry of State for Science and Technology, MOSST, 1981).

In the early 1970s, growing evidence demonstrated that the prospects for continued growth in an increasingly industrial economy depended upon a country's capacity for development of high technology capability in such areas as space age materials, microelectronics, robotics, and biotechnology (Science Council of Canada, 1977). Industries that were based upon high technology were found to have higher R&D intensities, triple the growth rate, twice the productivity, significantly greater rates of employment and lesser inflation in their prices, than industries not based upon high technology (Ministry of State for Science and Technology, MOSST, 1981).

In a world beset by shrinking energy resources, escalating health care costs, petroleum dependence, imbalances in food supply, and environmental pollution, one area of high technology was emerging which offered substantial promise towards alleviation of some of these ills. This area of high technology was identified as biotechnology.

In recognition of the enormous developmental potential that biotechnology could offer the Canadian economy, a Task Force on Biotechnology was struck in 1980. The terms of reference of this Task Force were:

- To advise the federal government on the possibility and suitability of instituting specific policies and programs designed to allow Canada to take advantage of the opportunities offered by biotechnology;
- To identify those areas of research and development of biotechnology in which Canada might most appropriately specialize, given its economic and social structure; and,
- To review possible ways of encouraging and promoting the required research and development, assuring that the results of this research and development would be used to meet economic and social development needs (Ministry of State for Science and Technology, MOSST, 1981).

The principle findings of this Task Force were that Canada's biotechnology efforts were characterized by a wide scattering of research activity in university and government laboratories and the absence of any major industrial biotechnology activity. The major recommendation was the need for a substantive government effort to stimulate growth of an industrial sector based upon biotechnology, and at the same time to promote a focusing of the country's research efforts in this area.

The results of the Task Force were the announcements in 1981 of a National Biotechnology Strategy - a long term federal government commitment to support and accelerate biotechnology development through a variety of activities involving government, university, and industry collaboration; increased funds to support basic research and development initiatives; establishment of major biotechnology institutes in Ontario, Quebec, and Saskatchewan; and recommendations for both manpower training and improvements in the regulatory infrastructure to accommodate indigenous biotechnology growth in Canada.

The principle area of focus for development of biotechnology in Canada was that of her resource sectors — long considered the mainstay of Canadian economic growth and prosperity. Within these sectors, priorities were assigned to the following areas:

- Human and animal health care products,
- Nitrogen fixation and plant strain development,
- Mineral leaching and metal recovery, and
- Cellulose utilization and waste treatment improvement.

Although environmental concerns were being raised elsewhere associated with the development of biotechnology, they were overshadowed in Canada (as well as her major trading partners). Recommendations of the Task Force on Biotechnology imparted a sense of immediate urgency towards an accelerated rate of economic development for Canadian biotechnology, so as to rapidly secure important niches in the international marketplace an omission that was to cause significant problems in later years as the technology evolved (Nader, 1986).

Finally, although the term "sustainable development" had yet to become vogue in the Canadian political arena, there had began to evolve an ever-broadening acceptance of the notion of sustainability at the grassroots level elsewhere. Associated with this acceptance was a growing recognition of the need for major changes to government policies and practices so as to integrate economic, environmental, and societal decision-making to maintain the welfare of people and the global ecosystem (United Nations, 1969; United Nations, 1972; IUCN, 1973; Science Council, 1977; World Conservation Strategy, 1980; CEAC, 1987).

2. THE NARROWING OF THE DIVERGENCE OF BIOTECHNOLOGY AND SUSTAINABLE DEVELOPMENT IN CANADA — THE FORMATIVE YEARS (1981–1991)

Although the recommendations from the Canadian 1980 Task Force on Biotechnology were quite overwhelming in their support for a strong federal government presence in the development and promotion of biotechnology, they fell short of providing direction as to how this support was to be manifested - particularly in the face of broader and increasing fiscal pressures from competing (and sometimes volatile) forces, as well as emerging sustainable development sensitivities. Some of these forces included widespread down sizing of state institutions, liberalization of markets, periodic political instability, increasing privatization, deregulation of prices, multiple demands for reduced government resources, lack of skilled human resources, and diminished public sector support for basic research (Allende, 1990; Jaffe, 1991; Cohen, 1994; Weiss, 1994). Further, there were a number of economic forces unique to the evolving biotechnology industry that were exerting strong influence on the rate, direction, and ability of the Canadian biotechnology industry to survive during this period. These forces are summarized in Table 1.

As the biotechnology industry evolved in other industrialized countries, a number of environmental forces evolved that would act to exert additional influence on the development of biotechnology in Canada. These forces are summarized in Table 2.

Primary amongst the areas of environmental concern associated with biotechnology development in Canada had been the adequacy of federal government regulatory oversight—particularly its ability to respond to a variety of environmental challenges raised by the introduction of novel substances into the Canadian environment. A summary of these environmental/regulatory concerns is contained in Table 3.

Table 1. Economic forces influencing biotechnology development in Canada (1981–1991)

- Moribund natural resource sector economy in Canada during this period
- Early absence of an indigenous biotechnology industry
- Limited availability of venture capital to support biotechnology development
- Increase in biotechnology activities amongst Canada's major trading partners (Nb USA, Japan, UK)
- Proprietary information and patenting concerns
- Concerns over availability of skilled human resources to support biotechnology development
- Development of Canada–United States Free Trade Agreement
- Evolution of North American Free Trade Agreement
- Diffuse industry location with disparate regional pockets of expertise
- Sporadic federal government support for basic research
- Limited cost and performance data on biotechnology products and processes
- Need for regulatory clarity by all levels of government regarding its applicability towards biotechnology development
- Low "glamour" of environmental bioindustries compared to applications of biotechnology in agriculture and pharmaceuticals

Source: McIntyre (1990).

Table 2. Environmental forces influencing biotechnology development in Canada (1981–1991)

• Public concerns over dual role of federal government as promoter and regulator
• Questions pertaining to adequacy of environmental regulations and diffuse regulatory responsibilities amongst Canadian federal government departments
• Growing importance of the precautionary principle in dealing with scientific uncertainty
• Increasing priority for capacity building and the need for integration of public participation into government decision making in Canada and at the international level
• Concerns over cumulative impacts from multiple environmental releases of micro-organisms
• Concerns over adequacy of existing pollution control technology to treat emission streams from biology based manufacturing facilities
• Questions over appropriateness of emergency planning techniques which evolved largely from experiences with the nuclear and chemical industries and their applicability to spills of biological materials
• Collapsed product development periods and reduced time frames to fulfill environmental assessment requirements
• "Adequacy" of environmental assessment methodology and its "suspect" ability to accurately discern impacts on biological community
• Fundamental knowledge gaps on microbial biodiversity, microbial ecology, and effects on ecosystem structure and function, biodegradation kinetics etc.
• Questions on relevance of much of the ecological risk assessment data which was based on experiences primarily with higher organisms (plants and animals)
• Suspect risk assessment methodologies which were based on experiences with the nuclear and chemical industries

Source: McIntyre (1990).

As a result of heightened international awareness of the importance of sustainable development and its applicability to biotechnology (World Commission on Environment and Development, 1987; Stewart, 1989; Daly and Cobb, 1990; World Conservation Union, 1991; Giampetro, 1995), the gap between economic and environmental consideration was narrowing considerably.

This flurry of international initiatives resulted in a number of excellent primers that both identified critical objectives for national environmental and development policies to follow the concept of sustainable development and recipes for maintenance of a sustain-

Table 3. Rationale for development of environmental regulatory oversight for biotechnology

• Obligatory requirements for review of regulations as outlined in the Canadian National Biotechnology Strategy, announced in 1983
• Stated federal government Green Plan commitments to make all federal departments more environmentally accountable
• Origin of previous legal framework in Canada based on experiences solely with nuclear and chemical industries, and questions regarding its appropriateness for dealing with biotechnology products and processes
• Rich history of introduction of new technology to society and several examples of untoward environmental experiences
• Enhanced environmental regulatory activity amongst Canada's major trading partners
• Inadequacy of Environment Canada's Environmental Contaminants Act at the time and its limitations for consideration of "animate" substances
• Unique attributes of microorganisms (microscopic in size, can reproduce and mutate, mobility, etc.)
• Exponential increase in biotechnology activities in Canada towards the end of the 1980s
• Increased public anxiety in Canada over the dual role of government as both regulator and promoter of biotechnology development
• Recognition of the importance of regulations as a catalyst for industry and technology development

Source: McIntyre (1990).

Table 4. Critical objectives for national environmental and developmental policies adhering to sustainable development

• Reviving growth
• Changing the quality of growth
• Meeting essential needs for jobs, food, energy, water
• Ensuring a sustainable level of population growth
• Conserving and enhancing the resource base
• Re-orienting technology and managing risk
• Merging environment and economics in decision making

Source: *World Commission on Environment and Development, 1987.*

Table 5. Operating principles for maintenance of a sustainable society

• Respect and care for the community of life
• Improve the quality of human life
• Conserve the Earth's vitality and diversity
• Minimize the depletion of non-renewable resources
• Keep within the Earth's carrying capacity
• Change personal attitudes and practices
• Enable communities to care for their own environments
• Provide a national framework for integrating development and conservation
• Create a global alliance

Source: *(World Conservation Union, United Nations Environment Program and World Wide Fund for Nature, 1991).*

able society—once achieved. Representative examples of these international primers can be seen in Tables 4 and 5.

3. THE COALESCENCE OF BIOTECHNOLOGY AND SUSTAINABLE DEVELOPMENT IN CANADA (1992–1994)

The two-year period from 1992–94 in Canada marked a watershed for sustainable development as a result of its emergence as an important global consideration, as evidenced in the 1992 United Nations Conference on Environment and Development, the World Bank's World Development Report of 1992, and the United Nations 1994 Agenda for Development. Calling for development that "meets the needs of present generations without compromising the needs of future generations," all three reports highlighted the need for governments world-wide, to simultaneously address developmental and environmental imperatives. In particular, the reports identified the need for integration of three major policy clusters into sustainable development decisions by government:

- *Ecological integrity*: The capability of natural systems to maintain their structure and functions in the face of natural pressures, imposed stress, anthropogenic activities, and irregular events.
- *Economic growth*: The improving of standards of living and the quality of life.
- *Social equity*: The fairer distribution of both income and opportunities among existing (intra-generational) and future (inter-generational) populations and the specification of what is to be sustained (Sadler, 1996).

The reports also recognized the global nature of many environmental problems and stressed the importance of major efforts to sustain both development and the quality of life in industrialized countries and in developing nations alike.

In concert with the newfound imperative for global sustainable development, Canada experienced a resultant ground swell of sustainable development activities at all levels of government and industry alike — activities that in the preceding ten years were largely non-existent (Baxter, 1994). Table 6 summarizes these activities.

In Canada, the environmental departments of governments were largely responsible for the job of leading the transition to sustainability, mainly because the initial focus was primarily on areas of environment concern. Within Environment Canada, sustainable development responsibilities, particularly in fulfilment of the department's dual role in biotechnology development, were manifested in a number of ways.

As chair of the Federal Government Working Group on Safety and Regulations in Biotechnology during this period, the department was instrumental in securing federal government consensus for development of a regulatory framework for biotechnology as well as the providing of benchmark legislation for biotechnology products under the Canadian Environmental Protection Act. This framework, developed in 1993, is featured in Table 7.

Although these were considered necessary and encouraging initiatives towards sustainable development in Canada, there still remained a wide gap between declaration and achievement, with the transition to a sustainable economy hamstrung by two important factors. In Canada's market economy, like those of its major competitors, the market place does not allow for internalization of environmental costs; that is, the costs of environmental pollution and resource depletion are not reflected in the prices that Canadians pay for products. This factor has the effect of leading society into pollution and resource de-

Table 6. A synopsis of Canadian sustainable development initiatives 1992–1994

Federal Government
- Signatory to Agenda 21 and Biodiversity Convention
- Establishment of a National Roundtable on Environment and Economy to promote sustainable development
- Signing by all federal ministers of a Guide to Green Government committing them to implement sustainable development practices and procurement policies
- Signing of the Environmental Accountability Partnership Agreement between Environment Canada and Treasury Board, focusing upon waste reduction, energy efficiency, water conservation
- Requirements under the Auditor General Act for individual federal departments to submit sustainable development strategies by 1997
- Laying of groundwork to create a Commissioner of Environment and Sustainable Development (Bill C-83)
- Commitment of the Canadian Council of Ministers of the Environment to rationalize the environmental management framework in Canada
- Undertaking side agreements under NAFTA and GATT for environmental protection
- Establishment of a Task Force on Economic Instruments and Barriers and Disincentives to Sound Environmental Practice

Provincial Government
- Establishment of institutes for sustainable development in Vancouver and Winnipeg
- Development of omnibus environmental legislation in Alberta, Nova Scotia, and Ontario
- Drafting of sustainable development legislation in Manitoba

Municipal Government
- Establishment of over 200 community round tables on sustainable development (NB Hamilton-Wentworth and Ottawa-Carleton Vision 2020, Atlantic Coastal Action Plan)

Professional Associations
- Strengthening of corporate environmental accountability by the Canadian Institute of Chartered Accountants
- Development of environmental standards by the Canadian Standards Association

Source: Government of Canada (1993).

Table 7. Canadian federal government regulatory framework for biotechnology

• Maintain Canada's high standards for the protection of human health and the environment
• Build upon existing legislation and institutions, clarifying jurisdictional responsibilities and avoiding duplication
• Develop guidelines, standards, codes of practice and monitoring capabilities for pre-release assessment of risks associated with release to the environment
• Develop a sound scientific database upon which risk assessments and evaluations of products can be made
• Promote development and enforcement of Canadian regulations in an open and consultative manner, in harmony with national priorities and international approaches
• Foster a favorable climate for development, accelerating innovation and adoption of sustainable Canadian biotechnology products and processes

Source: McIntyre (1996).

pletion. The second factor is considered more significant. Market prices in Canada, amongst its major competitors, are pushed even lower by a variety of government subsidies and other measures. These types of fiscal interventions can lead to a variety of untoward effects that include environmentally adverse processes and greater consumption of products, increased deficits, and reduced employment opportunities (Caccia, 1995).

In light of the foregoing, two federal government programs were initiated. First, in 1995 the federal government Standing Committee on Environment and Sustainable Development undertook a series of general and sectoral panel discussions on the matter of fiscal disincentives to sustainable environmental practices. With this report, the Committee was attempting to undertake the following:

- To understand what is implied in, and how to create a comprehensive baseline study of barriers and fiscal disincentives to sound environmental practices.
- To inject a sense of urgency into the need to conduct a comprehensive baseline study of federal taxes, grants, subsidies, and other fiscal disincentives to sound environmental practices.
- To highlight certain obstacles to achievement of sustainable development particularly evident in the energy and transportation sectors.
- To ensure that the federal government becomes an effective agent for development that is economically, environmentally, and socially sustainable (Caccia, 1995).

Second, Environment Canada began to explore mechanisms to remove barriers to sound environmental practices by examining the use of economic instruments as an adjunct to environmental regulations to assist industry in fulfilling their environmental responsibilities. A number of attractive advantages were found to be associated with the use of market forces in shaping industry behavior in the area of environmental protection - particularly in recognition that:

- Markets are the single most important indicator of technology change.
- Markets offer a very efficient forum for assimilating and processing information.
- Market mechanisms allow great flexibility in designing responses to publicly determined environmental aims.
- Economic mechanisms, particularly those focused on marketable permits allows polluters with relatively low abatement costs to treat their wastes, while allowing those with relatively high abatement costs, to buy permits and thus avoid abatement costs (Environment Canada, 1994).

An overview of those economic instruments is provided in Table 8.

Table 8. Overview of economic instruments for influencing environmental behavior

Environment charges	Includes user, polluter, input, product, and emission and effluent charges
Deposit–refund systems	Calls for the creation of incentives to return used products and packages
Financial incentives	Includes government grants and indirect subsidies to promote environmentally-responsible behavior
Marketable permits	Involves the establishment of a cap or ceiling by regulatory officials on total contaminant releases or on the production, importation, or consumption of a polluting product.

Source: *Environment Canada*, 1994.

Further, the department set about to draft the "environmentally sound management of biotechnology" chapter of the Canadian federal government response to Agenda 21—a synopsis of how Canadians were responding to the sustainable development challenges poised by biotechnology development in Canada in such areas as: increasing the availability of food, feed, and renewable raw materials; enhancing protection of the environment; enhancing safety and developing international mechanisms for co-operation; and establishing enabling mechanisms for the development and the environmentally sound application of biotechnology.

Finally, Environment Canada initiated a series of program elements in support of research and development for innovative environmental applications of biotechnology based largely on its potential for toxics reduction and elimination. The impetus towards wholehearted government support for environmental technology like biotechnology owed its origin to the recognition of the critical importance that this sector was to play in achieving sustainable development (Heaton, 1991 and Derzko, 1996). This increased support for environmental applications of biotechnology was based on a number of factors which are outlined in Table 9. Enhanced support for environmental applications of biotechnology within Environment Canada, although recent in nature, offered various segments of the Canadian environmental bioindustries community the opportunity to initiate research into a variety of innovative and cutting edge technology activities in such areas as biosensors, bioremediation, phytoremediation, biological gas cleaning, and process engineering and systems design.

Table 9. Rationale for Environment Canada support for innovative environmental applications of biotechnology

- Recognition that the environmental bioindustries component was lagging behind enhanced applications of biotechnology in agriculture and human health
- Acknowledgment of the increasing importance of environmental industries as a source of economic development in Canada
- Significant potential of bioremediation for reduction of clean-up costs
- Minimum site disruption offered by a variety of bioremediation techniques
- Relative simplicity of a variety of biological processes for pollution control
- Some regulatory encouragement for the use of biotechnology in Canada
- Possibility for complete contaminant destruction utilizing biological techniques in pollution control and waste treatment
- Demonstration that biosystems were less energy intensive
- Elevated public support for environmental applications of biotechnology in Canada
- Environmental applications of biotechnology can augment physical, chemical, and thermal techniques

Source: *Environment Canada*, 1996.

4. FROM PROMISE TO PRACTICE: RESPONDING TO THE SUSTAINABLE DEVELOPMENT CHALLENGES OF ENHANCED ENVIRONMENTAL APPLICATIONS OF BIOTECHNOLOGY IN CANADA (1994–96)

This period marked an intense effort by Environment Canada to exploit the incredible potential offered by environmental applications of biotechnology—many of which were in keeping with the thrust of the sustainable philosophy being espoused at the national level and amongst many of Canada's major trading partners. Major initiatives included those outlined in Table 10.

Parallel to these departmental initiatives were a series of federal government workshops designed to re-evaluate the effectiveness of the first fifteen years of the National Biotechnology Strategy. Current priority areas for consideration identified by the revitalized National Biotechnology Strategy exercise for 1997 include: improved provisions for public input into the decision-making processes, consideration of the federal government role in social and ethical issues associated with biotechnology, the need for establishment

Table 10. Environment Canada responses to the challenges of environmental biotechnology and sustainable development

- Development of New Substances Notification Regulations for biotechnology products (enzyme and microorganisms) under the Canadian Environmental Protection Act
- Initiation of a series of Interagency Agreements with the U. S. Environmental Protection Agency (USEPA) for test methods development, microcosm validation of soil microorganisms, bioremediation risk assessment techniques, etc.
- Increased in-house support for ecological risk assessment research in such areas as termination procedures for field releases of GEMs, *in situ* techniques for remediation of contaminated sediments, microbial ecological profiles, biological gas cleaning, phytoremediation, soil treatability protocols, and establishment of demonstration projects
- Establishment of an Environment Canada intra-departmental committee on biotechnology representing the key regulatory and promotional responsibility centers involved in all facets of biotechnology development
- Development of an environmental biotechnology curriculum for training sessions and to act as a template for possible utilization in community colleges across Canada
- Reconstitution of a national BIOQUAL network (biotechnology in pollution control sector) with BIOFOR (biotechnology in forestry sector) network and collaboration with BIOREM (biotechnology in agriculture sector) to support biotechnology initiatives in the renewable resources sector
- Development of centrally located demonstration sites for scientists from industry, academia, and government, to evaluate biotechnology products (planned for 1997)
- Joint collaborative project with Industry Canada to support development of environmental bioindustry commercialization projects
- Development of a certification procedure for bioremediation products based on USEPA California model (planned for 1997)
- Completion of an analysis of Canadian environmental biotechnology industries
- Extensive participation in the Senate Committee of Environment and Sustainable Development examination of regulatory issues associated with biotechnology development
- Completion of a study to gage public awareness/support/understanding of environmental applications of biotechnology in Canada
- Conduct a series of International Environmental Management Initiatives to assist host countries in capacity building for developing the necessary environmental infrastructure for utilizing biotechnology products and processes in a sustainable manner
- Secured agreements amongst a number of federal government clients with contaminated sites to establish demonstration projects involving the use biotechnology products and processes
- Initiated discussions amongst financial institutions to assist in targeting the more promising innovative environmental biotechnology products and processes for support

of a National Advisory Committee on Biotechnology responsible directly to federal cabinet, improvements in biotechnology regulations and international harmonization, intellectual property issues, and the acknowledgment of the need for greater emphasis on sustainability principles.

The environmental bioindustry in Canada is currently posed to experience the same explosive growth as that which has occurred with developments from enhanced applications of biotechnology in the agriculture and pharmaceutical sectors. It is only now beginning to make significant contributions to jobs and wealth creation in Canada, with the challenge being to maintain not only that which we have achieved to date, but to take advantage of the incredible potential that environmental applications of biotechnology can offer for future generations.

5. CONCLUSIONS

For the current and future success of environmental biotechnology development in Canada, the evolving federal government National Biotechnology Strategy revitalization initiative and Environment Canada current and evolving biotechnology program elements must be maintained and continued to be based on principles of sustainability. For this to occur, it will require development of stronger linkages and partnerships among stakeholders in industry, academia, environmental non-government organization, other levels of government, and of the public, to enhance overall government abilities to gain and share knowledge in the area of environmental biotechnology. It must acknowledge the need for all stakeholders to integrate sustainable environmental biotechnology knowledge whenever possible, at the local level, where people are more likely to be able to learn, feel, and be empowered to act. It means using a wide array of instruments that include economic incentives, equity considerations, ecoefficient technologies, and regulations. Finally, it means continued focus on narrowing the disciplinary gaps between environmental protection and economic development, while recognizing that economics and ecology can no longer be sensibly considered as separate academic endeavors.

Indeed, the last principle is perhaps the most important. Whereas traditionally in Canada, environmental protection was seen as occurring at the expense of economic development, the new paradigm proposes that environmental protection, economic development, and consideration of quality of life components are in reality complementary objectives that, when properly understood and pursued, are mutually reinforcing. If government does nothing else but reinforce this new way of looking at issues associated with biotechnology development, it will have gone a long way towards nurturing and achieving Canada's most important strategic objectives in advancing Canadian biotechnology priorities - particularly those involving innovative environmental products and processes - in a socially responsible and sustainable manner.

SUGGESTED READINGS

The following references have provided valuable information which has been incorporated into the ideas presented in this article.

Asian Development Bank, (1990) "Economic Policies for Sustainable Development," Times-Offset, Singapore, October.

Beckerman, W., (1995) "Small is Stupid: Blowing the Whistle on the Greens," Duckworth, U. K.

Biswas, A. K., (1989) "Environmental Aspects of Hazardous Waste Management for Developing Countries: Problems and Prospects," in *Hazardous Waste Management*, Maltezou, S. P., et al., eds., UNIDO, Vienna, Austria, pp.261–271.

Carley, M. and Christie, I., (1993) "Managing Sustainable Development," University of Minnesota Press, Minneapolis.

Clifford, G., (1994) "The Multistakeholder Process in Consensus-Building: Guiding Principles," *Optimum*, Autumn, pp.44–49.

Colwell, R. R., (1991), "Risk Assessment in Environmental Biotechnology" In: *Current Opinion in Biotechnology*, V. 2, pp. 470–475.

Current, D., (1994), "Forestry for Sustainable Development: Policy Lessons from Central America and Panama", The Environmental and Natural Resources Policy and Training Project, December, Madison, Wisconsin

Environment Canada, (1995a), "Guiding Principles for Ecosystem Initiatives", Ottawa, Ontario.

Environment Canada, (1995b), "Key Issues in the Transfer of Environmentally Sound Technology at Globe 1994", An OECD Special Session, Ottawa, Ontario.

Giampietro, M., (1995), "Sustainability and Technological Development in Agriculture-A Critical Appraisal of Genetic Engineering", in *BioScience*, V. 44, N. 10, pp. 677–694.

Goodland, R., and Ledoc, G., (1987), Neoclassical Economics and Principles of Sustainable Development," *Ecological Modeling*, V. 38

Goodland, R., Daly, H., and El Serafy, Salah, Eds. (1992), "Population, Technology, and Lifestyles: The Transition to Sustainability," Island Press.

Government Accounting Office, (1994), "Ecosystem Management-Additional Actions Needed to Adequately Test a Promising Approach" GAO/RCED-94–111, Report to Congressional Reporters, Washington, D. C. 20548. pp. 1–9.

Hambleton, P., et al, (1992), "Biosafety Monitoring Devices for Biotechnology Processes", In: *TIBTECH*, V. 10, June.

Haverkort, B. and Hiemstra, W. (1993), "Differentiating the Role of Biotechnology," *Biotechnology and Development Monitor 16*; The Netherlands

Inter-American Institute for Co-operation in Agriculture, (1994) "Common Regulatory Frameworks and Establishment of Biosafety Committees" In, *BINAS News*, V. 1, N. 1, p. 11.

James, C., and Krattiger, A. F., (1994), "The ISAAA Biosafety Initiative: Institutional Capacity Building Through Technology Transfer," In: *Biosafety for Sustainable Agriculture: Sharing Biotechnology Regulatory Experiences of the Western Hemisphere*, Krattiger, A. F., and Rosemarin, A., (eds.) ISAAA: Ithica and Stockholm. pp. 225–237.

King, R. B., et al, (1992), "Practical Environmental Bioremediation", Lewis Publishers Ltd., Ann Arbour

Krattiger, A. F., and Lesser, W. H., (1994) "Biosafety-An Environmental Impact Assessment Tool-and the Role of the Convention on Biological Diversity," In: *Widening Perspectives on Biodiversity*, Krattiger, A. F., et al., (eds.) IUCN: Gland and International Academy of the Environment: Geneva.

Lemon, J., (1994), "Proposals to Improve the Scientific Basis for Environmental Decision Making" In: *The Environmental Professional*, V. 16, pp. 93–192.

National Roundtable on the Environment and Economy, (1994), "Canada and Sustainable Development: Progress or Postponement," *National Roundtable Review*, Fall, Ottawa, Canada.

Nickerson, M., (1993) "Planning for Seven Generations: Guideposts for a Sustainable Future," Voyageur Publishing.

OECD, (1986), "Recombinant DNA Safety Considerations", Paris.

OTA (Office of Technology Assessment), (1991), "Biotechnology in a Global Economy," OTA-Ba-494, Washington D.C.

Parenteau, R., (1988) "Public Participation in Environmental Decision Making," Federal Environmental Assessment Review Office, Ottawa, Canada.

Pearce, D., (1988) "Optimal Prices for Sustainable Development," In D. Colard, D. Pearce, and D. Ulph (eds), *Economics, Growth, and Sustainable Environment*, London, Macmillan.

Potier, M., (1996), "Integratingm Environment and Economy", *The OECD Observer*, N. 198, February-March, pp. 6–10.

Repetto, R., (1990) "Promoting Environmentally Sound Economic Progress: What the North Can Do," *World Resources Inc.*, April, p. 6.

Sasson, A., (1988), "Biotechnologies and Development", UNESCO.

Serageldin, I., (1993), "Making Development Sustainable", In: *Finance and Development*, December, pp.6–11.

Sharples, F., (1993) "Ecological Aspects of Hazard Identification for Environmental Uses of Genetically Engineered Organisms," Publication No. 3554, Environmental Sciences Division, Oak Ridge National Laboratory, Oak Ridge, Tennessee.

Tolba, M. K., (1990), "Building and Institutional Framework for the Future", In: *Environmental Conservation,* V. 17, N. 2, pp. 105–110.

Tzotzos, G. T. (Ed), (1995), "Genetically Modified Organisms-A Guide to Biosafety", Centre for Agriculture and Biosciences International, UNIDO Secretariat.

United Nations, (1995) "Agenda for Development-Celebration of the 50th Anniversary of the United Nations," New York, New York.

United Nations Industrial Development Organization, (1985). "Safety Guidelines and Procedures for Bioscience Based Industry and Other Applied Microbiology," UNIDO Secretariat, Vienna, Austria.

United Nations Council on Sustainable Development, (1995) "Environmentally Sound Management of Biotechnology-Report of the Secretary General," Draft document of the Third Session-Commission on Sustainable Development, April 11–28, pp.2–37.

Visser, B., (1994) "The Prospects for Technical Guidelines for Safety in Biotechnology," in *Biotechnology and Development Monitor,* N. 20, September, pp.21–22,

Wilson, D. C., and Balkau, F., (1990), "Adapting Hazardous Waste Management to the Needs of Developing Countries-An Overview and Guide to Action," in, *Waste Management Research,* V.8, N.7, pp. 87–98.

World Bank (1995) "METAP Views Participation as Key to Capacity Building," *Environment Bulletin,* Winter, V. 7, N. 5, P. 6

Wu, R., (1986), "Building Biotechnology Research and Development Capability in Developing Countries", In: *Capability Building in Biotechnology and Genetic Engineering in Developing Countries,* Vienna: UNIDO, pp. 61–75.

REFERENCES

Allende, J. E., (1990), "Biotechnology Perspectives in Latin America." In: *Biotechnology-Science, Education, and Commercialization, Vasil,* I. K., Ed., Elsevier, New York, USA, pp 165–183.

Baxter, K. H., (1994). "Can We Get Past the Easy Part?" in, *Canada and Sustainable Development: Progress or Postponement?* National Roundtable Review, Fall, Ottawa, Ontario.

Canadian Environmental Advisory Committee, (1987). "Canada and Sustainable Development," Ministry of Supply and Services Canada, Catalogue Number En92–6,1987E.

Caccia, C., (1995). "Keeping a Promise: Towards a Sustainable Budget", *Report of the Standing Committee on Environment and Sustainable Development,* N. 85, December 7.

Cohen, J. I. (1994), "Selected Management Issues Regarding Biotechnology: Increasing Demands and Competing Resources," In: *Turning Priorities Into Feasible Programs.* Proceedings of a Regional Seminar on Planning, Priorities, and Policies for Biotechnology in South East Asia, ISB, SG., pp. 81–86.

Daly, H. E., and Cobb, J. B., (1990). "For the Common Good: Redirecting the Economy Toward Community, the Environment, and a Sustainable Future," Boston: Beacon Press.

Derzko, N. M., (1996). "Using Intellectual Property Law and Regulatory Processes to foster the Innovation and Diffusion of Environmental Technologies" in, *Harvard Environmental Law Review,* V. 20, pp. 3–59.

Environment Canada, (1994). "Economic Instruments", Minister of Supply and Services, Catalogue En-40–224/9–1994

Government of Canada, (1993). "Record of Decision of Canadian Federal Government Sub Group on Safety and Regulations in Biotechnology," Clean Technologies Advancement Division, Environment Canada, Ottawa, Canada.

Heaton, G. H. et al., (1991). "Transforming Technology - An Agenda for Environmentally Sustainable Growth in the 21st Century", *Worldwatch Institute.*

Jaffe, W. (1995), "Biodiversity and Harmonization of Biosafety in Central America and the Dominican Republic." The Inter-American Institute for Co-operation in Agriculture (IICA). *BINAS News,* 1(1), N 11.

McIntyre, T., (1990). "Asleep at the Switch-The Federal Government, Environment, and Planning for High Technology Development-The National Biotechnology Strategy 1981–1991" Unpublished Ph.D. thesis presented to the Faculty of Environmental Sciences, University of Waterloo, Waterloo, Ontario, Canada.

McIntyre, Terry, (1996). "Environment Canada Submission to National Biotechnology Strategy Review Steering Committee - A Strategic Assessment of the Canadian Environmental Biotechnology Industry," Clean Technologies Advancement Division - Environment Canada, Ottawa, Canada, 17 pp.

Ministry of State for Science and Technology, (1981). "Biotechnology: A Development Plan for Canada", Ministry of Supply and Services Canada. Cat. No. ST 31–9/1981E

Nader, C., (1986). "Technology and Democratic Control: The Case of Recombinant DNA", In: *The Gene Splicing Wars,* Zellinskas, R.A. and Zimmerman, B.K., (Eds.), McMillan Publishing Company, New York, pp. 139–167.

Sadler, B., (1996). "Environmental Assessment in a Changing World: Evaluating Practice to improve Performance" International Study of the Effectiveness of Environmental Assessment, Final Report. Canadian Environmental Assessment Agency.

Science Council of Canada, (1977). Uncertain Prospects-Canadian Manufacturing Industry 1971–1977, Minister of Supply and Services Canada.

Stewart, R., (1989). "International Aspects of Biotechnology and Its Uses in the Environment", *Environmental Law Report,* V. 19, N. 11, Harvard University, Cambridge, Massachusetts, pp. 10511–13.

United Nations, (1969). "Man and His Environment: Problems of the Human Environment", Report of the Secretary General-United Nations

United Nations, (1972). "United Nations Conference on the Human Environment", Stockholm.

Weiss, C., (1994), "Technology Modernization in Latin America," *Interciencia*, 19(5), Sept.-Oct., pp 229–238.

World Commission on Environment and Development (WCED), (1987). "Our Common Future," London: Oxford University Press.

World Conservation Union et al., (1991). "Caring for the Earth: A Strategy for Sustainable Living", Gland, Switzerland.

17

ENVIRONMENTAL BIOTECHNOLOGIES IN MEXICO

Potential and Constraints for Development and Diffusion

José Luis Solleiro and Rosario Castañón

CamBioTec (Canada–Latin America Initiative of Biotechnology for Sustainable Development)
Center for Technological Innovation
National University of Mexico
PO Box 20-103, 01000 México

1. INTRODUCTION

The emerging biotechnology revolution is giving rise to the hope that it can provide the basis for more sustainable development, and the field of environmental management is no exception. It is widely recognized that environmental biotechnology will play a major role in efforts aimed at preventing further damage to the environment and combating environmentally related diseases, mainly by generating food at a lower environmental cost, by providing less toxic chemicals or less polluting processes for chemical manufacture, by eliminating toxic chemicals from the environment, and by controlling infectious disease in overcrowded populations (Timmis, 1992).

Environmental biotechnology encompasses methods that use naturally occurring microorganisms to remove pollutants from the environment. Since it involves little energy input, has minor chemical needs in the form of nutrients, and operates at ambient conditions, the cost factors for these systems are believed to be low in relation to other methods for cleaning up environmental contamination (Sayler and Fox, 1991). Perception of all these advantages has provoked an increasing involvement of governments, companies and research institutions in extensive research and development programs in search of useful, cost-effective applications of biotechnology to conserve natural resources and improve environmental quality.

Mexico also has great interest in improving its environmental management sector. For this reason, it has recently adopted an ecological policy that reflects national objectives and also the obligations under international agreements such as those deriving from the United Nations Conference on the Environment and Development (Declaration of Rio, Agenda 21, Non-Binding Declaration on Principles for Forests, Framework Convention on Climatic Change, Framework Convention on the Protection of Biological Diversity). The North American Free Trade Agreement (NAFTA) establishes environmental management standards for the three partners as negotiated in the parallel agreement on environmental protection.

During the last administration, the Mexican government developed a policy framework for environment and sustainable development covering the following areas (Solleiro, 1995):

- inclusion of environmental considerations in economic and social policies;
- ecological policies and their instruments;
- legal and institutional framework;
- fight against poverty;
- social participation;
- protection and promotion of human health;
- education, increasing levels of awareness and training;
- science and technology for sustainable development; and
- international trade and environment.

Priority was given to biotechnology, as a relevant area for sustainable development, particularly for the prevention and control of soil pollution, the correct handling of dangerous substances, and preserving the quality of water and its optimal use.

The last government administration established ambitious programmes for education, research, training and diffusion of information on ecology and the environment. The present administration is trying to give continuity to the approach of the last one, and according to the current National Programme of the Environment (Diario Oficial, April 3, 1996), advances will be made in the following areas:

- to increase the private sector's participation in environmental management activities as a way to both protect the environment and improve the competitiveness of Mexican industry;
- to continue the Pilot Programme for Decentralizing and Strengthening Environmental Management aimed at creating capacities in the state governments for the conception and implementation of sound local environmental management strategies, coordinated with those of the federal government;
- to achieve coordination with environmental actions derived from other federal government programmes such as the National Urban Development Programme and the National Forest and Soil Programme;
- to make reforms to the General Law for Ecological Equilibrium and Environmental Protection with the purpose of consolidating measures for the sustainable use of natural resources, promoting an effective and rational process of decentralization of environmental management, incorporating economic criteria and instruments, reducing discretionary decisions by the authorities, and regarding as crime several activities affecting the natural resources of the country (La Jornada, March 29, 1996).

All the above provides evidence of the interest of the Mexican government in improving the environmental management in the country. This could become a very interesting market opportunity for environmental biotechnologies. It is, however, important to recognize that a number of obstacles must be overcome for the effective application of biotechnologies. This study therefore examines developments in environmental biotechnology against the background of Mexico's recent changes in macro-economic, science and technology and environmental policies. It also tries to identify incentives and constraints in the different phases of research, technology development and diffusion of biotechnology in this area, as well as the roles of different actors participating in the innovation process.

2. METHODOLOGY

This study is strongly based on the experience of the authors in managing technology development and transfer projects in the area of biotechnology. We have managed some of the most important developments in environmental biotechnology conducted by the National University of Mexico. Results of the projects have been transferred to private firms and technologies are currently in use in different locations.

Our follow-up to these experiences (e.g., Biodiscos - aerobic system for waste water treatment, Anaerobic Reactor for Waste Water Treatment, Gas Washer Equipment, Lactic Fermentation of Organic Waste, Market Assessment of Bioremediation in Mexico and Waste Management of the Metal Works Industry in Mexico City) has been complemented by the following activities:

a. We analyzed the main official documents that constitute the economic, legal and environmental background to the development of environmental biotechnologies in Mexico.
b. We also analyzed the recent evolution of Mexico's science and technology policies and the initiatives to implement specific national biotechnology programmes.
c. We identified the main institutions involved in research and development, production and commercialization of environmental biotechnologies in Mexico with the aim of drawing a picture of the possibilities for introducing biotechnologies in the environmental management sector.
d. With the objective of refining our assessment, we conducted interviews with six key actors in successful environmental biotechnology innovations in order to document cases and to obtain more precise information on the factors behind success and the obstacles for innovation in this field.

3. GENERAL OVERVIEW OF THE ENVIRONMENT IN MEXICO

In Mexico, there is currently an increasing awareness of the problems derived from pollution, and interest in preserving and restoring the environment is increasing mainly due to the following complementary factors:

a. The first legal instrument to protect the environment was introduced early in the 1970s, but in the second half of the 1980s, a number of reforms and new laws were added to this initial instrument, establishing a much wider coverage; strengthening regulations for the prevention and control of pollution, protection of health, improvement of safety and hygiene, and protection of the environment; and introducing effective mechanisms for the enforcement of this legal framework.
b. Mexican society, as a whole, has shown increasing concern over the high levels of pollution in some urban and industrial areas, as well as in agricultural regions in which chemical fertilizers and pesticides have been used since many years

back. This growing awareness of environmental problems is generating strong pressure on policy makers and company leaders to implement more and better measures for pollution abatement.

c. Participation in global markets and international trade establishes stricter rules and standards for production processes and products. For some of the most important export products of Mexico, non-tariff barriers can be erected, mainly based on the argument of a lack of sound environmental management practices. Moreover, the challenge for Mexican firms will grow very soon when the relatively new series of standards ISO 14000, establishing international standards for environmental management at the firm level, are implemented as a requirement to export.
d. In the case of Mexico, special attention should be paid to NAFTA. It establishes strict requirements for the use of more sustainable production methods that represent a powerful stimulus for transforming the technological basis of the Mexican productive sector.

Another important driving force for change is, of course, the size and degree of emergency of Mexico's pollution problems. Contaminants are spread throughout the country, in urban, industrial and rural areas. Pollution intensity (measured as the ratio between the annual volume of emissions and the total economic value of production) doubled between 1950 and 1989 (SEDESOL-INE, 1994), considering that the manufacturing industry grew 10 times during the same period and production of pollutants increased 20 fold. At the same time, taking fuel consumption as an indicator of change in the production of industrial emissions, intensity of energy consumption (expressed by unit of production) increased 5.7% between 1970 and 1990, unlike the trend of Organization for Economic Coordination and Development (OECD) countries, where this indicator decreased 35.3% over the same period.

According to a recent study (Mercado et al., 1995), in which intensity and volume of pollutants produced by the manufacturing industry were estimated, the states with the highest intensity of pollution are Tabasco, Chiapas and Veracruz (see Table 1) and the most polluting industrial sectors are oil and derivatives, basic petrochemical industry, basic chemical industry, plastic products, paper and pasteboard, basic iron and steel industries and other metallic products, except machinery (see Table 2).

Numerous factors can be mentioned as promoters of this deterioration of the Mexican environment. The following should be highlighted:

- a clear trend to centralize industrial activities in just a few locations (Mexico City and the state of Mexico alone concentrate 32.5% of the country's value added and 30.5% of total employment);
- indiscriminate application of subsidies to the consumption of resources such as water, electricity, gas and oil for promotion of industry and tourism, which was intensified after the oil "boom" at the end of the 70s;
- late and weak enforcement of environmental legislation;
- lack of financial resources for investment in environmental management issues;
- lack of economic and tax incentives for environment protection;
- lack of technical information for the assessment of alternative solutions to environmental problems;
- inadequate and inaccurate standards for different economic sectors;
- inadequate environmental education; and
- a very low level of investment in new cleaner production technologies.

Table 1. Mexico: Indirect estimate of the intensity and volume of the pollution produced by the manufacturing industry by State, 1993.

State	Intensity	Volume (tons p.a.)
Aguascalientes	3.3	6.1
Baja California	5.3	16.2
Baja California Sur	2.2	0.4
Campeche	3.7	0.6
Coahuila	6.9	52.0
Colima	2.2	0.4
Chiapas	39.1	65.6
Chihuahua	6.9	26.2
Federal District	6.7	173.7
Durango	4.5	7.2
Estado de México	8.8	236.2
Guanajuato	12.1	82.2
Guerrero	2.2	1.0
Hidalgo	5.9	29.2
Jalisco	6.7	78.5
Michoacán	18.2	40.4
Morelos	6.6	17.6
Nayarit	2.1	0.9
Nuevo León	8.0	116.1
Oaxaca	6.4	21.0
Puebla	9.2	62.3
Querétaro	10.1	36.9
Quintana Roo	5.4	1.1
San Luis Potosí	7.8	27.4
Sinaloa	3.9	6.0
Sonora	4.5	18.4
Tabasco	48.7	84.7
Tamaulipas	17.6	101.2
Tlaxcala	15.3	19.3
Veracruz	27.9	286.5
Yucatán	4.7	6.0
Zacatecas	1.9	0.5
National total	10.7	1621.18

Source: Mercado, A. et.al. (1995), "Contaminación Industrial en la zona metropolitana de la Ciudad de México" *Comercio Exterior*, October, 766-77.

3.1. The Main Environmental Problems in Mexico: Market Opportunities

There is no doubt about the fact that the most important environmental problem in Mexico is that of water pollution. SEMARNAP (Mexican Ministry of the Environment, Natural Resources and Fisheries) estimated that, in 1994, total investment in infrastructure for waste water treatment was about 1,091 million dollars. The goal for the year 2000 is to reach a level of investment of 2,700 million dollars (La Jornada, April 1, 1996). Another estimate points out that currently only 12% of the waste water generated in Mexico (around 200 cubic meters per second) is treated. To satisfy this potential demand would require investment in the order of one billion dollars per year. This represents a good market opportunity for companies derived from environmental biotechnologies, but it is important to bear in mind that very strong competition already exists. In 1993 alone, 70 new companies were created in this field, though most of them are partnerships between Mexican distributors and foreign firms.

Table 2. Indirect estimate by sector of the volume of pollutants emitted by the manufacturing industry in Mexico, including the metropolitan zone of Mexico City, 1993

Sector	Total (tons)
Meat and dairy products	10.0
Fruit and vegetables	4.8
Wheat grinding	3.6
Maize grinding	0.9
Sugar	1.7
Oils and edible fats	7.2
Animal feed	1.5
Other food products	2.8
Alcoholic drinks	2.2
Tobacco	1.7
Yarn and woven material made from soft fibers	27.6
Yarn and woven material made from hard fibers	0.8
Other textile industries	5.7
Clothing	14.5
Leather and footware	19.7
Sawmills, threeply and boards	9.2
Other wood and cork products	18.7
Paper and pasteboard	70.8
Printers and publishing houses	62.4
Oil and derivates	91.3
Basic petrochemicals	347.1
Basic chemical products	255.8
Synthetic resins and artificial fibers	30.3
Pharmaceutical products	27.2
Other chemical products	33.7
Rubber products	4.2
Plastic articles	68.5
Glass and glass products	5.4
Cement	4.9
Non-metallic mineral products	7.9
Basic iron and steel industries	68.0
Basic non-ferrous metal industries	38.5
Metal furniture	2.1
Structural metal products	14.1
Other metallic products excluding machinery	50.5
Non-electrical machinery and equipment	6.1
Electrical machinery and apparatus	8.8
Electro-domestic apparatus	4.0
Electronic equipment and apparatus	9.4
Electrical equipment and apparatus	2.8
Vehicles	11.9
Transport equipment and material	0.7
Other manufacturing industries	3.2
Total	1362.1

Regarding the generation of solid waste, Mexico's level of generation of municipal solid waste was estimated to be 29.47 million tons, of which 82.84% is thrown to open-air deposits with very little or no sanitary control. (It is worth mentioning that these data can be very unreliable. This figure is based on estimates of the generation of about 80,746 tons/day of solid municipal waste, while a previous study by the Ministry of Social Development, or SEDESOL, mentioned over 400,000 tons/day in 1992). Thirty percent of this waste is not biodegradable (this percentage was only 5% in 1950). In the case of hazardous wastes, there is not at present a reliable inventory, but projections carried out by the National Ecology Institute led to an estimate of 7.7 million tons of hazardous waste produced in 1994. The most generated hazardous wastes are solvents, oils and fats, paint and varnish, welding material, resins, acids and bases and oil derivatives (SEDESOL-INE, 1994). In these cases, evolution of the market for environmental biotechnologies will depend, on the one hand, on the potential of specific technologies to solve concrete problems and, on the other hand, on the enforcement of regulations. In 1993, seven national official standards were approved dealing with the management of hazardous wastes. These standards describe the characteristics of hazardous wastes, procedures for testing toxicity and the determination of compatibility of two or more residues, requirements for controlled confinement sites and for the design and construction of complementary facilities, requirements for design, construction and operation of confinement cells, and requirements for the operation of confinement facilities (Diario Oficial de la Federación, 1993). SEMARNAP estimates investment in solid-waste treatment facilities will be in the order of one billion dollars over the next four years (La Jornada, April 1, 1996).

Soil pollution represents another area of opportunity for the application of environmental biotechnologies. In this case, a market analysis we conducted recently (Castañón et al., 1995) has shown that even though there is very little information to estimate the actual dimension of the problem of polluted soils in Mexico, there is clear evidence of vast areas presenting serious problems of pollution from oil spills and excessive use of agrochemicals. These areas are spread all over the territory. Some positive signs open windows of opportunity for the development of a market for bioremediation of soils. Maybe the most important one is PEMEX, Mexico's largest enterprise, which is taking the lead and exploring technological options to apply bioremediation techniques to recover polluted areas. Given the economic power of PEMEX and that the company is under strong political and social pressure to come out with effective solutions to the pollution problems it has caused over the last 25 years, an interesting market opportunity is there for those companies with advanced technology and with the possibility of offering cost-effective solutions in the short run.

In the case of bioremediation of soils contaminated with agrochemicals, opportunities will arise only in the medium term, mainly because social and regulatory pressures are not strong enough to force greater investment. It is possible that export requirements will do the job of inducing cash-crop producers to invest in testing and implementing bioremediation techniques. This may occur over a 5–7 year period.

In both cases, for the application of bioremediation it is important to take into account some limitations. One of them is that the lack of qualified people to deal with these techniques makes it almost mandatory to include a capacity building strategy for technology transfer. Bioremediation is still an expensive alternative and Mexico's economic crisis will play against it. Costs have to be brought down for a more successful and rapid diffusion of this technology. Finally, to deal with the big client, PEMEX, establishing partnerships with at least one Mexican firm is mandatory.

Another area of opportunity for the application of environmental biotechnologies in Mexico is the biological treatment and filtration of air and gases. Biotechnology can have a positive influence on the production of equipment, reactive agents and services for measuring, identifying, abating and reducing gaseous contaminants present in combustion, production and ventilation gases as well as those used for hauling volatiles in remediation operations (Revah and Noyola, 1996). In Mexico, up to now, the principal applications deal with the biological treatment of emissions from fixed sources through biowashers and biofilters. These technologies have proven to be useful and cost-effective in handling flows between 1000 and 10,000 cubic meters per hour and concentrations of contaminants in the range of 0.3 to 1.5 grams per cubic meter. Important technological developments have been achieved to improve this equipment, making use of specialized microorganisms, using new supporting materials and combining biological operations with physicochemical ones to improve the absorption of hydrophobic compounds, to support microorganisms in selective membranes and to include previous steps of advanced oxidation in the process. Numerous industrial sectors can apply biofilter technologies in Mexico, but the main competition for this market is the physicochemical option. Biotechnologies have therefore to prove their advantages to make profit from interesting industrial niche markets. Some successful experiences promise to pave the way for biotechnologies. We will provide some details in the last section of this paper.

4. LEGAL AND INSTITUTIONAL FRAMEWORK FOR ENVIRONMENTAL MANAGEMENT

Until November 1994, SEDESOL (Ministry of Social Development) was the governmental body in charge of environmental policy of Mexico. The present administration has created a new ministry to deal with these issues: the SEMARNAP. SEMARNAP's duties are:

- to formulate and lead the general policy for environmental improvement;
- to establish ecological standards and criteria for the sustainable use of natural resources and for the preservation and restoration of environmental quality;
- to monitor and enforce the compliance of standards and programmes regarding environmental protection, defense and restoration, through the creation of ad hoc bodies and the establishment of actions, mechanisms and administrative procedures for such goals, according to applicable laws;
- to establish ecological criteria and general standards for waste water flows;
- to develop proposals for the establishment of natural protected areas of interest to the federation; and
- to assess environmental impact studies corresponding to development projects submitted by different sectors.

SEMARNAP's activities are conducted with the support of two institutions with technical and operative autonomy: the National Ecology Institute (INE) and the Federal Environment Attorney's Office (PFMA). INE is responsible for the development of environmental policies and to keep environmental standards updated. PMFA is in charge of overseeing legislation for the defense, protection and restoration of the environment.

SEMARNAP is also assisted by a National Advisory Council for Sustainable Development in which firms, universities, cooperatives, government institutions and nongovernmental organizations are represented. The Council develops proposals for improvement of legislation and regulatory frameworks. SEMARNAP has also anticipated the creation of a National Council for Environmental Investment, with the objective of generating new projects and informing on investment opportunities in this field. The development of environmental standards is assisted by a National Advisory Committee of Environmental Protection Standards.

Another action to promote private investment in environmental protection and clean-up is the establishment of a long-term agreement among SEMARNAP, the Ministry of Trade and Industrial Promotion (SECOFI) and the National Confederation of Industrial

Chambers (CONCAMIN) to put forward a joint Programme for Environmental Protection and Industrial Competitiveness with the following lines of action:

- development of a new approach to environmental regulation that concurrently promotes environmental efficiency and total quality in industrial processes, with a focus on preventing and minimizing generation of wastes and residues;
- promotion of voluntary environmental initiatives and programs in industry through the commitment of private sector organizations with environmental audits, input substitution, energy efficiency and technological modernization;
- administrative simplification aimed at minimizing bureaucratic procedures;
- development of a joint environmental information system to support decision-making in industry;
- education and training for environmental management in industry;
- promotion of cleaner technologies;
- decentralization of environmental management through the creation of joint regional centers for industrial environmental management;
- financial support for feasibility studies and projects through the promotion of international financial mechanisms; and
- promotion of private investment in environmental infrastructure for handling, recycling, treatment, transportation and destruction of residues, effluents and emissions as well as promotion of the integration of productive chains through sound environmental management.

4.1. Environmental Law

Currently, the most important legal instrument governing environment protection (among other laws such as the General Health Law, 1992; the Phytosanitary Law, 1992; the Federal Labour Law; the National Water Law, 1992; and the New Agrarian Law, 1992) is the General Law for Ecological Equilibrium and Environmental Protection. The Law establishes the framework for the interaction of the federal and local governments in environmental concerns; defines the general policy and the regulations for enforcement; establishes guidelines for the ecological order, for risk and environmental impact assessments, for protection of wild flora and fauna and the sustainable use of natural resources; defines also guidelines for ecological prevention and restoration for air, water and soil, patterns of social participation, and ecological education; and control and safety measures as well as penalties.

Bylaws exist for the following areas:

- environmental impact assessment;
- hazardous residues;
- prevention and control of pollution generated by vehicles circulating in Mexico City; and
- prevention and control of atmospheric pollution.

Table 3 presents a summary of the Law.

4.2. Environmental Policy Instruments

Table 4 presents an overview of the type of policy instruments implemented by the Mexican government to promote private sector participation and investment in environmental management. Table 5 shows the main economic instruments and their potential.

Table 3. General law governing ecological equilibrium and environment protection (summary)

Title	Chapter	Section
First: General provisions	I. Preliminary norms II. Agreement between the federation, the states and the municipalities. III. Powers of the ministry and coordination between the federal public administration agencies and offices ecological policy IV. Ecological policy instruments	1. Ecology planning 2. Ecological ordinance 3. Ecological criteria in promoting development 4. Ecological regulation of human settlements 5. Evaluation of environmental impact 6. Ecological technical norms 7. Measures of protection for natural areas 8. Ecological research and education 9. Information and monitoring
Second: Protected natural areas	I. Category, declarations, and ordinance of protected natural areas II. National system of protected natural areas III. Wild and water flora and fauna	1. Types and character of protected natural areas 2. Declarations to establish, conserve, administer, develop and monitor protected natural areas
Third: Rational use of the natural elements	I. Rational use of water and aquatic ecosystems II. Rational use of the soil and its resources III. Effects of the exploitation of non-renewable resources on the ecological equilibrium	
Fourth: Protection of the environment	I. Prevention and control of atmospheric pollution II. Prevention and control of water pollution and of aquatic ecosystems III. Prevention and control of soil pollution activities considered to imply risk IV. Dangerous materials and residues V. Nuclear energy VI. Noise, vibrations, light and thermal energy, smells and visual pollution	
Fifth (only): Social participation		
Sixth: Control and safety measures, sanctions	I. Observance of the law II. Inspection and supervision III. Safety measures IV. Administrative sanctions V. Appeals in case of dissent VI. Federal offenses VII. People's denouncement	

As can be seen, Mexico has a very complete policy framework and sound definition of legal and economic instruments for protecting the environment. However, the application of this framework is not yet tangible. Mexico has still a long way to go before it can adopt a modern environmental outlook. Environmental protection is still considered by most entrepreneurs and producers to be a luxury reserved for the advanced countries (Solleiro, 1995). The fact is that the productive sector still conceives environmental pro-

Table 4. Principal environmental policy instruments

Instrument	Principal characteristic
Moral persuasion	Seeks, through information, education and persuasion, that the agents change their behaviour
Economic instruments	Modifies agents' behaviour through changes in costs and benefits
Direct control instruments	Introduction of standards and technologies
Government investment	Attempts to make agents change their behaviour through direct investment in infrastructure and other types of support

Source: SEDESOL-INE (1994), "México. Informe de la situación general en materia de equilibrio ecológico y protección al ambiente," Secretaría de Desarrollo Social-Instituto Nacional de Ecología, Mexico.

tection as a matter of excessive cost in order to comply with certain regulations. Only a minority are making a transition to a modern view in which environmental protection is seen as a source of competitive advantage.

Besides this cultural factor, there are some other set obstacles to the proper use of the environmental legislation. According to our experience and the opinion of experts interviewed, the following are the main aspects:

Table 5. Principal economic instruments and the case of Mexico

Instrument	Present applications	Planned applications	Potential applications
Charges for the emission of pollutants	Levying of rights for emitting waste water	None	Charges for waste disposal Charges for emissions that pollute the atmosphere Charges for generating noise
Charges on products	Tax on the purchase of second-hand vehicles Increase in road tax	None	Charges on petrol Charges on packaging Charges on fertilizers and pesticides Charges on detergents
Permits for sale	None	Reduction of substances that affect the ozone layer in order to comply with the Montreal Protocol Reduction in SO_2 emissions	Reductions of emissions from mobile sources Reduction of industrial NO_X emissions
Deposit–reimbursement systems	The instrument is used but not for environmental reasons. The existing schemes generate profits for persons who take part in them	None	Treatment of used batteries Treatment and recycling of used car oil Recycling of solvent containers
Transferable development rights	None	None	Conservation of protected natural area adjacent to zones with a good tourist development, urban and suburban potential
Third party insurance cover	Civil liability for nuclear damage (by Law although it has not yet been applied) Article 153 of the LGEEPA—hazardous waste	None	Handling of highly dangerous substances and hazardous waste

Source: SEDESOL-INE (1994), "México. Informe de la situación general en materia de equilibrio ecológico y protección al ambiente," Secretaría de Desarrollo Social-Instituto Nacional de Ecología, Mexico.

- part of the regulatory framework has been developed as a response to political pressures exerted by different interest groups;
- most people are still unaware of existing programmes, their operation and the advantages they could offer;
- there is no adequate operative framework to implement economic incentives;
- financial support is very limited because of the huge rate of interest for loans and a very passive attitude of financial institutions; and
- lack of continuity of programmes as a result of a short-term planning approach.

5. OVERVIEW OF BIOTECHNOLOGY IN MEXICO

Development of biotechnology in Mexico has taken place in a changing economic context. Mexico used to have an economic development model based on import substitution. Under this model, strong protection to domestic enterprises was granted in the form of high tariffs for imported goods, considerable incentives and subsidies and strong State intervention in the economy. In 1984, Mexico began a drastic transformation, discarding the import-substitution model and the protected economy in favor of export promotion and insertion in the global market. An important part of the new economic strategy has been based on free trade and the Mexican government has placed great expectations on the action of market forces as the main incentives to foster the competitiveness of Mexican industry.

Biotechnology research in Mexico has its origins in the late 60s and early 70s, when a heavy investment in the development of infrastructure and research personnel was made at the main public universities. This laid a solid foundation for the creation of research capacity in some public institutions. Since then, important academic achievements have been made and the number of research groups dealing with biotechnology has grown steadily. Some of these groups have international prestige and are part of global collaborative networks. Recent government scientific policies have given strong stimulus to scientific research projects, training and creation of infrastructure.

However, the development of a biotechnology industry in Mexico has not followed the same path. The number of biotechnology firms is still very low and has not grown in the last five years. On the contrary, three important companies recently suspended their biotechnology operations in the country (Quintero, 1994). In sharp contrast to the scientific policy, Mexican industrial policy has not considered bioindustries as a special objective. That coincides with a sort of indifference by financial institutions towards the creation of biotechnology-based enterprises.

There are in Mexico some incipient programmes to support new technology-based ventures such as incubator facilities and technology parks, with the objective of helping new high-technology businesses by giving them access to facilities as well as managerial and financial support, but they are unspecialized, and one of their main weaknesses is that they are not well funded and there is no venture capital instrument for start-ups. Thus the climate for modern biotechnology companies is not very supportive (Solleiro, 1995).

In summary, Mexico does not have a specific biotechnology strategy and this has resulted in very uneven development dominated by academic achievements. Biotechnology industries are small and their growth has been very slow, in part, because there is no support or guidance for specific goals, since the government has been reluctant to define development priorities. However, important weaknesses are to be found at enterprise level and relate mainly to shortcomings in the definition of sound innovation and marketing strategies (Solleiro et al., 1996).

6. ENVIRONMENTAL BIOTECHNOLOGIES: EXISTING CAPACITIES AND NEEDS

Environmental biotechnology development in Mexico is driven by academic research, as well. Most institutions and research programmes rely on public funding. There are four or five well consolidated research centers performing environmental biotechnology research, all of them located in Mexico City, which gives a clear picture of the concentration of this research area (see Table 6). Another three to four institutions have small groups (or individual researchers) working in areas such as bioremediation of soils, biolixiviation of metals and waste water treatment. Other groups (about 15) work on the development of biopesticides and related subjects of agricultural biotechnology (Solleiro, 1995), but they do not have explicit environmental biotechnology programmes. Very little is being done to build, integrate and strengthen capacities in areas of research considered fundamental for environmental biotechnology such as microbiology, biochemistry, molecular ecology, environmental science, chemical/environmental engineering, physical chemistry and analytical chemistry (Sayler and Fox, 1991). Evidence of this statement are the results of a quick analysis we made of the projects registered in the data bases of UNAM's Environmental Research Programme (PUMA) and ARIES (Inventory of Research Programmes and Resources of Higher Education Institutions). This clearly shows that most projects deal with the diagnosis and evaluation of environmental problems and conditions of different regions; others relate to applied research mainly to adapt existing technologies to local conditions and resources.

Private industry has still very little presence in environmental biotechnology research. As mentioned before, there are important international companies introducing technologies and equipment from abroad, mainly waste water treatment systems. Just a few companies conduct research in this field and all of them do it under collaborative research agreements with the main research labs shown in Table 6.

Table 6. Main Mexican institutions and researchers carrying out projects in the environmental field

A. Universidad Autónoma Metropolitana–Iztapalapa	
Oscar Monroy	Anaerobic waste water treatment
Sergio Revah	Gas treatment by biofilters
Mariano Gutiérrez and Gustavo Viniegra	Solid waste treatment
B. Instituto Politécnico Nacional	
CINVESTAV–D.F.	
Mayra de la Torre	
C. Universidad Nacional Autónoma de México	
1. Instituto de Biotecnología	
	Strongest basic research capability
	Some applied studies in toxicity issues
2. Instituto de Ingeniería	
Adalberto Noyola	Anaerobic waste water treatment
Simón González	Aerobic waste water treatment
Susana Saval	Soil bio-remediation
D. Universidad Autónoma del Estado de Morelos	
Mario Trejo	Bio-remediation of soils polluted by oil
E. Instituto Tecnológico de Durango	
	Bioleaching of metals
F. Instituto Tecnológico de Sonora	
Oscar Cámara	Bio-remediation of soils polluted by agrochemicals
G. Instituto de Ecología–Jalapa	
Eugenia Olguín	Solid waste treatment

Moreover, only a small number of firms have infrastructure and production capacities in Mexico. An analysis of SEMARNAP's Directory of Service Suppliers in Environmental Issues (SEMARNAP, 1995) and of Kompass, a well known industrial directory, leads us to say that most players in the market of environmental products and services are consultants, engineering firms and dealers representing foreign firms. Even in the case of waste water treatment, where the biological option is gaining acceptance, the sector is still under the control of sanitary engineers, who have a strong background in the construction of hydraulic infrastructure, but very little knowledge of biological processes. Our experience with the transfer and diffusion of biological systems for waste water treatment has shown that sanitary engineers are opposed to the introduction of biotechnologies in this area. This suggests that an effort to educate them and to establish technological alliances with these professionals will be critical for biotechnology diffusion for water treatment.

7. LESSONS FOR TECHNOLOGY TRANSFER AND DIFFUSION

As mentioned above, we tried to complement the general analysis with some empirical evidence from successful environmental biotechnology innovations.

One of the first technologies for biological treatment of waste water adapted, improved and put on the market by Mexicans was the biological rotatory reactor. This is an aerobic system developed a long time ago. First attempts to test this technology in Mexico began in the late 1970s and early 1980s. UNAM purchased a system from an American company for water treatment on the campus and also for experimental purposes. In so doing, UNAM's researchers started a technology development project aimed at adapting these systems to locally available materials and improving their performance. An improved technology package was then ready for transfer. After a long search for potential industrial companies to take the new system to the market, which was quite difficult because of lack of credibility and the low level of national investment in pollution clean-up, a major milestone opened the door to technology diffusion: the 1988 environmental law was passed by Congress and the Mexican government promised strict measures to enforce it. It was like magic. Immediately, a company that specialized in hydraulic systems with whom contact had previously been made, signed a technology transfer agreement. Then the university-company realized that the biological part of the technology package was well designed, but the same could not be said for a number of issues related to the equipment. This let us learn our first lesson: strong equipment design capabilities must be available to achieve success. Thanks to the new law, some private companies and municipal governments became interested in these systems and a new problem came to the surface: potential users either do not have information on the quality of their waste water or they do not want to tell the truth. In both cases, a new lesson must be learned: you have to work with your potential client to characterize his/her problem. It does not work at all to offer "package" solutions. Currently, a major problem is that relationships between UNAM and the licensee of the patent have broken down, because of the inadequate terms of the licensing agreement and lack of good will on the part of the entrepreneur. License was granted in exclusivity, which is an obstacle for the university to look for new, more effective licensees. A new lesson for technology transfer is to include very clear exit clauses in the agreement.

A system for the anaerobic treatment of waste water based on the UASB concept was adapted and improved by a research consortium constituted by the Metropolitan Autonomous University (UAM), the Engineering Institute of UNAM and ORSTOM

(France). This technology has already resulted in two patents and has been transferred on a non-exclusive basis to eight Mexican consulting firms (Revah and Noyola, 1996). According to the research leader and one of the licensees, the factors behind the success of this technology have been

- the new environmental legislation along with the political will to enforce it;
- a big market potential where a large part of the required infrastructure is yet to be built;
- anaerobic technology is a mature, ready-to-use option for industrial waste water clean-up;
- capacity exists in the research team and companies to develop specific applications of the system; and
- experience in negotiating terms of licensing agreements let the consortium avoid past mistakes (see above).

A number of obstacles must be overcome, however, for the broad diffusion of these (and other) environmental biotechnologies. According to one of the licensees, the main obstacles are as follows:

- lack of credibility of Mexican firms offering advanced technology solutions;
- multinational firms constitute very powerful competition to Mexican firms which do not have the same resource endowment and lack a consolidated image;
- Mexican firms in this business are usually too small and this limits their financial and productive capacity;
- potential users still believe that previous studies to characterize their pollution problems were a waste of time and money which sets an obstacle for the development of effective solutions;
- there is no financial incentive or instrument to support the user wanting to clean-up waste water or the biotechnology firm;
- for the wide use anaerobic system, there is still a limitation because there is no supply of inoculating sludge.

UAM has developed another successful application: a biofilter system to eliminate H_2S and CS_2 emissions of industrial processes. The R&D project involved laboratory work, and building, starting-up and operating five pilot plants (one in university facilities and four in the company's). Since 1992, an industry-scale plant has operated in the user company. Technology is protected by patents owned by the user, a large Mexican firm in the business of chemical products (CYDSA). CYDSA has already designed and built two additional plants in the USA, which represents a successful technology export. Improvements and new applications of this technology are under way, keeping the collaboration model with the university. Key factors for this success, according to the research leader and the firm's scientist, have been the cultivation of excellent interpersonal relationships based on mutual trust; the high technical level of technology recipient; and adequate negotiation of intellectual property. We want to stress that no recipe for success exists. In this case the rights went to the sponsor firm. As limiting factors, again, the lack of experience and skills to deal with equipment technology can become a major issue. Changes of priorities and management on the firm's side caused discontinuities and delays. For the firm it has been difficult to make a big jump: in this case, from being an environmental biotechnology user to an enterprise developing and selling such technologies.

The last case we documented refers to PEMEX's efforts to select bioremediation technology. A research team led by Dr. Susana Saval (UNAM's Engineering Institute) is assessing alternatives and potential suppliers. The pattern of "innovation" we can see in this area is based on Mexican firms offering bioremediation services, without having internal capacities to adapt and/or improve processes or techniques, and only relying on commercial ties to foreign, mostly American, firms. There is no technology transfer process, mainly because of the lack of interest on the side of the technologist and of assimilation capacity on the side of the recipient. This lack of domestic capacities sets a large limitation for technology diffusion, because, unlike the case of waste water treatment, soil remediation is not amenable to generic methodologies and, for this reason, the way to proceed should begin with an assessment of the soil's problems on a case-by-case basis. Capacity building is therefore a priority, bearing in mind that multidisciplinarity in this case is mandatory, because the biotechnologist's effort must be complemented by those of geologists and hydrogeologists, for instance.

8. CONCLUSIONS

a. Without doubt, the most important factor in promoting the use of environmental biotechnologies is the existence of a broad, strict legislation and adequate instruments to bring it into force.
b. Mexico has made great progress in the development of an environmental policy and regulations framework. However, this is barely a first step, since successful implementation will depend on the presence of a series of factors which are not as yet in existence. One of the most important of these is the lack of funding for environmental investment and serious deficiencies, especially regarding trained human resources, on complying with oversight activities.
c. At a national level, it is important to advance and consolidate the environmental management decentralisation process. It would be a great mistake to think that the deficiencies we have mentioned will be overcome exclusively through measures laid down by the federal government.
d. Greater participation of the private sector in environmental management is essential, but an active environmental policy is needed that incorporates much more aggressive economic incentives and fiscal systems. Furthermore, we consider that this policy must be closely linked to industrial policy (now being drawn up) with the objective of using environmental management as a vehicle for industrial competitiveness.
e. With respect to environmental biotechnologies, it can be clearly observed that there are mature alternatives for treating home and industrial waste water and for washing and filtrating industrial emissions. In the case of these mature options, public institution and enterprise strategies should place emphasis on a wide diffusion that will lead to real applications.
f. There is a group of emerging environmental biotechnologies that, to be put on the market, still depend on the strengthening of research capacities. It is also important to have ease of access to advanced technologies developed in industrialized countries by means of transfer processes in which there is a strong component of domestic capacity building. It should be pointed out that success in technology transfer will depend on national capacity to explore, select and adapt technologies that are suitable for the environmental, economic and social conditions of the country.
g. Mexico's share of global markets leaves no room for doubt: it is essential that firms adopt new concepts of environmental management. This calls for activi-

ties to make firms aware of the matter and greater diffusion of the technological and commercial achievements of those firms that have used environmental management as a powerful tool for competitiveness.

h. Environmental biotechnology is rapidly evolving. Mexico will have to strengthen the academic sector involved in fundamental research in order to be able to follow this evolution. It is not possible to think that with just a few research groups of quality the enormous environmental problem of a country as large as Mexico can be confronted. Strengthening of research in environmental biotechnology demands immediate actions to train human resources and intensify cooperation with academic centres in other countries.

ACKNOWLEDGMENTS

The authors are grateful to Adalberto Noyola, Armando Roa, Sergio Revah and Susana Saval for the valuable information and the constructive comments they gave us.

The authors are exclusively responsible for the opinion expressed in this paper.

REFERENCES

Castañón, R. et al., (1995). "Mexico: business opportunities in soil treatment using bio-remediation techniques. Industry Canada.

Diario Official de la Federación (Official Gazette) dates indicated in text.

La Jornada (1996). Mexico City Daily Newspaper, dates indicated in text.

Mercado, A. et al., (1995). "Contaminación Industrial en la zona metropolitana de la Ciudad de México" Comercio Exterior, October, 766–774

Quintero, R. (1994). "Retrospectiva de la biotecnología en México", BIOCIT Siglo XXI, 3, 8, CIT-UNAM, México.

Revah, S. and Noyola, A. (1996). "El Mercado de la Biotecnología Ambiental en México y las Oportunidades de Vinculación Universidad-Industria" in E. Galindo (editor) "Fronteras en Biotecnología y Bioingeniería", Sociedad Mexicana de Biotecnología y Bioingeniería, México (in press)

Sayler, G. and Fox, R. (1991). "Environmental Biotechnology: Perceptions, Reality, and Applications", in Sayler et al., (eds.) "Environmental Biotechnology for Waste Treatment", Plenum Press, New York, 1–13

SEDESOL (1992). "Informe nacional del ambiente 1989–1991 para la Conferencia de las Naciones Unidas sobre Medio Ambiente y Desarrollo", Secretaría de Desarrollo Social, Mexico

SEDESOL-INE (1994). "México. Informe de la situación general en materia de equilibrio ecológico y protección al ambiente", Secretaría de Desarrollo Social-Instituto Nacional de Ecología, Mexico

SEMARNAP (1995). "Padrón de prestadores de servicios en materia de impacto ambiental". México.

Solleiro, J.L. (1995). "Biotechnology and Sustainable Agriculture: the Case of Mexico", *Technical Papers* 105, OECD Development Centre, Paris

Solleiro, J.L. et.al. (1996). "Innovation strategies for follower biotechnology firms: Business development under adversity" in Mason, Lefebvre and Khalil (editors) Management of technology V, Technology Management in a Changing World. Elsevier Science Ltd., England; 243–252

Timmis, K.N. (1992). "Environmental biotechnology. Editorial Overview", *Current Opinion in Biotechnology* 3, 225–226

18

ENVIRONMENTAL BIOTECHNOLOGY

The Japan Perspective

Osami Yagi and Minoru Nishimura

[1]Water and Soil Environment Division
National Institute for Environmental Studies
Japan Environment Agency, 16-2, Onogawa
Tsukuba, Ibaraki, 305 Japan
[2]Japan Research Institute Ltd., 16 Ichiban-cyo
Chiyoda-Ku, Tokyo, 102 Japan

1. INTRODUCTION

Soil and groundwater pollution by toxic chemicals has become a major issue in recent years in Japan. In 1994, 232 cases of soil contamination were reported, and major contaminants were organochlorine compounds such as trichloroethylene (TCE), tetrachloroethylene (PCE) and heavy metals (Environment Agency of Japan, 1995). The Environment Agency of Japan formulated the Environmental Quality Standards for soil in February, 1994, to protect human health and conserve the natural environment, and adopted the following values: TCE, 0.03 milligrams per liter (mg/l); PCE, 0.01 mg/l; and l,l,l-trichloroethane (TCA): 1 mg/l. The standards are basically applied to all kinds of soil.

Table 1 shows the survey of groundwater pollution by volatile chlorinated compounds (VOCs). Before 1988, TCE-contaminated wells (as determined by the drinking water quality standard above) accounted for 2.7%, PCE for 4.1%, and TCA for 0.2% of the wells which were surveyed. By 1993, these percentages had dropped to 0.3%, 0.5% and 0.0%, respectively. More than 2,000 wells polluted by volatile chlorinated organic compounds were detected.

Various attempts are now being made to develop and evaluate soil clean-up technologies in Japan. For heavy metals, solidification, sealing and excavation are the technologies commonly applied. For chlorinated compounds, soil vapor extraction and groundwater pumping followed by air stripping are very common methods. Physical and chemical methods for remediation are expensive. Therefore, less expensive technology leading to complete elimination of pollutants is required. Emphasis has been placed on in-situ bioremediation, because TCE and PCE are the major agents of soil and groundwater pollution in Japan. There are many reports about TCE degradation by aerobic bacteria, such as phenol degraders, toluene degraders, ammonium oxidizing bacteria, propane de-

Biotechnology in the Sustainable Environment, edited by Sayler *et al.*
Plenum Press, New York, 1997

Table 1. Survey of groundwater pollution by VOCs

Year	Chemicals	Surveyed wells	Polluted wells	Ratio (%)
1984–1988	TCE	26607	722	2.7
	PCE	26594	1078	4.1
	TCA	2657	46	0.2
1989	TCE	3388	30	0.9
	PCE	3388	42	1.2
	TCA	2569	2	0.1
1990	TCE	5817	44	0.8
	PCE	5817	79	1.4
	TCA	4515	1	0.0
1991	TCE	6158	27	0.4
	PCE	6518	44	0.7
	TCA	5135	0	0.0
1992	TCE	4762	18	0.4
	PCE	4762	35	0.7
	TCA	3952	3	0.1
1993	TCE	4480	15	0.3
	PCE	4480	24	0.5
	TCA	3960	0	0.0

Drinking water standard: TCE 0.03 mg/l, PCE 0.01 mg/l, TCA 0.3 mg/l.

graders and methanotrophs. A field pilot test for in situ bioremediation of TCE contaminated soil and groundwater was evaluated, which follows.

2. SITE CHARACTERIZATION

Characterization of the TCE contaminated area was carried out. TCE contamination of well water was detected in 1990. The contaminant was mainly TCE and the contaminated soil zone included the upper 1st aquifer, from a 14 meter (m) to 23 m depth. Contamination originated from an electric part factory. It was determined that groundwater pollution could be alleviated by pumping and air stripping, as well as the use of activated carbon.

To evaluate the application of bioremediation, the groundwater quality of the site was analyzed (Table 2). The pH value was 6.28, temperature was 16.2°C, and the amount

Table 2. Summary of site characterization (groundwater at C city)

Depth from surface (m)	14–23
pH	6.28
Temperature	16.2
Dissolved oxygen (mg/l)	4.94
TCE (mg/l)	5.26–6.5
C-DCE (mg/l)	0.042–0.05
VC (mg/l)	0
Total carbon (mg/l)	—
Total nitrogen (mg/l)	7.5
Total phosphorus (mg/l)	0.01
Copper (mg/l)	0.1
Aerobic heterotrophs (CFU/ml)	2×10^2
Methanotrophs (MPN/ml)	1×10^2

of dissolved oxygen was 4.9 mg/l. TCE concentration ranged from 5.3 to 6.5 mg/l. Total nitrogen and total phosphorus content were 7.5 mg/l and 0.01 mg/l, respectively. The number of methanotrophs was 10^2 cells/milliliter.

3. BIOTREATABILITY TESTING

A biotreatability test was conducted. Soil and groundwater samples collected from the contaminated layer were incubated in a mineral salt medium containing methane as a sole carbon source using serum bottles. A ten-day culture was examined for TCE degradation ability, and 4.5 mg/l of TCE decreased to 1.1 mg/l after 48 hours of incubation. Without bacteria 20% of the amount of TCE decreased after 48 hours (Figure 1). These results showed that TCE degrading methanotrophs were present in this contaminated area. Therefore, it was planned to conduct in situ bioremediation using methane and oxygen injection.

4. IN SITU BIOREMEDIATION

Figure 2 shows the in situ bioremediation design for this site. Remediation consisted of pumping up to 100 metric tons/day of groundwater from the extraction well (EW) and air stripping. At this site, three injection wells (IWl, IW2, IW3) were constructed; IW1 and IW2 were for oxygen and IW3 was for methane. Two monitoring wells (MWl, MW2) were also constructed. The purpose of in situ bioremediation was to stimulate the activity of the methanotrophs by methane and oxygen injection, and to reduce the TCE concentration in MW2. One hundred metric tons of water were continuously pumped up, and 50 metric tons of treated water were injected per day.

Figure 3 shows a diagrammatic representation of the in situ bioremediation facilities. The distance between IW and MW1 and MW1 and MW2 was 3.5 m. The distance from MW2 to EW was 5 m. Oxygen, methane, 7 mg/l of nitrogen, and 25 mg/l of phosphorus were applied to the injection water. The screen of the well was located at a depth from 14 m to 23 m where the groundwater and the soil layer were contaminated by TCE. Before bioremediation, the flow rate of groundwater was normally 10 centimeters per day (cm/day). However, after injection and pumping, the flow rate increased to 60 cm/day. Therefore, it took 10 days for groundwater to reach MW2 from IW.

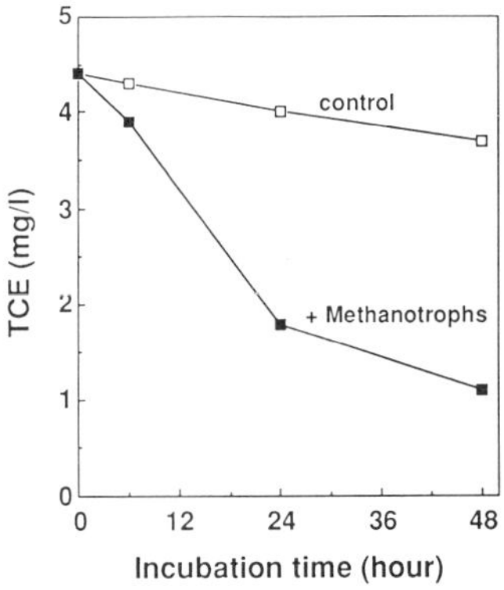

Figure 1. TCE degradation by methanotrophic enrichments from contaminated sites (flask test).

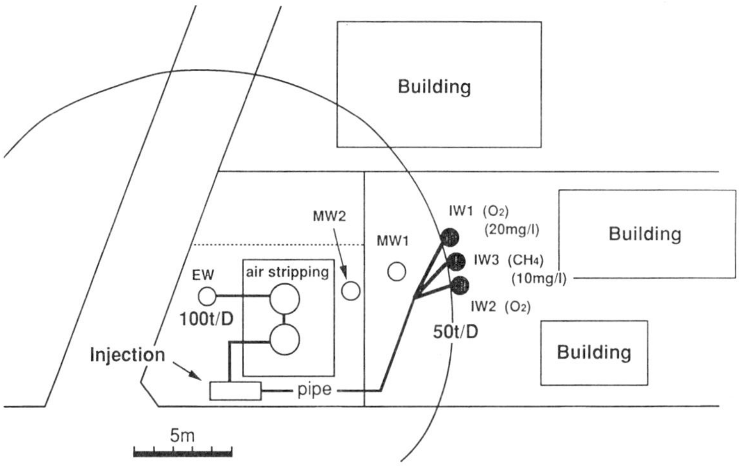

Figure 2. In situ bioremediation design.

Monitoring parameters for in situ bioremediation using methanotrophs included the determination of the number of methanotrophs, activity of soluble methane-monooxygenase in methanotrophs and number of aerobic heterotrophs. The amounts of trichloroethylene and TCE by-products such as 1,1-, cis- and transdichloroethylene and vinyl chloride were measured. Temperature, pH, dissolved oxygen, methane content, nitrogen and phosphorus were also determined.

Figure 4 shows the changes in the TCE concentration in IW3, MW1 and MW2. Before bioremediation, the TCE concentration in IW3 was about 7 mg/l. At first, oxygen-amended water was injected using IW1 and IW2. After 10 days, methane-amended water was injected using IW3. After 35 days, oxygen and methane injection was stopped. Therefore, the period from time 0 to 35 days corresponded to the growing phase of the methanotrophs, and after 35 days the degradation phase started. After oxygen injection, the TCE concentration decreased to less than 0.02 mg/1.

Methane emission is considered to play a major role in global warming. However, at the concentrations used in this trial, methane was completely metabolized by the methanotrophs. This low concentration was not associated with TCE degradation.

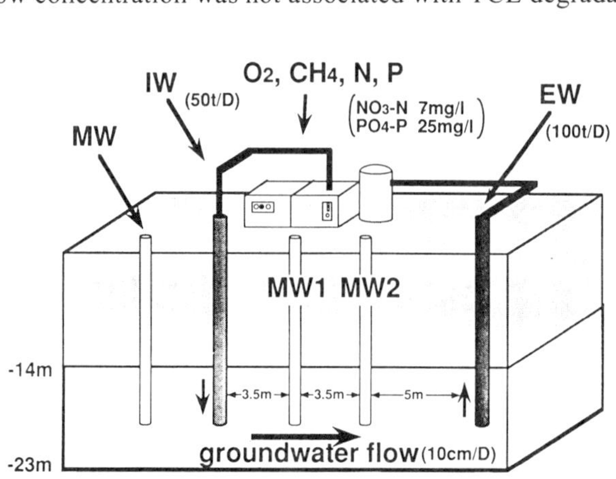

Figure 3. View of in situ bioremediation facilities for TCE contaminated groundwater.

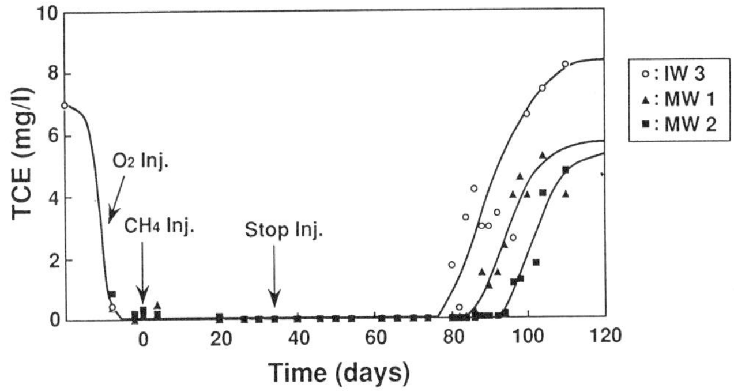

Figure 4. TCE concentration in injection and monitoring wells.

Injection of 50 metric tons of water per day increased the water level of the monitoring wells. However, contaminated groundwater did not reach MW1 and MW2 during the injection phase. After the discontinuation of the oxygen and methane injection, the contaminated water would reach MW1 and MW2 within 10 days, theoretically. Therefore, if methanotrophs did not grow, the TCE concentration in MW2 should have increased at 45 days. However, the TCE concentration in MW2 did not increase until after 90 days. Therefore, it appeared that the TCE degradation activity continued for about 40 days after the interruption of the injection.

Numbers of methanotrophs in the injection and monitoring wells were determined by the most probable number (MPN) method using methane as a sole carbon source (Figure 5). In this groundwater, methanotrophs were naturally present at numbers ranging form 10 to 10^3. After oxygen and methane injection, the number methanotrophs increased rapidly to 10^5 and 10^6, and did not decrease immediately after the interruption of the injection. When the number of methanotrophs decreased to 10^4 levels, the TCE concentration increased gradually in the injection wells. The density of methanotrophs in MW1 and MW2 increased within 20 days after the methane injection. The numbers of methanotrophs in MW1 and MW2 were significantly smaller than in the injection wells. After 80 days, the number decreased to 10^3 and the TCE degradation activity stopped. Methanotroph density was presumed to be indicative of TCE degradation activity.

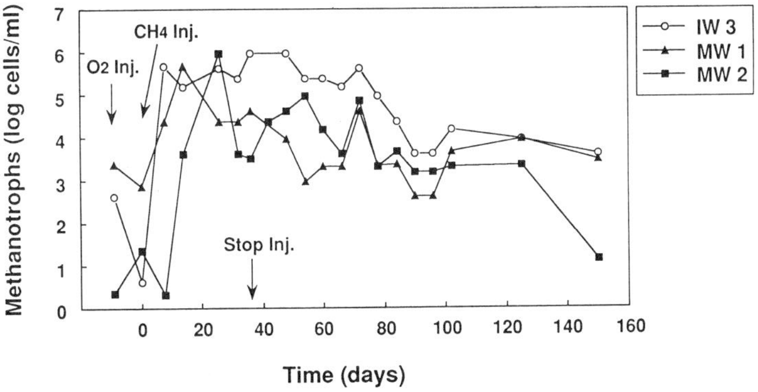

Figure 5. Change of methanotrophs.

5. RISK AND EFFICACY EVALUATION

In the application of bioremediation, risk assessment, in order to determine the impact of metabolite concentration on ecosystems, is very important. In the soil environment, TCE was transformed to 1,1-, cis-, and trans-dichloroethylene, vinyl chloride and chloroacetic acid (Vogel 1985, Nakajima 1992). Dichloroethylene concentration in the injection and monitoring wells was usually less than 0.02 mg/l during the 100 day period. The concentration of vinyl chloride was always less than 0.001 mg/l. Since these concentrations of dichloroethylene and vinyl chloride were so low, it thus appeared that the aquifer was maintained under aerobic conditions.

Efficacy measurement was carried out by determining the changes in the TCE concentration of the injection, monitoring and extraction wells during the 40 day period of degradation. TCE concentration in MW1, before bioremediation, was 6.7 mg/l. During the degradation period, the average TCE concentration was 0.025 mg/l and 99.6% of TCE was removed through the 3.5 m soil layer.

The amount of TCE degradation was calculated based on the changes of TCE in the extraction well. One hundred metric tons per day of groundwater were continuously extracted and the average TCE concentration ranged from 0.8 mg/l to 0.5 mg/l. Before bioremediation, the amount of TCE in the groundwater was calculated as follows:

$$0.8 \text{ mg/l} \times 40 \text{ days} \times 100 \text{ metric tons/day} \times 1000 \text{ l/metric ton} \times 1 \text{ kg}/10^6 \text{ mg} = 3.2 \text{ kg TCE.}$$

During bioremediation, the amount of TCE in the groundwater was:

$$0.5 \text{ mg/l} \;\; 40 \text{ days} \times 100 \text{ metric tons/day} \times 1000 \text{ l/metric ton} \times 1 \text{ kg}/10^6 \text{ mg} = 2.0 \text{ kg TCE.}$$

The difference of 1.2 kg represents the amount of degradation.

6. CONCLUSION

Based on this field test, it is suggested that biostimulation technology using methane and oxygen injection is effective in the remediation of TCE-contaminated soil and groundwater (Little, 1988; Nelson, 1990; Yagi, 1994). It is important to develop a bioaugmentation technology for the cleanup of soil and groundwater contaminated with volatile chlorinated compounds. We are developing a bioaugmentation technology using methanotrophs which were isolated and identified as *Methylocystis* sp. strain M. The strain M can degrade 30 mg/l of TCE (Uchiyama, 1989). We place emphasis on the development of effective evaluation methods, such as the determination of toxicity, movement and survival of methanotrophs, presence of metabolites such as di- and trichloroacetic acids, amended nitrogen concentration, and the fate of TCE and its influence on ecosystems. These data are necessary to win public acceptance.

REFERENCES

Environment Agency of Japan (1995). Environmental White Paper, Printing Bureau of the Finance Ministry, Tokyo.

Little, C. D., A. V. Palumbo, S. E. Herbes, M. E. Lidstrom, R. L. Tyndall, and P J. Gilmer (1988). Trichloroethylene Biodegradation by a Methane-oxidizing Bacterium., *Appl. Environ. Microbiol.* 54:951–956.

Nakajima, T., H. Uchiyama, O. Yagi, and T. Nakahara (1992). Novel Metabolite of Trichloroethylene in a Methanotrophic Bacterium, *Methylocystis* sp. M and Hypothetical Degradation Pathway., *Biosci. Biotech. Biochem.* 56:486–489.

Nelson, M. J., J.V. Kinsella, and T. Montoya (1990). In Situ Biodegradation of TCE Contaminated Groundwater, *Environ. Prog.* 9:190–196.

Uchiyama, H., T. Nakaiima, O. Yagi, and T. Tabuchi (1989). Aerobic Degradation of Trichloroethylene by a New Methane-Utilizing Bacterium Strain M, Type 2., *Agric. Biol. Chem.* 53:2903–2907.

Vogel, T.M. and P.O. McCarty (1985). Biotransformation of Tetrachloroethylene, Dichloroethylene, Vinyl Chloride and Carbon Dioxide under Methanogenic Conditions, *Appl. Environ. Microbiol.* 49:1080–1083.

Yagi, O., H. Uchiyama, and K. Iwasaki (1994). *Bioremediation of Trichloroethylene Contaminated Soils by a Methane-utilizing Bacterium Methylocystis* sp. M., *Bioremediation of Chlorinated PHA Compounds*, Lewis Publishers, 28–36.

19

ENVIRONMENTALLY ACCEPTABLE ENDPOINTS

The Scientific Approach to Clean-up Levels

Hon Don Ritter

National Environmental Policy Institute
Washington, DC

The following question has baffled many in making environmental decisions: *How clean is clean?* Decision-makers engaged in answering this question ranged from the White House to the State House to the school house—with Congress in between—and was especially an issue during the reauthorization of Superfund, the Clean Air Act, and the Resource Conservation and Recovery Act (RCRA). In the process of answering this question, another question is raised: *What will it cost?* In other words, how many hundreds of billions of dollars must this nation spend to meet the requirements of these acts? This, in turn, brings about a third question: *What government or societal programs will suffer if we spend these huge amounts of money on environmental cleanup?* Originally, most of these costs were borne by industry. With the requirement that federal facilities meet the same requirements, billions and billions of dollars of costs are now allocated to the taxpayer. The Department of Energy (DOE) alone had a $6.6 billion dollar remediation budget for 1996. They are forecasting that remediation of their sites will take 30 to 75 years. In times of limited budgets, do these funds come from Social Security, Medicare, education, crime control, or defense? If not, where do the funds come from, when many feel that the citizenry is already overtaxed?

The United States is at a crossroads in environmental policy formulation and implementation. If the US is to capture greater environmental benefit from vast investments in regulation and cleanup, the politicians, regulators and the general public must ask how clean is clean, or what is an environmentally acceptable end point (EAE). Good science, risk assessment, economics and decisionmaking flexibility can help provide the answers. An involved and better educated citizenry, working together to build trust and understanding, are crucial to the answer. If science shows that many of the compounds of concern are complexed in the soil and do not represent a true risk to human health, wildlife, plants or trees, then the process of risk assessment must take this into account.

In 1993, the National Environmental Policy Institute (NEPI) was formed. NEPI is a non-profit bipartisan organization devoted to advancing new ideas for developing environ-

Biotechnology in the Sustainable Environment, edited by Sayler *et al.*
Plenum Press, New York, 1997

mental policies based on sound science, consideration of risks, costs and benefits, and the involvement of new constituencies. NEPI's early efforts focused on systemic policy reform through the "Reinventing EPA/Environmental Policy Working Group" chaired by Professor Bruce Piasecki of the Rennsselaer Polytechnic Institute in New York State. Within this working-group NEPI formed a "Science and Risk Assessment" sector chaired by Don Elliott (former General Counsel at the US Environmental Protection Agency) and Dr. Jack Moore (former Assistant Administrator for Prevention, Pesticides and Toxic Substances). The group progressed towards developing a consensus opinion on the key elements of good risk assessments and the whole issue of ensuring integrity of science in the regulatory process. NEPI became involved with EAEs through work with Dr. Tom Roose of the Gas Research Institute (GRI) and Dr. Dave Nakles of Carnegie Mellon University. Currently, NEPI is examining the policy implications of the EAE issues through the help of some of the most knowledgeable experts in the field, who are solution-oriented and are capable of distinguishing the political agenda from the science.

Other problems in risk assessments have involved the selective use of only the worst case scientific data while at the same time ignoring other data that in many cases showed a lesser or zero effect. In some cases human data were ignored and only animal data used. When EAE considerations are taken into account, it becomes apparent that a far lower level of conservatism is required. Full acceptance may not come easily. The Dioxin Risk Assessment was sent back for further review by the US Environmental Protection Agency's (USEPA) own Scientific Advisory Board based on the fact that its summary did not match the detailed scientific content of the assessment. This assessment was probably the most extensive risk assessment ever conducted. In addition, more scientific research has been conducted on dioxin than probably any other compound.

Some of the preliminary conclusions of the NEPI Science and Risk study are summarized below.

- All available data should be included in the assessment, including data that shows no effect. An emphasis should be placed on data that has been peer reviewed in the scientific literature or developed in accordance with good laboratory practice.
- The assessment should be open and transparent, meaning that sources should be identified and confidence in the data discussed. If data is ignored, the reasons for ignoring it should be stated.
- When conflicts in the scientific data exist, the risk assessment shall include alternative interpretations of the data with an emphasis on the most likely conclusion.
- An iterative process should be used starting with a relatively simple screening analysis and progressing to a more rigorous analysis if the risks justify it.
- In determining the need to proceed to a more detailed analysis, consideration should be given to whether additional data is likely to significantly change the estimate of risk.
- Realistic exposure assessments should be used. An explanation of the exposure scenarios should be provided including an estimate of the population at risk and the likelihood of the exposure scenario.
- Default assumptions, inferences, models, or safety factors should not be used when good scientific data and scientific understanding, including-site specific data, are available.
- The levels of conservatism used in the assessment, as well as the compounding effect of multiple levels of conservatism, should be discussed.
- Substitution risks should be taken into account in the assessment.

- Scientific peer review of the assessment should be conducted and opportunity provided for public participation and comment.

Following early work under the Reinventing EPA/Environmental Policy Working Group, it was recognized that the issue of *how clean is clean?* is central to the reinvention of environmental policy. Environmental policy cannot be reinvented without addressing the approach to cleanup and cleanup standards, a rational treatment of chemicals and more importantly, the underlying science, risk and cost determinations. On this basis NEPI formed the "How Clean is Clean Working Group," chaired nationally by Lynn Scarlett, Vice President of the Reason Foundation with Dr. Winston Porter as project director. Dr. Porter was formerly Assistant Administrator for Solid Waste and Emergency Response at the USEPA, where he also served as National Program Manager for the Superfund and RCRA programs, and is one of the foremost experts in designing cost-effective solutions to real clean up problems.

Under this leadership, the How Clean is Clean? project involved approximately 100 public and private experts in various fields including industry, the environmental and scientific communities, academia, and government at all levels. Four major areas are being addressed:

- Clean Up and Corrective Action
- Federal Facilities
- Brownfields
- Chemical Emissions

A primary objective of the work is to uncover common themes and principles across these individual areas and to involve new constituencies in the decision process. Environmentally Acceptable Endpoints is one of these themes. It is an issue in the first three categories and in some components of the fourth.

"*How Clean Is Clean? First Phase Report*" (NEPI, 1995) highlighted the following principles:

- The key decision maker should be closest to the work to be done, most likely at a state or local level. Unlike some problems such as air pollution, which can effect large regions of the country, waste sites are generally local problems and cleanup decisions are strongly influenced by local environmental, economic, and land-use considerations. Local citizens have shown much common sense when given information on risk and land use.
- Consideration of costs, benefits and risk assessment should be given a central role in determining how clean is "clean enough," and for what purpose. Once again sharing such information with concerned citizens is quite often helpful to making rational decisions.
- Voluntary and other methods to accelerate cleanup activities should be facilitated.
- The role of the at-risk stakeholders should be increased in the cleanup process as opposed to the role of national intervenors.
- The focus should be shifted from ever increasing lists of bad chemicals to the more relevant approach of assessing chemical toxicity along with concentration and exposure in soil and water, or looking at facility-wide or "bubbling" approaches for air. In other words, all chemicals can be benign or harmful depending on their location, concentration, bioavailability and usage. Adding EAE considerations for soils certainly changes the nature of the discussion vis a vis concentration.

- With respect to chemical lists, it is time to compare the reduction in risks and the gains in health and environmental quality due to the recent sharp reductions in emissions before simply adding more chemicals to the lists. It is time to look at whole facilities' emissions rather than the myriad of operations going on inside.

The approaches summarized above also tie in well with the guide for risk-based corrective action that was developed by the American Society for Testing and Materials (ASTM, 1994) for application to petroleum release sites. This guidance, more commonly known as RBCA (or "Rebecca") standardizes existing Superfund risk assessment methods for use at other locations and promotes the concentration of resources at sites that pose the greatest threat to human health and the environment. The RBCA approach recognizes the diversity of parameters affecting risk, such as contaminant complexity, physical and chemical characteristics. It uses a tiered approach that tailors assessment and remediation activities to site-specific conditions and risks. RBCA is only one example of a flexible, tiered risk-based framework. Others can be developed as needed.

The political debate on the issues discussed above has become very polarized, particularly when risk started out as a bipartisan issue. A number of risk bills were introduced in the 103rd Congress. Senator Patrick Moynihan (D-NY) introduced a Risk Reduction Act. Senator Bennett Johnston (D-LA) introduced a risk amendment to the USEPA Cabinet Elevation bill that passed in the Senate by a vote of 93 to 3. In the same time frame, President Clinton issued an Executive order requiring all government agencies to conduct risk assessments and cost-benefit analyses on all new regulations that would have an impact on the economy of greater than $100 million per year. More and more risk assessment input is becoming a part and parcel of the regulators' modus operandi. For example, the President's Commission on Risk issued its report in June, 1996 and the USEPA issued its Risk Guidance this same year.

With the arrival of the new Republican Majority in the 104th Congress, risk and cost benefit analysis were incorporated into the Contract with America. The House risk bill HR 9 passed with a vote of 353 to 142. Senator Dole incorporated risk and cost benefit analysis in his Regulatory Reform bill S-343. With the Presidential election still less than eighteen months away, the vote on the bill was very close to a party line vote. Senator Dole was unable to obtain the 60 votes required for cloture. After that, Senator Robb (D-VA) made a number of unsuccessful attempts to move forward with a less controversial reform bill.

Despite the lack of progress on the legislative front, progress was made in other areas. The National Academy of Public Administration (NAPA), at the request of Barbara Mikulski (D-MD) (previous Chair of the Senate VA/HUD and Independent Agencies Appropriations subcommittee), completed the report "Setting Priorities, Getting Results, A New Direction For EPA" (NAPA, 1995). This report was well received by the Senate Appropriations subcommittee, now led by former ranking member, Sen. Kit Bond. who is taking an active role in follow-up.

Recommendations included the use of comparative risk analyses to inform the selection of priorities and the development of specific program strategies within USEPA. It also recommended that the USEPA refine and expand its use of risk analysis and cost-benefit analysis in making decisions.

The USEPA is currently incorporating the recommendations of the NAPA report into their operations. They have reorganized their laboratories along risk lines and have formed a new central planning and budget office. It is intended that this new office will give science, risk and cost-benefit a stronger role in setting USEPA priorities. In addition, the

USEPA has also made assignments to strengthen the role of scientific peer review within the agency.

The NEPI book "Reinventing the Vehicle for Environmental Management" (NEPI, 1995) highlights the need for a systemic and holistic reinvention of the USEPA and reinvigorate our national environmental policies. In the next century, NEPI envisions an era in which environmental quality is enhanced using approaches that are far more effective and less expensive than so many of today's "command-and-control" rules.

NEPI recommends that the architecture be built around a new model that is based on States having greater authority to allow more flexible regulations, partnerships resulting in increased decision making at the local level, and increased use of flexible, voluntary and free market approaches to compliance. To enhance this transition, an Integrating Environmental Statute is being developed.

In the other agencies, DOE is doing major risk assessment work. They have issued a report entitled "Risks and the Risk Debate-Searching for Common Ground" (USDOE, 1995). The report addresses risk assessment, risk management, risk communication and priority setting with DOE's remediation programs. Also, the US Department of Agriculture has formed an Office of Risk Assessment and Cost-Benefit Analysis. Again, while Congress did not score, the debate that has emerged has had a very significant impact.

So how does all of this relate to Biotechnology in the Sustainable Environment? One of the most expensive components of remediation is when soil removal and/or treatment is required. Bioremediation, natural attenuation or a combination of the two down to EAE levels offers a way to remediate soil in-situ at a far lower cost as compared to conventional soil removal and/or treatment technologies. It reduces the potential of increased exposure due to soil removal and shipping. It avoids the painful "Not in My Backyard" (or NIMBY) issue involved with soil incineration.

A detailed understanding of bioavailability is key to *the how clean is clean?* issue. The conventional wisdom assumes that the chemical concentration of a contaminant in a medium defines toxicity and, given a certain exposure, health risk. The scientific underpinnings of remediation and the engineering to get the job done all are based on the premise that the relative risk to a receptor is directly proportional to the chemical concentration present in the soil. Much research has been done in recent years that shows that premise may be wrong, resulting in a national cleanup strategy possibly costing tens of billions of dollars more than necessary.

NEPI has worked with the GRI consortium effort, University of Texas Professor Ray Loehr and others who are at the cutting edge in trying to take the latest research on the availability of chemicals in soil and bring it forward into the public policy debate. The hypothesis is that chemicals that are present in a soil become less available for uptake by living organisms due to interactions between the chemical and the soil, and that this reduction in availability reduces the risk associated with these chemicals—known as "bioavailability". It is necessary to communicate to key decision makers that the availability of a chemical is not simply equivalent to its measured concentration, but is more realistically related to the type of soil, the chemical properties of the contaminants and the extent to which the contaminant has weathered in the natural environment. The type and extent of treatment to which it has been subjected also plays a role. Environmentally Acceptable Endpoints, contaminant concentration levels remaining in soils deemed safe by virtue of their lack of availability, then become the critical targets for scientific investigation. This changes the way that risk assessment is developed and carried out, shifting the research and development focus from detecting minute amounts of chemicals to studying the actual, site specific impact of chemicals in soils on biological organisms. The determination of EAEs thus requires further understanding the complex chemistry of soil - contaminant interactions.

Finally, the EAE concept can vastly reduce the price of cleanup. From old gas stations to brownfields to Superfund sites, the opportunity exists to get much more bang for the buck, many more cleanups and the return of useful lands to their rightful purpose. In some instances, it may actually be better to leave things alone. The USEPA, DOE, DOD, the state, and even local governments, are learning more about this and acting on this as this article is written.

NEPI has actively sought to educate key members of Congress and their staff on this issue, trying to get it into the middle of the Superfund and brownfields debate. Bioavailability language was part of the improved technical language in HR 2500, the last Congress' Superfund reauthorization. This kind of information must also be shared with additional members of Congress serving on relevant committees in this Congress, DOE, USEPA science and policy leadership and funding agencies looking at better and less costly ways to remediate contaminated media.

REFERENCES

National Academy of Public Administration, April, 1995. "Setting Priorities, Getting Results: A New Direction for EPA," 1120 G street, NW, Suite 850, Washington, DC, 20005–3801.

NEPI, Summer 1995. "Reinventing the Vehicle for Environmental Management."

NEPI, 1995. "How Clean is Clean?" September 1995 First Phase Report, National Environmental Policy Institute, 1100 17th NW, Suite 330, Washington, DC 20036.

ASTM, 1994. "Emergency Standard Guide for Risk-Based Corrective Action Applied at Petroleum Release Sites, ES 38–94, American Society for Testing and Materials, Philadelphia, PA.

USDOE, June 1995. Office of Environmental Management," Risks and the Risk Debate: Searching for Common Ground."

20

ENVIRONMENTAL RISK ASSESSMENTS AND THE NEED TO COST-EFFECTIVELY REDUCE UNCERTAINTY

Robin D. Zimmer

IT Corporation
312 Directors Drive
Knoxville, Tennessee 37923

1. INTRODUCTION

Rachel Carson's *Silent Spring* (1962) was instrumental in the dawning of environmental awareness by alerting the public to the hazards of chemical stressors such as DDT. This classic work also focused on the need to guard against unnecessary degradation of the natural environment and sparked a new era of environmental activism and regulation. The "green movement" was borne, and the legislative ground swell has been overwhelming. The environmental pendulum swung from extreme leniency linked to ignorance, to extreme intolerance and mistrust. Impractical terms such as "zero discharge" and "remediation to non-detect" became the ultimate goal of public policy. Fortunately however, the environmental policy pendulum is swinging back to a center point following the exertion of forces such as national and international economics, and a more sound understanding of environmental processes.

Authorities are now carefully evaluating the severity of site-specific environmental stress caused by anthropogenic constituents, and weighing costs for containing or eliminating the stress versus the significance of the environmental threat. Accordingly, federal and state regulatory agencies are becoming more receptive to risk-based standards for establishing hazardous waste cleanup levels as well as air emission and wastewater discharge limits. Scientists, however, continue to struggle with the issue of uncertainty. Reduction of uncertainty in a cost-effective manner is of paramount importance to not only practitioners of risk theory, but also to policy makers and the regulatory authorities tasked with implementation of environmental protection programs.

Environmental risk analysis has matured into a commonly used means of evaluating the likelihood that hazards, imposed by chemical or physical stressors, will exhibit unacceptable environmental effects in the near term or at some point in the future. Defining "unacceptable effects" is of course a critical starting point in the process. Once a clear determination has been made regarding the resource to be protected, assessors are tasked with identifying the means to *measure* whether these adverse effects have occurred or are likely to occur in the future. Herein lies the issue of uncertainty. What is the likelihood that this adverse response will take place, given present or future stressor levels? Although the science of risk and uncertainty analysis has matured, it is still in its infancy at a time when regulators and policy makers are becoming increasingly dependent upon it.

This paper details some of the uncertainties and relative costs associated with field and laboratory practices presently employed in conducting "quantitative" assessments of ecological hazards and risk.

2. ENVIRONMENTAL RISK ANALYSIS

Environmental or ecological risk analysis is commonly conducted via a phased or tiered approach. The initial phase (Tier I) includes a simple analytical screening of compounds or analytes which may be present based on historical use of a site. Site-specific concentrations of these analytes are then compared to levels documented in the literature as having the potential to cause adverse effects on natural biotic communities. This initial screening phase is not the appropriate level to assess risk from stressors that fall somewhere between "clear threat" and "no threat." If concentrations of site-specific constituents are significantly elevated and pose a clear risk to the surrounding biota, remedial or interim actions should be taken. Conversely, if site concentrations present no possible threat, the "risk evaluation" can be discontinued and remedial actions, if any, will be driven by factors other than ecological risk. If, however, the results of the conservative Tier I screening risk assessment are unclear as to the presence or absence of a potential ecological risk, more detailed quantitative measures should be employed to further evaluate potential risks and to reduce the associated uncertainties. Because of site-specific variables such as bioavailability, most efforts require more definitive assessments of hazards when screening data indicate elevated constituent levels.

Figure 1 illustrates a simplified risk decision tree. If the risk is properly characterized following a given level or phase of analysis, management of that risk, via remedial

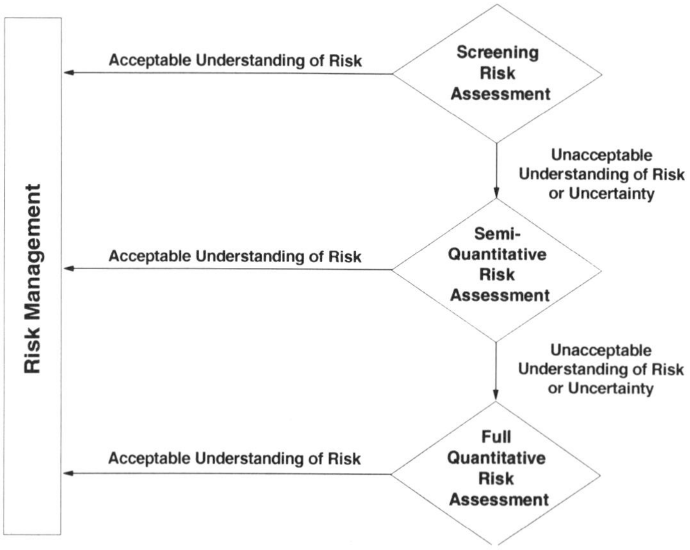

Figure 1. Risk analysis decision tree.

measures or possibly no action at all, is pursued. However, if the risk to the natural biota is unclear, more quantitative approaches are pursued.

The more quantitative approaches require direct measurements of effects within the receiving waters or hazardous waste site of concern. It is during these quantitative phases that assumptions are tested and uncertainties associated with the assumptions are reduced coincident with increasing tests or field measurements and modeling. The measurement of effects, either in the laboratory or in the field, is really an assessment of potential hazards. Laboratory tests or direct field measurements provide an investigator with more confidence in assessing the hazard. Suter's (1993) hazard assessment paradigm (Figure 2) displays decreasing confidence intervals (CI) coincident with an increase in the sequential tests for hazards. Overlapping CIs, as shown in the cross hatched area within Figure 2, represents an unclear hazard potential. The highest expected environmental concentration could exceed the documented effect concentration. Sequential tests such as acute and chronic bioassay testing, and on-site ecological health measurements, reduce the "uncertainty" of overlapping CIs and allow the investigator to more clearly evaluate the margin between the expected site concentration and adverse effect levels. It is important to note however, that Suter's paradigm also shows that at some point increasing the number of tests does little to reduce the CIs or increase the level of confidence. The point at which the investigator receives an acceptable quantitative "picture" of the hazard without realizing a diminishing return on effort and costs is critical.

Presently, while reduction in uncertainty is important, it is often cost intensive. Coincident with each sequential test of hazards is a concomitant increase in cost. Increasing the quantitative assessment of a constituent of potential concern (COPC) or multiple COPCs should reduce the uncertainty associated with its effect on a site. However, costs are generally inversely proportional to the more quantitative efforts and uncertainty reduction as shown in Figure 3. It is the increasing costs of characterizing ecological effects via exhaustive field measurements and laboratory testing that is of growing concern to scientists, decision makers and policy makers. Adoption of risk-based approaches to site evaluations frequently increases the demand for more definitive and quantifiable data to

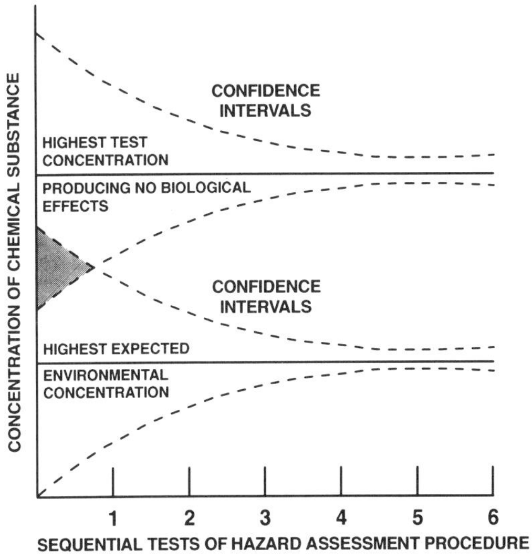

Figure 2. Tiered testing and assessment in the hazard assessment program. (Adapted from Suter, 1993).

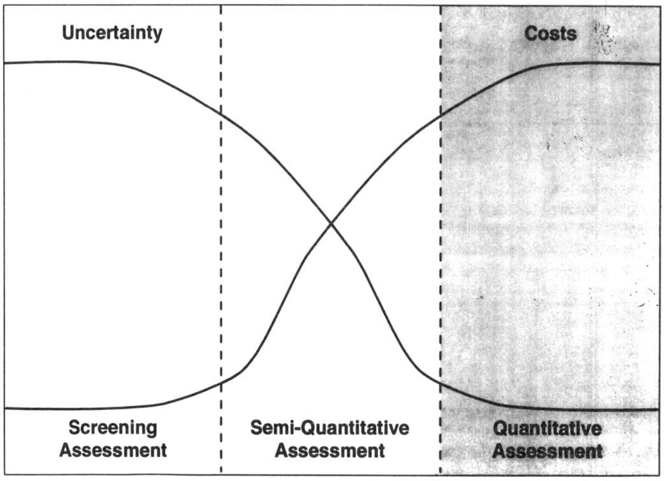

Figure 3. Costs and uncertainty relationship in ecological risk assessments. (Adapted from Landis and Yu, 1995).

measure damage to aquatic and terrestrial ecosystems. The balance of costs for these quantitative efforts then becomes a driving factor related to uncertainty.

3. ADVERSE EFFECT MEASUREMENTS

Before a measurement of effects can be initiated, it is critical that the investigator identify the natural resource or biotic component of a specific site or receiving water body that must be protected or restored. In other words, the *goal* of the assessment must be clearly defined. Assessment endpoints and measurement endpoints are then identified to support the assessment goal. An example of the relationship between these variables is shown below for protection of a freshwater pond ecosystem:

Assessment Goal

- Protection of the food web integrity for aquatic species (both recreationally important and protected species).

Assessment Endpoints

- No likelihood for a reduction in population abundance of prey species for higher trophic level species due to acute or chronic adverse effects resulting from exposure to site-related COPCs.
- No contamination of prey items for higher trophic level receptors resulting from bioaccumulation of COPCs.

Measurement Endpoints

- Mortality in individuals representing higher trophic levels such as fish, or surface and wetland species such as muskrats, ducks or herons.
- Benthic community structure.
- Elevated COPC tissue concentrations in high trophic level species.
- Reproduction or growth impairment leading to reduction of individuals of a prey species for critical food organisms.

Data Needs

- Aquatic fish and invertebrate—No Observable Effect Levels (NOEL).
- Aquatic fish and invertebrate—Lethal Concentrations of water or sediment causing 50 percent lethality (LC_{50}).
- Benthic community diversity, abundance, species evenness, and presence/absence of pollutant sensitive species.
- Waterfowl and fish tissue analysis.
- Food chain modeling.

A simplified illustration of a pond food web is shown in Figure 4. Note that target organisms which are representative of various levels within the food web would be selected as "sentinels" upon which hazard assessments can be conducted. This differs of course from a human health assessment in which the obvious goal is protection of a single species. The data needs to support measurement endpoints dictate the level of analytical sophistication and subsequent costs.

In this case, acute and chronic bioassay testing, under laboratory conditions, is required along with direct field measurements of the benthic community's health and tissue analyses of representative higher trophic level organisms. Although these types of laboratory and field-based data reduce the uncertainty of characterizing ecological effects, they fall short of eliminating uncertainty. In regard to sediment quality criteria for example, investigators continue to struggle with the issue of bioavailability. Under laboratory conditions, are organisms being exposed via direct sediment uptake, interstitial water, or at the sediment/overlying water interface? One of the key challenges is the extrapolation of ef-

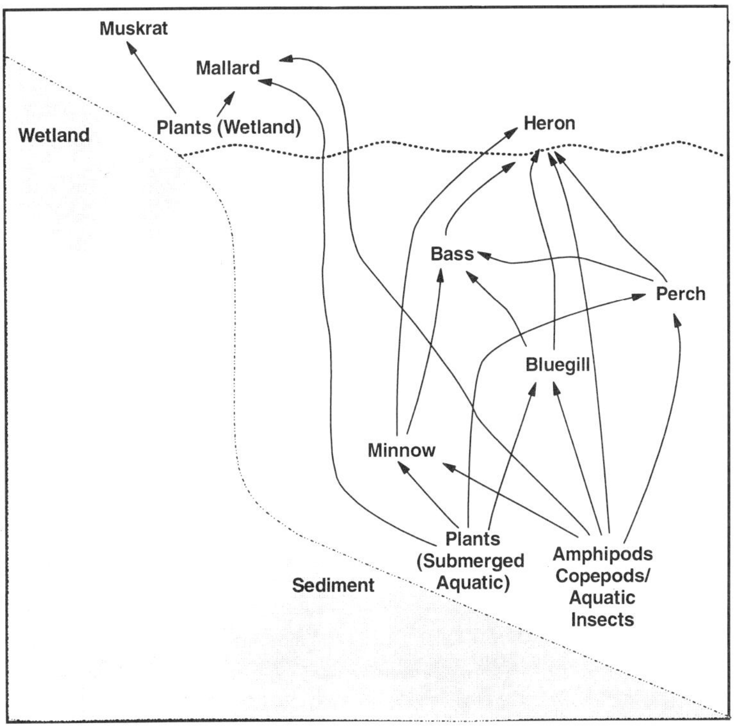

Figure 4. Aquatic (pond) food web.

fects observed in the laboratory, to a prediction of effects in the field. Is exposure in the laboratory comparable to exposure in the field? Are the test organisms reacting to stress in the laboratory in a manner which is comparable to the field? An additional extrapolation relates to the complexities of ecosystems and the reaction of individuals, populations, and communities to the physical or chemical stress of concern. Although these extrapolations contribute to uncertainty, they are certainly more quantitative then the conservative literature based assumptions indicative of Tier I assessments. Their value is undeniable when combined with formal uncertainty analysis. However, they are far from inexpensive when done properly.

The rising costs of detailed ecological damage assessments has been recognized as a limiting factor in conducting extensive investigations at numerous sites. A number of researchers are working frantically to develop less costly means to assess damage on a macroscopic and microscopic scale. The US Environmental Protection Agency (USEPA), for example, has introduced *Rapid Bioassessment Protocols For Use In Streams And Rivers* (Plafkin, et al., 1989). The USEPA protocols recommend a phased approach in measuring damage to streams or rivers in an attempt to reduce field related costs. There are also significant gains being made in developing micro scale bioindicators.

4. MICRO-BIOINDICATORS AND THEIR RELEVANCE

Bioindicators can simply be defined as "measurable" impairment indices reflecting environmental effects ranging from subcellular damage to lethality. Indicators can therefore exist on the micro or macro scale. Observance of mortality rates on fathead minnows or water fleas within a laboratory over a 48–96 hour exposure period is a macro bioindicator, as is the measurement of a test water's effect on growth or reproduction of these organisms over an extended exposure period. The death of test organisms or statistically significant inhibition on growth or reproduction as compared to controls, is the exhibition of stress induced by chemical or physical forces present within the test media. The macro effects, such as these, are observed after internal cellular damage has occurred.

Earlier detection of stressor damage on the cellular or subcellular level may provide a clearer dose - response relationship, and be more cost-effective. The key is measuring damage as close to the origin of the stressor's action as possible. Identification of this "action site" is critical because compound classes often differ in their site of action. Identification of the site can therefore be useful in identifying the stressor type. The interaction of a xenobiotic stressor or compound is portrayed by Landis and Yu (1995) within Figure 5. The figure provides a dose or time scaled effect line of a xenobiotic from its introduction into a biological system to exhibition of ecosystem-wide effects. Again, a xenobiotic, which may or may not undergo biotransformation, will have some subcellular site of action resulting in possible effects on a target organism's nucleic acids, enzymes, or general biochemical integrity, which will ultimately be exhibited in physiological and behavioral effects and possibly mortality. Effects on individuals within a population may result in an upset in population dynamics, community structure, and ultimately the stability of an entire ecosystem.

Common macro indications of stress can be observed as physiological and behavioral effects on individuals or populations; however it is often difficult to attribute the adverse response to specific stressors, or for that matter, the time and duration of exposure. Moreover, measurements of population, community, or ecosystem effects represent gross responses to stress, and may be exhibited well after serious damage has occurred.

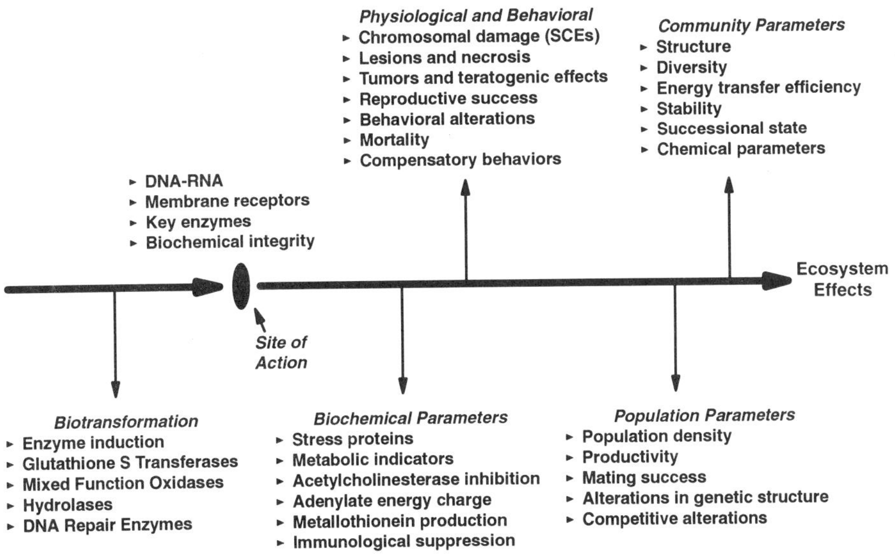

Figure 5. Interaction of a xenobiotic with the ecosystem.

5. CONCLUSIONS

Micro indicators may offer a more sensitive means to detect damage earlier in the "effects process", and a number of researchers are working to develop new micro biomarker tools. However, in order for these new tools to be useful they should offer more sensitivity, be more quantitative, offer more specificity regarding the source of stress, and be more cost effective. It is also important to note that new sub-cellular biomarkers must be developed with the ultimate intent of being able to extrapolate observed or measured results to effects on populations, communities, and possibly ecosystems. If a biomarker cannot be used to predict larger scale effects without some degree of confidence, its usefulness will be limited.

The ultimate goal is to conduct quantitative assessments of effects more cost-effectively, while improving confidence in extrapolating effects observed on the subcellular or individual organism level to effects on the larger scale of populations, communities and ecosystems. Given the complexities of the structure and dynamics of these larger systems, extrapolations can be encumbered with too much uncertainty. Development of cost-effective bioindicator tools which will link effects in the organism or subcellular level to larger scale implications represents the challenge at hand for the research community.

REFERENCES

Carson, R., 1962. *Silent Spring*, Houghton Mifflin, New York.

Suter, G.W., 1993. *Ecological Risk Assessment*, Lewis Publishers, Ann Arbor, Michigan.

Plafkin, J.L., Barbour, M.T., Porter, K.D., Grass, S.K., and Houghs, R.M., 1989. *Rapid Bioassessment Protocols For Use In Streams and Rivers*, USEPA. Office of Water (EPA/444/4-89-001). Washington, D.C.

Landis, W.G., and Yu, M., 1995. *Introduction To Environmental Toxicology - Impacts of Chemicals Upon Ecological Systems*, Ann Arbor, Michigan.

21

ACCURATELY ASSESSING BIODEGRADATION AND FATE

A First Step in Pollution Prevention

Thomas W. Federle

Environmental Science Department
The Procter and Gamble Company
Ivorydale Technical Center
Cincinnati, Ohio 45217

1. INTRODUCTION

An important first step in pollution prevention is accurately assessing the biodegradation and fate of a chemical prior to its manufacture and widespread use. With the exception of pesticides and agricultural chemicals, which are deliberately introduced into the environment, most chemicals enter the environment as a secondary result of their manufacture, transport and disposal. In the case of consumer product chemicals, the major routes of entry into the environment result from their disposal by the household down the drain or in the solid waste. In the United States, most domestic wastewater is treated prior to release to natural waters. Approximately 75% of total wastewater flow is treated in publicly owned sewage treatment works, while the remaining 25% receives some type of on-site treatment. The most common form of on-site treatment involves a septic tank and tile field.

Depending upon the route of disposal and the effectiveness of treatment, a chemical can ultimately reside in the atmosphere, soil, groundwater or surface waters. Pollution prevention is contingent on accurately predicting: 1) where a chemical will go (partitioning & transport), 2) how much of it will get there (environmental loading & concentration), and 3) how long it will remain there (persistence). The answer to these questions is governed by the interaction of five major fate processes: dilution, volatilization, sorption, chemical degradation and biodegradation.

Dilution results in a decrease in concentration, but at the expense of increased dispersion through the environment. Volatilization is movement into the atmosphere, which is another form of dilution. Once again a decrease in concentration comes at the expense of increased dispersion through the environment. Chemicals that enter the atmosphere can be transported and redeposited at a global level. Sorption is the binding of a chemical to a solid such as sludge, sediment or soil. Sorption generally slows transport and limits dispersion, but concentrates the chemical on the solid. While these three processes affect the transport and concentration of chemical, none lead to loss of chemical identity.

Chemical degradation includes a variety of hydrolytic, oxidative and photolytic reactions. These reactions are often slow and occur in a narrow range of pH and light intensity conditions, which may not exist in the environmental compartments in which a chemical resides. Chemical degradation results in loss of chemical identity and often function, but the chemical's constituent elements are not fully returned to natural mineral cycles. Biodegradation is mediated by microorganisms, primarily bacteria and fungi, which are ubiquitous and can increase in concentration in response to a chemical's presence. Biodegradation is usually faster and more complete than chemical processes and occurs over a wide range of environmental conditions. It results in a loss of chemical identity and function as well as return of the chemical's constituents to natural mineral cycles.

2. FATE ASSESSMENT

To predict a chemical's impact on the environment, the various fate processes need to be integrated and incorporated into a fate assessment. A qualitative fate assessment can be conducted based upon a chemical's inherent physical chemical properties, which include vapor pressure or Henry's Law constant, water solubility and K_{ow} (octanol water partitioning coefficient). In addition, a QSAR (quantitative structure activity relationship) prediction or screening level biodegradation test can provide an indication of the likelihood that a chemical will undergo biodegradation. A qualitative assessment provides an indication of whether a chemical will partition into the atmosphere, soil and sediments or water column and if it will persist. A more quantitative assessment, in which actual environmental concentrations are estimated, requires the use of mathematical models that incorporate kinetic descriptions of loading, transport and losses.

Unfortunately, fate models are often inaccurate. They fail because 1) they do not accurately depict the various fate processes and their interactions, 2) they are formulated incorrectly from a mathematical perspective, or 3) they are incorrectly parameterized. In the last case, they may be parameterized with K_{ow} versus a K_d for the relevant particulate (clay, sludge etc) or a biodegradation rate that is based upon a QSAR or biodegradation test that was performed at the wrong concentration, in the wrong matrix, etc. It is sometimes a consequence of poor parameterization that well formulated models make poor predictions and bad models make good predictions, thus providing a false sense of security. Because of the ease of reformulation, improvement often focuses on the models rather the parameters used in the models. As an end result, individual models are not fairly evaluated, little understanding is gained and overall model reliability is not substantially improved.

3. BIODEGRADATION ASSESSMENT

Biodegradation is the major loss mechanism for many synthetic organic chemicals in the environment, and consequently the biodegradation rate is among the most important parameters for any exposure model. A comprehensive biodegradation assessment needs to address two questions: 1) Is the chemical completely biodegradable? and 2) Is the chemical practically biodegradable? The first question relates to the potential for a chemical to undergo complete biodegradation as evidenced by extensive mineralization and no persistent metabolites. When a chemical undergoes complete biodegradation, the impact of the parent rather than any metabolite remains the focus of the safety assessment. The second question relates to the realization of this potential in the environmental compartment in which a chemical resides and focuses on the kinetics of the biodegradation process. A chemical that is practically biodegradable not only degrades in the compartment in which it resides but also biodegrades at a rate sufficient to reduce exposure concentrations and prevent accumulation over time. Practical biodegradation results in tangible environmental benefits, which include less dispersion through the environment and lower environmental concentrations, which can translate into higher safety factors.

To accurately parameterize exposure models and conduct comprehensive biodegradation assessments, there is a need for biodegradation data that is obtained under conditions resembling those occurring in situ. For the kinetic data to have relevance, it is key that the data be generated at the concentrations, which the chemical occurs in the environment. In addition, the environmental form of the chemical should simulate that in situ. In some circumstances the environmental form of a chemical has a profound effect on its rate and extent of biodegradation. In a like manner, the level, composition, and physiological status of the microbial community should reflect that occurring in situ. Adaptation of a microbial community plays a key role in determining the rate and extent at which a chemical might degrade. The decision to utilize a pre-exposed or acclimated community should be based upon the discharge pattern of a chemical into the environment. If a chemical is emitted in a random and discontinuous fashion, the use of a pre-acclimated community for testing can be misleading. On the other hand, if a chemical will be discharged continuously, the community will become acclimated and remain acclimated since the chemical will represent a new ecological niche in the receiving compartment. In this case, the relevant biodegradation rates are those of an adapted community. Finally, the test matrix should itself be relevant. If one is estimating rates in activated sludge or soil, the test should consist of activated sludge or soil with all their relevant attributes including solids level, pH and oxygen concentration.

In addition to testing under realistic conditions for the chemical, the biology and the matrix, accurate predictions of fate depend upon the measurement of all relevant endpoints and robust kinetic analysis of the data. Biodegradation testing ideally should include concurrent analyses of parent loss, metabolite formation and disappearance as well as mineralization. Measuring only parent disappearance or mineralization can be misleading. Finally, the resulting data should be analyzed in such a way to not only determine a rate but also identify the most appropriate mathematical functions describing the biodegradation process.

4. BIODEGRADATION TESTING

Many existing biodegradation test methods do not simulate the field and are limited to selected end points. Dosing of the test chemicals to a test is often based solely upon practical considerations and expediency rather than how a chemical normally enters or resides in an environmental matrix. Furthermore, kinetic analysis is often not performed or the data are fitted to a preconceived model such as the first-order. These combined errors are subsequently propagated in the exposure models.

At present, most biodegradation data are generated in screening tests and occasionally in more realistic test systems that utilize radiolabeled test materials. Examples of standardized screening level tests include the Sturm CO_2, closed bottle, and MITI tests. These test procedures are applicable mainly to sewage and utilize nonspecific detection methods, such as total CO_2 evolution, DOC loss or oxygen uptake. These tests were designed to achieve a sufficient signal above noise to show loss of chemical utilizing a nonspecific analytical techniques. They are characterized by low levels of biomass to minimize background noise and high concentrations of test chemicals to maximize the signal. These tests are very useful for demonstrating completeness of biodegradation, and due to their stringency, are useful for differentiating chemicals that will easily degrade from those that may be more problematic. As a consequence, they play important roles in registering new chemicals in Europe and Japan. Nevertheless, the unrealistically high ratio of test material to biomass can at times lead to false negative results, and the kinetics observed in these systems do not necessarily reflect those that occur in situ.

This latter point is illustrated by a comparison that was conducted by Federle et al., (in review), which examined the first-order biodegradation rates of nine diverse chemicals in a Sturm CO_2 test and more realistic ^{14}C tests with activated sludge, river water and soil. To limit variability, the Sturm and river water tests were inoculated with microorganisms from the same sludge used in the activated sludge test. Table 1 shows the correlation between the mineralization rates for the various chemicals in the different test systems. Surprisingly, the rates observed in the screening test were nearly perfectly noncorrelated with the rates for these same chemicals in the more realistic ^{14}C tests. Instead, the rates in the screening test were highly correlated with the solubility of the individual test chemicals. These relationships clearly demonstrate that while results from screening tests are very useful for qualitative assessments, the rate in the test was more governed by solubility, which appeared to decrease in importance at environmentally relevant concentrations.

The most common realistic test systems are batch tests in which a radiolabeled test chemical is incubated with actual environmental samples. The ability to monitor radioactivity rather than a nonspecific endpoint makes it possible to test realistic concentrations of a chemical. Such systems can be used to determine completeness of biodegradation as well as actual kinetics. They are applicable to and have been used for activated sludge, river, estuary and ocean waters, freshwater and marine sediment, groundwater and surface and subsurface soils. While these tests have many advantages, they also have some limitations as currently employed. Often, the only definitive endpoint is evolution of $^{14}CO_2$. This is an easy endpoint to measure since CO_2 can easily be trapped in base and quantified by liquid scintillation counting (LSC). With CO_2 as a sole endpoint, these tests are only useful for determining the fate of chemicals that undergo mineralization. Thus, they have little application for chemicals that are biotransformed rather than mineralized. Furthermore, disappearance of the parent rather than mineralization is the relevant parameter for exposure models used for safety assessments. Low recovery in some systems, especially soil, can

Table 1. Correlation of first-order mineralization rates for nine diverse chemicals in a Sturm CO_2 test with mineralization rates in realistic ^{14}C tests and chemical solubility

	r
Mineralization rate	
Activated sludge	–0.06 ns
River water	0.04 ns
Soil	0.12 ns
Chemical	
Solubility	0.92**

ns = not significant
**$p \leq 0.01$.

leave one uncertain about the fate of a significant fraction of a material. Besides being mineralized, a chemical can be incorporated into biomass or natural humic materials. Such low recovery of CO_2 can make it difficult to ascertain the biodegradation of individual components within a mixture and detect persistent metabolites. Furthermore, analysis of only one endpoint yields little fundamental understanding on the mechanism of biodegradation.

5. THE IMPORTANCE OF TEST CHEMICAL DOSING AND KINETIC ANALYSIS

An important aspect of experimental design that can have a profound effect on the observed biodegradation even in realistic ^{14}C test systems is how the chemical is dosed to the system. While the mode of dosing is not likely to have much effect on the outcome observed with highly soluble compounds, this aspect of a test takes on greater importance as test materials become increasingly less soluble and sorptive. The potential for generating artifactual findings with such compounds is immense. Figure 1 shows mineralization in activated sludge of a long chain amine ion paired with an anionic surfactant used in laundry applications. In one case, the chemical was dosed to the test system in isopropanol and in the other it was dosed in 20% isopropanol-water. In the former, it was dosed drop-wise; in the latter as more of a bolus. The effect on mineralization was dramatic. In one case, mineralization was extensive and rapid; in the other slow and incomplete. An even more dramatic illustration of the effect of dose form is shown in Figure 2, in which stearic acid was dosed into soil as an aqueous solution or pre-associated with various soil constituents. This experiment was originally reported by Knaebel et al., (in press). Once again, the rate as well as the extent of mineralization was highly affected by how the chemical was introduced into the test system. The problem with such disparate results is establishing what represents reality. In the first example, neither dosing scheme accurately represents how the chemical enters wastewater. This sensitivity of biodegradation test results to dosing procedures demonstrates how important it is to think critically about the physical/chemical form of a chemical in the environment, and the need to recreate this form in the actual test.

Another important aspect of experimental design is the treatment of the data. Often no kinetic analyses are performed or the data are fitted to a preconceived model, which is commonly the first-order model. A major reason for this tendency is that many models utilize a first-order function to describe biodegradative and chemical losses. While first-order models often provide very reasonable fit to many sets of biodegradation data, a

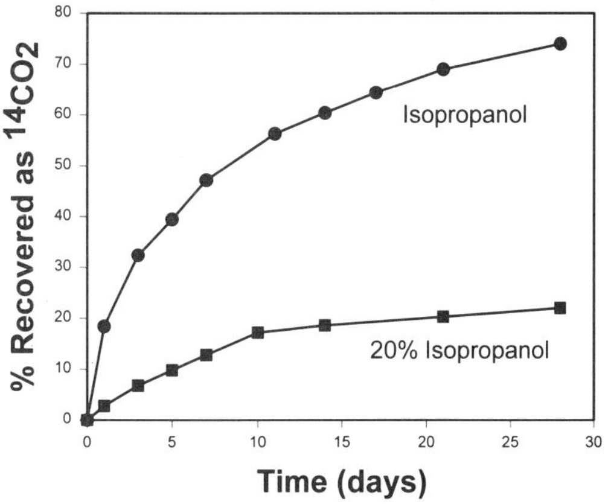

Figure 1. Mineralization of a long-chain amine ion paired with an anionic surfactant in activated sludge as a function of dosing.

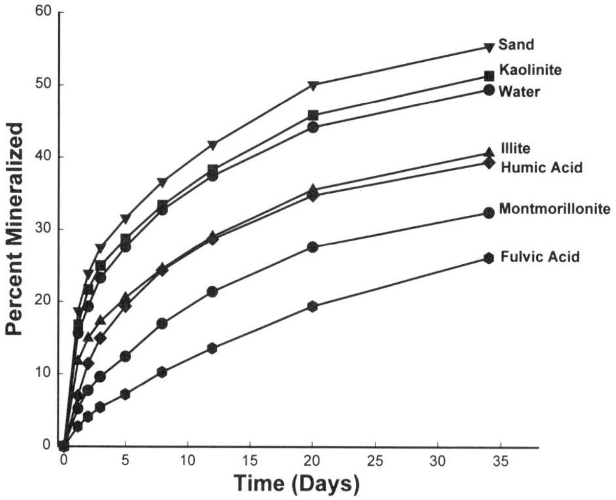

Figure 2. Effect of dose form on the mineralization of sodium stearate in soil. (Redrawn from Knaebel et al.).

problem can arise when a first-order model data provides an approximate but not exact fit for the data, and the error is propagated in the exposure model. Figure 3 shows primary biodegradation of a surfactant in activated sludge. Alkyl ethoxylate sulfate (AES) is an important anionic surfactant, which consists of an alkyl chain, condensed with a series of ethoxylate groups, and possessing a sulfate on the terminal alcohol. Based upon statistical considerations and visual inspection, primary biodegradation of AES is well described by a first-order decay model, and AES appears to have a half-life of less than 5 minutes. However, closer inspection of the data and resultant fit (inset) indicates that while the model predicts no AES remaining after 30 min, in reality a low level (<1%) is still detectable even after three hours. Use of a simple first-order loss function and a half-life of a few minutes would significantly overestimate removal during sewage treatment, which is indeed the case based upon actual monitoring data (McAvoy et al., 1993).

6. IMPROVED BIODEGRADATION TEST METHODOLOGY

Based upon the previous discussion, it is clear that three major opportunities exist for maximizing the value of ^{14}C tests currently utilized for biodegradation and exposure

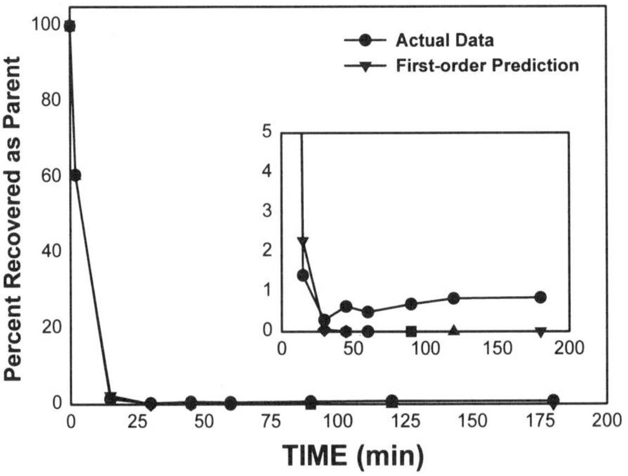

Figure 3. Primary biodegradation of a homologue of alkyl ethoxylate sulfate (AES) in activated sludge: comparison of first-order fit to actual data.

assessments. These include 1) increasing the number of endpoints examined to include parent, metabolites and dissolved CO_2, 2) introducing the chemical into the test system in a realistic fashion and 3) performing more robust kinetic analyses of the data to identify both the correct biodegradation rates and identify the most appropriate model describing the biodegradation process.

To extend the value of ^{14}C batch tests, existing methods were modified with the goal of determining the levels of parent and metabolites and specifically measuring incorporation into biomass constituents (i.e., protein, lipid, nucleic acids, cell walls). To accomplish these goals, existing methods were coupled with specific radioanalytical techniques including Rad-HPLC, Rad-TLC and Rad-GC/MS as well as sequential extraction schemes to recover parent, metabolites and biomass constituents. The key challenge was to develop generic analytical approaches, which had wide application and were efficient and cost effective. The solutions to these challenges were to use flash-freezing and lyophilization for sample preparation and concentration, Rad-TLC for analysis of parent and metabolites and sequential extraction and biochemical fractionation for measurement of biomass incorporation. The latter was made efficient by performing the extractions in microcentrifuge tubes.

By combining these techniques, several new test procedures were developed. Figure 4 shows a generic outline of these new die-away tests. The major variable is the matrix, which can be activated sludge, raw sewage, river water, effluent diluted into river water or anaerobic digester sludge. In the last case, the incubation and most manipulations are performed in an anaerobic chamber. In brief, the test chemical is incubated with samples freshly obtained from the environment. An abiotic control is prepared through autoclaving and addition of mercuric chloride. This treatment serves as a control for analytical recovery of the parent, nonspecific recovery of radioactivity into the various biomass fractions, abiotic losses resulting from hydrolysis, and sorption to the test vessel or volatilization. Periodically, subsamples are removed from both treatments and lyophil-

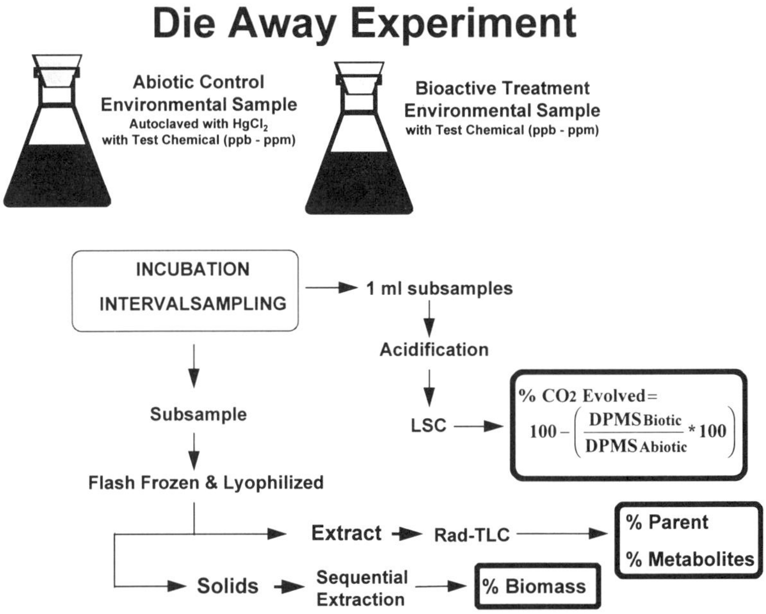

Figure 4. Schematic diagram of a generic die-away test for assessing the primary and ultimate biodegradation of a ^{14}C test chemical in an environmental sample.

ized. The lyophilized solids are then extracted with an appropriate solvent(s) to recover parent and likely metabolites. These extracts are analyzed by LSC to determine total radioactivity and Rad TLC to determine the relative abundance of parent and various metabolites. The extracted solids are analyzed directly or quantitatively transferred to microcentrifuge tubes for biochemical fractionation of the biomass. These solids are sequentially extracted with cold trichloracetic acid (TCA) to recover low molecular cytoplasmic components, ethanol/ether to recover lipids, hot TCA to recover nucleic acids, and 10 N NaOH to recover proteins. After each extraction step, the microcentrifuge tubes are centrifuged and the supernatant is recovered and counted by LSC. The final step involves combustion of the fully extracted solids to determine incorporation into cell walls. Data for each fraction is corrected using the abiotic control. $^{14}CO_2$ is determined by acidifying subsamples from the bioactive and abiotic treatments and comparing the difference. Alternatively, evolved $^{14}CO_2$ is trapped in base and quantified by LSC, and dissolved $^{14}CO_2$ is determined by acidifying subsamples in biometer flasks and quantifying the evolved radioactivity following trapping in base.

Figure 5 shows the type of data that results from such a test design. This experiment shows the biodegradation of AES in activated sludge. The test chemical was a $^{14}CE_3S$ homologue, uniformly radiolabeled in the 1 and 3 ethoxylate groups, and dosed at a final added concentration of 1 milligram per liter (mg/L). Disappearance of parent started im-

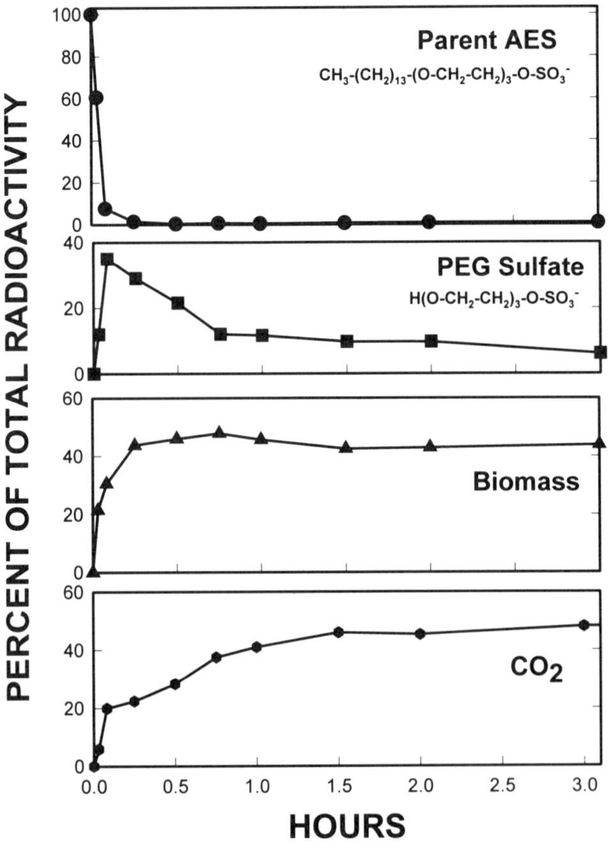

Figure 5. Initial events in the biodegradation of AES ($^{14}CE_3S$) in activated sludge.

mediately and was rapid. Concurrent with the disappearance of parent, was uptake of radioactivity into biomass, evolution of $^{14}CO_2$, and the transient appearance of a polar metabolite, which was identified as polyethylene glycol sulfate (PEG sulfate). This sequence of events indicated that AES was initially cleaved at its internal ether, resulting in the release of PEG sulfate and fatty alcohol, which were subsequently incorporated into biomass or mineralized. A similar central cleavage has been reported with cultures of bacteria (Hale et al., 1982). Notably, within 3 hours, approximately 95% of the radioactivity derived from the parent was equally distributed between biomass and $^{14}CO_2$. This distribution is consistent with a growth yield of approximately 0.5. Continuing the test much longer only would yield information of the conversion of biomass carbon to $^{14}CO_2$. Thus, the use of mineralization as the sole endpoint would grossly overestimate the time needed to degrade the parent and its primary metabolite.

7. IMPROVED TEST DESIGN AND DATA ANALYSIS

Besides the analytical aspects of a test, another area that has received attention is the introduction of test chemicals. In the case of highly soluble chemicals, the chemical is added as an aqueous solution. However when the material is marginally soluble or sorptive, careful attention is now paid to the probable mode of entry of the chemical into the wastewater. In the case of laundry ingredients, these chemicals are discharged into the sewer associated with a variety of other cleaning agents, particularly surfactants. The preferred means of dosing such compounds is as a component of gray water. This approach not only mimics how chemicals normally enter wastewater but also results in homogenous distribution of a chemical through the test system, which results in improved replication during subsampling. In the case of surface soil, most laundry ingredients reach soil as a component of sludge during the sludge amendment of soils. The preferred procedure for dosing a test chemical to soil is to place a small sample of sludge into the test vessel, add the test chemical directly to the sludge in solvent or preferably water, add the soil to test vessel and mix the sludge with the test chemical into the soil. This procedure therefore more accurately simulates how a laundry ingredient normally enters a topsoil than dosing it to the whole soil in water or an organic solvent. This philosophy if not these same procedures can be used for dosing other chemicals, which enter soil or wastewater by other routes.

The final area for improvement relates to the kinetic analysis of the data. Recently, regression software (Jandel Table Curve 2D) has become commercially available that will fit data to multiple equations using nonlinear regression. Within seconds, this software generates a rank ordered listing of the fits based upon r^2. In addition, at the click of a mouse each actual fit and its residuals can be visually inspected on the computer screen. As a matter of practice, each set of biodegradation data are fitted to a variety of decay and production equations including zero-order, first-order, logistic first-order, first-order with lag or three-half order with growth or three half-order without growth. The most appropriate model is then identified based upon statistical considerations and visual inspection. Figure 6 shows the ability of this approach to properly discern the best model. This graph shows the relative ability of a first-order and three-half order model to fit data on the primary biodegradation of an AES homologue in activated sludge. While the r^2 indicated that a first-order decay model accurately described the loss of AES, closer inspection of the actual fit and residuals indicated that the 3/2-order model is much more appropriate and accurate.

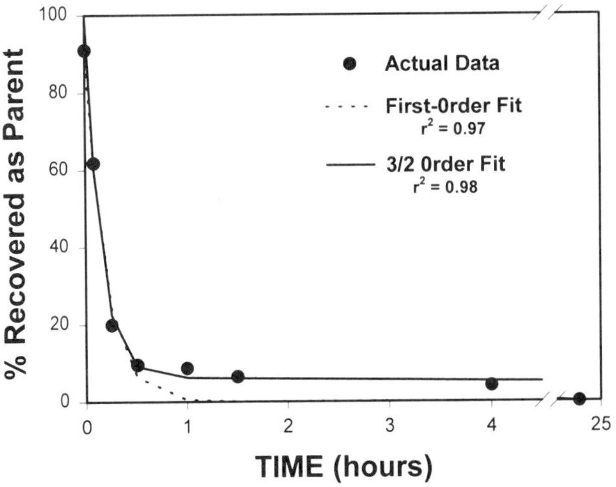

Figure 6. Primary biodegradation of a homologue of alkyl ethoxylate sulfate (AES) in activated sludge: comparison of first-order and 3/2-order model fit to actual data.

8. CONCLUSIONS

Accurately determining the biodegradation kinetics of a chemical is critical to assessing its fate and likely effects in the environment. New testing approaches utilizing simple radioanalytical methods and realistic simulations of environmental conditions are key to establishing the complete and practical biodegradation of a chemical in a particular environmental compartment. Such testing approaches combined with realistic dosing of test materials and robust kinetic analysis of the data can provide 1) realistic kinetic descriptions of primary and ultimate biodegradation, which are critical for exposure modeling, 2) an understanding of the biodegradation mechanism and the metabolites that might be formed in the environment, which is important for conducting a holistic risk assessment on the chemical as well as its products, and 3) a detailed accounting of disappearance of parent, formation and disappearance of metabolites, uptake into biomass and mineralization. The net result is less uncertainty regarding the fate of a chemical, more accurate exposure assessments and more confident risk assessment.

REFERENCES

1. Federle, T.W. Gasior, S.D. and Nuck, B.A., 1996. Extrapolating mineralization rates from the ready CO_2 screening test to activated sludge, river water and soil, *Environ. Toxicol Chem.* (in review).
2. Hales, S.G., Dodgson, , K.S., White, G.F., Jones, N. and Watson, G.K. 1986. A comparative study of the biodegradation of the surfactant sodium dodecyltriethoxy sulfate by four detergent-degrading bacteria. *J. Gen. Microbiol.* 132: 953–961.
3. Knaebel, D.B., Federle, T.W., McAvoy, D.C. and Vestal, J.R., 1996. Microbial mineralization of organic compounds in an acidic agricultural soil: Effects of preadsorption to various soil constuents, *Environ. Toxicol. Chem.* (in press).
4. McAvoy, D.C., Fendinger, N.J., Norwood, K.T. and Dyer S.D., 1993. Removal of alkyl ethoxylates and alkyl ethoxylate sulfates during wastewater treatment, *In Abstracts of the Annual Meeting of the Society of Environmental Toxicology and Chemistry.*

22

MODELING TO PREDICT BIODEGRADABILITY

Applications in Risk Assessment and Chemical Design

Robert S. Boethling

U.S. Environmental Protection Agency
Office of Pollution Prevention and Toxics 7406
401 M St., SW
Washington, DC 20460

1. BACKGROUND

In the US, the safety of specific chemical substances is evaluated primarily under three statutes. Substances used as food additives, drugs and cosmetics are regulated by the Food and Drug Administration (FDA) under the Federal Food, Drug and Cosmetic Act (FFDCA). Chemical substances proposed for use as pesticides are regulated by the US Environmental Protection Agency (USEPA) under the Federal Insecticide, Fungicide and Rodenticide Act (FIFRA), which imposes a host of data requirements for any submitter seeking to register the substance as an active ingredient. Industrial chemicals are regulated under the Toxic Substances Control Act (TSCA; Public Law 94–469), which was enacted by Congress in 1976 in response to a perceived need to limit exposure to "environmental chemicals" such as polychlorinated biphenyls (PCBs). As stated in the Act, its primary purpose is "to assure that...innovation and commerce in...chemical substances and mixtures do not present an unreasonable risk of injury to health or the environment" (TSCA, section 2(b)). TSCA requirements are different for existing substances and substances not yet in production ("new" chemicals). One of the first tasks of the newly created Office of Toxic Substances (now the Office of Pollution Prevention and Toxics; OPPT) was to assemble and publish a list of chemical substances already in commerce. This was accomplished in July, 1979 as the TSCA Chemical Substance Inventory, which listed approximately 50,000 substances then in production or being imported into the US. Since that time the Inventory has grown to include over 70,000 substances by the addition of new chemicals.

Anyone who wishes to manufacture or import into the US for commercial purposes a substance not listed on the Inventory and not otherwise excluded by TSCA (e.g., pesticides, drugs) must submit formal notice of their intent to do so. Such a submission is called a Premanufacture Notice (PMN), and for most new chemicals it must be submitted to the USEPA at least 90 days prior to manufacture or import. Substances approved by the USEPA are then added to the Inventory and become existing chemicals as soon as the

Biotechnology in the Sustainable Environment, edited by Sayler *et al.*
Plenum Press, New York, 1997

USEPA receives the required Notice of Commencement (NOC) declaring the submitter's intent to commence manufacture. Since 1979, the USEPA has received more than 30,000 valid PMNs (Figure 1), and submissions currently average well over 2,000 per year.

Receipt of a PMN sets in motion a process of review that has evolved over time to meet the unique requirements established by TSCA. The fundamental purpose of PMN review is stated in the law and is to determine whether "the manufacture, processing, distribution in commerce, use, or disposal [of a new chemical substance] or any combination of such activities presents or may present an unreasonable risk of injury to health or the environment" (TSCA section 5(b)).

However, TSCA imposes serious challenges to the USEPA's ability to accomplish this. The most significant are, first, that PMN submitters are only required to furnish data already in their possession (if any), and are not required to conduct a battery of tests as a precondition for approval; and second, that the review must be conducted in 90 days or less. The burden of proof that a substance presents or may present an unreasonable risk rests on the Agency's shoulders, which often must make sound decisions based on few or no submitted test data, within a very short period of time. This situation is unlike that in the European community, where a series of tests that supply the "minimum premarket data set" are prescribed by law.

By its nature the PMN process provides a powerful impetus for developing estimation methods for the many parameters needed in the assessment process. But the need to rapidly screen existing chemicals to identify substances of priority concern has been equally important. Here again TSCA provided much of the early impetus, since it created the Interagency Testing Committee (ITC) and charged it with the responsibility of screening the Inventory and establishing priorities for further review. Now, virtually every

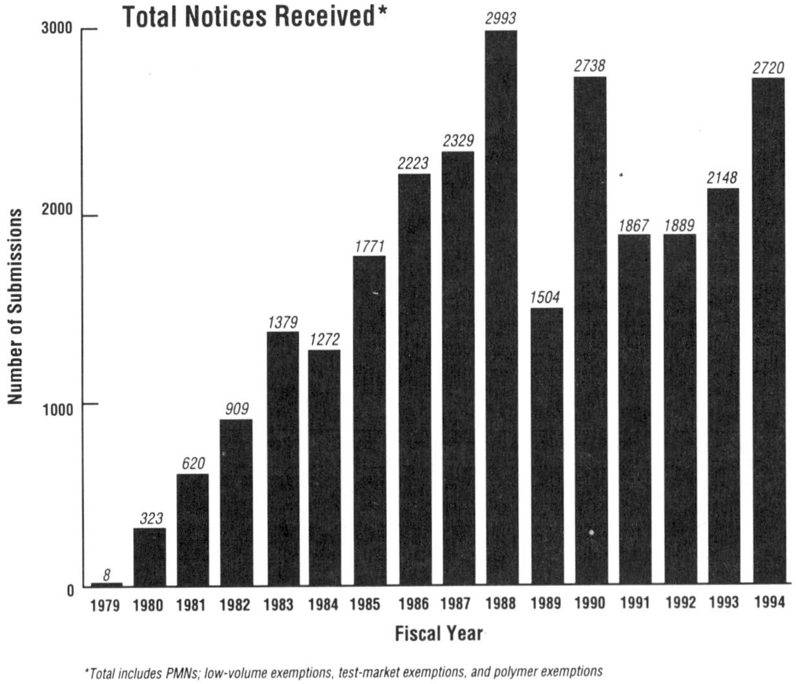

Figure 1. Total number of valid premanufacture notices received by USEPA for the period 1979–1994.

USEPA program is involved in chemical scoring in one way or another. Examples of chemical prioritization systems include Reportable Quantity (RQ) adjustment methodology under the Comprehensive Environmental Response, Compensation and Liability Act (CERCLA, or "Superfund"); the Superfund Hazard Ranking System; the USEPA Office of Pesticide Programs' Inerts Ranking Program for "inert" components of pesticide formulations; and OPPT's Use Cluster Scoring System (UCSS).

A recent SETAC workshop (Chemical Ranking and Scoring; Destin, FL, 2/95) identified more than 100 different systems[1]. In 51 systems subjected to a comparative analysis there were at least 12 different endpoints used to measure environmental persistence, at least 15 to estimate mobility, partitioning or bioaccumulation, and 34 kinds of data used to estimate exposure. It is axiomatic that the required input data simply are not available for many of the chemical substances of interest. Thus, a common need exists for ways to furnish the missing data short of expensive laboratory testing.

2. PREDICTING BIODEGRADABILITY

Most evaluation systems in current use, whether for premanufacture/premarket review or prioritization of existing chemicals, include explicit consideration of environmental persistence, bioconcentration and ecotoxicity. Persistence is primarily a function of biodegradability for the majority of chemicals released to soil and water. This creates a problem for chemical screening because experimental biodegradation data are typically lacking entirely or do not exist in a form that can be easily incorporated into automated screening methods. We responded to this problem by first developing a weight-of-evidence procedure for collecting and evaluating available data, and subsequently using these and other data to develop models for predicting biodegradability. Before discussing the BIODEG database and models, however, it is necessary to review briefly what is known about the effects of chemical structure on biodegradability.

2.1. Chemical Structure and Biodegradability

More than 40 years of experience have shown that relatively small changes in molecular structure can appreciably alter a chemical's susceptibility to biodegradation. These studies have resulted in several "rules of thumb"[2,3,4,5] about the effects of chemical structure on biodegradability. The following molecular features generally increase resistance to aerobic biodegradation:

- Halogens; especially chlorine and fluorine
- Chain branching, especially quaternary C and tertiary N
- Nitro, nitroso, azo, arylamino groups
- Polycyclic residues (such as in polycyclic aromatic hydrocarbons or PAHs), especially with more than 3 fused rings
- Heterocyclic residues; e.g., pyridine rings
- Aliphatic ether bonds

For the most part, these features affect the ability of a chemical compound to serve as an inducer or substrate, or both, of degradative enzymes and cellular transport systems. For example, addition of a chlorine atom to a phenyl ring makes the ring less susceptible to attack by oxygenase enzymes, which utilize a form of electrophilic oxygen as a cosubstrate. This list is not exhaustive, nor should it be inferred that the presence of only a single atom or group from the list necessarily renders a compound recalcitrant. Moreover, in most cases the mechanism by which increased resistance to biodegradation is conferred is not known in detail.

In contrast, biodegradability is generally enhanced by the presence of potential sites of enzymatic hydrolysis (e.g., esters, amides); by the introduction of oxygen in the form of hydroxyl, aldehydic or carboxylic acid groups; and by the presence of unsubstituted linear alkyl chains (especially ≥4 carbons) and phenyl rings, which represent possible sites for attack by oxygenases. The second of these three factors is particularly important because the first step in the biodegradation of many compounds (e.g., hydrocarbons) is the enzymatic insertion of oxygen into the structure, an activity carried out solely by bacteria, and this step is almost always rate limiting.

The number and position of substituent groups appended to a base structure such as a phenyl ring also seem to have some bearing on biodegradability, but it is more difficult to apply these generalizations to specific substances. For some polymers, such as modified cellulosics (e.g., methylcellulose), degree of substitution is a relatively precise concept and has predictive value. But for most nonpolymeric structures this is not true. An effect of substituent position has been noted for many classes of compounds including biphenyls, phenols, phenoxy herbicides, benzoates and anilines, but there are presently no consistent rules that are useful for predicting relative biodegradability. The greater resistance of meta- than of ortho- or para-disubstituted benzenes has become so familiar as to achieve the status of maxim, but may in fact apply only to degradation of substances in a few specific classes in soil.

2.2. Group Contribution Method for Predicting Biodegradability

An ability to predict relative rates of biodegradation from chemical structure alone would not only be useful for scoring and priority setting, but would also greatly facilitate the design of safer chemicals. In this section I describe the approach we used to develop mathematical models capable of such predictions. These models utilize the molecular features listed in the previous section as the basis for prediction.

Fragment contribution methods such as that used in our models have been used for many years in chemical engineering, but only more recently in environmental chemistry. The basic premise is that the activity of interest is a function of the contribution of one or more molecular substructures or fragments of which the molecule is composed, and that the contribution of each fragment does not vary from compound to compound (i.e., there is no interaction between fragments). The ideal situation is that each fragment in a model has a clear mechanistic relationship to the activity of interest, which is understood at the molecular level. This situation is rarely if ever realized. Fortunately, it doesn't really matter as long as (i) there exists for model development a set of measured values ("training set") of adequate size, and (ii) a reasonably comprehensive set of structural fragments associated in some way with activity can be identified.

We used this approach to develop a set of 4 models for predicting biodegradability. Two of the models (based on linear and nonlinear algorithms) classify chemicals as easily or not easily biodegradable, and the other two (for primary and ultimate biodegradation) make semi-quantitative estimates of aquatic biodegradation rates[6,7]. The rate models and classification models are based on two distinct and independent training sets. For purposes of model development the positive and negative molecular features listed in the previous section were formally defined and constituted the independent (predictor) variables.

A training set of biodegradability data for the two classification models was developed using data from the literature that are well documented and widely available. To take advantage of all available data for the largest possible universe of chemicals we instituted several years ago a data evaluation procedure[8] that utilizes biodegradation data from all types of studies other than pure cultures. The objective of this "weight of evidence" approach is to increase confidence that model predictions reflect chemical structure rather than experimental conditions. Each test result, whether for biochemical oxygen demand (BOD), CO_2 production, loss of parent or something else, is assigned a qualitative biodegradation code such as BR (Biodegrades Rapidly) or BSA (Biodegrades Slowly even with Acclimation). Aspects of biodegradation such as acclimation, microbial toxicity and temperature are considered in the evaluation process. Summary biodegradation codes are then assigned to each chemical for each of several endpoints, and for overall evaluations of aerobic and anaerobic biodegradability based on the summary evaluations for the individual endpoints. The latter reflect basic categories of test protocol. For aerobic biodegradation there are screening studies, soil grab sample studies, water grab sample studies, biological treatment simulations, and field tests. Finally, a reliability code, reflecting the amount and consistency of available data (1, three or more consistent test results available; 2, two test results available; 3, only one test result, or two or more inconsistent results), is assigned for each biodegradation summary code.

Biodegradation data and evaluations are entered into a file called BIODEG, which is a component of the Environmental Fate Data Base[9]. BIODEG presently contains extracted biodegradation data on more than 800 organic chemicals. The records for each chemical constitute a comprehensive assessment of existing biodegradation data for the chemical.

We used the overall aerobic biodegradation summary codes with reliability evaluations of 1 or 2 to develop the two biodegradability classification models. What the models predict is the probability that a chemical is in the BR category. The models predicted biodegradation category correctly for approximately 90% of the 295 chemicals in the training set[7]; an earlier study showed similar results for an independent validation set[6]. This level of accuracy is on par with other published biodegradability models that used different training sets and statistical methods[10,11,12,13], but we consider it an advantage of our approach that the predictor variables explicitly reflect generally accepted rules of thumb. That these rules are so accepted has been confirmed by carefully conducted surveys of expert knowledge in the field[7,14]. Fragments and their coefficients are listed in Table 1. Both the signs and relative magnitudes of the coefficients are generally consistent with expectation.

The other two models, for semi-quantitative estimation of primary and ultimate biodegradation rates, use the same molecular fragments but were developed from a completely independent training set. That training set was derived from a survey in which a panel of 17 experts estimated rates of primary and ultimate biodegradation under aerobic conditions in aquatic environments for 200 organic chemicals[7]. Each expert rated the biodegradability of each chemical using the terms hours, days, weeks, months and longer than months to indicate the approximate time they thought would be required for the process to proceed to completion. These responses were encoded as integers ("hours" = 5; "days" = 4; etc.), and the mean of all expert responses was then used as the dependent variable in multiple regressions against the 36 structural fragments and molecular weight. The 200 survey chemicals covered a very wide range of structure and molecular weight, and the majority were multifunctional. In general, chemicals were selected to be included in the survey for the specific purpose of testing hypotheses regarding the effects of certain substructures on estimated biodegradability.

The primary and ultimate biodegradation survey models both calculated biodegradation rates for chemicals in the training set with $R^2 = 0.7$, and > 90% of the residuals ≤ 0.5.

Table 1. Structural fragments and coefficients from the BIODEG model

Fragmentor parameter	BIODEG models			Survey models		
	Freq[a]	Linear coeff	Non-linear coeff	Freq[a]	Primary coeff	Ultimate coeff
Equation constant	—	.748	3.01	—	3.848	3.199
M_w	295	−.000476	−.0142	200	−.00144	−.00221
Unsubstituted Aromatic (≤ 3 rings)	2	.319	7.191	1	−.343	−.586
Phosphate ester	5	.314	44.409	6	.465	.154
Cyanide/nitrile (−C≡N)	5	.307	4.644	11	−.065	−.082
Aldehyde (−CHO)	4	.285	7.180	5	.197	.022
Amide (−C(=O)−N or −C(=S)−N)	9	.210	2.691	13	.205	−.054
Aromatic −C(=O)OH	24	.177	2.422	6	.0078	.088
Ester (−C(=O)−O−C)	23	.174	4.080	25	.229	.140
Aliphatic −OH	34	.159	1.118	18	.129	160
Aliphatic $-NH_2$ or −NH−	13	.154	1.110	7	.043	.024
Aromatic ether	11	.132	2.248	11	.077	−.058
Unsubstituted phenyl group ($-C_6H^5$)	25	.128	1.799	22	.0049	.022
Aromatic −OH	46	.116	.909	21	.040	.056
Linear C4 terminal alkyl ($-CH_2CH_2CH_2CH_3$)	44	.108	1.844	26	.269	.298
Aliphatic sulfonic acid or salt	4	.108	6.833	4	.177	.193
Carbamate	4	.080	1.009	6	.194	−.047
Aliphatic −C(=O)OH	33	.073	.643	10	.386	.365
Alkyl substituent on aromatic ring	36	.055	.577	36	−.069	−.075
Trizine ring	5	.0095	−5.725	4	−.058	−.246
Ketone (−C−C(=O)−C−)	12	.0068	−.453	10	−.022	−.023
Aromatic −F	1	−.810	−10.532	1	.135	−.407
Aromatic −I	2	−.759	−10.003	2	−.127	−.045
Polycyclic aromatic hydrocarbon (≥ 4 rings)	6	−.657	−10.164	2	−.702	−.799
N−nitroso (−N−N=O)	4	−.525	−3.259	1	.019	−.385
Trifluoromethyl ($-CF_3$)	1	−.520	−5.670	2	−.274	−.513
Aliphatic ether	11	−.347	−3.429	16	−.0097	−.0087
Aromatic $-NO_2$	14	−.305	−2.509	13	−.108	−.170
Azo group (−N=N−)	2	−.242	−8.219	3	−.053	−.300
Aromatic $-NH_2$ or −NH−	32	−.234	−1.907	23	−.108	−.135
Aromatic sulfonic acid or salt	11	−.224	−1.028	8	.022	.142
Tertiary amine	10	−.205	−2.223	10	−.288	−.255
Carbon with 4 single bonds & no H	9	−.184	−1.723	32	−.153	−.212
Aromatic −Cl	40	−.182	−2.016	27	−.165	−.207
Pyridine ring	18	−.155	−1.638	8	−.019	−.214
Aliphatic −Cl	12	−.111	−1.853	14	−.101	−.173
Aromatic −Br	5	−.110	−1.678	4	−.154	−.136
Aliphatic −Br	5	−.046	−4.443	2	.035	.029

[a]Number of compounds in the training set containing the fragment.

For interpretation, calculated biodegradability values and residuals should be compared to the integer scale used to codify the experts' responses. Figure 2 shows the distribution of residuals for the primary biodegradation survey model.

There is no need to use these particular models, or any other model, to make an educated guess about biodegradability of an untested compound. But the fragment contribution models do provide a rapid, convenient, and systematic way to accomplish this, in a way that is consistent with knowledge in the field. Molecular structure is the only input

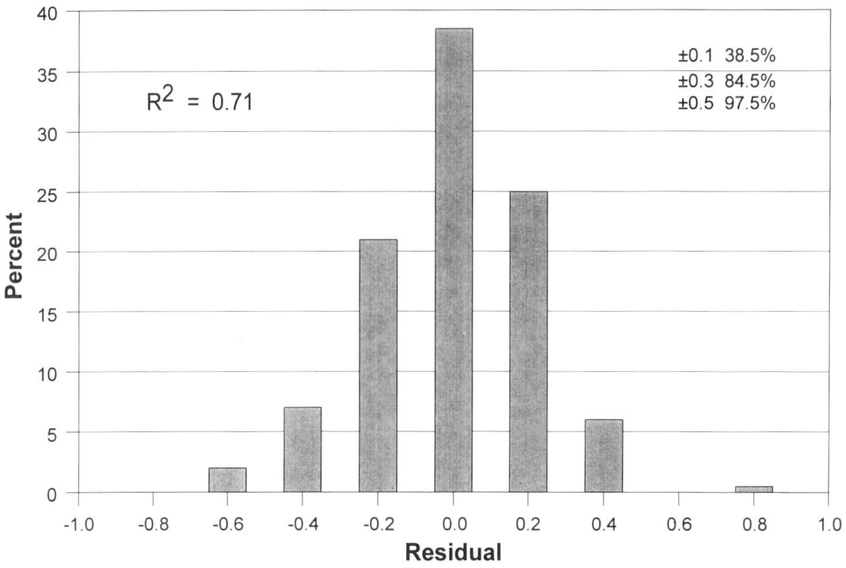

Figure 2. Distribution of residuals for the primary biodegradation survey model.

needed and is entered from the PC keyboard via the chemical's SMILES (Simplified Molecular Information and Line Entry System[15]) notation. The models clearly identify and list for the user relevant fragments and their predicted contributions.

3. APPLICATIONS OF BIODEGRADABILITY MODELS

Assumptions about the potential utility of these models have focused mainly on scoring exercises aimed at prioritizing long lists of chemicals for more detailed assessment. However, other applications can be envisioned. In the discussion that follows special emphasis is placed on premanufacture screening of chemicals, in the design phase of development.

3.1. USEPA's Use of the Cluster Scoring System (UCSS)

The UCSS has been used by the USEPA's Office of Pollution Prevention and Toxics (OPPT) since 1992 to screen chemicals on the TSCA Inventory, for the purpose of setting priorities for further review under the RM (Risk Management) process. The UCSS is the outgrowth of years of frustration with the inefficiency of the "single-chemical" approach to risk assessment/risk management. To regulate a substance under TSCA the USEPA must show that it does or may present an *unreasonable risk*, and this in effect necessitates an assessment of not only the risks from exposure to the chemical being scrutinized, for all possible uses, but of the risks of actual or potential substitutes as well. Obviously this is an extremely slow process when carried out on a chemical-by-chemical basis.

The UCSS aims at creating efficiencies by first grouping substances by common or closely related uses, and then subjecting the chemicals within the cluster to a procedure designed to evaluate risk at the screening level. Use clusters can subsequently be compared to one another as a means of identifying the highest priorities for full assessment. Moreover, this approach facilitates the identification of safer substitutes as a means to pollution prevention. Table 2 lists the chemicals in a typical use cluster, in this case for flame retardants used in plastics. Substances in this and other clusters are assigned scores for human exposure, environmental exposure, human toxicity, environmental toxicity and pollution prevention potential, by algorithms that are consistent across all clusters. Scores for these basic elements are then combined to yield an overall score for each chemical, and individual chemical scores are combined to yield an overall score for the cluster.

Use volume, total releases to the environment, number of use sites and environmental persistence are the four elements currently used to estimate the potential for environmental exposure. Since the emphasis at this stage of the existing chemical process is on screening and not full assessment, a decision was made early in the development of the UCSS to automate the calculation of persistence scores using components of the BIODEG model. Scores of high, medium or low for persistence are assigned as follows:

BIODEG ultimate degradation survey model	BIODEG non-linear probability model	Score
≤ 2	—	H
$> 2; \leq 3$	—	M
$> 3; \leq 4$	0.5	M
	≥ 0.5	L
> 4	—	L

Experience with this scheme is still limited, but the results thus far seem to be consistent with expectations. Also listed in Table 2 are the UCSS persistence scores calculated as above for flame retardants used in plastics.

3.2. Pollution Prevention: Designing Biodegradable Chemicals

One way to reduce pollution at the source is to design safer chemicals. Chemicals that persist in the environment remain available to exert toxic effects and may bioaccumu-

Table 2. Typical use cluster and persistence scores: Flame retardants used in plastics

Chemical	BIODEG model[a]	Suggested interpretation[b]	Persistence score
1,2–bis(2,4,6–tribromopenoxy) ethane	0.78	recalcitrant	high
2–ethylhexyl diphenyl phosphate	2.89	weeks	medium
Phosphonium bromide	—[c]	—[c]	—[c]
Ammonium fluoroborate	—[c]	—[c]	—[c]
Antimony pentoxide	—[c]	—[c]	—[c]
Antimony trioxide	—[c]	—[c]	—[c]
Isodecyl diphenyl phosphate	2.83	weeks	medium
Triphenyl phosphate	2.70	weeks–months	medium
Tris(2,3–dibromopropyl) phospate	1.98	months	high
Alumina trihydrate	—[c]	—[c]	—[c]
Tetrabromobisphenol A	1.31	≥ months	high
Decabromodiphenyl oxide (ether)	–0.34	recalcitrant	high
Chlorendic anhydride	0.70	recalcitrant	high
Isopropylphenyl diphenyl phosphate	2.51	weeks–months	medium
Tert–butylphenyl diphenyl phosphate	2.34	weeks–months	medium

[a]Direct output of BIODEG ultimate biodegradation survey model.
[b]Suggested interpretation of BIODEG predictions using the following scale: 5 = hours; 4 = days; 3 = weeks; 2 = months; 1 = longer than months (approximate time required for complete ultimate biodegradation).
[c]Inorganic chemical; no BIODEG prediction possible.

late. Since microbial degradation is the major loss mechanism for most organic chemicals in soil, water and sewage treatment, biodegradability should be included as a factor in product design along with toxicity and the ability of the chemical to function in the desired application. By incorporating greater biodegradability to nontoxic products into a molecule's structure, the risk of environmental damage is reduced. In fact biodegradability has been an important design consideration for down-the-drain consumer products like laundry detergents for more than 40 years, but this has not been the case for industrial chemicals with mainly non-consumer uses.

The following examples are drawn from high-volume chemicals in current use and show how enhanced biodegradability has—or might have—avoided unnecessary environmental damage. Further, the examples offer straightforward illustrations of how the BIODEG models may be used to make predictions about the relative biodegradability of chemical substances.

3.2.1. Linear Alkylbenzene Sulfonates. The development of laundry detergents based on linear alkylbenzene sulfonate (LAS) is a brilliant success story, and a case can be made that this is still the best illustration to date of molecular engineering to enhance biodegradability and thus environmental acceptability. The replacement of soap as the workhorse surfactant in household laundry products occurred as early as the 1940s[16], with the development of manmade alkylbenzene sulfonate (ABS) surfactants. At first the alkyl chains were derived from a kerosene fraction, but these products were soon replaced by ABS produced from propylene tetramer. Tetrapropylene alkylbenzene sulfonate (TPBS) was a more efficacious and economical product, obtained via a one-step Friedel-Crafts process involving addition of benzene at the double bond of the olefin feedstock to yield the branched alkylbenzene, followed by sulfonation of the benzene ring. As manufactured TPBS is actually a complex mixture with a typical structure as shown in Figure 3.

Environmental problems with these highly branched products appeared almost immediately, as they were found to be incompletely biodegraded in municipal sewage treatment systems. According to Painter[17] TPBS was degraded by only about 50% in sewage treatment units, and as a result excessive foaming occurred in activated sludge aeration tanks and receiving waters. Because of its incomplete biodegradation levels of TPBS in river waters in Britain and elsewhere were as high as 2 mg L^{-1}, and water tended to foam when coming out of the tap.

Among the results of this foaming were impaired efficiency of the treatment plants and increased dispersal of potentially pathogenic bacteria[17]. Eventually methods were developed that permitted the economical manufacture of a more environmentally acceptable product, LAS (Figure 3). This technology involved use of molecular sieves to obtain predominantly linear alkanes from petroleum, followed by any of several methods for producing the olefin. Voluntary changeover from TPBS to LAS was complete by the early 1960s in the US[16].

LAS surfactants are almost completely biodegradable in sewage treatment, and this has been amply demonstrated in hundreds of studies, including numerous monitoring studies conducted at full-scale treatment plants[18]. Of course we now take it for granted that the lower biodegradability of TPBS is due to its highly branched alkyl group, and that the greatly enhanced biodegradability of LAS is due to the absence of such branching. The enhanced biodegradability of linear LAS is easily predicted from information in Table 1 because the molecule now contains an additional fragment positive for biodegradability that is not present in TPBS, the linear terminal alkyl with $\geq$4 carbons.

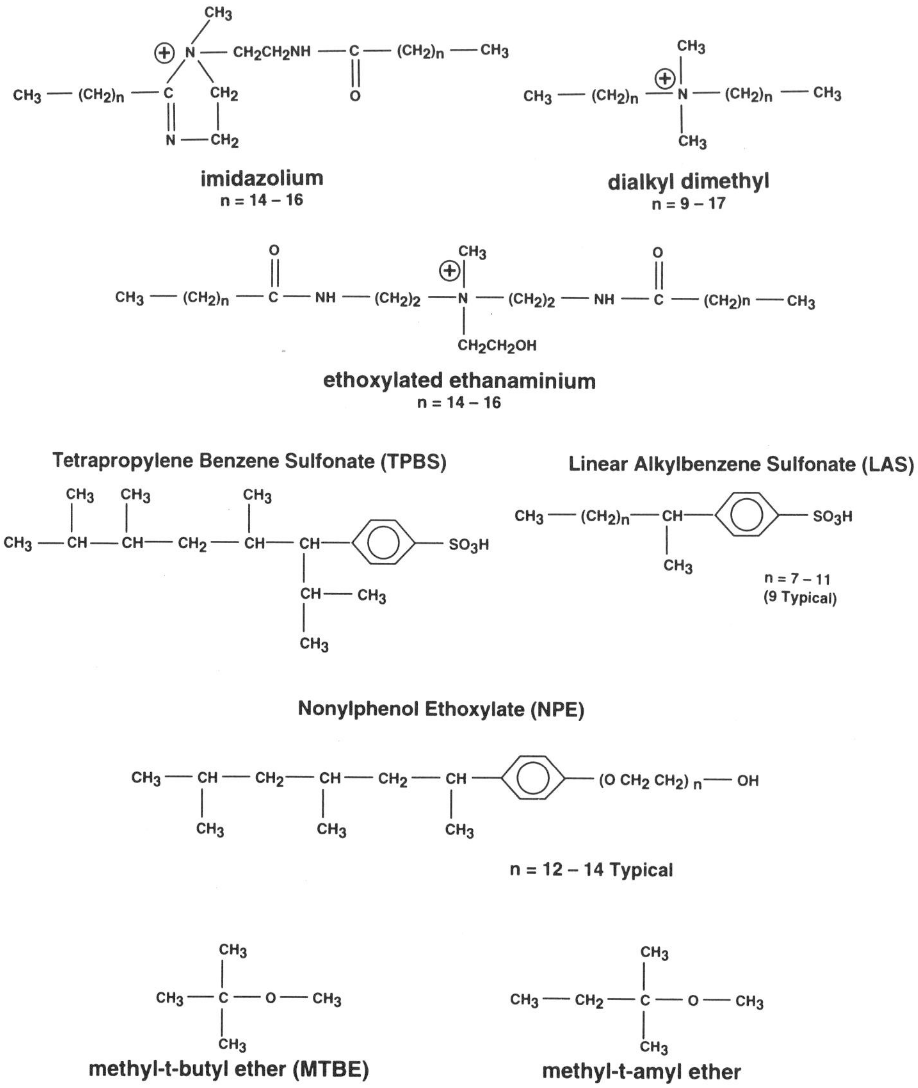

Figure 3. Structures of chemicals mentioned in this paper.

3.2.2. Dialkyl Quaternaries. Surface-active quaternary ammonium compounds (generally referred to as QACs or quats) first gained prominence more than 50 years ago, with Domagk's discovery that the biocidal properties of simple quaternary ammonium compounds were greatly enhanced by the presence of a long alkyl group[19]. There are still many QAC-type biocides in use, but household fabric softeners presently constitute the largest market by far for QACs. Other applications are mainly industrial and include multiple uses in textile processing, road paving, oil well drilling and mineral flotation, to name just a few. According to Cross[20], 66% of the market for QACs is dominated by three classes, all of which are dialkyl quaternaries, meaning that hydrophobicity is imparted to the molecule by two linear alkyl chains in the C_{10} to C_{18} range. The three classes are dialkyl dimethylammonium salts, imidazolium quaternary ammonium salts, and ethoxylated ethanaminium quaternary ammonium salts, with typical structures as shown in Figure 3. Most uses of QACs, and especially the high-volume fabric softeners, lead to their release to municipal wastewater treatment systems.

Until recently the fabric softener market was dominated by a QAC of the first type, dihydrogenated tallow dimethyl ammonium chloride (DHTDMAC). The long alkyl groups in DHTDMAC are derived, as the name implies, from purified animal fat (tallow), and consist of a mixture chiefly in the C_{16}–C_{18} (tallow fatty acids) range. The true aqueous solubility of DHTDMAC is exceedingly low, and the chemical sorbs strongly to solids in wastewater treatment and the environment. Removal in treatment is therefore high (>95%), unlike TPBS, but does not necessarily correspond to ultimate biodegradation[21]. More importantly, the heavy use of DHTDMAC before 1990, its relatively low rate of biodegradation in aquatic sediments, and its high intrinsic ecotoxicity led to a public perception in some parts of the world, particularly Europe, that DHTDMAC was placing a critical load on surface waters. As a result, DHTDMAC has been phased out of the European market, with consumption declining from a peak of about 50,000 metric tons per year before 1990 to 4,500–9,000 tons in 1993[22]. Similar changes are occurring in the US market as manufacturers are voluntarily reformulating their products.

DHTDMAC is being replaced by dialkyl QACs in the other two classes listed above, the imidazolium and ethoxylated ethanaminium QACs. Although the database on environmental fate and removal in wastewater treatment of these compounds is less extensive than for DHTDMAC, the new dialkyl QACs seem to biodegrade more rapidly due to the linkage of the alkyl groups to the remainder of the molecule via hydrolyzable amide bonds[21]. These new, more biodegradable fabric softening agents thus represent another illustration of how safer surfactants can be developed by rational molecular design. The approach here may appear to be different from that for TPBS/LAS since it involves direct incorporation of a molecular feature known to be positive for biodegradability (Table 1), not the deletion of an impediment (branching) as in TPBS. But this distinction is really not very meaningful since the replacement of tetrapropylene with a linear alkyl group in ABS/LAS also facilitates beta-oxidation by competent microorganisms. Attack on the alkyl group is in fact the principal degradation pathway for LAS[16].

3.2.3. Alkylphenol Ethoxylates. Alkylphenol ethoxylates (APEs) are one of two major classes of nonionic surfactants. APE uses are mainly industrial and cover a wide range, including applications in textile processing, emulsion polymerization, printing, metal cleaning, oil well drilling and papermaking. According to the Chemical Manufacturers Association (CMA), 450 million pounds of APEs were sold in the US in 1988[23]. Nonylphenol ethoxylates (NPEs) represent the largest class of APEs, accounting for nearly 75% of US APE production in 1980. A good example of NPE use in industry is in printing. In screen reclamation NPEs are used in ink, emulsion and haze removal formulations, according to the USEPA's draft *Cleaner Technologies Substitutes Assessment* (CTSA) for the screen reclamation use cluster[24]. And in a new CTSA being prepared for the lithographic blanket wash process, NPEs appear in 7 of 37 product formulations.

Unlike linear alcohol ethoxylates (the other major class of nonionic surfactants), APEs are mostly branched[25]. The industrial synthesis of APEs and evolution of the manufacturing process are somewhat parallel to those of alkylbenzene sulfonates[26]. As with alkylbenzene sulfonates the alkylphenyl portion of the molecule is synthesized by addition of an aromatic feedstock (phenol in this case) to the double bond of an olefin. At first the olefins were formed from polymerization of butene and isobutene, which resulted in highly branched alkyl groups with an abundance of quaternary carbons. These products long ago were replaced by APEs derived from propylene oligomers, which yield structures without quaternary carbons but still containing branched alkyl groups as in TPBS (Figure 3).

But the substitution of linear for branched olefins, which would logically constitute the next step in the evolution of safer APEs, has not occurred on a large scale. There may be various reasons for this, but the fact that the stringent biodegradability criteria adopted by the US Soap and Detergent Association apply only to consumer products[27] would seem to be important, since APE uses are mainly commercial. This and the absence of obvious environmental problems such as foaming in waterways have translated into a lack of impetus for change.

The environmental risks associated with APEs and especially NPE are a complex and contentious issue. Most attention has focused on the mono- and diethoxylated nonylphenol adducts (NP1EO and NP2EO, respectively), which have been reported to be relatively stable intermediates in NPE biodegradation[26]. NP1EO, NP2EO and nonylphenol itself are highly toxic to aquatic organisms, whereas the parent NPEs (the number of ethoxylate groups may be as high as 30–50, but 12–14 is more typical) are much less toxic. Recently added to this is a new controversy, as nonylphenol, NP2EO and related compounds have been reported to be estrogenic in fish[28].

The actual margins of safety under environmental conditions for the above effects are a subject of intense debate, but the current situation might not have arisen in the first place if APEs were manufactured mainly from linear olefins. Branched APEs biodegrade initially by stepwise shortening of the polyethoxylate hydrophile one ethoxylate group at a time[16], and attack on the branched alkyl group does not seem to be a significant pathway. Substitution of a linear alkyl group would eliminate an impediment to biodegradation and provide another site for microbial attack, by analogy to TPBS and LAS. This should lead to faster biodegradation and make transient accumulation of toxic intermediates unlikely. Experimental data support this contention (Table 3). Indeed, studies[16] suggest that a fundamental shift in the principal route of breakdown occurs such that degradation by beta-oxidation of the alkyl group (not attack on the polyethoxylate chain) now predominates. As with LAS, the enhanced biodegradability of linear APEs is easily predicted from information in Table 3.

3.2.4. Oxygenates: Methyl-Tert-Butyl Ether. The phaseout of lead-based octane enhancers in gasoline began in 1973 as a result of growing environmental and health concerns[29]. To replace the octane lost with the removal of lead the aromatic content of gasoline was increased from about 20 % by volume before the lead phaseout to as much as 40% in some cases, but the resulting high volatility created problems with emissions of volatile organic compounds (VOCs). This led to the USEPA's vapor pressure reduction program in 1989 and since then, oxygenates have been added to boost the octane rating of

Table 3. Biodegradability of linear and branched alkylphenol ethoxylates

APE[b]	% Biodegradation Linear[c]	Branched	Method[d]	Analysis[d]
C_8APE_9	71	46	In, 28 d	Wt
	51	49	In, 20 d	Wt
C_9APE_9	65	25	RW, 15 d	CT
	65	30	SF, 5 d	CT
	88	55	CAS, 4 hr	CT
	57	33	In, 9 d	CT
	66	32	In, 9 d	ST
	75	0	In, 9 d	F
	62	10	SF, 7 d	CT
	60	18	SF, 7 d	ST
	0–50	0	SF, 7 d	F
	89	75	RW, 10 d	CT

[a]Adapted from Swisher[16].
[b]APE = alkylphenol ethoxylate. C_8APE_9 signifies octylphenol with 9 ethoxy groups and C_9APE_9 signifies nonylphenol with 9 ethoxy groups.
[c]Linear seconday.
[d]Abbrevations: In=natural or synthetic medium inoculated with acclimated or unacclimated microorganisms; RW=river water die-away; SF=shake-flask culture; CAS=continuous flow activated slude; Wt=weight of soluble organics in cell-free medium; CT=cobalt thiocyanate; ST=surface tension; F=foaming properties.

fuel while lowering its vapor pressure. Since the USEPA's oxygenated fuels program went into effect in November 1992, oxygenates have also been added seasonally to gasoline to reduce wintertime generation of carbon monoxide. Oxygenates in current use include ethanol, methyl-*tert*-butyl ether (MTBE), ethyl-*tert*-butyl ether (ETBE) and *tert*-amyl methyl ether (TAME) (Figure 3), but MTBE is by far the most common. Unlike ethanol it can be produced and blended into gasoline in existing refineries, and reformulated gasoline can be transported through existing pipelines. US production of MTBE has increased rapidly in the last several years and was reported to be nearly 9 billion kg in 1993[29].

Unfortunately, research has shown MTBE to be quite persistent. MTBE resists biodegradation under aerobic conditions in both soil and water, and several recent studies have extended these findings to sulfate-reducing, nitrate-reducing and methanogenic anaerobic conditions in soil and water microcosms[30] and aquifer materials[31,32]. In view of the many possibilities for fugitive release that result from the huge infrastructure required to store, deliver and use gasoline, it is therefore not surprising that a recent study detected MTBE with high frequency in shallow ground water[33]. Among other results, this study by the US Geological Survey's (USGS) National Water Quality Assessment program found MTBE in 27% of 211 wells in urban areas, with Denver, CO yielding the highest frequency of detection at 79% of the wells sampled.

A report issued by the USEPA in 1993 concluded that MTBE is unlikely to pose "a substantial risk of acute health symptoms among healthy members of the public receiving typical exposures," but acknowledged that "more subtle health risks, especially among susceptible subpopulations" may still exist[34]. The USEPA considers MTBE to be a possible human carcinogen, and the USEPA's Office of Drinking Water is planning to issue a final health advisory for MTBE in drinking water. The levels of MTBE found in the USGS survey are low, but that study was limited in scope and did not sample deeper aquifers. Beyond this, it seems doubtful that MTBE's persistence could have been seriously considered in any risk assessments conducted prior to its widespread use, because if it had been

Table 4. Summary evaluations of experimental data from the BIODEG database, and BIODEG model predictions of biodegradability, for a series of oxygenates

	BIODEG model[a]		Database summary evaluation[b]	
Chemical	Linear prob	Nonlinear prob	Aerobic	Anaerobic
Methanol	0.89	0.98	BR 1	BR 3
Ethanol	0.88	0.97	BR 1	BR 3
Isopropanol	0.88	0.96	BR 1	—[c]
Tert–butanol	0.53	0.56	BST 1	—[c]
Methyl ethyl ketone	0.72	0.82	BR 1	—[c]
Acetone	0.73	0.85	BR 1	—[c]
Methyl isobutyl ketone	0.71	0.76	BRA 1	—[c]
Ethyl acetate	0.88	1.00	BR 1	BR 3
Methyl–*tert*–butyl ether	0.17	0.03	BSA 1	BSA 1
Methyl–*tert*–amyl ether	0.17	0.03	BSA 3	BSA 3
Ethyl–*tert*–butyl ether	0.17	0.03	BSA 3	BSA 1
Disopropyl ether	0.35	0.13	BS 2	BSA 2
Diethyl ether	0.36	0.19	BS 1	BSA 3[d]
Di–n–propyl ether	0.35	0.13	BRA 3	BSA 2
Di–n–butyl ether	0.56	0.86	BSA 2	BSA 2
Methyl–n–butyl ether	0.47	0.54	—[c]	BRA 3[d]
Ethyl–n–butyl ether	0.46	0.49	BSA 3	BSA 2

[a]Probability of rapid biodegradation; generally, if p > = 0.5 the prediction is BR (biodegrades rapidly).
[b]Biodegradability and reliability codes; 1 = high reliability; 2 = intermediate reliability; 3 = low reliability; BR = biodegrades rapidly; BRA = biodegrades rapidly with acclimation; BS = biodegrades slowly; BSA = biodegrades slowly even with acclimation; BST = biodegrades sometimes; see text for details.
[c]No BIODEG database record available.
[d]No BIODEG database record available; evaluation is based on data in Suflita and Mormile[31].

it would have been obvious to any expert that the compound's tertiary carbon and aliphatic ether functionalities (Figure 3) would render it difficult to degrade. It is noteworthy that a 1987 consent order for MTBE testing that was negotiated by the USEPA did not include any environmental fate or effects testing.

Table 4 contains summary evaluations from the BIODEG database for several low-molecular weight ethers, alcohols, ketones and esters, including the major gasoline oxygenates. BIODEG model predictions are also given for the same compounds. It is clear that the model predictions accurately reflect relative biodegradability as demonstrated by the experimental data.

4. THE FUTURE

Absent specific knowledge of a chemical's environmental behavior, the way to design more biodegradable chemicals is to incorporate positive features like ester linkages and hydroxyl groups and exclude halogens, quaternary carbons, nitro groups and the like. The positive and negative features highlighted earlier in the text are a good starting point. Table 1 contains a more extensive list of relevant fragments and coefficients from the BIODEG models[7]. The BIODEG models provide a convenient way for chemists in research and development to quickly compare alternative designs.

Product performance and economics obviously are as closely linked as biodegradability is to molecular structure. This makes the task of following these recommendations potentially very difficult, and it further suggests that chemical design is properly a multidisciplinary process. It's easy to see the need to consider biodegradability for high-volume, down-the-drain chemicals used in consumer products. But manufacturers, processors and users of chemicals whose uses are mainly commercial must also try harder to make this a part of their economic equation. Often the balancing process may demand compromises in economics or function that were not previously necessary. But this change should be viewed as merely one part of the process of internalizing environmental costs. The desirability of integrating environmental considerations into business decisions and designing products to minimize their environmental impact are nothing new and have been acknowledged in CMA's Responsible Care® program.

Persistent chemicals are of special concern because they may remain available to biota for long periods of time, and thus may build up in exposed organisms to levels that cause toxic effects even if releases to the environment are small or infrequent. Through the process of bioaccumulation levels may be achieved that appear safe on the basis of acute toxicity criteria, but which ultimately prove harmful to the organisms in chronic exposures, or even to other organisms that are consumers of the exposed organism. Human populations may be exposed to such chemicals through consumption of contaminated fish, and history contains numerous examples of cases in which accumulation of chemical substances from the ambient environment by fish has led to decreases in reproductive success in fish-eating birds.

Halogenated hydrocarbon insecticides like aldrin and DDT and their metabolites are now ubiquitous global contaminants. The obvious lesson from this is that we should not design, manufacture and release to the environment new substances that are so persistent. The less obvious but equally compelling message to be derived from the various examples given above is that all industrial chemicals that may be released in substantial amounts should reflect the principles of safe design to the extent practicable. There are four reasons why we must pursue this goal:

1. We cannot know in advance all of the possible toxic effects of released chemicals. This is certainly true for new substances that have a minimum of test data, but it is also true for well-studied existing chemicals as witnessed by controversy over nonylphenol ethoxylates;
2. Consumption of a chemical, and therefore, potentially, its release to the environment, may increase significantly if the product is successful;
3. New uses may develop over time, and with them, a greater possibility of environmental release. This is especially true for new substances because the original manufacturer typically envisions one or more specific uses, but only in the marketplace is a chemical's utility fully revealed. For example, a surfactant with a very limited market niche may find its way into broader uses in textile processing, papermaking or another industry that generates large volumes of wastewater;
4. Recognizing that we live in an era of global markets and of concern for the global environment, chemical substances produced in the US may be exported to other nations where environmental controls are less stringent. Due to the possibility of long-range atmospheric transport, environmental damage may not be confined to the site of release.

ACKNOWLEDGMENT

The author gratefully acknowledges Prof. J.M. Suflita's suggestion of oxygenates as an example to be included in Section 3 of this paper.

REFERENCES

1. *Chemical Ranking and Scoring: Principles and Guidelines* (draft), Proceedings of the workshop: Chemical Ranking and Scoring, Destin, FL, 11–16 Feb 1995; SETAC Press: Pensacola, FL.
2. Chapman, P. J. In *Workshop: Microbial Degradation of Pollutants in marine Environments*; Bourquin, A. W.; Pritchard, P. H., Eds.; USEPA report no. 600/9–79–012; US Environmental Protection Agency: Gulf Breeze, FL, 1979; pp 28–66.
3. Gibson, D. T. In *The Handbook of Environmental Chemistry*; Hutzinger, O., Ed.; Springer-Verlag: Berlin, Germany, 1980, Vol. 2A; pp 161–192.
4. Dagley, S. *Ann. Rev. Microbiol.* 1987, *41*, pp 1–23.
5. Scow, K. M. In *Handbook of Chemical Property Estimation Methods*; Lyman, W. J.; Reehl, W. F.; Rosenblatt, D. H., Eds.; McGraw-Hill: New York, NY, 1982; pp 9–1 to 9–85.
6. Howard, P. H.; Boethling, R. S.; Stiteler, W. M.; Meylan, W. M.; Hueber, A. E.; Beauman, J. A.; Larosche, M. E. *Environ. Toxicol. Chem.* 1992, *11*, pp 593–603.
7. Boethling, R. S.; Howard, P. H.; Meylan, W. M.; Stiteler, W. M.; Beauman, J. A.; Tirado, N. *Environ. Sci. Technol.* 1994, *28*, pp 459–465.
8. Howard, P. H.; Hueber, A. E.; Boethling, R. S. *Environ. Toxicol. Chem.* 1987, *6*, pp 1–10.
9. Howard, P. H.; Sage, G. W.; LaMacchia, A.; Colb, A. *J. Chem. Inf. Comput. Sci.* 1982, *22*, pp 38–44.
10. Niemi, G. J.; Veith, G. D.; Regal, R. R.; Vaishnav, D. D. *Environ. Toxicol. Chem.* 1987, *6*, pp 515–527.
11. Gombar, V. K.; Enslein, K. In *Applied Multivariate Analysis in SAR and Environmental Studies*; Devillers, J; Karcher, W., Eds.; Kluwer: Boston, MA, 1991; pp 377–414.
12. Klopman, G.; Balthasar, D. M.; Rosenkranz, H. S. *Environ. Toxicol. Chem.* 1993, *12*, pp 231–240.
13. Tabak, H. H.; Govind, R. *Environ. Toxicol. Chem.* 1993, *12*, pp 251–260.
14. Boethling, R. S.; Gregg, B.; Frederick, R.; Gabel, N. W.; Campbell, S. E.; Sabljic, A. *Ecotoxicol. Environ. Saf.* 1989, *18*, pp 252–267.
15. Weininger, D. J. *Chem. Inf. Comput. Sci.* 1988, *28*, pp 31–36.
16. Swisher, R. D. *Surfactant Biodegradation*, 2nd ed.; Surfactant Science Series, Vol. 18; Marcel Dekker: New York, NY, 1987.
17. Painter, H. A. In *The Handbook of Environmental Chemistry*; Hutzinger, O., Ed.; Springer-Verlag: Berlin, Germany, 1992, Vol. 3F; pp 1–88.
18. Rapaport, R. A.; Eckhoff, W. S. *Environ. Toxicol. Chem.* 1990, *9*, pp 1245–1257.
19. Fredell, D. L. In *Cationic Surfactants*; Cross, J.; Singer, E. J., Eds; Surfactant Science Series, Vol. 53; Marcel Dekker: New York, NY, 1994; pp 31–60.
20. Cross, J. In *Cationic Surfactants*; Cross, J.; Singer, E. J., Eds.; Surfactant Science Series, Vol. 53; Marcel Dekker: New York, NY, 1994; pp 3–28.
21. Boethling, R. S. In *Cationic Surfactants*; Cross, J.; Singer, E. J., Eds.; Surfactant Science Series, Vol. 53; Marcel Dekker: New York, NY, 1994; pp 95–135.
22. European Centre for Ecotoxicology and Toxicology of Chemicals (ECETOC). *DTDMAC: Aquatic and Terrestrial Hazard Assessment. CAS No. 61789–80–8*; Technical Report No. 53; ECETOC: Brussels, Belgium, 1993.
23. Chemical Manufacturers Association (CMA). *Alkylphenol Ethoxylates. Human Health and Environmental Effects*; Interim Report; CMA: Washington, DC, 1991.
24. US Environmental Protection Agency (USEPA). *Cleaner Technologies Substitutes Assessment. Industry: Screen printing. Use Cluster: Screen Reclamation*; USEPA report no. 744R-94–005; USEPA: Washington, DC, 1994.
25. Naylor, C. G. *Environmental Fate of Alkylphenol Ethoxylates*; paper presented at the Annual Meeting of the US Soap and Detergent Association, Boca Raton, FL, 1992.
26. Holt, M. S.; Mitchell, G. C.; Watkinson, R. J. In *The Handbook of Environmental Chemistry*; Hutzinger, O., Ed.; Springer-Verlag: Berlin, Germany, 1992, Vol. 3F; pp 89–144.
27. Cahn, A.; Lynn, Jr., J. L. In *Kirk-Othmer Encyclopedia of Chemical Technology*, 3rd ed.; Grayson, M.; Eckroth, D., Eds.; Wiley: New York, NY, 1983, Vol. 22; pp 332–432.
28. Jobling, S.; Sumpter, J. P. *Aquat. Toxicol.* 1993, *27*, pp 361–372.
29. Peaff, G. *Chem. Eng. News* 1994, *72* (September 26), pp 8–13.
30. Yeh, C. K.; Novak, J. T. *Water Environ. Res.* 1994, *66*, pp 744–752.
31. Suflita, J. M.; Mormile, M. R. *Environ. Sci. Technol.* 1993, *27*, pp 976–978.
32. Mormile, M. R.; Suflita, J. M. *Environ. Sci. Technol.* 1994, *28*, pp 1727–1732.
33. Newman, A. *Environ. Sci. Technol.* 1995, *29*, p 305A.
34. Anderson, E. V. *Chem. Eng. News* 1993, *71* (September 20), pp 9–12, 16–18.

23

ANALYTICAL MICROSYSTEMS

Emerging Technologies for Environmental Biomonitoring

Kenneth L. Beattie[1,2]

[1]Health Sciences, Research Division
Oak Ridge National Laboratory, P.O. Box 2008
Oak Ridge, Tennessee 37831-6123
[2]Center for Environmental Biotechnology
University of Tennessee

1. INTRODUCTION

Widespread environmental biomonitoring, involving analysis of very large numbers of biological samples, is not readily attainable with today's technology. New processes and devices must be developed for collection, processing and analysis of very large numbers of biological samples in industrial and natural ecosystems. Environmental surveillance activities that would be enabled by technological advances include (i) monitoring of microbial populations in industrial bioprocesses, such as activated sludge treatment and fermentations; (ii) detection and control of biological contamination in industrial chemical processes; (iii) pathogen surveillance in soil, water, food products and animal vectors; (iv) studies of genetic diversity in wild animal and plant species; (v) genotyping of large populations of domesticated animals and plants; (vi) assessment of biological effects of toxic exposure in natural ecosystems through molecular endpoint surveillance; and (vii) toxicity testing in natural ecosystems and wastewater effluents. The wide scale monitoring activities listed above will require development of new devices that will enable sample manipulation and analysis at low cost and high throughput.

2. ANALYTICAL MICROSYSTEMS

2.1. Miniature Sample Processing Devices

In the fields of DNA diagnostics and genome analysis there is a trend toward automation and miniaturization of analytical systems, to provide rapid, ultrahigh throughput genetic analysis. The ultimate form of miniaturized systems, ideally suited for the future needs of environmental biomonitoring, would be a fully integrated "laboratory on a chip," capable of multiple microscale sample preparation and analysis. Microfabrication technol-

Biotechnology in the Sustainable Environment, edited by Sayler *et al.*
Plenum Press, New York, 1997

ogy, which has revolutionized the electronics industry during the past two decades, promises to do the same for the analytical instrument field in the coming decade. Availability of low cost mass-produced microscale analytical devices should greatly improve the prospects for widespread environmental monitoring.

The "front end" of the bioanalytical process is sample processing, consisting of cell disruption, biomolecule extraction, and purification. Current strategies for automation of sample preparation typically involve robotic manipulation of samples, whereby multiple micropetting tips move samples from one location to another at each step in the sample processing, such as from one microtiter plate to another, or from a microtiter plate to a multisample extraction manifold, and ultimately from the sample processing module to the analytical system such as multiple wells in a gel or capillary electrophoresis apparatus[1,2]. In this traditional "anthropomorphic" approach the sample manipulations are carried out by a robotic device that mimics the operations that a technician performs in the laboratory, except that numerous samples are manipulated simultaneously. In the future a more robust strategy for multiple microscale sample processing, providing a higher degree of miniaturization and parallelism, will be needed. Recent advances toward development of a "laboratory on a chip[3–5]" exemplify the kind of technological innovation that is needed for multiple microscale sample processing.

An array of microelectromechanical systems (MEMS array), conceptualized in Figure 1, may be the ideal front end component of an integrated analytical microsystem for environmental biomonitoring. Numerous biological samples are introduced at the upper face of the MEMS array and purified molecular components emerge from the lower face, ready for delivery to an interfaced analytical module such as a biosensor array. The front end MEMS array would contain all of the reagents, solid phase support materials and fluidic flow and switching means that are necessary for the sample processing. Adaptation of advanced MEMS technology[6–8] to the task of parallel bioprocessing would appear straightforward, and eventual mass production of MEMS arrays, ultimately containing disposable components, should enable a reasonable cost per sample.

2.2. Genosensors

The "back end" of the analytical microsystem, interfaced with the MEMS array, would consist of a detection system for specific analytes, such as a biosensor array. One such device, under development for nucleic acid sequence analysis, is a miniature two-di-

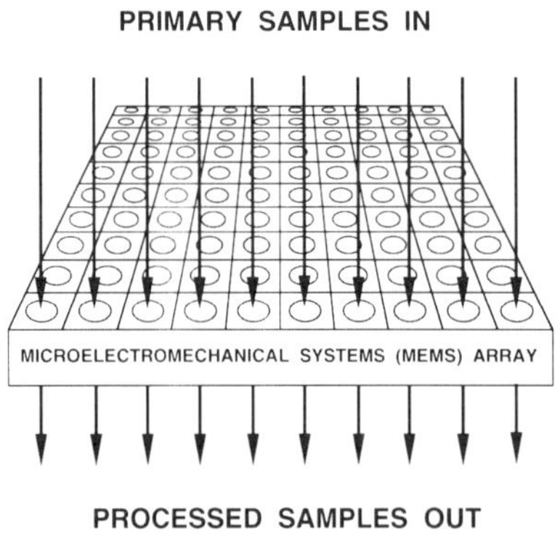

Figure 1. Microelectromechanical Systems Array (MEMS Array) concept.

mensional array of DNA probes, termed a "genosensor" or "DNA chip"[9–13]. As illustrated in Figure 2, each site in the genosensor array contains a specific DNA sequence (typically a short oligonucleotide), tethered to the surface at one end. Each immobilized DNA "probe" serves as a specific recognition element that binds to a complementary sequence in the DNA or RNA "target" strand through specific A•T and G•C base pairing. Thus, when a nucleic acid sample is hybridized to the genosensor array, a pattern of binding is obtained which reflects the nucleotide sequence of the target strand. The ultimate form of array hybridization, originally envisioned for complete *de novo* DNA sequence determination, is known as "sequencing by hybridization" (SBH)[15–33]. This approach requires a comprehensive set of arrayed probes, such as a 256×256 array of all 65,536 octamers. In principal, if a DNA target were hybridized in parallel with the complete set of octamers, the hybridization pattern would reveal the octamer content of the target, and a computer program searching for 7-base overlaps among the hybridizing octamers could then generate the DNA sequence from the nested set of octamers. However, because of technical challenges associated with SBH (mainly, lack of absolute fidelity of hybridization and multiple chance occurrence of short probe sequences within the target strand), most researchers in the DNA chip field are focusing on simpler DNA diagnostic applications of oligonucleotide arrays, requiring tens to a few thousand probes.

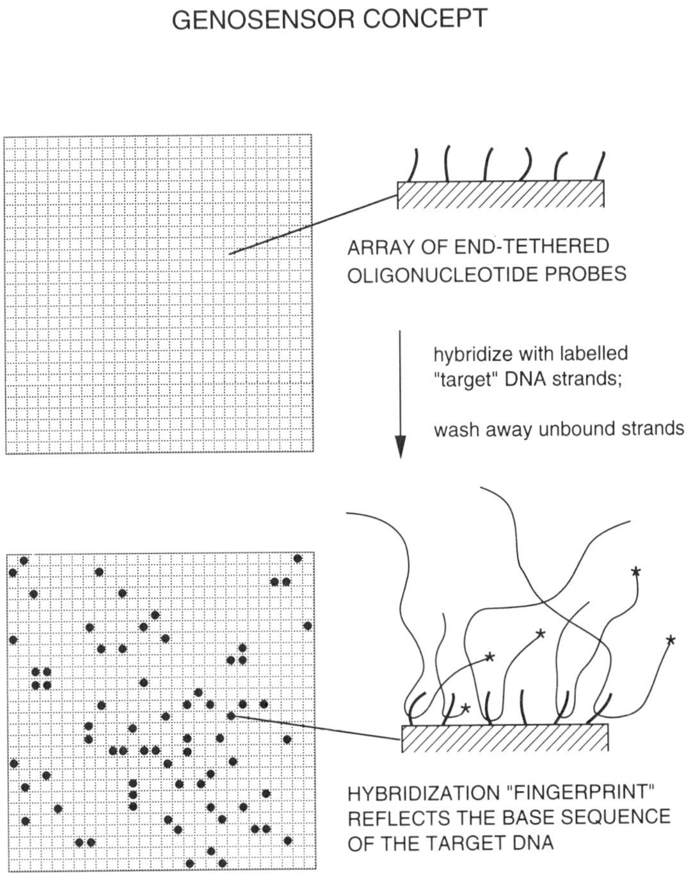

Figure 2. Genosensor concept.

2.2.1. Genosensor Array Fabrication. Researchers in the SBH field have employed two approaches to create genosensor arrays: (i) the in situ synthesis of numerous sequences onto a support surface[16,34–40], and (ii) the attachment of presynthesized oligonucleotides at each site in the array[9–14,23]. Both strategies employ the standard phosphoramidite method of solid phase chemical synthesis of oligonucleotides[41]. The author's laboratory has followed the post-synthesis attachment approach, in which the desired set of oligonucleotides is first synthesized, then chemically immobilized at specific sites on the surface of the chip. Synthesis of thousands of oligonucleotides, in quantities sufficient for millions of chips, is readily accomplished by an efficient segmented synthesis strategy initially developed in the author's laboratory[42–46] and further developed by Genosys Biotechnologies, Inc. (The Woodlands, Texas, USA). In the Genosys process, cylindrical porous Teflon synthesis wafers, containing controlled pore glass support material held in place by Teflon mesh at the upper and lower faces of the wafer, are stacked to form a sealed column through which reagents are flowed from below, permitting simultaneous addition of A, G, C or T to all oligonucleotides within a stack. Following each cycle of base addition, wafers are sorted into a new configuration, resulting in synthesis of a unique base sequence within each wafer. A single Genosys instrument is capable of synthesizing 200–300 oligonucleotides simultaneously, and about a thousand oligonucleotides per day. Thus, the full set of 65,536 octamer probes could be synthesized in a week using ten of the Genosys instruments.

Robotic protocols have been developed in the author's laboratory for arraying probes onto a surface, using a Hamilton Microlab 2200 fluid delivery system. The instrument is capable of precisely delivering droplets as small as 10 nanoliters (nL) (8 at a time) to the surface of glass slides at 500 micrometer (μm) pitch across the surface[12,13], and has a working area consisting of two 14″×14″ platforms (capacity of 8 microtiter plates or 60 microscope slides each), separated by a washing station. Theoretically, the Hamilton robot could array about 7,500 probes onto a single microscope slide. At 1 millimeter (mm) pitch the practical limit is about 1,500 oligonucleotides per slide. For higher density arrays (anticipated in future microfabricated genosensors) the probes can be placed about 100 μm apart on the surface using piezoelectric microjet tips developed at Microfab Technologies, Inc. (Plano, Texas, USA)[11].

The author's laboratory previously reported a relatively simple, reliable procedure for covalent linkage of oligonucleotide probes to silicon dioxide surfaces, involving specific condensation of a primary amine on the 5′- or 3′-terminus of the oligonucleotide with an epoxysilane group on the glass[12,13]. A more convenient and reliable procedure has recently been developed in the author's laboratory for covalent linkage of oligonucleotide probes at one end to glass surfaces[14,47]. The covalent linkage involves the direct reaction of 3′-propanolamine functions on synthetic oligonucleotides to silanol groups on underivatized glass. The proposed ester linkage, illustrated in Figure 3, is suggested by the following findings[14,46]: The linkage is (i) stable in hot water, enabling multiple cycles of hybridization; (ii) stable in mild acid but labile in mild base (favoring the ester linkage over the amide linkage); (iii) not formed with 5′-hexylamine-derivatized oligonucleotides (primary amine alone is insufficient); (iv) inhibited by pretreatment of glass with propanolamine but not propylamine; and (v) blocked by acetylation of primary amine on oligonucleotide (indicating that the amine function is necessary but insufficient for the linkage reaction). Phosphorimager quantitation of [5′-^{32}P]probe immobilization has shown that the surface density of glass-immobilized oligonucleotide probes is 10^{10}–10^{11} probes/mm^2, as with the previously employed epoxysilane-amine linkage chemistry[12,13]. However, the new method is much faster, more convenient, and provides a

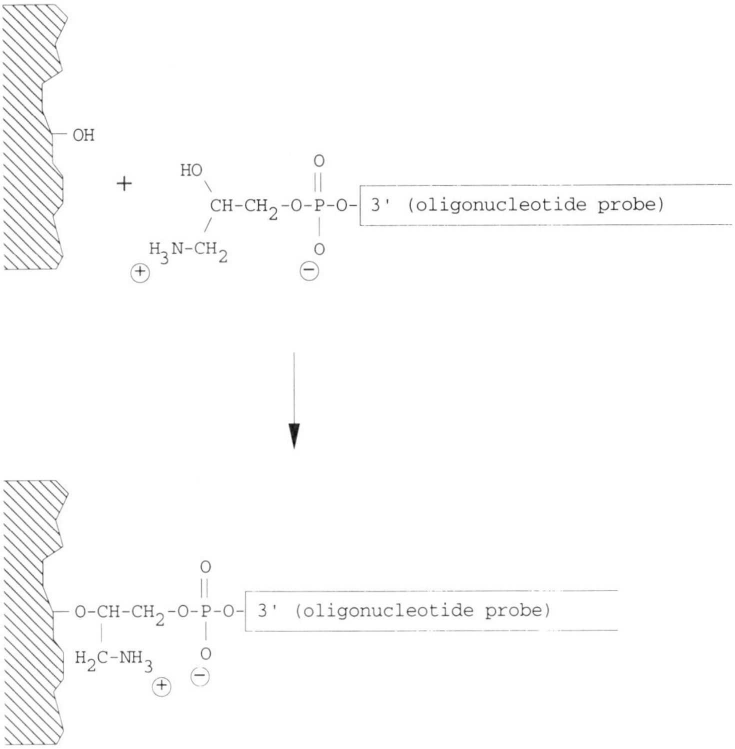

Figure 3. Proposed reaction scheme for covalent linkage of 3′-propanolamine-derivatized oligonucleotides to glass (SiO_2) surfaces.

lower background of nonspecific target DNA binding than the previous epoxysilane-amine method.

2.2.2. Flowthrough Genosensors. An advanced genosensor configuration is under development in the author's laboratory, in which the hybridization reactions occur within three-dimensional volumes of porous silicon dioxide or channel array glass, rather than on a two-dimensional surface[12,14,47]. The flowthrough configuration provides about 100-fold greater surface area per unit cross section, compared with the flat surface design, thus increasing the binding capacity per unit cross sectional area. The 100-fold increase in binding capacity per site should result in improved detection sensitivity, and should enable a higher degree of miniaturization of the array, since a given number of bound molecules would occupy a ten-fold smaller cross sectional area in the porous configuration. This configuration also enables 100 times as many hybridization reactions to be carried out per unit cross sectional area, compared with a flat surface. The flowthrough genosensor concept is illustrated in Figure 4. In the design shown here an array of square regions of porous silicon is first formed in a silicon wafer, then the porous silicon array is bonded to a second wafer in which square holes (aligned with porous patches) have been acid etched in a layer of silicon or silicon dioxide. The thickness of the silicon wafer and geometry of the porous patches and sample wells can be varied at will. For detection of hybridization using a phosphorimager, 200 μm diameter porous cells, spaced 500–1000 μm apart in 200 μm thick silicon, is a suitable geometry. For optical detection of hybridization (using a

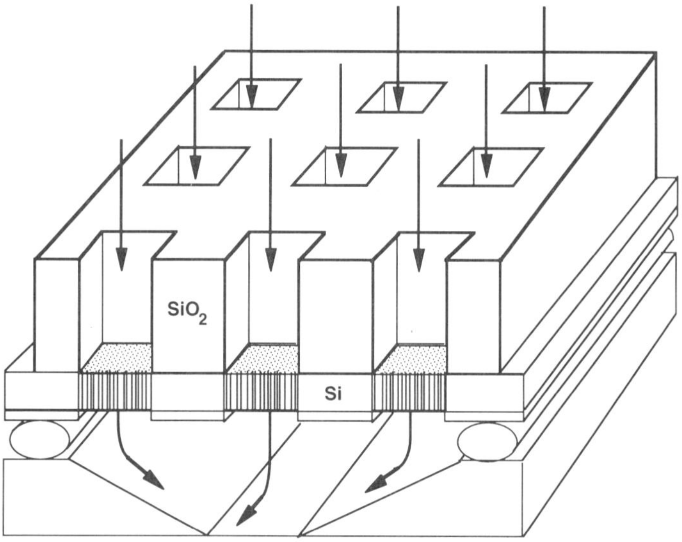

Figure 4. Flowthrough genosensor concept.

charge-coupled device, or CCD, camera), 50 m diameter porous cells at 100–200 μm pitch would be appropriate.

Shown below (Figure 5) are scanning electron micrographs (kindly supplied by Dr. Peter McIntyre, Physics Dept., Texas A&M University and Accelerator Technology Corp.) of a regular array of square pores formed in pure silicon using an electrochemical acid etch process[48], following an initial photolithography step which defines the position of each pore. The appearance of individual pores (3 μm diam. at 10 μm pitch) as viewed obliquely from an upper corner of the wafer is shown on the right, and a cross sectional view showing the straight channels opening onto the lower face of the wafer is shown on the left.

The advantages of porous glass over flat glass genosensors have recently been demonstrated in the author's laboratory, using commercially available glass capillary arrays (Galileo Electro-Optics Corp., Sturbridge, MA). In these proof-of-concept experiments[14,47] 9mer probes were immobilized within 0.5 mm thick wafers containing dense arrays of 10-

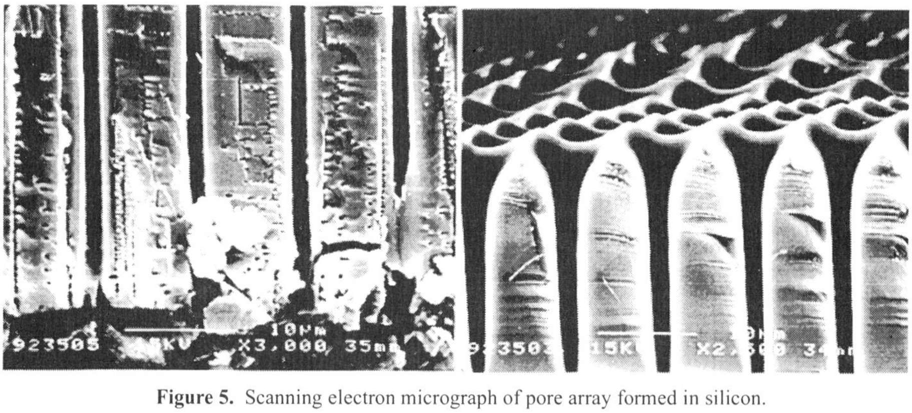

Figure 5. Scanning electron micrograph of pore array formed in silicon.

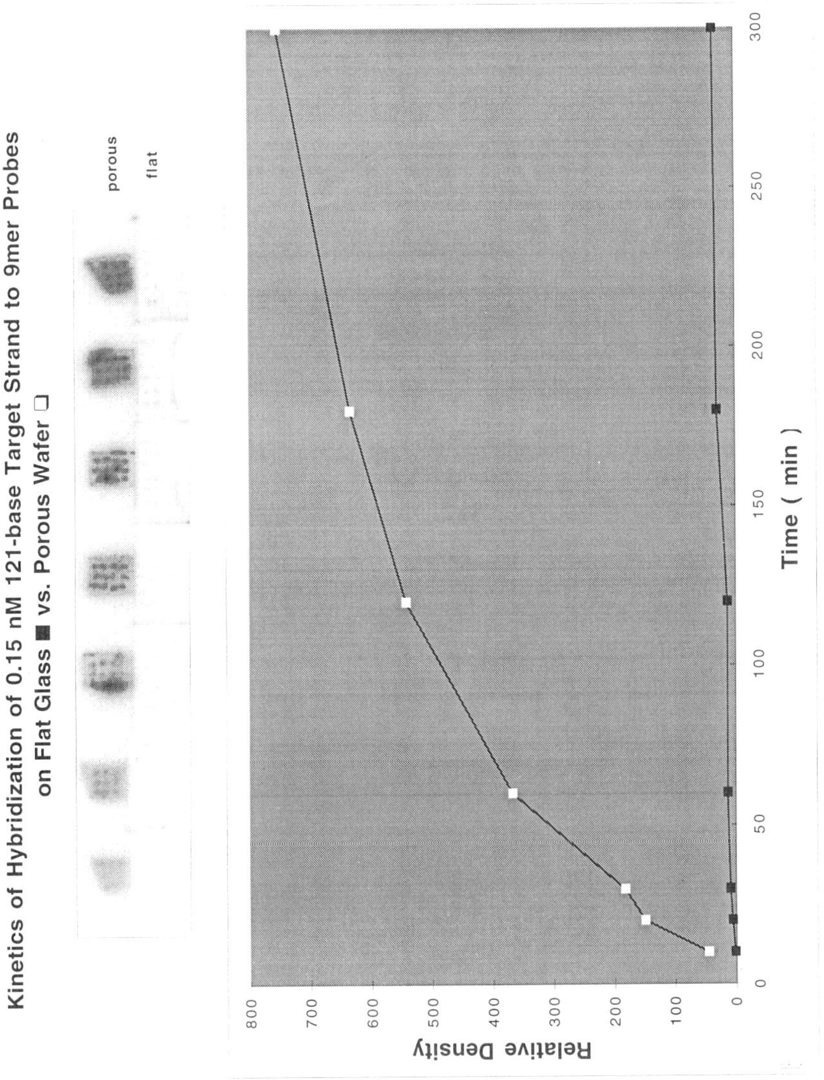

Figure 6. Hybridization kinetics with microchannel glass versus flat glass.

μm diam. pores. Figure 6 shows the results of an experiment in which groups of Cystic Fibrosis wild-type (CF wt) 9mer probes, immobilized on flat glass versus capillary array glass (250 nL of 10 µM probe applied to each spot), were bathed in a dilute solution of 5′-^{32}P-labeled 121-base CF wt target strand (0.15 nM in 2 milliliters 3M tetramethyl ammonium chloride at room temp.). After the indicated times, hybridization supports were removed and washed for 2 hours with 3M tetramethyl ammonium chloride (TMAC) at 4°C, then analyzed using a phosphorimager. As seen in Figure 6, the rate and extent of hybridization to the porous glass arrays was very much greater than with the flat glass supports.

The results shown in Figure 7 demonstrate the reusability of the flowthrough genosensor as well as mismatch discrimination. A model experiment was carried out using immobilized 9mer probes specific for a region of the cystic fibrosis (CF) gene and representing the normal (wild-type, designated wt) sequence and two different single base substitution mutations (designated 549 and 551). Three different 5′-^{32}P-labeled synthetic 21mer targets (representing wt, 551 and 549 sequences) were used in successive cycles of

Probe: CF 9mer (3’-aminopropyl, 10 µM); 2 X 200 nL/spot applied
wt = wild type; 551 & 549 = mutant (single base substitution)

Array Key:

551	549	wt	wt	wt
o	o	o	o	o
o	o	o	o	o
o	o	o	o	o

Hybridization: 20 µL of 10nM [5’-^{32}P]21mer target in 3M TMAC, room temp.

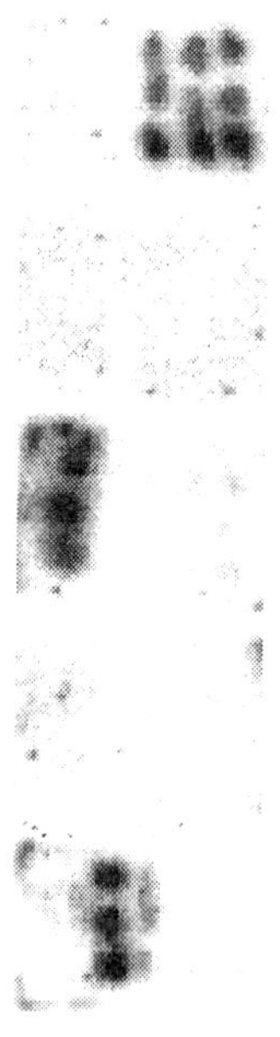

Figure 7. Demonstration of multiple cycles of hybridization within microchannel array glass.

hybridization and regeneration (removal of bound target by hot water wash) as depicted in the figure, and the capillary array wafer was analyzed using a phosphorimager after each step. As expected, the wt target produced a hybridization signal only in the three rows containing wt probe. After stripping of the hybrids by flowthrough of hot water, the wafer was hybridized with the 551 target, which bound to the corresponding row of 551 probes, then the wafer was stripped with hot water and hybridized with the 549 target, which bound to the row of 549 probes. Each cycle of hybridization, washing and regeneration required about 30 minutes with the porous glass genosensor. Using flat glass hybridization arrays, each cycle of analysis requires several hours. These results demonstrate the reusability of porous glass genosensors, the high specificity of hybridization, and fast analysis time, which should enable increased sample throughput.

Figure 8 summarizes the results of an experiment in which single-stranded and denatured double-stranded 212-base CF polymerase chain reaction (PCR) fragments were hybridized to complementary glass-tethered 9mer probes, immobilized on flat microscope slides versus in capillary array glass supports. Single-stranded PCR fragments, isolated using a streptavidin affinity minicolumn (with one of the two PCR primers labeled with biotin) were hybridized to the 9mer arrays, in parallel with an equimolar quantity of heat-denatured double-stranded PCR fragments. The relative hybridization signals are shown in the bar graph below the phosphorimager data. Dark bars represent hybridization on flat glass slides and light bars represent hybridization within porous wafers. Hybridization was carried out at 4°C for 4 hr in 1.0 mL of hybridization buffer containing 3M

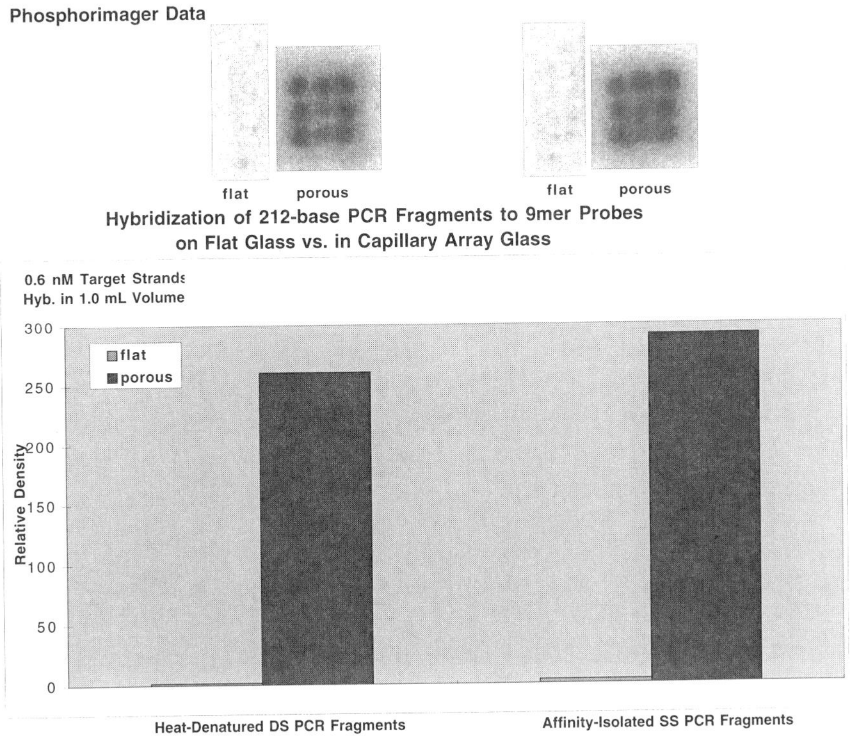

Figure 8. Hybridization of single-stranded versus heat-denatured double-stranded PCR fragments within microchannel glass and flat glass.

Table 1. Environmental applications of flowthrough genosensors

- Molecular endpoint surveillance
  - Assessment of toxic exposure in natural ecosystems
    - Quantitation of DNA adduct levels
    - Detection of cellular stress responses through mRNA profiling
      - probes targeted to known stress- and DNA damage-inducible genes
      - arbitrary sequence arrays for discovery of new stress response genes
    - Toxicity testing in wastewater effluents
- Microbial population analysis
  - Industrial microbial population monitoring
    - Biodegradation process control
    - Fermentation control
    - Contamination control in manufacture of chemicals, foods, pharmaceuticals
  - Pathogen surveillance
    - Agricultural soil samples
    - Streams, lakes and municipal water supplies
    - Insect and mammalian vectors
- Genetic diversity surveillance
  - Assessment of genetic diversity in wild animal and plant species
  - Genotyping of large populations of domesticated animals and plants
    - Gene discovery
    - Marker-assisted breeding

TMAC and 0.6 nM strands. The supports were then washed in 3M TMAC at 4°C for 2 hours and analyzed using a phosphorimager. The results demonstrate two important advantages of the porous glass hybridization supports over the flat glass genosensor: (I) extent of hybridization is superior with porous glass supports; and (ii) porous supports enable direct analysis of heat-denatured double-stranded PCR fragments, eliminating the cumbersome step of strand isolation.

2.3. Practical Applications of Flowthrough Genosensors in Environmental Diagnostics

The imminent availability of analytical microsystems will enable widespread biological monitoring in three categories of environmental science: (i) measurement of molecular endpoints for toxicology testing and risk assessment; (ii) monitoring of natural microbial ecosystems and industrial bioprocesses; and (iii) genetic analyses related to biodiversity and agricultural productivity. Anticipated applications of flowthrough genosensors in environmental surveillance are listed in Table 1.

ACKNOWLEDGMENTS

This work was performed at the Houston Advanced Research Center with support from National Institute of Health Grant P20 HG00665 and a grant from the Houston Endowment. The author is grateful to Dr. Peter McIntyre for permission to reproduce scanning electron micrographs of porous silicon substrates.

REFERENCES

1. http://www.perkin-elmer.com:80/ga/280602/280602.html

2. Meier-Ewert, S., Maier, E., Ahmadi, A., Curtis, J. & Lehrach, H. (1993). An automated approach to generating expressed sequence catalogues. *Nature 361*:375–376.
3. Jacobson, S.C. & Ramsey, J.M. (1996). Integrated microdevice for DNA restriction fragment analysis. *Anal. Chem. 68*:720–723.
4. Jacobson, S.C., Hergenroder, R., Moore, A.W. & Ramsey, J.M. (1994). Precolumn reactions with electrophoretic analysis integrated on a microchip. *Anal. Chem. 66*:4127–4132.
5. Woolley, A.T. & Mathies, R.A. (1994). Ultra-high-speed DNA fragment separations using microfabricated capillary array electrophoresis chips. *Proc. Natl. Acad. Sci.*, U.S.A. *91*:11348–11352.
6. Sniegowski, J.J. (1996). Moving the world with surface micromachining. *Solid State Technol.*, Feb., 1996, pp. 83–90.
7. Gabriel, K.J. (1995). Engineering microscopic machines. Scientific American *273*:118–121.
8. Bryzek, J., Petersen, K. & McCulley, W. (1994). Micromachines on the march. IEEE *Spectrum*, May, 1994, pp. 20–31.
9. Beattie, K.L., Eggers, M.D., Shumaker, J.M., Hogan, M.E., Varma, R.S., Lamture, J.B., Hollis, M.A., Ehrlich, D.J. & Rathman, D. (1992). Genosensor technology. *Clin. Chem. 39*:719–722.
10. Lamture, J.B., Beattie, K.L., Burke, B.E., Eggers, M.D., Ehrlich, D.J., Fowler, R., Hollis, M.A., Kosicki, B.B., Reich, R.K., Smith, S.R., Varma, R.S. & Hogan, M.E. (1994). Direct detection of nucleic acid hybridization on the surface of a charge coupled device. *Nucl. Acids Res. 22*:2121–25.
11. Eggers, M., Hogan, M., Reich, R.K., Lamture, J., Ehrlich, D., Hollis, M., Kosicki, B., Powdrill, T., Beattie, K., Smith, S., Varma, R., Gangadharan, R., Mallik, A., Burke, B. & Wallace, D. (1994). A microchip for quantitative detection of molecules utilizing luminescent and radioisotope reporter groups. *BioTechniques 17*:516–525.
12. Beattie, K., Beattie, W., Meng, L., Turner, S., Bishop, C., Dao, D., Coral, R., Smith, D. & McIntyre, P. (1995). Advances in genosensor research. Clin. Chem. *41*:700–706.
13. Beattie, W.G., Meng, L., Turner, S., Varma, R.S., Dao, D.D. & Beattie, K.L. (1995). Hybridization of DNA targets to glass-tethered oligonucleotide probes. *Molec. Biotechnol. 4*:213–225.
14. Doktycz, M.J. and Beattie, K.L. 1996. Construction and use of genosensor chips. In Beugelsdiik, A. (Ed), *Advanced DNA Sequencing Technologies*. J. Wiley, in press.
15. Bains, W. & Smith, G.C. (1988). A novel method for nucleic acid sequence determination. *J. Theor. Biol. 135*:303–307.
16. Southern, E.M. (1988). Analyzing polynucleotide sequences. International patent application PCT GB 89/00460.
17. Drmanac, R., Labat, I., Brukner, I. & Crkvenjakov, R. (1989). Sequencing of megabase-plus DNA by hybridization: Theory of the method. *Genomics 4*:114–128.
18. Khrapko, K.R., Lysov, Y.P., Khorlyn, A.A., Shick, V.V., Florentiev, V.L. & Mirzabekov, A.D. (1989). An oligonucleotide hybridization approach to DNA sequencing. FEBS Lett. *256*:118–122.
19. Strezoska, Z., Paunesku, T., Radosavlijevic, D., Labat, I., Drmanac, R. & Crkvenjakov, R. (1991). Sequencing by hybridization: First 100 bases read by a non gel-based method. *Proc. Natl. Acad. Sci.*, USA *88*:10089–10093.
20. Drmanac, R., Drmanac, S., Labat, I., Crkvenjakov, R., Vicentic, A. & Gemmell, A. (1992). Sequencing by hybridization: Towards an automated sequencing of one million M13 clones arrayed on membranes. *Electrophoresis 13*:566–573.
21. Drmanac, R. & Crkvenjakov, R. (1992). Sequencing by hybridization (SBH) with oligonucleotide probes as an integral approach for the analysis of complex genomes. *Int. J. Genome Res. 1*:59–79.
22. Drmanac, R., Drmanac, S., Strezoska, Z., Paunesku, T., Labat, I., Zeremski, M., Snoddy, J., Funkhouser, W. K., Koop, B. & Hood, L. (1993). DNA sequence determination by hybridization: A strategy for efficient large-scale sequencing. *Science 260*:1649–1652.
23. Khrapko, K.R., Lysov, Y.P., Khorlin, A.A., Ivanov, I.B., Yershov, G.M., Vasilenko, S.K., Florentiev, V.L.& Mirzabekov, A.D. (1991). A method for DNA sequencing by hybridization with oligonucleotide matrix. *DNA Sequence 1*:375–388.
24. Pevzner, P.A., Lysov, Y.P., Khrapko, K.R., Belyavsky, A.V., Florentiev, V.L. & Mirzabekov, A.D. (1991). Improved chips for sequencing by hybridization. *J. Biomol. Struct. Dynam. 9*:399–410.
25. Bains, W. (1991). Hybridization methods for DNA sequencing. *Genomics 11*:294–301.
26. Cantor, C. R., Mirzabekov, A., and Southern, E. (1992). Report on the sequencing by hybridization workshop. Genomics *13*:1378–1383.
27. Bains, W. (1993). Characterizing and sequencing cDNAs using oligonucleotide hybridization. DNA Sequence *4*:143–150.
28. Beattie, K.L. (1994). Report on the 1993 international workshop on sequencing by hybridization. *Human Genome News*, Jan., 1994.

29. Mirzabekov, A.D. (1994). DNA sequencing by hybridization - a megasequencing method and a diagnostic tool? *Trends Biotechnol. 12*:27–32.
30. Lysov, Y.P., Chernyi, A.A., Balaeff, A.A., Beattie, K.L., Florentiev, V.L.& Mirzabekov, A.D. (1994). DNA sequencing by hybridization to oligonucleotide matrix. Calculation of continuous stacking hybridization efficiency. *J. Biomolec. Struct. Dynam. 11*:797–812.
31. Broude, N.E., Sano, T., Smith, C.L. & Cantor, C.R. (1994). Enhanced DNA sequencing by hybridization. *Proc. Natl. Acad. Sci.*, USA *91*:3072–3076.
32. Hoheisel, J.D. (1994). Application of hybridization techniques to genome mapping and sequencing. *Trends Genet. 10*:79–83.
33. Drmanac, S. & Drmanac, R. (1994). Processing of cDNA and genomic kilobase-size clones for massive screening, mapping and sequencing by hybridization. *BioTechniques 17*:328–336.
34. Fodor, S.P.A., Read, J.L. , Pirrung, M.C., Stryer, L., Lu, A.T. & Solas, D. (1991). Light-directed, spatially addressable parallel chemical synthesis. *Science 251*:767–773.
35. Jacobs, J.W. & Fodor, S.P.A. (1994). Combinatorial chemistry - applications of light-directed chemical synthesis. *Trends Biotechnol. 12*:19–26.
36. Pease, A.C., Solas, D., Sullivan, E.J., Cronin, M.T., Holmes, C.P. & Fodor, S.P.A. (1994). Light-generated oligonucleotide arrays for rapid DNA sequence analysis. *Proc. Natl. Acad. Sci.*, U.S.A. *91*:5022–5026.
37. Doktycz, M.J., Jacobson, K.B., Beattie, K.L., & Foote, R.S. (1995). Optical melting as a tool for optimizing SBH analysis. *In* Cohn, G.E., Lerner, J.M., Liddane, K.J., Scheeline, A. & Soper, S.A. (Eds.), *Ultrasensitive Instrumentation for DNA Sequencing and Biochemical Diagnostics*, Proc. SPIE 2386, pp. 30–34.
38. Southern, E.M., Maskos, U. & Elder, J.K. (1992). Analyzing and comparing nucleic acid sequences by hybridization to arrays of oligonucleotides: Evaluation using experimental models. *Genomics 13*:1008–1017.
39. Maskos, U. & Southern, E.M. (1992). Parallel analysis of oligodeoxyribonucleotide (oligonucleotide) interactions. I. Analysis of factors influencing oligonucleotide duplex formation. *Nucl. Acids Res. 20*:1675–1678.
40. Matson, R.S., Rampal, J.B. & Coassin, P.J. (1994). Biopolymer synthesis on polypropylene supports. *Anal. Biochem. 217*:306–310.
41. Matteucci, M.D. & Caruthers, M.H. (1981). Synthesis of deoxyoligonucleotides on a polymer support. *J. Am. Chem.* Soc. *103*:3185–91.
42. Beattie, K.L. & Frost, J.D. III. (1992). Porous wafer for segmented synthesis of biopolymers. U.S. Patent # 5,175,209.
43. Beattie, K.L., Logsdon, N.J., Anderson, R.S., Espinosa-Lara, J.M., Maldonado-Rodriguez, R. & Frost, J.D. III. (1988). Gene synthesis technology: Recent developments and future prospects. *Appl. Biochem. Biotechnol. 10*:510–521.
44. Beattie, K.L. & Fowler, R.F. (1991). Solid phase gene assembly. *Nature 352*:548–549.
45. Beattie, K.L. & Hurst, G.D. (1994). Synthesis and use of oligonucleotide libraries. *In* Epton, R. (Ed.), *Innovation and Perspectives in Solid Phase Synthesis*, Proc. 3rd International Symposium on Solid Phase Synthesis, Mayflower Worldwide Ltd., Birmingham, U.K., pp. 69–76.
46. Beattie, Wanda G., Raia, Gina C., Pratt, J. Darrell, Budowsky, Edward I., Kumar, Sudhir, Varma, Rajender S. & Beattie, Kenneth L. (1996). Oligonucleotide array hybridization: improvements in probe attachment to glass. *Nucleic Acids Res.*, submitted.
47. Beattie, K.L., Zhang, B., Pratt, J.D. & Beattie, W.G. (1996). Flowthrough genosensors employing porous glass hybridization substrates. *Nucl. Acids Res.*, submitted.
48. Lehmann, V. (1993). The physics of macropore formation in low doped n-type silicon. J. Electrochem. Soc. *140*:2836–2843.

24

BIOREPORTERS AND BIOSENSORS FOR ENVIRONMENTAL ANALYSIS*

R. S. Burlage,[1] J. Strong-Gunderson,[1] C. Steward,[2] and U. Matrubutham[2]

[1]Environmental Sciences Division
Oak Ridge National Laboratory
Oak Ridge, Tennessee 37831-6036
[2]Center for Environmental Biotechnology
University of Tennessee
Knoxville, Tennessee 37932

1. INTRODUCTION

The study of microbial ecology is necessary for a full understanding of the natural world. Microorganisms affect every aspect of life, from global nutrient cycling to organismal physiology to specific commensal relationships that fill diverse niches. The investigation of microbial ecology is made more difficult by the extremely small size of the organisms, as well as by the multiplicity of species that characterize a microbial community. In fact, most of the microorganisms have not yet been described, so that any examination of a natural community will, of necessity, involve only a minority of species. Yet even these limited investigations can yield important information. It is thus useful to be able to examine individual species as they establish in a community, and as they interact with other members of that community. To accomplish this task, both bioreporters and biosensors have been developed. These tools have already increased our knowledge of normal physiological processes, and will undoubtedly contribute much more in the future.

Bioreporters and biosensors are closely related, yet have distinct differences. Bioreporters are gene sequences that produce a product that is easy to detect (i.e., is assayable). Biosensors are biological molecules (e.g., antibodies, enzymes, microorganisms) that respond to the presence of an analyte by producing an effect that can be detected by electronic means. Both of these rely on a biological sensing of an environmental change, and

* Research sponsored by the Office of Health and Environmental Research, US Department of Energy. Oak Ridge National Laboratory is managed by Lockheed Martin Energy Research Corp. for the US Department of Energy under contract number DE-AC05–96OR22464.

Biotechnology in the Sustainable Environment, edited by Sayler *et al.*
Plenum Press, New York, 1997

the subsequent production of a detectable change. However, the bioreporter is a specific fragment of DNA that is genetically introduced into the cell, while the biosensor utilizes a variety of biological molecules.

The subjects of bioreporters and biosensors are too broad for a comprehensive treatment here, although other references will be suggested. Instead, a description of some useful examples that have been used for environmental analysis will demonstrate the strong and weak points of this technology, and how they can properly be used.

2. BIOREPORTERS: STRENGTHS AND LIMITATIONS

Bioreporters are a powerful means of understanding the transcriptional activity of genetic expression in any cell (Bronstein et al., 1994). Among them are those employing the genes for light production derived from fireflies (*luc*) and from microorganisms (*lux*), chloramphenicol acetyltransferase (CAT), and β-galactosidase (*lac*Z) activity. Detection by bioreporters is based on one or a combination of methods inclusive of colorimetric, bioluminescence, chemiluminescence, fluorescence assays, or production of a specific enzymatic or immunological product. Of these methods, the bioreporter systems based on bio- or chemiluminescence have become popular replacements of conventional analytical tools for environmental study largely due to increased sensitivity and quicker detection of the gene activities (Bronstein et al.,1994). They are especially useful for environmental analysis since such bioluminescent reporter systems provide the capability of real-time, on-line monitoring of biological activities through the use of opto-electronic devices (Matrubutham et al., in press). This non-invasive monitoring through the use of bioreporters provides data which is relevant to the spatial and time components of the system being investigated.

Bioreporter use for environmental studies is not without certain inherent problems. The genetic construct contains reporter genes which have been incorporated into an organism to track properties of interest. This makes the strain a recombinant organism, and certain regulatory processes must be followed for the safe and secure handling of the strain. This is especially true if a release to the environment is planned. Regulatory issues must be taken into account when considering environmental release of a genetically engineered bioreporter strain. This includes an extensive review for issue of a consent order from the US Environmental Protection Agency (USEPA) (Sayler et al., in press, USEPA). Permission for release experiments is a long and technically involved process, requiring substantial knowledge of the organism and its probable effects in the wild. While these experiments are rather rare at the present time, the pace of experimentation is expected to increase dramatically in the near future, as more information on possible effects becomes known. Present data indicate that environmental effects are usually small, transient, or non-existent.

Production of light or other bioreporter functions as well as growth and maintenance of the strain have metabolic requirements. In addition, genetic regulation of the reporter system can come at some expense to the growth of the organism. Genetic constructs, particularly those which are plasmid-based, have been postulated to decrease the fitness of a microorganism. However, this is a difficult property to quantify, and in the majority of cases a well-made gene fusion does not appear to have any great effects on the health of the organism. A greater concern may be the ability of the microorganism to survive and establish in the wild, although this is a difficult property for native bacterial strains that are grown under laboratory conditions. The effect of a genetic fusion on these processes has also been difficult to quantify.

Systems such as the *lux* bioreporter have growth constraints, requiring oxygen, adenosine triphosphate (ATP), reduced nicotinamide adenine dinucleotide phosphate

($NADPH_2$) and an aldehyde substrate which may limit their applicability or diminish their performance. The luciferase (light-producing enzyme) is also labile at temperatures above 30 °C. A bioluminescent *lux* reporter system would not be effective in a high temperature or anaerobic system. Such considerations for growth requirements, fitness, and genetic stability are all factors which must be taken into account when utilizing a genetically engineered bioreporter strain. Other bioreporters, such as the *lacZ* or β-galactosidase assay, require that a cellular fraction containing the enzyme be obtained and used in an assay system. The probability of obtaining a characteristic sample of organisms from the microbial community may limit the usefulness of this technique. In addition, this type of biochemical assay is destructive of a portion of the system under study, and thus requires perturbation of the community.

In addition to growth requirements, application or emplacement and monitoring of the bioreporter strain can affect its utility. Means must exist to deliver the bioreporter to the necessary location, which can be problematic in subsurface systems. Monitoring of bioluminescent bioreporters requires the use of light detection and amplification devices adjacent to the bioreporters to effectively monitor light production. Biological and physical limitations as well as technological limitations must be overcome to be able to effectively monitor light production. Recent advances in fiber optic and light monitoring technology have lead to improvements in monitoring of bioluminescent bioreporters (Matrubutham et al., In press). As discussed above, bioreporters that depend on biochemical assays have sampling requirements that make them useless in most large-scale test systems.

3. ENVIRONMENTAL APPLICATIONS OF BIOREPORTERS

The *lux* bioreporter has found its greatest application in environmental systems. Studies of the bioluminescence phenomenon in the naturally occurring bacteria *Vibrio fischeri* have led to the construction of engineered strains containing *lux*, the light producing genes. Engineered strains can serve as a bioanalytical tool for measurement of biological and non-biological processes. Bioreporters indicate the expression of a gene of interest by producing an easily assayable or detectable gene product (such as light in the example above). *Lux* fusions have been utilized as bioreporters for the specific detection of heavy metals in the environment (Corbiser et al., 1993; Selifonova et al., 1993). Bioluminescence markers have also been used to study root colonization by rhizosphere bacteria (DeWeger et al., 1991). An *algD-lux* bioluminescent strain has been constructed to monitor the production and formation of alginate in biofilms (Wallace et al., 1994).

Light detection is the essential means of measuring and reporting the expression of this bioreporter. Current technologies have lead to the construction of a multichannel automated fiber optic-based monitoring system. This system provides higher sensitivities in detection of light from multiple locations in an environment. The system has been coupled to a computer containing custom automated data acquisition software. The automated light monitoring device is currently undergoing laboratory testing, prior to an upcoming field deployment.

4. BIOREMEDIATION ACTIVITIES

Bioluminescent bioreporters are currently being used in bioremediation studies with *Pseudomonas fluorescens* HK44 (Heitzer and Sayler, 1993). *Pseudomonas fluorescens*

HK44 is a bioluminescent bioreporter which has been constructed for the purposes of monitoring and optimizing the degradation and bioremediation of polyaromatic hydrocarbons (PAH) (King et al., 1990). It responds to the presence of naphthalene and salicylate, and the light response has been shown to be linear in relation to the concentration of naphthalene (Heitzer et al., 1990). This organism contains a catabolic plasmid (pKA1) that encodes naphthalene degradation and which is similar to the archetypal NAH7 plasmid. Plasmid pKA1 carries a promoterless *lux* gene cassette due to transpositional insertion (Tn4431; Shaw and Kado, 1987) in the *nah*G gene of the lower (salicylate) operon of the naphthalene degradation pathway (Figure 1). The modified strain (HK44) has a Nah+, Sal^+, Lux^+, Tet^+ phenotype.

The efficacy of this microbial bioreporter has been extensively tested in laboratory studies since its construction (King et al., 1990; Heitzer et al., 1992,1994). The light response has been shown to be linear in response to a wide range of contaminant concentrations. The bioreporter can be applied in a variety of systems such as aqueous flow-cells, column bioreactors, and through direct application to contaminated soils (Heitzer et al., 1990,1994; Huang et al., 1993, Sayler et al., in press). Bioreporter bacteria can be added directly to soil to augment existing polyaromatic hydrocarbon degrading bacteria and provide direct bioluminescent on-line light measurements. HK44 can function as a direct bioreporter or as a biosensor. The bioreporter function is accomplished by direct measurement of light production emitted from cells as naphthalene or salicylate is degraded. The biosensor function is accomplished by immobilizing HK44 in strontium alginate beads. The beads are placed in a porous biosensor module which is attached to fiber optic monitoring devices. The biosensor module can then be buried in subsurface soils or aqueous flowing system. This system works best under saturated conditions, but will also work in the presence of naphthalene vapors. The biosensor function allows HK44 to serve as sentinel monitors for the chemical constituents in an advancing plume. They also act as *in situ* surrogates to evaluate the efficacy of engineering approaches to simulate and maintain degradative performance in bioremediation. Encapsulated HK44 have been shown to be responsive to repeated salicylate induction over a period of 35 days in groundwater at various pH regimes (Sayler, G.S., 1994).

A large-scale field study using the genetically modified *Pseudomonas fluorescens* HK44 for polyaromatic hydrocarbon bioremediation of contaminated soils will be per-

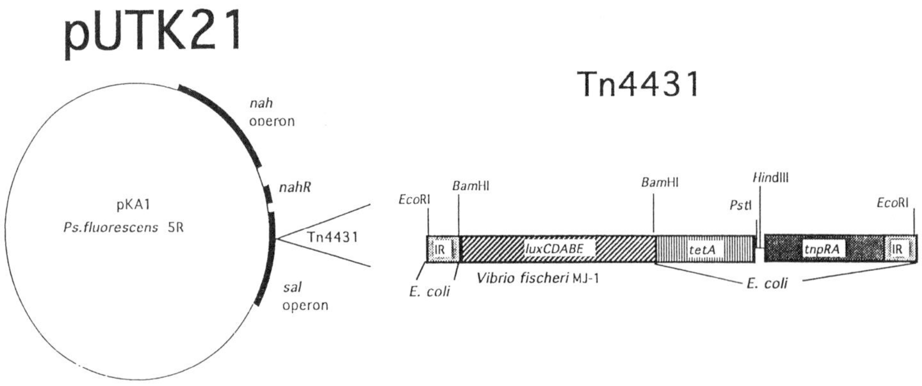

Figure 1. Construction of plasmid pUTK21.

formed. As a preparation for the study, preliminary investigations were initiated to test fundamental hypotheses whether the pre-existing structure of the microbial community, fitness of the engineered strain and selection imposed by the contaminant interact to either permit or exclude the engineered strain as an agent of bioremediation. Laboratory microcosm studies indicate that HK44 is able to establish a population in naphthalene contaminated soils. Presence of other potentially competitive *nahA*-containing microorganisms did not inhibit introduction, growth and maintenance of the genetically engineered strain (Sayler et al., in press, Proceedings 7th International Symposium on Microbial Ecology). In addition, since the engineered strain functions as a bioreporter for similar catabolic genotypes at large in the community, control of the activity and dynamics of the strain can be used to manage or enhance the bioremediation process.

As powerful as the bioluminescent bioreporters are for describing genetic expression, they do have substantial disadvantages as well. They are dependent on the addition of co-factors (e.g., oxygen, aldehyde) that might not always be present in the cell. This could lead to false negative results. For this reason the gene for the Green Fluorescent Protein (GFP) was developed as a bioreporter. GFP is an extremely bright fluorescent protein that was isolated from the jellyfish, *Aequorea victoria*. The fluorescence is so strong that individual cells expressing it can easily be seen under the epifluorescent microscope. This is in contrast to bioluminescent cells, which produce a relatively weak signal. GFP addresses many of the disadvantages of the bioluminescent bioreporters, including the cofactor dependency (although it does require small concentrations of oxygen). In addition, the fluorescence is inherent in the protein, and does not require continual functioning of an enzyme for its activity. It is also a highly stable molecule, even at high temperatures. Its stability, however, is also a drawback, because once the protein is made it persists. The bioluminescent system is far more responsive to rapid changes in conditions. The presence of GFP in bacteria does not appear to have deleterious effects on the host, although a comprehensive analysis has not been performed.

The GFP gene has been cloned and sequenced, and the protein has been extensively characterized (Perozzo et al., 1988; Prasher et al., 1992). The fluorescence makes the cell easy to detect using ultraviolet light, which has an excitation wave length of 395 nanometers (nm) and an emission wave length of 509 nm as well as conventional light-gathering equipment. As with the measurement of bioluminescence, fluorescence can be measured accurately and with great sensitivity. Detection is dependent on the ability of the researcher to expose the GFP molecule to the excitation wavelength, and this can be performed with flexible fiber optic cables that are introduced into a microbial ecosystem. The intact GFP gene has been inserted into a derivative of Tn5, and therefore random mutations with GFP are possible (Burlage et al., 1996). This transposon, Tn5GFP1, can be introduced into a variety of Gram-negative species using electroporation.

The stability of GFP makes it an ideal bioreporter for some applications, such as for tagging bacteria for a transport experiment (Burlage et al., 1996). A standard fluorescence spectrometer can be used to detect concentrations of GFP-tagged down to 10^4 milliliter^{-1}, and smaller concentrations can be detected by microscopy. The fluorescence signal is proportional to the number of bacteria present in the sample. Using established methods, it is possible to detect even single cells in small samples. It was shown in these experiments that the fluorescence signal was an excellent index of conventional colony forming unit (CFU) counts on selective agar plates. A dual bioreporter system has been developed by Khang et al., (manuscript in preparation). This system makes use of the proven *mer-lux* bioreporter of mercury cations (Selifonova et al., 1993) and the expression of GFP by a constitutive promoter. Using this microorganism, the number of cells can be calculated based on the fluorescent signal, and used to normalize results for the bioluminescent (mercury-sensing) response. This construction has proven valuable in biofilm systems, and will enable three-dimensional maps of the biofilm to be created.

Variants of GFP have been described which offer alternative bioreporters. The Red Shifted - Green Fluorescent Protein (RS-GFP) (Delagrave et al., 1995) is a GFP mutation in which the excitation wavelength has shifted towards the red end of the spectrum. The protein fluoresces at approximately the same wavelength (the maximum is at 505 nm instead of 510 nm), but excites at 490 nm instead of 395 nm. This shift is expected to be helpful, since the 490 nm excitation wavelength is beyond the wavelengths of excitation for cellular proteins fluorescence (due to their aromatic amino acids). A mutant GFP developed by Heim et al., (1994) results in the production of a blue color instead of green.

5. BIOSENSORS

A biosensor is a type of probe in which a biological component (e.g., enzyme, antibody, or nucleic acid) interacts with an analyte, and in which this reaction is then detected by an electronic component and translated into a measurable (electronic) signal. Biosensor probes are possible because of a fusion of two technologies, microelectronics and biotechnology. Biosensors are useful for measuring many analytes (e.g., gases, ions, organic compounds) or even bacteria, and are suitable for studies of complex microbial environments. Although a comprehensive treatment of biosensors is not possible here, a recent review on the application of biosensors for environmental study is available (Rogers, 1995).

The specificity of the biological component for an analyte (or group of related analytes) is the essential ingredient in making a good biosensor. For example, a single strand of DNA will hybridize only to its complementary strand under the appropriate conditions. An enzyme may have one or a few substrates that it complexes with. It is important to note that the assay conditions are especially important in whether the biosensor will be functional. Biological molecules are generally labile, and will not withstand field conditions. In many cases, the biological component will be degraded as a food source by microorganisms in the test solution. However, short exposures may be efficacious.

The electronic components, or transducers, generally fall into distinct categories: electrochemical, optical, piezo-electric, and calorimetric (Sethi, 1994), each of these is based on established technologies, although they all measure different attributes. For example, piezo-electric biosensors measure changes in mass, while electrochemical transducers report on changes in voltage when current is held constant (potentiometric) or report changes in current when voltage is held constant (amperometric). It is necessary to match the appropriate biological and electronic components in order to measure a relevant event. Since the piezo-electric transducer detects a change in the attached mass, it is effective with antibodies, which will bind tightly with specific antigens. Both electrochemical and optical electrodes are very useful for the detection of signals from attached enzymes. The enzymatic reaction can cause a potential change that is detected by the electrochemical electrode, or a change in one of the components of the enzyme system that can be detected by the optical electrode. Nucleic acid biosensors depend on the ability of a single-stranded nucleic acid to hybridize with another fragment of DNA by complementary basepairing. A nucleic acid biosensor that utilizes evanescent wave technology has been described by Graham et al., (1992). They used short fragments of nucleic acids that are small enough to reside within the field of the evanescent wave. They were able to detect fluorescein-labeled DNA hybridizing to their complementary immobilized probes in a flow cell. Fluorescence was monitored and reported as a change in the output voltage. The biosensor described by Eggers et al., (1994) integrates microelectronics, molecular biology, and computational science in an optical electrode format. This type of biosensor may be very valuable in characterizing the DNA isolated from complex microbial communities, and for comparing the change of that community over time.

6. SUGGESTIONS FOR TECHNOLOGY DEVELOPMENT

Improvements in the ability to utilize bioluminescent bioreporters and biosensors is expected to continue in the area of opto-electronic monitoring equipment. Development of photodiodes, more sensitive cameras and photon counting devices, and in general smaller devices to improve applicability will lead to enhanced monitoring of bioluminescent bioreporters. One area which could use more improvement is the development of improved methods to send the produced light signal from the source to the monitoring devices. Current use of fiber-optic and photo-diode based systems is limited by the physical distance between the monitoring device and source, and the loss of signal over increasing transmission distances. It is also important to improve on the resolution of these light-detecting devices, so that single cells can be identified. This would stand in marked contrast to the present technique of examining thousands of cells to detect an average response of a community. Study of single cells will be vital to understanding the interplay of ecological principles in a real-world setting. Fortunately these are technological areas which receive considerable attention by various research groups and industries which should lead to improved monitoring of bioreporter and biosensors in the future.

Development of additional bioreporter molecules is an ongoing process. One of the most unusual, and potentially most useful, bioreporters is the ice nucleation gene of *Pseudomonas syringae.* Expression of this gene creates a protein which permits freezing of liquid samples at a higher temperature. Detection of frozen droplets is a phenomenon that is easy to demonstrate and measure. This is a technique that requires the collection of representative samples, but the amounts needed are so small that it would not disrupt the experiment unduly. It is reported that sensitivity to a single bacterium can be achieved (Strong-Gunderson, et al., 1995). Other bioreporters are being developed by private industry that utilize different colors of visible light, or that fluoresce at different wavelengths of excitation and emission.

REFERENCES

Bronstein, I., J. Fortin, P.E. Stanley, G.S.A.B. Stewart, and L.J. Kricka. 1994. Chemiluminescent and bioluminescent reporter gene assays. Anal. Biochem. 219:169–181.

Burlage, R.S., Z. Yang, and T. Mehlhorn. 1996. A Tn5 derivative labels bacteria with Green Fluorescent Protein for transport experiments. *Gene* 173: 53–58.

Corbiser, P., L. Diels, G Nuyts, M. Mergeay, and S. Silver. 1993. *Lux*AB fusions with the arsenic and cadmium resistance operons of *Staphylococcus aureus* plasmid p1258. *FEMS Microbiol. Lett.* 110: 231–238.

Delagrave, S., R.E. Hawtin, C.M. Silva, M.M. Yang, and D.C. Youvan. 1995. Red-shifted excitation mutants of the green fluorescent protein. *Bio/Technology* 13: 151–154.

DeWeger, L.A., P. Dunbar, W.F. Mahafee, B.J. Lugtenberg, and G.S. Sayler. 1991. Use of bioluminescence markers to detect *Pseudomonas* spp. in the rhizosphere. *Appl. Environ. Microbiol.* 57: 3641–3644.

Eggers, M., M. Hogan, R.K. Reich, J. Lamture, D. Ehrlich, M. Hollis, B. Kosicki, T. Powdrill, K. Beattie, S. Smith, R. Varma, R. Gangadharan, A. Mallik, B. Burke, and D. Wallace. 1994. A microchip for quantitative detection of molecules utilizing luminescent and radioisotope reporter groups. *BioTechniques* 17: 516–524.

Graham, C.R., D. Leslie, and D.J. Squirrell. 1992. Gene probe assays on a fibre-optic evanescent wave biosensor. *Biosensors and Bioelectronics* 7: 487–493.

Heim, R., D.C. Prasher, and R.Y. Tsien. 1994. Wavelength mutations and posttranslational autoxidation of green fluorescent protein. Proc. Natl. Acad. Sci. USA 91: 12501–12504.

Heitzer, A., and G.S. Sayler. 1993. Monitoring the efficacy of bioremediation. *Trends in Biotech.* 11:123–142.

Heitzer, A., K. Malachowsky, J.E. Thonnard, P. Bienkowski, D.C. White, and G.S. Sayler. 1994. Optical biosensor for environmental on-line monitoring of naphthalene and salicylate with immobilized bioluminescent catabolic reporter bacteria. *Appl. Environ. Microbiol.* 60:1487–1494.

Heitzer, A., O.F. Webb, J.E. Thonnard, and G.S. Sayler. 1990. Specific quantitative assessment of naphthalene and salicylate bioavailability using a bioluminescent catabolic reporter bacterium. *Appl. Environ. Microbiol.*, 58:1839–1846.

Huang, B., T.W. Wang, R. Burlage, and G. Sayler. 1993. Development of an On-line Sensor for Bioreactor Operation. *Appl. Biochem. Biotech.* 39/40:371–382.

King, J.M.H., P.M. DiGrazia, B.M. Applegate, R. Burlage, J. Sanseverino, P. Dunbar, F. Larimer, and G.S. Sayler. 1990. Rapid, sensitive bioluminescent reporter technology for naphthalene exposure and biodegradation. *Science* 249:778–781.

Matrubutham, U., D. Hueber, C. Steward, J. Thonnard, T. Vo-Dinh, and G.S. Sayler. Novel, automated fiber optic device for environmental biosensor technology detecting *in situ* bioluminescent signals. (In press).

Perozzo, M.A., K.B. Ward, R.B. Thompson, and W.W. Ward. 1988. X-ray diffraction and time-resolved fluorescence analyses of *Aequorea* green fluorescent protein crystals. *J. Biol. Chem.* 263: 7713–7716.

Prasher, D.C., V.K. Eckenrode, W.W. Ward, F.G. Prendergast, and M.J. Cormier. 1992. Primary structure of the *Aequorea victoria* green-fluorescent protein. *Gene* 111: 229–233.

Rogers, K.R. 1995. Biosensors for environmental applications. *Biosensors and Bioelectronics* 10: 533–541.

Sayler, G.S. 1994. Biodegradation process analysis: molecular application in simulations and environmental verification. *Proceedings EC-US Task Force on Biotechnology Research*, pp. 47–54.

Sayler, G.S., C. Steward, U. Matrubutham, G. Huxel, J. Thonnard, and J. Drake. A species invasion paradigm for managing biodegradative microbial communities. *Proceedings 7th International Symposium on Microbial Ecology.* (In press)

Sayler, G.S., U. Matrubutham, C. Steward, A. Layton, C. Lajoie, J. Easter, and B. Applegate. Towards field release of engineered strains for bioremediation. *Proceedings USEPA Symposium on Environmental Release.* (In press)

Selifonova, O., R. Burlage, and T. Barkay. 1993. Bioluminescent sensors for the detection of bioavailable Hg(II) in the environment. *Appl. Environ. Microbiol.* 59:3083–3090.

Sethi, R.S. 1994. Transducer aspects of biosensors. *Biosensors and Bioelectronics* 9: 243–264.

Shaw, J.J. and C.I. Kado. 1987. Direct analysis of the invasiveness of *Xanthomonas campestris* mutants generated by Tn4431, a transposon containing a promoterless luciferase cassette for monitoring gene expression, *in* "Molecular Genetics of Plant-Microbe Interactions", Verma, D.P.S. and N. Brisson (eds.) Martinus Nijnoff, Dordrecht.

Strong-Gunderson, J.M., S. Carroll, B. Sanford, and A.V. Palumbo. 1995. Demonstration of an environmental tracer for groundwater, contaminant, nutrient, and microbial transport. *Abstracts of the 95th General Meeting of the American Society of Microbiology*, Washington, D.C., May 21–25.

Wallace, W.H., J.T. Fleming, D.C. White, and G.S. Sayler. 1994. An *algD*-bioluminescent reporter plasmid to monitor alginate production in biofilms. *Microb. Ecol.* 27:225–239.

25

RISK ASSESSMENT FOR A RECOMBINANT BIOSENSOR

Philip Sayre

USEPA, Office of Pollution Prevention and Toxics
Mail Code 7403
401 M Street, SW
Washington, DC 20460

1. THE TOXIC SUBSTANCES CONTROL ACT AND THE RISK ASSESSMENT PROCESS

Under the Toxic Substances Control Act (TSCA), intergeneric microorganisms intended for commercial purposes are reviewed for their safety. TSCA-subject microorganisms include those in several use categories such as fermentation applications for specialty chemical production (detergent enzymes, etc), biofertilizers, ore mining, oil recovery, biomass conversion, and bioremediation. Such intergeneric microorganisms are subject to Premanufacture Notification (PMN) review prior to commercialization, and voluntary PMNs are requested if the microorganisms will be used in field tests to examine issues such as efficacy prior to full commercialization. In either case, the US Environmental Protection Agency's (USEPA) Office of Pollution Prevention and Toxics (OPPT) reviews the PMN and legally-binding conditions on the microorganism's production, and use are negotiated between the proponent and OPPT in order to address any risk concerns identified in the USEPA's review. In the case of field tests proposed for *Pseudomonas fluorescens* Strain HK44 (pUTK21), the decision took the form of a Consent Order under Section 5(e) of TSCA.

2. A FRAMEWORK FOR RISK ASSESSMENT OF BIOREMEDIATION

The risk assessment process used by OPPT consists of eight reviews, as noted in Table 1. In addition, two other technical documents are produced. An economics report identifies the cost of complying with TSCA, the benefits of the product to the economy and society, and other issues relevant to the risk-to-benefit analysis. Second, a TSCA Inventory Listing document notes how the microorganism will be listed on the Inventory. Finally, the risk management document, termed a Consent Order issued under Section 5(e)

Biotechnology in the Sustainable Environment, edited by Sayler *et al.*
Plenum Press, New York, 1997

Table 1. OPPT risk assessment reports prepared for biotechnology submissions

Name of report	Focus of report
Taxonomy report (Segal, 1995)	Identifies genus and species of recipient microorganisms. May address donor microorganisms also.
Chemistry report (Tou, 1995)	Identifies genetic manipulations made to construct intergeneric microorganism. May include a flow diagram for construction process and final construct illustration.
Construct analysis (Sayre, 1995)	Identifies hazard and gene transfer issues associated with introduced DNA used to construct the intergeneric microorganism. Identifies any inserted DNA whose function is uncertain. May include a flow diagram for construction process and final construct illustration.
Ecological hazard assessment (McClung, 1995)	Identifies potential environmental impacts of the recombinant microorganism and its products on environmental receptors such as aquatic and terrestrial vertebrates, invertebrates, and plants.
Human health assessment (Source: SRA Technologies)	Identifies potential impacts of the recombinant microorganism and its products on human health. Pathogenic and toxic effects are considered.
Engineering report (Source: Radian Corporation)	Identifies releases of microorganisms and their products to environmental media and estimates worker exposure to the subject microorganisms.
Exposure assessment	Identifies concentrations of microorganisms in receiving air, water and soil.
Risk assessment (Broder, 1995)	Balances hazard and exposure concerns to arrive at an overall determination for the field test.

of TSCA, identifies risk issues for the field test, restrictions placed on the field test, and may note any remaining concerns for use of the microorganism at other sites. In order to organize this information in a way that can be evaluated more rapidly, the information in these reports will be presented in accordance with the draft risk assessment scheme for bioremediation products developed at the EPA/Environment Canada Bioremediation Risk Assessment Workshop (Figure 1). This approach was used earlier to present a preliminary risk assessment of a recombinant pseudomonad intended to degrade polychlorinated biphenyls (Sayler and Sayre, 1995). The draft risk assessment scheme for bioremediation is discussed in the next section.

3. APPLICATION OF THE RISK ASSESSMENT FRAMEWORK TO A FIELD TEST WITH *Pseudomonas fluorescens* STRAIN HK44 (pUTK21)

Pretest information for *Pseudomonas fluorescens* Strain HK44 (pUTK21), or Strain HK44, was provided to OPPT in the Premanufacture Notice PMN P95–1601 from the University of Tennessee (Sayler, 1995a). The PMN requested a release of the strain to soils contained in large lysimeters located at the Department of Energy's Y-12 site in Oak Ridge, Tennessee. Strain HK44 has introduced DNA that allows it to both degrade polynuclear aromatic hydrocarbons (PAHs) and serve as a biosensor which produces visible light in the presence of bioavailable PAHs.

The PMN containing pretest information on Strain HK44 was structured according to the USEPA "Points to Consider" guidance document (USEPA, 1994). The "Points to Consider" document lists the types of information needed to complete the individual USEPA reviews listed in Table 1. The risk assessment for Strain HK44 can be summarized by referring to the major "Steps" in Figure 1: the risk based on pretest information, the exposure assessment, the hazard assessment, and the risk characterization/regulatory decision.

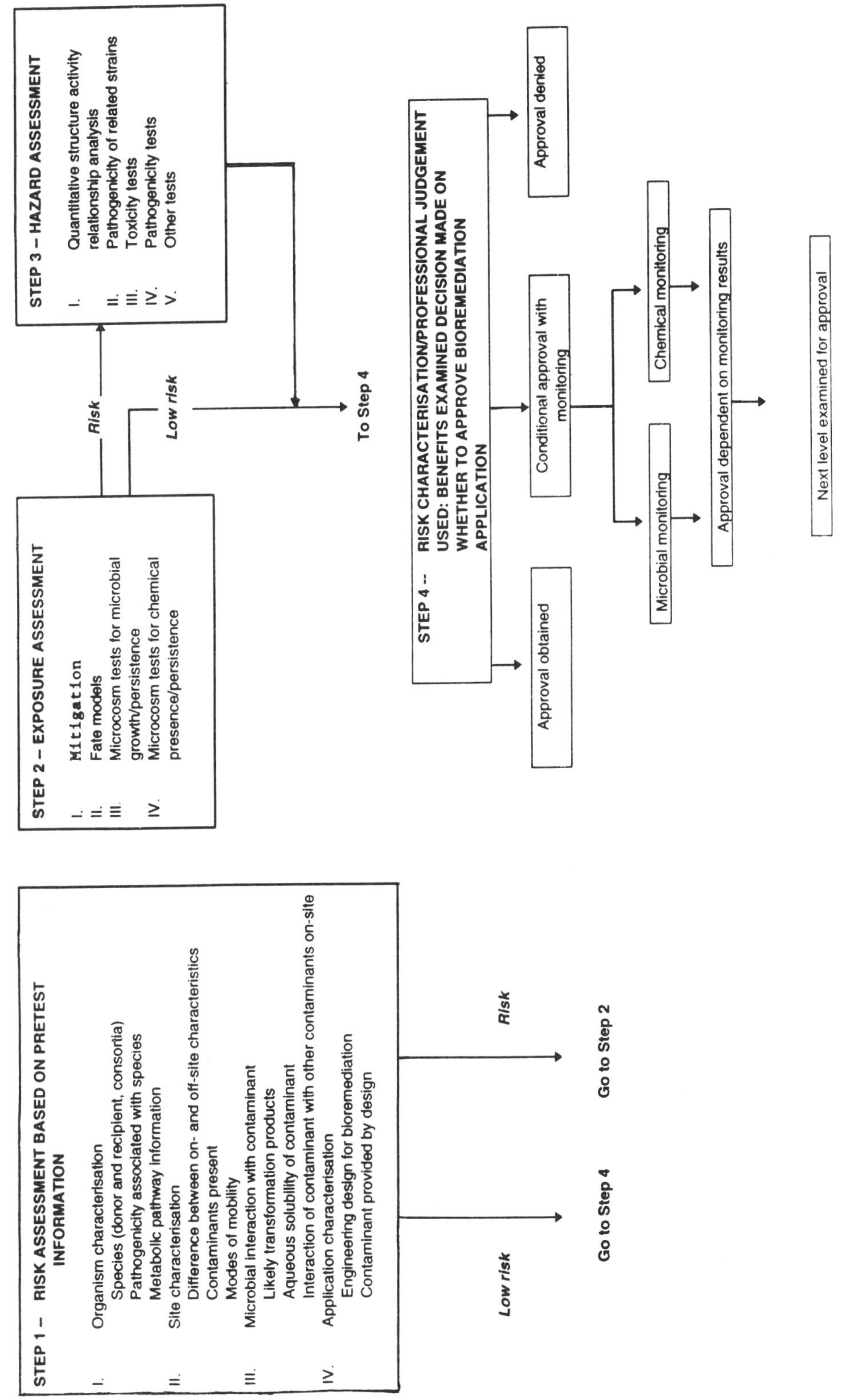

Figure 1. Bioremediation risk assessment scheme from the 1993 EPA/Environment Canada bioremediation risk assessment workshop.

3.1. Review of Pretest Information

The pretest information addressed under Step 1 of Figure 1 consists of information submitted in the PMN and as a result of follow-up requests by the USEPA. It characterizes the (1) recombinant organism, (2) the site where the organism will be applied, (3) interaction of the microorganism with any contaminants on the site, and (4) engineering design for application of the microorganism at the site and any physical containment afforded as a result of the engineering design.

3.1.1. Organism Characteristics. Strain HK44 consisted of the recipient strain *Pseudomonas fluorescens* 18H (also referred to as Strain HK9) which contained the 116 kb (kilobase) plasmid pUTK21. Plasmid pUTK21 (Figure 2) was constructed by adding a *lux* cassette to the naturally-occurring plasmid pKA1. Plasmid pKA1 was identified in *P. fluorescens* 5R, and Strain 5R was obtained from a mixed slurry reactor inoculum used to treat manufactured gas plant soils contaminated with polynuclear aromatic hydrocarbons (Sanseverino et al., 1993). Strain 18H is an obligate aerobe originally isolated from a manufactured gas plant soil. Strain 18H is ampicillin resistant, carries two cryptic plasmids retained in Strain HK44, and is able to degrade salicylates but not naphthalene (Sayler, 1995a; Sanseverino et al., 1993). The identities of both strains were supported by fatty acid methyl ester analyses. Strain 5R had a similarity index of 0.547 for *P. fluorescens* Biovar II, and an index of greater than 0.540 for *P. chlororaphis*. Strain 18H had an index of 0.394 for *P. chlororaphis* and one of 0.319 for *P. fluorescens* Biovar II. The USEPA's initial analysis indicated that Strain 18H appears to be a strain intermediate between *P. chlororaphis* and *P. fluorescens* Biovar II. The DSM-German National Collection of Type Cultures accepted both strains as *P. fluorescens*. Strain HK44 also produced the green fluorescing compound pyoverdin, typical of the species *P. fluorescens* (but not the yellow-green phenazine carboxylate pigment chlororaphin characteristic of *P. chlororaphis*) (Palleroni, 1984b). Therefore, the USEPA agreed with the identification of Strain HK44 as a *P. fluorescens* Biovar II.

Plasmid pUTK21, present in *Pseudomonas fluorescens* Strain HK44, consisted of plasmid pKA1 with Tn*4431* inserted into the *nahG* gene. Tn*4655* sequences are considered likely to be present based on similarities between plasmid pKA1 and NAH7; *nah* are naphthalene-degradative genes; Tn*1721* IR are inverted repeats from *Escherichia coli* Strain D1021; *lux* genes are derived from *Vibrio fischeri* Strain MJ-1 (now a *Photobacterium fischeri*). The following genes were all derived from Tn*1721*: *tetR* and *tetA*, which are the tetracycline repressor and tetracycline resistance genes; and *tnpR* and *tnpA*, which encode the resolvase and transposase enzymes.

Pathogenicity and toxicity information on the species *P. fluorescens* (Ballows et al., 1991) indicated that clinical cases have been documented for this species including empyema, urinary tract infections, postoperative infection, pelvic inflammatory disease and fatal transfusion reactions due to contaminated blood. Palleroni (1984) noted that *P. fluorescens* was not prevalent in clinical laboratories and hospitals; its ability to grow at refrigerator temperatures can lead to contamination of clinical samples, but it may not grow at body temperature (37°C). Although *P. fluorescens* has beneficial effects in that it inhibits the growth of some microbial plant pathogens, *P. fluorescens* Biovar II also causes soft rot of onions (Wright and Hale, 1992), alfalfa (Turner and Van Alfen, 1983), broccoli (Canaday et al., 1991), lettuce (Miller, 1980), and other plants. Further, it can cause blight of cucumbers (Ohta et al., 1976), and has been associated with opportunistic pathogenicity in fish which are under stressed conditions (Bullock, 1964).

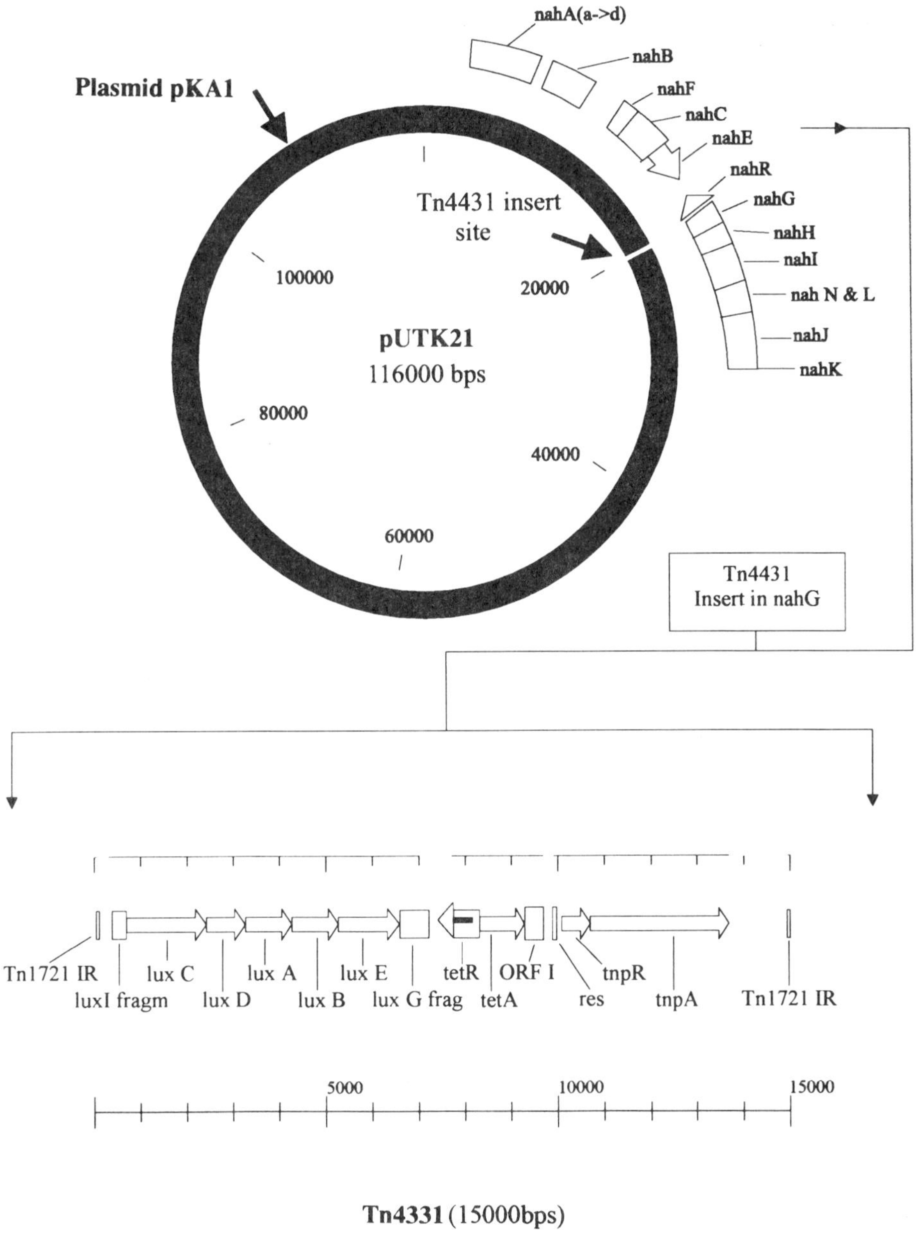

Figure 2. Map of pUTK21. Tn*4431* containing the promotorless *lux* cassette, inserted into the salicylate hydroxylase gene of the lower *nah* pathway.

A description of the construction of Strain HK44 occurs in King, et al., (1990). The *lux* cassette, contained within transposon Tn*4431*, was inserted into the *nahG* gene in the *sal* pathway of plasmid pKA1 to create plasmid pUTK21. Tn*4431* (Shaw et al., 1988) consisted of seven *lux* genes and gene fragments from *Photobacterium fischeri*, tetracycline resistance from *Escherichia coli Strain D1021*, and all genes and sequences from Tn*1721* which allows Tn*4431* to be a fully functional transposon.

In the presence of naphthalene, or the regulatory inducer metabolite salicylate, the pUTK21 bioluminescent reporter plasmid *lux* cassette mediates light production. The construct was intended to function as an indicator of the bioavailability of PAHs in sediments so that the ability to degrade the PAHs using microorganisms could be assessed. Secondly, Strain HK44 should be capable of degrading low to high molecular weight PAHs in the presence of the natural microbial community. For naphthalene, the *nah* pathway degrades naphthalene to salicylate, which then serves as an inducer for further naphthalene degradation. Salicylate itself would normally be degraded by *sal* pathway genes, except that the first gene in that pathway — *nahG,* which encodes salicylate hydrolase — has been inactivated by the insertion of Tn*4431* which contains the *lux* genes. According to Sayler (1995a), Tn*4431* contains translation stop codons in the two insertion sequences located on either end of the transposon which prevent translation in all three reading frames. Therefore, there is no fusion protein resulting from *nahG* and *lux* genes, and there is no translation of the *sal* genes downstream of Tn*4431*. The salicylate is, however, degraded further by an *ortho*-degradative pathway located on the chromosome. The *lux* genes lead to light production.

The construction of pUTK21 was well described. Only a limited number of sequences were unidentified, primarily nondegradative genes associated with the pKA1 backbone of plasmid pUTK21. Plasmid pKA1 is approximately 101 kb, and shares extensive homology with the well-described plasmid NAH7 in the 25 kb region which encodes the *nah* and *sal* pathways to the extent that these two pathways in pUTK21 and NAH7 can be considered homologous (Sanseverino et al., 1993). However, NAH7 is only 83 kb (Yen and Serdar, 1988), and Sanseverino et at. (1993) found significant differences in the restriction patterns of pKA1 as compared to NAH7 in the nondegradative portions of the plasmid.

In addition to unknown genes in the nondegradative portion of pUTK21, there is some uncertainty regarding the substrate range of pUTK21, based on the lack of information on the full substrate range of NAH7. It is known that pUTK21 is able to degrade the three-ring PAHs anthracene and phenanthrene to salicylate intermediates (Sanseverino et al., 1993), which should be further degraded by the chromosomal *ortho*-degradative pathway of Strain HK44. Sayler (1995b) noted that compounds from naphthalene to high molecular weight aromatics could be degraded, but not heavy tar residues. Hydroxylated and carboxylated intermediates could be expected from higher molecular weight PAHs, but other microbial populations present in soils are likely to further degrade these compounds. The issue of toxic metabolites which are generated from Strain HK44 is moot since pUTK21 would likely produce no different metabolites than pseudomonads which naturally bear pKA1 and the related NAH7 plasmids.

There is, however, a concern for salicylate-like metabolites if the *nah/sal* pathway of pUTK21 is transferred to other bacteria which may not bear the *sal* operon. In this case, metabolites structurally analogous to salicylate may be produced in naturally-occurring pseudomonads which receive the *nah/sal* pathway from HK44 via gene transfer. *P. fluorescens* Strain 5RL (donor of pUTK21 and original host for pKA1) does not have the *ortho*-degradative pathway and accumulated 1-hydroxy-2-naphthoic acid and 2-hydroxy-3-naphthoic acid from degradation of phenanthrene and anthracene, respectively (Menn et al., 1993). Other pseudomonads in the environment which acquire the degradative genes from pUTK21 may also be unable to degrade salicylate analogs. Again, such intermediates are likely to be further degraded by other bacteria in the environment, since pseudomonads containing NAH7 or plasmids similar to pKA1 could mineralize all three compounds (Sanseverino et al., 1993). Transfer of the *nah/sal* pathway could occur by conjugation since plasmid HK44 is a fully conjugative single copy plasmid (Sayler, 1995a). Transfer could also occur through mobilization of Tn4655-like transposon which may bracket the *nah/sal* pathway (as it does in NAH7). Tn*1721* sequences present in Tn*4431* could assist in the movement of the Tn4655-like transposon. Other gene transfer mechanisms are also theoretically possible.

A final concern identified with the introduced DNA is the presence of the tetracycline resistance gene. Risk issues arise when there is potential for an antibiotic resistance gene to spread (from an introduced microorganism) to microbial pathogens which are controlled (in clinical, agricultural, or veterinary settings) by the antibiotic against which the resistance is active (Neu, 1992). Expert panels convened by both the USEPA (USEPA, 1989) and by Health Canada (1995) found that tetracycline resistance is among the least desirable resistance markers to include in microorganisms released to the environment. This finding was made based on clinical and veterinary use of major antibiotics, and on the transmissibility of the replicons carrying the resistances. The presence of the tetracyline resistance gene imbedded in two transposons carried by a conjugative plasmid increases the potential for it to spread from Strain HK44 to other taxa.

3.1.2. Site Characterization. Strain HK44 will be released to contained lysimeters situated on the 2-acre Y-12 site located at the Oak Ridge National Laboratory in Oak Ridge, Tennessee. This site is moderately sloped, has uncontaminated soils and sediments, is close to electrical power, and reasonably secure. The lysimeters were originally designed for use in uranium leaching experiments connected with the disposal of uranium wastes generated at the DOE's Y-12 nuclear plant.

3.1.3. Microbial Interaction with Contaminants. The interactions of Strain HK44 with specific contaminants in the soils to be added to the lysimeters were undetermined since the contaminated soils to be treated in the lysimeters had not been selected. Following completion of the USEPA risk assessment it was decided that an uncontaminated, nonsterile, loamy soil would be spiked with napthalene, phenanthrene, and anthracene. This soil would be placed in the lysimeters for treatment with Strain HK44. Therefore, no metabolites would be expected from these three PAHs after degradation with HK44. If the *nah/sal* pathway was transferred to another microorganism that does not have the ability to degrade salicylate and its analogs, these intermediates would be present in the soils. However, degradation of these intermediates by other microorganisms present in the soils is thought to be likely. Microorganisms may generate toxic waste metabolites which are more water soluble and, therefore, have increased mobility as compared with the parent waste. However, this factor plays little role in assessing overall risk of PAH metabolites due to the degree of physical containment provided by the lysimeter design.

Other contaminants could undergo partial degradation if present in test soils. However, none of these contaminants are present in the soils in the lysimeters or at the Y-12 site. Dibenzofuran, in the presence of strains carrying NAH7, is converted to a dead-end product 4-[2′-(3′-hydroxy)benzofuranyl]-2-keto-3-butanoic acid (Selifonov et al., 1991). Naphthalene-related compounds can be converted by pseudomonad dioxygenases to oxygenated products which on steric grounds would not be anticipated. Chapman (1978) cited data indicating that pseudomonads which initially convert naphthalene to 1,2-*cis*-dihydrodiol encounter a related naphthalene waste — 1,5-dimethyl-naphthalene — which is a methyl substituent that is oxygenated to the primary alcohol and then converted to 1-methyl-5-naphthoic acid with neither of the aromatic rings being oxidized. Chapman noted other studies in which acenaphthene is converted by a naphthalene-grown pseudomonad to 1-acenaphthenol and then to 1-acenaphthenone. The stability and toxicity of these compounds was not noted.

3.1.4. Application Characterization. Application characterization consisted of two components: (1) the engineering design which detailed how, and under what conditions, the microorganisms will be applied to waste material and (2) the physical containment provided by the engineering design.

The test is being conducted in four 8-ft diameter by 10-ft deep lysimeters arrayed in a circle around a central core. The core is used for observation and gathering of samples and leachates. The treatments consist of (1) one lysimeter with PAH-contaminated soil only, (2) one lysimeter with Strain HK44 in the absence of PAHs, and (3) two lysimeters with Strain HK44 and PAH-contaminated soil. The test duration will be approximately two years and involve the use of 10^{14} cells of Strain HK44. The cells will be added to the lysimeters by mixing the cells into 2 cubic yards of moist contaminated soil. The cells will be grown in fermentors at the Oak Ridge National Laboratory, and will be introduced into the soil using a cement mixer. The 2 cubic yards of contaminated soil containing Strain HK44 will be sandwiched between an upper and lower layer of the same amount of clean soil.

The containment provided by the lysimeters and inactivation procedures for environmental media containing Strain HK44 allow minimal releases to the environment. Each of the lysimeters consists of a vertical 1/8″ thick corrugated steel pipe which is fitted with a steel lid, rests on a concrete apron, and has a leachate collection system which empties into a 55-gallon drum. All leachate and soils will be decontaminated prior to disposal if viable colony forming units are detected in the materials. Equipment will also be sanitized, and air monitoring will be conducted during periods when dispersal of the microorganism could occur such as during filling of the lysimeter.

3.2. Exposure Assessment

If the information provided in the PMN under Step 1 of Figure 1 was insufficient to complete the risk assessment for the field test, then further exposure assessment information would be considered. Exposure assessment information noted in Step 2 of Figure 1 includes information that would allow prediction of the fate of Strain HK44 and any PAH metabolites in the lysimeter soils, as well as any modifications of the field test design that would mitigate exposure to Strain HK44 and metabolites. Exposure assessment information included the field test design, detection limits for Strain HK44, data assessing the ability of Strain HK44 to survive in various contaminated soils, gene transfer information on Tn*4431* and pUTK21, proposals for worker protective gear, and sensitivity of Strain HK44 to hypochlorite.

Strain HK44 (Sayler, 1995c) was able to establish in a variety of nonsterile microcosms containing contaminated soils and sediments. Contaminants included naphthalene alone, diesel fuel, and a mixture of PAHs and other organic contaminants. In naphthalene-contaminated microcosms, the population of Strain HK44 increased over time with simultaneous degradation of naphthalene. After introduction at approximately 10^4 cells/gram, concentrations reached 10^7 colony forming units per gram (cfu/g) at day 10, then declined to 10^4–10^5 cfu/g at day 17 (with 4% of naphthalene remaining) and less than 10^3 cfu/g after 6 months. The declining concentration of Strain HK44 with decreasing PAH concentrations is considered beneficial from a risk assessment standpoint since the microorganism should decline in a similar fashion in the field.

Data relevant to assessing the stability of Strain HK44 and its ability to transfer pUTK21 and Tn*4431* were noted by Sayler (1995a). In nonselective chemostat experiments with Strain 18H, the Strain HK44 population experienced a 99% loss of the pUTK21 plasmid: 39 generations and a dilution rate of 0.086/hour resulted in 1.58 x 10^4 cfu per milliliter (ml) which retained the plasmid. The approximately 1 x 10^4 cfu/ml concentration was maintained for another 23 generations. Plasmid preparations showed that Tn*4431* remained stable in pUTK21, and all samples of isolates which lost pUTK21 were Tet^s, showing that the transposon did not insert into the chromosome.

Only skilled workers familiar with microbiological techniques will be involved in the field test, and no workers with open cuts or sores will be allowed on site. Gloves, and possibly respirators to protect against organic vapors, will be worn. The field test was scheduled for October, 1996.

All instruments, equipment, soils, and other samples will be sanitized. Strain HK44 can be detected by several techniques including a bioluminescence most probably number (MPN) procedure (detection limit approximately 10 cfu/g), agar plates with tetracycline and salicylate (detection limit 10^2–10^3 cfu/g), and by using *nah* and *sal* probes. In order to show the efficacy of hypochlorite inactivation, Sayler (1995c) provided data that showed colonies on plates cultured with yeast extract/peptone/glucose were unable to form colonies after treatment with 1–2% hypochlorite.

3.3. Hazard Assessment

According to Step 3 of Figure 1, if uncertainty regarding risk still exists after evaluation of pretest and exposure information, then hazard information should be examined. No quantitative structure activity relationship analyses were done to assess the toxicity of PAH metabolites due to the contained nature of the test and the likely further degradation of any salicylate-like analogs. Environmental and human health pathogenicity were addressed in part by data requested by the USEPA: growth curve information (Sayler, 1995c) showed that Strain HK44 does not increase in numbers at 37°C so it is unlikely to grow at mammalian body temperatures. This finding indicated that it was extremely unlikely that HK44 posed any human health or other mammalian toxicity/pathogenicity issues.

3.4. Regulatory Decision

Conditional approval of the field test (see Figure 1, step 4) was given in a TSCA 5(e) Consent Order which was agreed upon by the University of Tennessee and the USEPA on March 27, 1996. The Strain was only to be used at the Oak Ridge Y-12 site. Introduction of the Strain by means of a soil slurry which was sandwiched in the lysimeters between clean soil layers as suggested in the PMN was required. Sanitization of soils and other contaminated samples, equipment, and instrumentation was required. Such sanitization will be considered effective when there are no colony forming units at the limit of detection, considered to be 10 cfu/gram of soil or liter of water. Routine monitoring in the area around the lysimeters was requested, particularly during periods when aerosol generation is more likely (such as during introduction of Strain HK44 into the contaminated soils, and the soil's subsequent introduction into the lysimeters).

Quarterly reports on the status of the experiment generated for University of Tennessee and the Department of Energy will be forwarded to the USEPA. Reports will include operation evaluation of the lysimeters, operation evaluation of the monitoring equipment, analysis of data from sampling and monitoring, sampling schedule, and environmental safety and health evaluation (including accidents and injuries). Separate records of the progress of the field test will be kept by the University of Tennessee for several topics including production volume, standard operating procedures, sampling information, routine monitoring activities for detection of Strain HK44, and effectiveness of sanitization techniques.

The Consent Order also identified three issues which require resolution prior to allowing the use of the same microorganism at any site or under less stringent containment conditions. Since transfer of the tetracycline resistance to microbial pathogens could be of concern, data on the frequency of transfer of pUTK21 and Tn*4431* should be examined. Second, the presence of persistent toxic metabolites may need to be addressed prior to commercialization.

Finally, the Consent Order noted that plant and animal pathogenicity concerns may need to be addressed in more detail prior to commercialization. Following the USEPA decision which identified Strain HK44 as a *P. fluorescens* biovar II (which evolved from biotype B of Stanier et al., 1966), USDA reached a conclusion that this strain is not a plant pest and identified Strain HK44 as a member of the "*Pseudomonadaceae* RNA Group I biotype D" (USDA, 1996). Biotypes D and E of Stanier et al., (1966) have no current equivalent biovar designation. They were replaced by *P. chlororaphis* and "*P. aureofaciens*", and ultimately were combined as *P. chlororaphis* (Johnson and Palleroni, 1989). Since *P. chlororaphis* and *P. fluorescens* are so closely allied, it is expected that isolates will be found that are transitional between the two species. Due to the inadequacies of current methods for separating closely related strains within or between biovars of fluorescent *Pseudomonas* species, attempts to place strains within species or biovars of *P. fluorescens* and allied species may result in different designations using different criteria. Therefore, the USDA determination that Strain HK44 is a biotype D using one set of valid criteria may not be inconsistent with the taxonomic designation of Strain HK44 as a member of *P. fluorescens* biovar II. Even if Strain HK44 does belong to *P. fluorescens* biovar II, the similarity with biotype D organisms may alleviate some plant pathogenicity concerns: pseudomonads included in the former biotype D such as *P. chlororaphis* and *P. aureofaciens* have no noted adverse effects on plants, as opposed to some members of *P. fluorescens* biovar II (such as *P. marginalis*) which can be plant pests (Palleroni, 1984).

In summary, two conclusions may be reached from the information presented on Strain HK44. First, the review of this microorganism under TSCA indicated that its use in the proposed lysimeter field test poses minimal risks to humans and the environment. Second, this review also presents a structured and reasonable review process for recombinant bioremediation products.

Disclaimer: This document has been reviewed by the Office of Pollution Prevention and Toxics, USEPA, and approved for publication. Approval does not signify that the contents necessarily reflect the views and policies of the Agency, nor does mention of trade names or commercial products constitute endorsement or recommendation for use.

REFERENCES

Ballows, A., W.J. Hauser, and H.J. Shadomy (eds), 1991. *Manual of Clinical Microbiology*, 5th edition, ASM, Washington, D.C.

Broder, M., 1995. Risk assessment for PMN P95–1601. Office of Pollution Prevention and Toxics, Washington, D.C.

Bullock, G., 1964. *Pseudomonadales* as fish pathogens. *Devel. Industrial Microbiol.*, 5:101–108.

Canaday, C.H., J. E. Wyatt, and J.A. Mullins, 1991. Resistance in broccoli to bacterial soft rot caused by *Pseudomonas marginalis* and fluorescent *Pseudomonas* species. *Plant Disease*, 75:715–720.

Chapman, P., 1978. Degradation mechanics, in: *Microbial degradation of pollutants in marine environments*, USEPA, EPA-600/9–79–012, pp. 28–66.

Health Canada, 1995. Workshop on the assessment of microorganisms containing antibiotic resistance genes - Ottawa, January 27–28, 1993, Health Canada.

Johnson, J.L., and N.J. Palleroni, 1989. Deoxyribonucleic acid similarities among *Pseudomonas* species, *Intl. J. Syst. Bact.*, V39(3):230–235.

King, M.,H., P.M. diGrazia, B. Applegate, R. Burlage, J. Sanseverino, P. Dunbar, F. Larimer, and G.S. Sayler, 1990. Rapid, sensitive bioluminescent reporter technology for naphthalene exposure and biodegradation, *Science*, 249:778–780.

McClung, G., 1995. Ecological hazard assessment for PMN submission P95–1601, Office of Pollution Prevention and Toxics, Washington, D.C.

Menn, F., B. Applegate, and G. Sayler, 1993. NAH plasmid-mediated catabolism of anthracene and phenanthrene to naphthoic acids, *Appl. & Environ. Microbiol.*, V59(6):1938–1942.

Miller, S., 1980, Susceptibility of lettuce cultivars to marginal leaf blight caused by *Pseudomonas marginalis* (Brown 1918) Stevens 1925, *New Zealand J. Experimental Agriculture*, 8:169–171.

Neu, H., 1992. The crisis in antibiotic resistance, *Science*, 257:1064–1078.

Ohta, K., H. Morita, K. Mori, and M. Goro, 1976. Marginal blight of cucumber caused by a strain of *Pseudomonas marginalis* (Brown) Stevens, *Ann. Phytopath. Soc. Japan*, 42:197–203.

Palleroni, N, 1984. Family I: *Pseudomonadaceae*, in: *Bergey's Manual of Systematic Bacteriology*, Volume 1, Williams and Wilkins, Baltimore pp 156–165.

Sanseverino, J., B. Applegate, J. Henry King, and G. Sayler, 1993. Plasmid-mediated mineralization of naphthalene, phenanthrene, and anthracene, *Appl. & Environ. Microbiol.*, V59(6):1931–1937.

Sayler, G.S. 1995a. 14 June Premanufacture Notice for PMN P95–1601, The University of Tennessee, Knoxville, TN.

Sayler, G.S., 1995b. Personal communication.

Sayler, G.S., 1995c. August 25 memorandum from to EPA from the University of Tennessee's Center for Environmental Biotechnology, Knoxville, TN.

Sayler, G.S. and P. Sayre, 1995. Risk assessment for recombinant pseudomonads released into the environment for hazardous waste degradation, in: *Bioremediation: the Tokyo '94 Workshop*, OECD, Paris, pp. 263–272.

Sayre, P., 1995. Construct analysis for PMN P5–1601. Office of Pollution Prevention and Toxics, Washington, D.C.

Segal, M., 1995. P-95–1601 Recipient/donor identities, Office of Pollution Prevention and Toxics, Washington, D.C.

Selifonov, S., A. Slepenkin, V. Adanin, M.Nefedova, and I. Starovoitov, 1991. Oxidation of dibenzofuran by pseudomonads harboring plasmids for naphthalene degradation, *Microbiology* (English translation MiKrobiologiya), V60:714–717. As cited in Menn, et al., 1993.

Shaw, J.J., L.G. Settles, & C.J. Kado, 1988. Transposon Tn*4431* mutagenesis of *Xanthamonas campestris pv. campestris*: characterization of a nonpathogenic mutant and cloning of a locus for pathogenicity, *Molecular Plant-Microbe Interactions*, 1:39–45.

Stanier, R.Y., N.J. Palleroni, and M. Doudoroff. 1966. The aerobic pseudomonads: a taxonomic study. *J. Gen. Microbiol.*, 43:159–271.

Tou, J., 1995. ETD/ICB biotechnology PMN chemistry report, Office of Pollution Prevention and Toxics, Washington, D.C.

Turner, V. and N.K. Van Alfen, 1983. Crown rot of alfalfa in Utah, *Phytopath.*, 73:1333–1337.

USDA. 6 March 1996 letter. Written to Dr. Gary Sayler of the University of Tennessee from John H. Payne, USDA Animal and Plant Health Inspection Service.

USEPA, 1989. Summary of the Biotechnology Science Advisory Committee's subcommittee on antibiotic resistances, US EPA, Washington, D.C.

USEPA, 1994. Points to consider in the preparation and submission of TSCA premanufacture notices (PMNs) for microorganisms, USEPA, Office of Pollution Prevention and Toxics, Washington, D.C.

Wright, P. and C. Hale, 1992. A field and storage rot of onion caused by *Pseudomonas marginalis*. *New Zealand J. Crop and Horticul. Sci.*, 20:435–438.

Yen, K.M. and Serdar, C.M., 1988. Genetics of Naphthalene Catabolism in Pseudomonads. *CRC Critical Reviews in Microbiology*, 15:247–267.

26

RISK-RELATED ISSUES AFFECTING BIOIMPLEMENTATION

Kate Devine

DEVO Enterprises, Inc.
1003 K Street, NW; Suite 570
Washington, DC 20001-4425

1. INTRODUCTION

Laws and regulations drive the supply of and the demand for many commodities, including technologies used for industrial waste treatment or control. There are many risk-related issues underlying or embodied in environmental statutes, regulations and policies/programs affecting the immediate or potential future commercial deployment of bioremediation. These issues and their effect on the implementation of bioremediation are discussed below within the context of the applicable environmental statute. The term "risk" is used broadly to mean potential health or environmental effects due to exposure to chemicals considered to be detrimental to a pre-existing state of human health or ecological quality.

2. THE RESOURCE CONSERVATION AND RECOVERY ACT

Subtitle C of the Resource, Conservation and Recovery Act (RCRA) of 1976 was enacted to ensure that those wastes determined to be hazardous are managed in a manner that protects human health and the environment. This act, and later amendments to the act, required the US Environmental Protection Agency (USEPA) to define hazardous wastes, set treatment standards prior to land disposal and issue permits for facilities that store, treat, and/or dispose of such wastes. The USEPA has defined as hazardous those chemical substances that: either have been listed by name in rulemakings; exhibit any of the characteristics of ignitability, corrosivity, flammability, or toxicity; or have been mixed with, derived from or contain listed hazardous waste (40 CFR Part 261, Subparts C and D; 40 CFR 261.3(a)(2)(iv) and (c)(2)(I)). The definition of hazardous waste, therefore, is quite broad and encompasses a significant portion of the universe of all chemical substances.

2.1. The Land Disposal Restrictions

In 1984, amendments made to RCRA, termed the Hazardous and Solid Waste Amendments (HSWA), stipulated the restriction of land disposal of hazardous wastes un-

Biotechnology in the Sustainable Environment, edited by Sayler *et al.*
Plenum Press, New York, 1997

less such wastes had been treated according to treatment standards that HSWA directed the USEPA to develop. Wastes meeting these standards, promulgated through a series of rules known as the Land Disposal Restrictions (LDRs), would not be subject to the constraints on land disposal. These regulations specified either contaminant concentration levels or methods of treatment which substantially reduced the toxicity or mobility of RCRA hazardous wastes and their hazardous constituents (40 CFR Parts 268.30 through 35). According to the USEPA, a recurring debate during the Agency's development of the land disposal restrictions was whether treatment standards should be technology-based or risk-based (Biotreatment News, 1993). The USEPA had begun developing risk-based treatment standards before HSWA's passage and continued to promote a risk-based approach after HSWA. The USEPA states, however, that through hearings and written comments on the proposed LDRs, Congress made clear its desire for technology-based treatment standards. Therefore, the USEPA reversed its initial standard development decision and issued technology-based treatment standards in its final rule (USEPA, 1990a).

The standards, that required either a method of treatment or a specific concentration level for the target contaminant in the treated waste, were based on the concept of "best demonstrated available technology," or BDAT. BDAT was defined as follows:

- A treatment technology is *demonstrated* for a given waste if it is currently used to treat the waste (i.e., bench and pilot scale technologies are not demonstrated);
- A technology is *available* if it can be purchased or leased and in addition, the technology must substantially diminish the toxicity of the waste or substantially reduce the likelihood of migration of hazardous constituents from the waste; and
- After determining the technologies that are demonstrated and available, a statistical analysis is used which incorporates a variability factor to account for normal process variations in well-designed and well-operated treatment systems to determine the *best* technology (Jonesi, 1989).

At the time the LDRs were promulgated (1986–1990), many of the more innovative types of cleanup technologies, such as bioremediation, had been commercially implemented at full scale for a relatively short period of time. For example, in the case of bioremediation, the first controlled application of oxygen and nutrients to stimulate indigenous bacteria to significantly degrade soil contaminants took place in 1972 (Brown et al., 1995). Even by the late 1980s, bioremediation's potential was largely untapped. For example, although bioremediation was being used at petroleum underground storage tanks (UST) sites, where such technology is particularly amenable to petroleum hydrocarbon contamination, USEPA UST data collected in 1990 and 1991 indicated bioremediation was being used at less than ten percent of such sites in the US (USEPA, 1992a).

The USEPA felt insufficient data were available to consider using such treatment technologies as a basis for treatment standard development. Instead, the BDAT levels established for many contaminants were based on data from treatment of pure industrial process wastes, which are typically less difficult to treat than those that are highly variable (USEPA, 1990b). Therefore, while bioremediation offered significant risk reduction potential for many hazardous wastes, it could not be utilized as a treatment, either in instances where other technology had been explicitly dictated or in instances where the standard was a concentration level based on that attainable only by incineration. Additionally, because states can choose to implement standards more stringent than those mandated at the federal level, concentrations as restrictive as nondetectable (ND) or level of detection (LOD) can be required for cleanup under the RCRA program. Such non risk-based standards also can preclude use of bioremediation at a given site.

2.2. RCRA's Hazardous Waste Identification Rules

Recognizing the need for redefinition of hazardous, the Agency first proposed its Hazardous Waste Identification Rule (HWIR) in 1992. Significant public comment caused reassessment of this initial proposal (Biotreatment News, 1992) with two hazardous waste identification rules being created. "Hazardous Waste Management System; Identification and Listing of Hazardous Waste; Hazardous Waste Identification Rule (HWIR)," referred to as HWIR-Waste, which focuses on newly generated process waste, was issued in December 1995 (USEPA, 1995a). "Requirements for Management of Hazardous Contaminated Media (HWIR-Media)," referred to as HWIR-Media, was issued on April 29, 1996 and focuses on contaminated media managed in cleanups overseen by the USEPA or authorized states (USEPA, 1996). The USEPA has indicated that its ultimate policy preference is to establish risk-based levels that represent minimal threat levels and, in doing so, cap the extent of hazardous waste treatment. The USEPA claims that due to the generally low concentrations of hazardous constituents in soil, incineration is not uniformly appropriate for hazardous soil (USEPA, 1995a).

2.2.1. HWIR-Waste. HWIR-Waste proposes to amend regulations under RCRA by establishing specific exit levels for low-risk solid wastes that are designated as hazardous waste. The primary purpose of the rule is to address those hazardous wastes that either: have been listed as hazardous or are defined as hazardous because they are mixtures of listed hazardous wastes and solid wastes (as per the 1980 Mixture Rule; 40 CFR 261.3(a)(2)(iv)), or are residues derived from managing listed hazardous waste (as per the 1980 Derived-from Rule; 40 CFR. 261.3(c)(2)(I)) that under current rules continue to be designated as hazardous waste although they are either generated with constituent concentrations that pose low risks or are treated in a manner that reduces constituent concentration to low levels of risk (USEPA, 1995a).

Under the proposal, generators of listed hazardous wastes that meet this rule's proposed self-implementing exit levels would no longer be subject to the hazardous waste management system under Subtitle C of RCRA as listed hazardous waste. The USEPA states that the HWIR proposes to establish a risk-based "floor" to hazardous waste listing that will encourage pollution prevention, waste minimization and the development of innovative waste treatment technologies. Additionally, the USEPA also states that the exit levels developed are risk-based concentrations at which a human or wildlife species could be directly or indirectly exposed to the exempted waste and would be unlikely to suffer adverse health effects. The USEPA has developed exit levels for 376 constituents (USEPA, 1995a).

The Agency's intention with this self-implementing rule is to create incentives for effective and innovative waste minimization and waste treatment, and to reduce unnecessary demand for RCRA disposal capacity without comprising needed environmental protection. The Agency states that the risk-based floor for listed wastes is to give a strong incentive to generators of listed hazardous waste to apply pollution prevention to their processes, to avoid Subtitle C control. The Agency believes action should also give incentive for the development of innovative treatment technologies to render wastes less risky (USEPA, 1995a).

In this proposed rule, the USEPA claims that wastes that are subject to RCRA because of the Mixture Rule or the Derived-from Rule that cannot meet the exit level and remain hazardous typically will pose risk that warrant regulation under Subtitle C. However, HWIR is purported to allow rapid exemption for wastes that are either mixed with or derived-from wastes that present no significant threats to human health and the environment. The Agency states that the delisting process (in which the USEPA can be petitioned to delist a hazardous waste, based on data submitted) will remain available to exempt wastes with constituents at more site and waste-specific levels (USEPA, 1995a).

The Agency proposed to re-evaluate the basis for some of the existing performance standards established for listed wastes, given that the state of the art in making quantitative determinations of risk has advanced and available methods have improved significantly. In addition, the increased sensitivity of analytical methods has lowered achievable detection limits, better bioassays exist than in the past, and more extensive biological data is available for extrapolation. So, the universe of available health-based and ecological data has grown significantly and the reliability of this information has improved. The Agency now believes that these data can be used to establish levels that minimize threats to human health and the environment. The "minimize threat levels" proposed, so-termed from the statutory requirements that threats to human health and environment be minimized, would substitute for the current treatment standards where the minimize threat levels are higher (USEPA, 1995a).

The public comment period for this rule closed on April 22, 1996. Among those to comment was the Biotechnology Industry Organization (BIO), representing about 50 companies and organizations offering biotreatment services and/or products in biotreatment. Among BIO's comments were that: (1) the USEPA should develop exit levels based on an analytical method that measures the "bioavailability" of the chemicals rather than utilizing a theoretical risk assessment process; (2) the USEPA should develop regional and conditional exit levels; and (3) if the USEPA uses a theoretical risk assessment to set exit levels, more realistic assumptions should be used, taking into account, for example, natural biodegradation, less extreme assumptions, a probabilistic assessment of the uncertainty in the risk, and the latest scientific information (Biotechnology Industry Organization, 1996).

2.2.2. HWIR-Media. Proposed in April 1996, the HWIR-Media rule is addressing waste management issues relating to soil, groundwater and sediments by defining hazardous constituents concentrations levels below which a waste is no longer considered hazardous. The rule will address major RCRA Subtitle C management requirements: Land Disposal Restrictions (see above), Minimum Technological Requirements (MTRs, these requirements apply to landfills and surface impoundments, e.g., protective liners in surface impoundments) and RCRA permitting issues. One of the main themes of HWIR-Media is to redefine hazardous constituents by means of a "bright line" to distinguish highly contaminated media from other less contaminated media. Media above the bright line would be subject to Subtitle C requirements. The rule proposes alternative treatment standards to the current universal treatment standards (UTSs) required under RCRA. UTSs for organics and metals are based primarily on incineration and stabilization, respectively. The alternative standards proposed are: either 90% of the original concentration or 10 times the UTS level, whichever is higher. Provisions for site-specific variances are also proposed. A streamlined permitting process is also proposed and is intended to expedite the permitting process for remedial activities concerning contaminated media (Forlini, 1995).

Media below the bright line would be subject to more flexible site-specific management requirements as specified by approved state or federal cleanup programs (*Biotreatment News*, 1994). That is, overseeing agencies are given the opportunity to exempt the less contaminated media (i.e., that below the bright line) from Subtitle C management and set new requirements based on site-specific conditions (*Bioremediation in the Field*,

1995). Such regulatory action, based on risk-related issues, could positively affect commercial implementation of bioremediation at a RCRA site.

2.3. RCRA Underground Storage Tank Program

2.3.1. The Program's History and Bioremediation's Past Role. In 1984, Congress enacted Subtitle I of RCRA as a means of controlling and preventing leaks from USTs. The USEPA, along with states under cooperative agreements, was given the authority to clean up releases from UST systems or to require their owners and operators to do so. Later, as part of the amendments to the Comprehensive Environmental Response, Compensation and Liability Act (CERCLA; more widely known as Superfund; see below) in 1986, Congress added a trust fund to Subtitle I to aid in cleanup of select UST sites. The Leaking Underground Storage Tank Fund is financed through a tax on gasoline, diesel, and aviation fuels, and is used under certain conditions, such as when cleanup costs exceed coverage requirements of the financially responsible party, when the owner or operator refuses to comply with a corrective action order; when a solvent owner or operator cannot be found, or when an emergency exists (USEPA, 1990b).

As stated above, 1990 and 1991 USEPA data indicated that bioremediation was not being used to a great extent in the UST program. Risk-related factors contributed to this less than maximum implementation of the technology for USTs. That is, a few years ago many states based their cleanup goals specifically for benzene, toluene, ethylbenzene and xylenes (BTEX) constituents and total petroleum hydrocarbon (TPH) compounds on the Leaching Potential Analysis Method which led to typical parameters of 1 part per million (ppm) BTEX or 100 ppm TPH for soils at petroleum-contaminated sites (Liptak and Lombardo, 1994). Because few scientific aspects of subsurface contamination are known with absolute certainty, agencies may rely on policy to support decisions about soil and groundwater cleanup levels (Walsh, 1990). In the early days of the UST program, states set such UST site standards without a scientific approach or guidelines and these numbers were adopted across the country in an effort to quickly establish cleanup goals for their state (Liptak and Lombardo, 1994). Although from a risk reduction perspective, bioremediation may be an acceptable means of cleanup at a given UST site, it could be precluded from use because bioremediation cannot universally achieve levels of 1 ppm BTEX or 100 ppm TPH.

In order to ascertain the current state of use of innovative technologies in the UST program, a study was conducted that included a survey of state representatives in the fall of 1994 and winter of 1995. The survey revealed that all state representatives had knowledge of innovative technologies, various application techniques and their potential in the UST cleanup market. Several state representatives said that they had seen a pronounced increase in the number of proposals for cleanup that involved bioremediation, utilized alone or with other technologies. Many states indicated that whatever technology could meet their standards would be implemented in cleanup. Also, many states had the requirement of some type of permit to be obtained when conducting bioremediation. But, there was a great degree of variation in permit-related factors, including the actual methodology employed that would warrant a permit (recirculation of groundwater and/or the injection of nutrients or oxygen), the actual state agency or department involved in handling the permitting, the length of time to issue the permit (which varied from 2 to 6 months) and, whether or not cleanup activity could commence before the permit was actually issued. Some states indicated that the addition of microorganisms was not viewed highly and would lead to higher scrutiny of a project or additional permitting requirements. No formal policy on bioremediation or any other cleanup technology was seen in most states at the time of the survey (Devine and Graham, 1995). Another more recent study of select state regulatory programs that focused on the implementation of in situ bioremediation revealed that the degree to which in situ bioremediation was implemented was directly correlated with the regulatory flexibility that was observed (Colorado Center for Environmental Management, 1996).

2.3.2. Financial Crisis and Its Impact: Risk-Based Site-Specific Approach. This study also revealed that many states have been faced with a growing financial crisis as the cost of cleanup increases faster than the growth in petroleum restoration funds (see above for more on the Leaking Underground Storage Tank Trust Fund). The USEPA passed regulations in 1988 that required petroleum UST owners, predominantly commercial gas stations and operators of car and truck fleets, to upgrade USTs by December 1998. As of 1995, less than one-quarter of an estimated 463,000 contaminated UST sites had been cleaned up. Of the 45 states for which financial information was available in 1994, only 13 states would be able to pay on all claims submitted for cleanup reimbursement (Environmental Information Ltd., 1995). Many states are recognizing the need to move beyond traditional cleanup approaches as the cost of cleanup increases faster than the growth in petroleum restoration funds. Thus, the need for maximally cost-effective and cleanup solutions is growing ever more imperative in order for such funds to be used in the most effective manner.

The state UST representative survey showed that states were beginning to realize that past treatment standards adopted by many states that embodied the concept of "one size fits all," that is, one treatment standard (e.g., 100 ppm TPH) for all sites, was not a cost-effective means of site cleanup. Generic risk-based cleanup goals are an advance over non-risk-based remediation goals but are still not always appropriate for a particular site. Site-specific factors can greatly influence the potential for exposure and the use of generic values means that these factors are not considered. Consequently, the site may be cleaned up to meet standards or guidance that are inappropriate for the particular site. The most familiar example is the use of criteria based on residential use of a property for a site that cannot be used for this purpose. More accurate and less conservative but still health-protective risk-based cleanup goals can be developed to allow consideration of site-specific factors such as future site use, chemical availability, and natural attenuation. Increased use of risk assessment and true, site-specific risk-based cleanup levels should lead to more appropriate and efficacious remediation. And for non-fully-destructive technologies, such as bioremediation, risk-based criteria may be required to allow the use of the technology at certain sites (Devine and LaGoy, 1996).

The UST state representative survey showed that a more progressive attitude towards site cleanup has developed due to the rising financial crisis of depleting state cleanup funds. While many states recognized the need for cleanup standards or program modifications, there were differences among states in the degree of current or immediate future direction of regulatory activity concerning bioremediation. Standards based on chemical-specific risk-based approaches, as well as issuance of guidance that may call for the explicit consideration of certain technologies in site remedy selection has been initiated in some states (Devine and Graham, 1995).

The growing environmental financial crisis has led the USEPA to encourage states to consider alternative means of allocation of constrained resources. In 1995, the USEPA issued a directive, "Use of Risk-Based Decision-Making in UST Corrective Action Programs" (USEPA, 1995a). Additionally, by 1996, almost all 50 states have requested training on the American Society for Testing and Materials (ASTM) Emergency Standard Guide for Risk-Based Corrective Action Applied at Petroleum Release Sites (ASTM ES

38–94), more commonly referred to as "RBCA" (pronounced "Rebecca") (Devine and Graham, 1995). This guidance standardizes existing Superfund (see below for more on Superfund) risk assessment methods and promotes the concentration of resources for sites that pose the greatest threat to human health and the environment. The RBCA approach recognizes the diversity of parameters affecting risk, such as contaminant complexity, physical, and chemical characteristics, and uses a tiered approach that tailors assessment and remediation activities to site-specific conditions and risks (ASTM, 1994). The first two tiers are similar, with tier one generally consisting of fairly stringent numeric criteria for a selected group of chemicals, and tier two involving the development of values for chemicals not on the tier one list, using formulas that establish tier one values. In most states, the third tier generally allows the development of values based on a site-specific risk assessment. This last approach gives regulators the flexibility to consider factors that decrease the risks posed by chemicals at a particular site, such as known future site use, chemical behavior at the site, and soil cover (Devine and LaGoy, 1996).

In the state UST representative survey, many states indicated that a risk-based approach was being considered and some specifically mentioned the possibility of following the state of Texas' lead in its risk-based approach that was adopted into regulation in late 1993 (Devine and Graham, 1995). In 1995, however, the state of Michigan, cautioned that there has been difficulty in providing regulatory guidance concerning implementation of RBCA due to lack of training for the regulated and consulting communities. In other words, the shift from cleaning up sites to a specific number to cleaning up to a standard based on a site-specific risk level is not an easy change to make (*Underground Tank Technology Update*, 1995).

In addition to the UST state representative survey, a 1994 survey of state cleanup standards showed that 33 states may use a risk/health parameter and/or a site-specific approach in determining cleanup standards for soil and/or groundwater for certain contaminants in certain instances (*Soils*, 1994). It is likely this number will increase over time.

2.3.3. Financial Crisis and Its Impact: Interest in Natural Attenuation Grows. As concern for the cost of cleanup of the US's industrial contamination continues to grow, one site cleanup approach currently being considered and implemented, particularly for petroleum hydrocarbon contaminants, is intrinsic remediation or natural attenuation. Intrinsic remediation does not entail active remedial effort. Instead, contamination is left to the natural processes of biodegradation, chemical transformation, dispersion, sorption and volatilization.

Various public and private sector activities have taken place in the past year or so that have focused on assessing intrinsic remediation's benefits as well as educating those responsible for choosing and implementing the most appropriate remedial alternative at a given site. The following describes some of these activities to date.

One of the first proponents of intrinsic remediation as a cost-effective solution to select contamination problems was Dr. John T. Wilson of the USEPA Office of Research and Development's National Risk Management Research Laboratory (NRMRL; Ada, Oklahoma). Wilson and NRMRL's Dr. Fran Kremer organized "Symposium on Intrinsic Bioremediation of Ground Water", held in the fall of 1994 in Denver, CO. The symposium was a joint effort of the USEPA's Biosystems Technology Development Program, the US Geological Survey (USGS), with additional sponsorship from the US Air Force. The Agency held another meeting on intrinsic bioremediation for chlorinated compounds on Sept. 10–13, 1996 in Dallas, TX.

In 1996, the U.S. Air Force produced a two-volume protocol regarding the intrinsic remediation of fuel hydrocarbons. Volume 1 of "Technical Protocol for Implementing Intrinsic Remediation With Long-Term Monitoring for Natural Attenuation of Fuel Contamination Dissolved in Groundwater" (Wiedemeier, 1996) is a field guide for consultants and contractors. Volume 2 of the protocol consists of two examples of natural attenuation being utilized as the remedial option of choice. The Air Force Center for Environmental Excellence (AFCEE) intends to distribute the protocol to all Air Force bases nationally and internationally, as well as to Air Force consultants, the Army and the Navy (Haas, 1996). Other organizations have begun to promote the Air Force's work. For example, in January 1996, the International Ground Water Modeling Center, affiliated with the Colorado School of Mines (Golden), offered the course "Computer Implementation of the Air Force Intrinsic Remediation (Natural Attenuation) Protocol."

Prior to that protocol, AFCEE issued "United States Air Force Guidelines for Successfully Supporting Intrinsic Remediation with an Example From Hill Air Force Base", by Wiedemeier, Wilson, Miller and Kampbell. This document summarizes the technical protocol for implementation of intrinsic remediation (Wiedemeier, 1996).

Another AFCEE natural attenuation initiative entails documentation of 45 full scale sites for implementing long-term monitoring of dissolved phase fuel contamination in ground water. The information collected from this initiative is now being complied into a report, intended for regulatory personnel. The data shows that intrinsic was the cleanup option chosen either alone or in conjunction with source removal (Haas, 1996).

The AFCEE is now writing a protocol for chlorinated solvents natural attenuation, which is being co-authored by Wilson. The Air Force also sponsored "Intrinsic Bioremediation of Chlorinated Solvents Symposium" on April 2, 1996 in Salt Lake City, UT and the Air Force currently is conducting microbial-related research as pertains to intrinsic remediation.

A 1995 state survey of UST programs, shows that only six states would not consider natural attenuation as a stand-alone remedial option primarily for groundwater. Several states have informal policies and New Jersey and North Carolina have written policy that address natural attenuation of petroleum hydrocarbons as a stand-alone option (Ritz, 1996). This 1995 survey shows a dramatic difference in the acceptance of procedures, such as natural attenuation, from the state UST representative survey conducted less than a year previous (see above). In other recent state action the State of California issued a December 1995 policy for its underground leaking tank program (LUFT) specifying closure of low risk soil sites and monitoring of low risk groundwater sites. This policy is based on a study, funded in part by the USEPA and Shell Oil, and conducted by the Department of Energy's Lawrence Livermore National Laboratory (Livermore, California) and the University of California. The study recommended to utilize "passive bioremediation" as a remediation alternative whenever possible, to minimize actively engineered LUFT remediation processes, and not to use the UST cleanup fund to implement pump-and-treat remediation unless its effectiveness can be demonstrated (Lawrence Livermore National Laboratory and University of California, 1995).

Also, the state of Wisconsin issued an emergency rule in February 1993 stipulating that three cleanup options be submitted to the state prior to site remediation if the responsible party will seek reimbursement from the state's petroleum cleanup fund. One of these options must be natural biodegradation. Additionally, there is a formal task group in the Wisconsin Department of Natural Resources to assess how state internal procedures may be limiting the use of innovative technologies (Biotreatment News, 1993b). Since this regulatory code change became effective in May 1993, of the 900 sites reviewed for reimbursement, only 2 or 3 had selected intrinsic bioremediation as the cleanup option as of October 1995. The state had hoped that the code change would lead to a significant number of sites utilizing intrinsic bioremediation and, subsequently, considerable cost savings. Wisconsin postulates that the lack of utilization of this option may be due to several factors: the degradation rate for the contaminants may be too slow for property owners; data to support intrinsic bioremediation is rarely collected; many sites do not meet state criteria of concentration limits or no groundwater contamination; consultants do not understand the mechanics or actual costs of intrinsic bioremediation; and generic cleanup levels are difficult to attain with intrinsic bioremediation. The state has taken steps to address the situation (See Table 1; Evanson, 1995).

Additionally, the American Society for Testing and Materials (ASTM; West Conshohocken, PA) currently is working on guidelines for implementation of natural attenuation. Guidelines are expected either in late 1996 or in early 1997 (Barden, 1996).

The American Petroleum Institute (API) has initiated and partially funded a project designed to evaluate and compare sampling and analytical methods used to characterize intrinsic bioremediation. The ultimate goal is to provide guidance on recommended sampling and analytical procedures. The project was undertaken because of the poor quality data that can result from conventionally employed methods. Two reports were due in 1996. One report contrasts sampling and analytical techniques and the other will be a guide to different sampling and analytical methods (Piontek, 1995 and 1996).

In addition to API's involvement, individual oil companies have also developed their own protocols for evaluation of natural attenuation. Chevron Research and Technology Company's Health and Safety Group (Richland, CA) issued, "Protocol for Monitoring Intrinsic Bioremediation in Groundwater." The work was funded by Chevron Corporation Long Range Research and authored by staff hydrologist Tim Buscheck and Kirk O'Reilly, Lead Research Biochemist. Chevron recently produced a workshop, "A Practical Guide to Monitoring Intrinsic Bioremediation in Groundwater," in conjunction with the Sixth West Coast Conference on Contaminated Soils and Groundwater, held in March in Newport Beach, CA. In March 1995, Mobil Oil published "A Practical Approach to Evaluating Intrinsic Bioremediation of Petroleum Hydrocarbons in Groundwater." The document was prepared by the Groundwater Technology Group of the Environmental Health and Safety Department and the Environmental Health Risk Assessment Group, Stony Brook Laboratories. Amoco Corporation also published "Natural Attenuation as a Remedial Alternative Technical Guidance" in 1995. In 1994, P.M. McAllister and C.Y. Chang of Shell published "A Practical Approach to Evaluation Natural Attenuation of Contaminants in Ground Water" (*Ground Water Monitoring and Remediation*, 1994).

As the financial and regulatory climate continues to foster use of the most cost-effective measures for addressing contamination, active remediation could become a less prominent tactic in cleanup approach. If such happens, contractors' services will be more of a

Table 1. The State of Wisconsin's proposed solutions to low number of intrinsic bioremediation sites

Obstacle	Solution
Information needs	Develop guidance for assessment of intrinsic
Cleanup standards	Encourage site-specific closure standards
Long-term monitoring	Site by site basis, with many on a reduced monitoring schedule
Site close-out criteria	Revise

Source: Evanson, 1995.

consulting nature, providing design and implementation, sampling and analysis. While in the short run the revenues from active remediation would decline, a solid client-contractor relationship would be built for the long run as the contractor continues to offer the most cost-effective solution. Although revenues on a per site basis are higher for active remediation sites, monitoring numerous intrinsic sites also could produce a viable business.

2.4. Environmentally Acceptable Endpoints

As stated above, the Biotechnology Industry Organization's comments on a recent proposed RCRA rule advised the USEPA to develop constituent exit levels from the hazardous waste system that would be based on an analytical method measuring the bioavailability of the chemicals, that is, that takes environmentally acceptable endpoints (EAEs) into account, rather than the theoretical risk assessment process. This concept of EAEs could save billions in cleanup costs for private industry and federal and state governments, if utilized. An EAE is a risk-based concept and can be defined as the concentration of chemical substances in soil or other medium that does not adversely affect human health or the environment (*Biotreatment News*, 1995). One of the most important site-specific factors for risk evaluation of contaminated soil is the availability of the chemicals held within the soil. It has been reported that substantial reduction in the toxicity of soil contaminants has been achieved with varying degrees of chemical removal. These results support the concept that the risk associated with a contaminated soil is independent of the total concentration of specific chemicals and depends on the fraction of the total concentration that is available to either human or ecological receptors. Therefore, a scientifically-based approach to determine "How clean is clean?" should not rely solely on the analytical detection limit of each contaminant. Instead, it should consist of a combination of toxicological and biological tests that can directly assess the concentration of chemicals in soil or water that will not adversely affect human health or the environment (Gas Research Institute, 1995).

Over the past decade, numerous laboratory and field studies and actual cleanup operations have been conducted on hydrocarbons in soil. The results of these efforts indicate that (1) hydrocarbons are biodegraded by indigenous soil microorganisms to a plateau concentration, i.e., a concentration which no longer decreases or decreases very slowly with continued treatment, (2) reduction below the plateau concentration is limited by the availability of the hydrocarbons to the microorganisms, (3) residual hydrocarbons that remain after biological treatment, regardless of the extent of treatment which has occurred, exhibit significantly reduced leachability to the aqueous phase and significantly reduced toxicity, and (4) residual hydrocarbons that are present in an aged and/or weathered soil, depending upon the environmental condition to which they were exposed, also exhibit a reduction in leachability and toxicity compared to virgin hydrocarbons (Gas Research Institute, 1995).

It is believed that the hydrocarbons that remain in the treated soil are no longer available ecological and human receptors, and represent an EAE. Therefore, bioremediation that can reduce hydrocarbon concentrations to a level that no longer pose an unacceptable risk to the environment, i.e., that no longer leach to the groundwater or harm ecological receptors or human health, could be employed as a risk reduction technology in many instances (Gas Research Institute, 1995).

The Gas Research Institute (GRI; Chicago, IL) has announced that a collaborative research effort on environmentally acceptable endpoints has been launched with the Petroleum Environmental Research Forum. According to GRI, the program will: define the relative importance of availability of contaminants for establishing cleanup levels, characterize potential toxicity of complex hydrocarbon mixtures, and develop frameworks for establishing EAEs (Gas Research Institute, 1996).

3. THE COMPREHENSIVE ENVIRONMENTAL RESPONSE, COMPENSATION AND LIABILITY ACT (SUPERFUND)

In 1980, Congress passed the Comprehensive Environmental Response, Compensation and Liability Act in response to public concern over the large number of abandoned, leaking, hazardous waste sites threatening human health and the environment (USEPA, 1992b). A major difference in CERCLA and RCRA (see above) is that CERCLA was designed to remedy mistakes made in hazardous waste management in the past while RCRA is focused on avoiding such mistakes through proper management in the present and future (USEPA, 1990b).

CERCLA created a trust fund to pay for cleanup and imposed cleanup liability on those responsible. This "Superfund" was comprised primarily of tax assessment on oil and designated chemicals (USEPA, 1992a). CERCLA response is triggered by a release or substantial threat of release of a dangerous substance. CERCLA authorizes removal action and remedial actions. Removals are short term cleanup actions conducted in response to an emergency situation, such as cleaning up a spill. Remedial actions are intended to provide permanent solutions to hazardous substance threats. The USEPA takes remedial action at sites on the National Priorities List (NPL), which are placed on this list after being evaluated through the Hazard Ranking System (HRS), a model that determines the relative risks to public health and the environment posed by hazardous substances (USEPA, 1990b).

The Superfund process consists of several steps briefly outlined as follows. After site discovery, a site assessment, beginning with a preliminary assessment (PA), is conducted to determine if the site poses a potential hazard. If the PA shows a contamination problem, the USEPA will do a site inspection and use the HRS to evaluate the relative risks to human health and the environment. For remedial responses, that is, longer term actions to eliminate or substantially reduce threat from contamination, there are two main phases: remedial investigation and feasibility study (RI/FS) and remedial design and remedial action (RD/RA). During the RI, USEPA, the state or the potentially responsible party (PRP) collects and analyzes information to determine the nature and extent of the contamination. This information is used in a site-specific risk assessment, prepared following procedures outlined in the USEPA's manuals entitled "Risk Assessment Guidance for Superfund" (USEPA, 1989).

If a site is determined to pose a risk, remediation typically is needed and a FS is prepared. During the FS, specific alternative remedies are evaluated by the USEPA. A preferred remedy is identified from this list of remedies in what is termed a Record of Decision (ROD). In the remedial design and action stage, the recommended cleanup is undertaken (USEPA, 1992a).

3.1. CERCLA Applicable or Relevant and Appropriate Requirements under the Superfund Amendments and Reauthorization Act (SARA) of 1986

Congress stipulated that response actions conducted under CERCLA must at least attain all "applicable or relevant and appropriate requirements" (ARARs) of other federal environmental laws and more stringent state environmental laws. ARARs are identified on a site-by-site basis for all on-site response actions where CERCLA authority is the basis for cleanup (USEPA, 1992b). However, the use of ARARs as the dominant cleanup standard, particularly the reliance on drinking water standards, makes it difficult to implement a technology such as bioremediation. ARARs can raise the threshold of acceptance so high that bioremediation, is excluded in favor of more familiar technologies (Brown, 1995b)

3.2. CERCLA Remedy Selection

Proponents of bioremediation have often criticized the lack of flexibility in the Superfund remedy selection process. As in the RCRA program, there is a predisposition to familiar technologies, with pump and treat and incineration often the remedies selected. Often when a ROD is finalized, it specifies performance criteria, such as attaining acceptable levels for drinking water standards. If these standards are not met, the ROD frequently specifies that the site would have to be addressed with another technology, such as excavation or incineration. Additionally, the length of time that elapses between the selection and implementation of the remedy is often long enough that significant innovation in a technology can occur. However, the current system does not allow for a systematic reopening of the ROD so that new or improved technologies can be evaluated and implemented (Brown, 1995b).

Additionally, the applicable regulatory agency focus often seems to place as much emphasis on the volume of the contaminant as on its mobility. One of the benefits of bioremediation is that it removes the more mobile constituents of a waste first reducing the mobility of and, consequently, the potential for exposure to the contaminant (Brown, 1995b). Therefore, as a risk reduction technology, bioremediation would be an appropriate means of cleanup at many Superfund sites.

3.3. Proposed Modifications to CERCLA

Two significant bills were introduced in 1995 and 1996 in an attempt to reform the Superfund Law. In the House of Representatives, Rep. Michael Oxley (R-OH), Chair of the Trade and Hazardous Subcommittee of the House Commerce Committee, authored H.R. 2500. Among other things, H.R. 2500 proposed to: replace joint and several liability with a "fair share" system in which cleanup costs are in proportion to the amount of pollution at a site caused by each party; eliminate ARARs and a preference for permanence; introduce a risk range for cleanup; base remedy selection on site-specific evaluations; and consider cost-effectiveness of cleanup and reopen RODs. Oxley claimed to have eliminated liability for 90 of private industries responsible for cleanup. In the Senate, Bob Smith (R-NH), Chair of the Environment Public Works Committee's Subcommittee on Superfund, Waste Control, and Risk Assessment, introduced S. 1285, which contained remedy selection modifications similar to H.R. 2500 (Ruffin, 1995). Such proposed modifications to the remedy selection process, if realized in future legislation, could increase utilization of innovative technologies, such as bioremediation. In March 1996, the Senate introduced a new S.1285, which also contained language pertaining to significant repeal of retroactive liability (*Biotreatment News*, 1996). While the remedy selections of both bills remain favorable to the use of innovative technologies, President Clinton threatened to veto any bill that contained provisions for retroactive liability repeal. As a result of the inability of the Administration and Congress to compromise on the liability issue, a Superfund reform bill was not passed in 1996.

4. THE TOXIC SUBSTANCES CONTROL ACT

4.1. Intergeneric Microorganisms

Section 5 of the Toxic Substances Control Act (TSCA) of 1976 authorizes the USEPA to review new chemicals not under the scope of other legislative jurisdiction (e.g, the Food, Drug and Cosmetics Act) for potential human health and environmental effects prior to manufacture or import. It further dictates to limit any unreasonable risk which the new chemical may present by taking action (USEPA, 1987). In the summer of 1994, the USEPA published a proposed rule, "Microbial Products of Biotechnology; Proposed Regulations Under the TSCA". This rule proposes, among other things, that persons intending to manufacture, import or process intergeneric microorganisms for commercial or field experimentation purposes in the US must notify the USEPA, providing certain information for review, before such activity (*Biotreatment News*, 1994b). In doing so, the USEPA has made the determination to treat intergeneric microorganisms as new chemical substances. Thus, any genetically engineered microorganism used for bioremedial field research or commercial purposes will be subject to reporting requirements not imposed on those utilizing naturally occurring microorganisms.

In 1996, the Clean Technologies Advancement Directorate of Environment Canada (Ottawa), conducted a public perception survey on the use of biotechnology for environmental purposes. Results showed that those surveyed accepted the use of naturally occurring microorganisms for such purposes but did not approve of the use of genetically engineered microorganisms (GEM) (Environment Canada, 1996). Release of genetically engineered organisms is perceived as posing significant risk to health or the ecology. Due to factors, such as data requirements and public perception, the implementation of GEMs is not expected to comprise any significant portion of US commercial bioremediation revenues for some time.

4.2. PCBs

Under TSCA Section 6, Congress singled out PCBs for immediate regulation and phased withdrawal from the market (USEPA, 1987a). TSCA mandated that approved disposal methods be used and that Agency approval be sought by obtaining a permit. In 1986, the Agency issued guidelines for persons applying to the USEPA for approval of PCB disposal by methods alternative to incineration because under the PCB regulations, incineration is the standard for PCB destruction. USEPA approval using technologies other than incineration or alternatives to incineration can be sought but equivalency to incineration must be demonstrated (Blake, 1994). Therefore, as with RCRA and CERCLA programs, a familiar, traditional method of treatment is given preference in remedy selection, although from a cost-effectiveness risk-based perspective, other technologies could be prime candidates for use.

In order to issue a permit for bioremediation as an alternative disposal method, the USEPA must make the finding that the process not only destroys PCBs, but produces no toxic by-products or toxic emissions, and that any microorganisms used as inoculum pose no unreasonable risk to human health or the environment. At this time, incineration is the only PCB disposal process approved to remove PCBs from soils and sediments (Blake, 1994).

In December 1994, the USEPA issued proposed PCB rule amendments, that should moderate these regulatory restrictions pertaining to the use of innovative technologies.

Over 4000 comments were received on this proposal and the Agency is in the process of preparing its final rule which is scheduled for release in late 1996. This rule provides for greater flexibility at both the R&D and commercial use level. R&D treatability studies will be exempt from permitting. Cleanup standards will be risk-based as opposed to the current incineration-based standards (Blake, 1995).

5. CONCLUSIONS

While many risk-related factors pertaining to environmental statutes and regulations historically have presented barriers to the implementation of bioremediation, a more moderate, flexible regulatory atmosphere currently is in development. This development is due to several factors, including more and better data, a growing lack of cleanup funds for treatment of all sites to artificially low contamination levels, and the realization that a risk-based, site-specific basis for treatment is appropriate for adequate protection of human health and the environment. Such an atmosphere should bode well for commercial implementation of bioremediation in the future.

REFERENCES

Barden, M., 1996, Wisconsin Dept. of Natural Resources, Madison, personal communication with K. Devine, March.

Bioremediation in the Field, 1995, EPA to release proposed hazardous waste identification rule for contaminated media, EPA/540/N-95/500, No. 12, pp. 1 and 4, August.

Biotechnology Industry Organization, 1996, Comments on the Biotechnology Industry Organization on the proposed Hazardous Waste Identification Rule (HWIR Rule), docket Number F-95-WHWP-FFFFF, 60 Fed. Reg. 66344, Dec. 21.

Biotreatment News, 1992, EPA proposes revised hazardous waste identification rule, *2*(7): 3 and 4.

Biotreatment News, 1993a, EPA proposes long-awaited contaminated soils rule. *3*(11): 7,8 and 10.

Biotreatment News, 1993b, Wisconsin issues new rule requiring consideration of passive bioremediation, *3*(6):1 and 7.

Biotreatment News, 1994a, EPA rolls contaminated soil treatment standards into HWIR *4*(7): 4 and 19.

Biotreatment News, 1994b, Second phase of LDRs issued, Superfund and CWA die, TSCA rule proposed, *4*(12): 10–12.

Biotreatment News, 1996, Reg review: Superfund reform, *6*(5): 20.

Blake, J., 1994, Approach to the regulation of bioremediation of polychlorinated biphenyls, in: *Bioremediation of Chlorinated and Polycyclic Aromatic Hydrocarbon Compounds*, (R.E. Hinchee, A. Leeson, L. Semprini, and S.K. Ong, eds.), Lewis Publishers, Boca Raton, FL, pp. 432–435.

Blake, J., 1995, USEPA, Office of Pesticides and Toxic Substances, Washington, DC, personal communication with K. Devine, December.

Brown, R.A., Hinchee, R., Norris, R.D., and Wilson, J., 1995, Bioremediation of petroleum hydrocarbons: a flexible, variable speed technology, presented at: National Ground Water Association Petroleum Hydrocarbons and Organic Chemicals in Ground Water: Prevention, Detection and Restoration, Nov. 29–Dec.1, Houston, TX.

Brown, R.A., 1995, Regarding the advantages of using bioremediation to clean up Superfund sites, testimony of Dr. Richard A. Brown, Vice President, Remediation Technology, Groundwater Technology, Inc., testifying on behalf of the Biotechnology Industry Organization, Inc. before the Senate Environment and Public Works Committee, Subcommittee on Superfund, Waste Control and Risk Assessment, Washington, DC.

Devine, K. and Graham, L.L., 1995, States' attitudes on the use of bioremediation, in: *Applied Bioremediation of Petroleum Hydrocarbons,* (R.E. Hinchee, J.A. Kittel, and H.J. Reisinger eds.), Battelle Press, Columbus, OH, pp. 61–71.

Devine, K. and LaGoy, P., 1996, Regulatory issues applying to bioremediation as a risk reduction technology, In *Bioremediation: Principles and Practice*, (S. Sikdar and R. Irvine eds.), Technomic Publishing Co., Inc., Lancaster, PA, in press.

Environment Canada, Clean Technologies Advancement Directorate, 1996, Environmental applications of biotechnology: focus groups, draft, unpublished.

Environmental Information, Ltd., 1995, *Underground Storage Market: Cleanup: Status and Outlook*, Minneapolis, MN.

Evanson, T., 1995, Wisconsin's efforts to encourage naturally occurring biodegradation as a cleanup option, paper presented at: Intrinsic Bioremediation: Strategies for Effective Analysis, Monitoring and Implementation, Oct. 16–17, Annapolis, MD, International Business Communications (Southboro, MA).

Forlini, M., 1995, USEPA, Office of Solid Waste and Emergency Response, Technology Innovation Office, The hazardous waste identification rule and its effect on bioremediation, presented at bioremediation policy panel session at SUPERFUND XVI, Washington, DC, November 8.

Ground Water Monitoring and Remediation, 1994, *XVI*(2): 161–173.

Gas Research Institute, 1995, *Environmentally Acceptable Endpoints in Soil: Risk-Based Approach to Contaminated Soil Management Based on Availability of Chemicals in Soil*, compilation of working papers for May expert meeting, draft, Washington, DC, April.

Gas Research Institute, 1996, How Clean is clean?: Research initiative on environmentally acceptable endpoints (EAEs) for contaminated soils, issue 3, Dec. 1995.

Haas, P., 1996, Brooks Air Force Base, TX, personal communication with K. Devine, March.

Jonesi, G., 1989, Impact of RCRA land disposal restrictions on bioremediation, USEPA Office of Enforcement and Compliance Monitoring, presented at: SUPERFUND X, Washington, DC.

Lawrence Livermore National Laboratory and University of California, 1995, *Recommendations to Improve the Cleanup Process for California's Leaking Underground Fuel Tanks (LUFTs)*, Oct. 16. Liptak, J.F. and Lombardo, G., 1994, The development of chemical-specific risk-based soil cleanup guidelines results in timely and cost effective remediation, New Hampshire Department of Environmental Services, Groundwater Protection Bureau, Concord, NH.

Piontek, K., 1995, An evaluation of field methods for intrinsic bioremediation measurements, presented at: Intrinsic Bioremediation: Strategies for Effective Analysis, Monitoring and Implementation, Oct. 16–17, Annapolis, MD, International Business Communications (Southboro, MA).

Piontek, K., 1996, CH2M Hill, St. Louis, MO, personal communication with K. Devine, March.

Ritz, S., 1996, States speak out on natural attenuation, *Soil and Groundwater Cleanup*, Jan./Feb., pp. 18–27.

Ruffin, M., 1995, Biotechnology Industry Organization, Superfund reauthorization and the environmental biotechnology industry's perspective, presented at: bioremediation policy panel session at SUPERFUND XVI, November 8, Washington, DC, December.

Soils, 1994, State cleanup standards for hydrocarbon contaminated soil and groundwater, Dec.: 14–60.

Underground Tank Technology Update, 1995, Michigan's cleanup program, Dept. of Engineering Professional Development, the College of Engineering, University of Wisconsin-Madison, *9*(6): 10, Nov./Dec.

USEPA, 1987, *The Layman's Guide to the Toxic Substances Control Act*, Office of Pesticides and Toxic Substances, Washington, DC.

USEPA, 1989, *Risk Assessment Guidance for Superfund. Volume I. Human Health Evaluation Manual (Part A)*, interim final, EPA/540/1–89/002, December, Office of Emergency and Remedial Response, Washington, DC.

USEPA, 1990a, *RCRA Orientation Manual*, EPA/530-SW-90–036, US Environmental Protection Agency, Office of Solid Waste, Washington, DC.

USEPA, 1990b, *Quality Assurance Project Plan for Characterization Sampling and Treatment Tests Conducted for the Contaminated Soil and Debris (CS&D) Program*, Office of Solid Waste, Washington.

USEPA, 1992a, *Technologies and Options for UST Corrective Actions: Overview of Current Practice*, EPA/542/R-92/010, Office of Solid Waste and Emergency Response Technology Innovation Office and Office of Underground Storage Tanks, Washington, DC.

USEPA, 1992b, *CERCLA/Superfund Orientation Manual*, EPA/542/R-92/005, Office of Solid Waste and Emergency Response, Washington, DC.

USEPA, 1993, Land disposal restrictions for newly identified and listed Hazardous wastes and hazardous soil, 58 *Fed. Reg.* 48092–48204, Sept. 14, 1993.

USEPA, 1995a, Hazardous waste management system: Identification and listing of hazardous waste: hazardous waste identification rule (HWIR), 60 *Fed. Reg.* 66344–66469, Dec. 21.

USEPA, 1995b, OSWER directive 9610.17: Use of risk-based decision-making in UST corrective action programs, memorandum from U.S. Environmental Protection Agency, Office of Solid Waste and Emergency Response Assistant Administrator to Regional Administrators, Washington, DC.

USEPA, 1996, Requirements for management of hazardous contaminated media; Proposed rule, 61 *Fed. Reg.* 18780–18864, April 29.

Walsh, W.J, 1990, Making science, policy, and public perception compatible: A legal/policy summary, *In Groundwater and Soil Contamination Remediation: Toward Compatible Science, Policy, and Public Perception*, National Academy Press, Washington, DC., pp. 206–249.

Western Governors Association Develop On-Site Innovative Technology Committee's Interstate Technology and Regulatory Work Group, and Colorado Center for Environmental Management. 1996, *A Study of Selected In Situ Bioremediation Across the United States*, draft, Jan. 4.

Wiedemeier, T.H., Wilson, J.T., Miller, R. and Kampbell, D., 1995. *United States Air Force Guidelines for Successfully Supporting Intrinsic Remediation with an Example From Hill Air Force Base.*

Wiedemeier, T.H., Wilson, J.T., Kampbell, D.H., Miller, R. and Hansen, J. 1996. *Technical Protocol for Implementing Intrinsic Remediation with Long Term Monitoring for Natural Attenuation of Fuel Contamination Dissolved in Groundwater*, Vols. I and II, Air Force Center for Excellence, Technology Transfer Division, Brooks AFB, Texas.

27

BIOREMEDIATION

The Green Thumb in Brownfields Management

Maureen Leavitt*

SAIC
P. O. Box 2502
Oak Ridge, Tennessee 37831

1. INTRODUCTION

During the industrial revolution, cities were built to support complete communities. Many residents worked in the same neighborhood they lived in, and few owned cars. With the advancement of technology, suburban property became more attractive, causing rampant development of former farm and natural resources. While this urban sprawl flourished, the inner cities suffered neglect severe enough to cause urban "blight"; city blocks stood abandoned and deteriorating. The impact of such damage reached not only property, but socioeconomic factions as well. These properties, now termed brownfields, have to be reconciled into productive use to save remaining natural resources and revitalize the inner city. The environmental regulating community has developed programs to entice such efforts. The following paper describes these efforts and how biotechnology can contribute to the movement.

2. DEFINITION OF BROWNFIELDS

The primary issues that define brownfields are described below:

- Perception is reality

Brownfields are properties that are perceived to be tainted, either by contamination or by urban blight. These properties are underutilized or even abandoned, often causing further deterioration of the property.

- Presence of hazardous contaminants

* Paladin International / 1946 Oak Ridge Turnpike / Oakridge, TN 37830

Biotechnology in the Sustainable Environment, edited by Sayler *et al.*
Plenum Press, New York, 1997

Hydrocarbons, metals and solvents, three of the most prevalent contaminants, may be present due to spills during past use of the site, such as a gasoline fueling station. Other contaminants such as polychlorinated biphenyls (PCB) or asbestos may have been associated with the surface structures built on the site. In fact, the structures themselves could be considered contamination.

• Redevelopment barriers

Developers don't consider brownfields sites. If all of their needs can be met with "green fields" (suburban areas that have not been developed), they have no incentive to assume ownership of potential liability. This issue of liability is in itself the major barrier to redevelopment of brownfields.

3. THE US ENVIRONMENTAL PROTECTION AGENCY'S BROWNFIELDS ACTION AGENDA

The mission of the US Environmental Protection Agency (USEPA) has always been to protect human health and the environment. Recently, the agency has realized that many of its practices, specifically those related to the Comprehensive Environmental Response, Cleanup and Liability Act (CERCLA) have caused sites to be abandoned, even when no further remediation was planned. Site owners or perspective buyers could not overcome the burden of environmental liability. To alleviate this negative impact, the USEPA has taken several actions under the theme of the Brownfields Action Agenda that produce a cooperative, attractive potential to many sites throughout the nation.

3.1. "NFRAP" Sites Removed from Tracking System

Some sites placed on the CERCLA Information System (CERCLIS) list have been investigated and classified as "No Further Remedial Action Required" (NFRAP). Most sites in this category were not contaminated, or were under state-run cleanup programs. Even though these sites were "blessed" with this classification, they remained on the Superfund tracking system list (CERCLIS). Simple association with this list caused these sites to be undesirable. In 1995, USEPA set a goal to remove 25,000 NFRAP sites of the 38,000 sites on the CERCLIS list, and to better communicate their positive status to potential developers.

3.2. Brownfields Pilot Programs Funded

USEPA funded 50 pilot projects at an approximate cost of $200,000 each. These funds are not intended to be used to clean up brownfields sites, but are instead to be applied toward effort to develop a strategy for redevelopment. These pilots were funded in several batches, with programs located throughout the nation. In addition to aiding the community receiving the funding, this program was intended to provide guidance to many other municipalities that can learn from these experiences.

3.3. Personnel Training

To encourage facilitation of brownfields progress, selected USEPA staff members were temporarily assigned to brownfields programs. Additional efforts were placed in training initiatives at community colleges to provide a skilled job pool in the vicinity of the primary brownfields programs.

3.4. Technical Guidance

The USEPA has issued several guidance documents that are intended to limit liability to those agents that caused the contamination and not to those that acquire contaminated property or receive contaminated groundwater from another site. Other guidance includes rules and processes to limit cleanup costs. These guidance documents include:

- Prospective Purchaser Agreements
- Municipal Acquisition Liability
- Lender Liability Rule
- Presumptive Remedy Guidance
- Common Sense Initiative
- Soil Screening Guidance

4. BIOREMEDIATION TOOLS AS THEY PERTAIN TO BROWNFIELDS

Once contamination has been identified, steps toward protecting human health and the environment must be initiated, preferably without interfering with redevelopment. Brownfields classification offers several advantages that can result in significant cost savings. In some instances, brownfields properties have a designated end-use or at least a desired use category at this stage of the process. This allows more site-specific risk assessment which can produce realistic remediation targets and remediation alternatives. In some cases, it prevents excessive remediation.

Another brownfields advantage is flexibility to consider innovative alternatives. Since the action is under a partnership and not a strict enforcement philosophy, alternatives that are cost-effective but require extended treatment periods could be used to meet treatment needs and allow development without delay. Bioremediation is one of many tools that should be utilized at brownfields sites. Bioremediation offers effective protection of human health and the environment under a variety of site conditions and often in obscure configurations allowing development to progress.

4.1. Passive Bioremediation

The basic premise of most bioremediation systems is that naturally-occurring bacteria in an impacted environment develop the means to degrade or tolerate the presence of organic contaminants. This indigenous activity can be considered passive or intrinsic bioremediation. Intrinsic bioremediation has been recognized most frequently for groundwater (saturated systems) contaminated with hydrocarbons. If it is known that the source of contamination has been terminated and that the site poses no immediate threat to human health, the possible improvement of the site due to intrinsic bioremediation is often explored. This activity includes collecting site-specific data for parameters that are known to impact or be impacted by biological activity. Once data are collected, an estimation of the type of bacterial activity occurring in the subsurface can be made. From this model, a prediction of the long-term contaminant fate is developed and considered for regulatory acknowledgment.

Table 1. Intrinsic bioremediation analytes

Analyte	Soil/Water	Field- or lab-based analysis
pH	S/W	field
Conductivity	W	field
Oxidation-reduction potential	W	field
Oxygen gas/dissolved oxygen	W	field
Carbon dioxide	W/S	field
Methane	S/W	field
Iron	W	lab
Manganese	W	lab
Total Kjeldahl nitrogen	S	lab
Ammonia	S/W	lab
Nitrate	W	lab
Sulfate	W	lab
Sulfide	W	lab
Total organic carbon	S/W	lab
Biomass estimate (method varies)	S/W	lab

Table 1 lists analyses that are typically associated with intrinsic bioremediation cases. Many of the analytes in this table are potential electron acceptors; they are used as a source of energy by bacteria. Other analytes such as pH and conductivity are measured to confirm that the conditions are not adverse to bacterial activity. By comparing these analytes in samples from within the contaminant plume and outside the contaminated area, it may be possible to suggest that bacterial biodegradation of the contaminants is occurring.

In addition to those analytes, contaminant concentrations and speciation can also provide evidence for biodegradation. For example, chlorinated solvents such as trichloroethylene (TCE) and perchloroethylene (PCE) are transformed into distinct byproducts such as dichloroethylene and vinyl chloride (Vogel and McCarthy, 1985, Wilson and Wilson, 1985). The presence of these byproducts is additional evidence for biodegradation (Wackett and Gibson, 1988, Wackett et al., 1989). For organic solvents, the evidence is not as direct; however, there has been some success by comparing relative proportions of compounds that are readily degraded to those that are resistant to degradation (Butler et al., 1994). A sample with low quantities of degradable compounds and high resistant compounds would suggest biodegradation has occurred.

If evidence suggests that a site is improving due to natural attenuation, the environmental regulating agency may grant that site "monitoring only" status. A common resolution to these sites is a long term monitoring plan to confirm that water quality is improving. It is this incentive that has made natural attenuation so popular. Most brownfields sites will not have an organized history of groundwater quality over an extended period of time, so intrinsic bioremediation applications may require focused sampling in conjunction with the site characterization phase.

Further development of this alternative is needed. The specific fate of chemicals in the environment needs further study, which would benefit from the development of user-friendly field kits documenting bacterial degradation. While most effort has been applied to groundwater systems, further research should be applied towards the computer modeling of chemical fate in unsaturated soil, as well as bacterial activity. Finally, this application should be extended to address metals that are influenced by biological activity.

4.2. Semi-Passive Bioremediation Systems

When natural attenuation is not enough to ensure protection of human health and the environment, a more active remediation system must be considered. Semi-passive bioremediation systems are defined as those in situ treatments that induce favorable conditions for accelerated biodegradation with minimal operational needs. Semi-passive systems can satisfy regulatory requirements for treatment while remaining obscure to the public and to development efforts. Combined with the low cost of a minimal operation requirement, semi-passive bioremediation systems are a desirable alternative to off-site treatment or disposal.

4.2.1. Groundwater Systems. One remedy within bioremediation applications that can be considered semi-passive is the microbial fence (see Figure 1). Such a system sets up a bioactive zone on the down gradient edge of a contaminated groundwater area. As impacted groundwater enters the bioactive area, contaminants are biodegraded. The water leaves the bioactive zone clean, or at least meeting treatment requirements. In ideal cases, minimal maintenance and monitoring is required.

This approach does have specific requirements including:

- relatively homogeneous subsurface formation,
- readily biodegradable contaminants present in soluble form, and
- groundwater flow sufficient to induce treatment.

This approach has been demonstrated in pilot- and full-scale applications (Abou-Rizk et al., 1995, Abou-Rizk and Leavitt, 1997). The disadvantages of such an approach

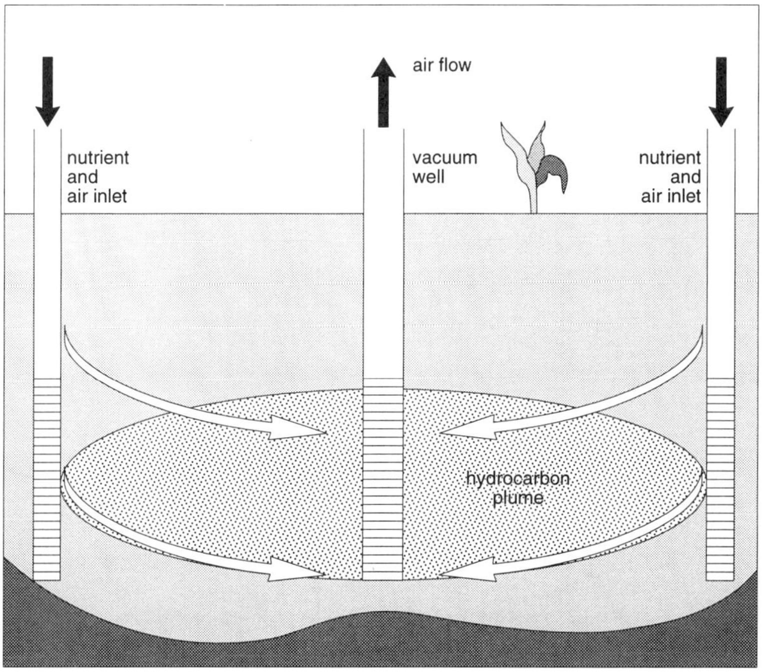

Figure 1. Conceptual illustration of a microbial fence.

include an extended treatment period, potential need for additional treatment and equipment, and the risk of unexpected new contamination.

4.2.2. Unsaturated Soil Systems. In situ treatment in unsaturated soil systems is a well-known application known as bioventing. This application stimulates aerobic biodegradation by circulating air, and therefore oxygen through the subsurface. There are some specific considerations that impact the success of this approach. First, if contamination is shallow, the remediation system must include a sealed surface (such as asphalt) to maximize air flow. Second, air will only move through the most permeable areas causing treatment to occur only in limited areas. For this reason, bioventing is generally limited to homogeneous subsurface formations. Heterogeneous formations can be treated, but excessive wells can be costly to maintain.

4.3. Aggressive Bioremediation

Bioremediation can be engineered to produce optimal treatment in both in situ and ex situ systems. Applications have been demonstrated to be unobtrusive at dozens of underground storage sites nation-wide. Examples of in situ groundwater treatment and soil treatment are shown in Figures 2 and 3. Site attributes include permeable formations, predictable water flow, ample water yield, and the absence of any free-phase organic layer (Lee et al., 1988). Engineering controls are more prevalent at these sites, resulting in a more optimized system, and in most cases, increased costs (both capital and operations and maintenance). The advantage is a much shorter treatment period, alleviating any construction limitations sooner than with passive systems.

While aggressive bioremediation is an excellent remedial approach, it is unlikely that these systems will be prevalent at brownfields sites. In essence, it can be considered paying full-price for a remedy when other alternatives may produce the same result in time with lower cost.

4.4. Challenges for the Future

The question remains, "Why aren't we seeing bioremediation implemented at more sites?" There is no single answer, but there are several underlying influences.

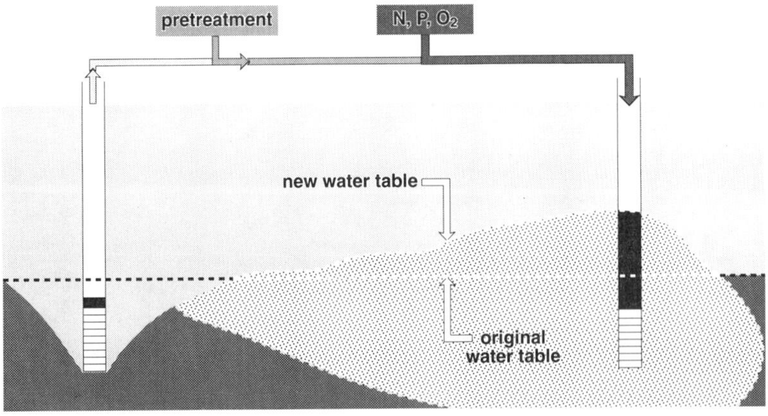

Figure 2. Aggressive groundwater bioremediation conceptual approach.

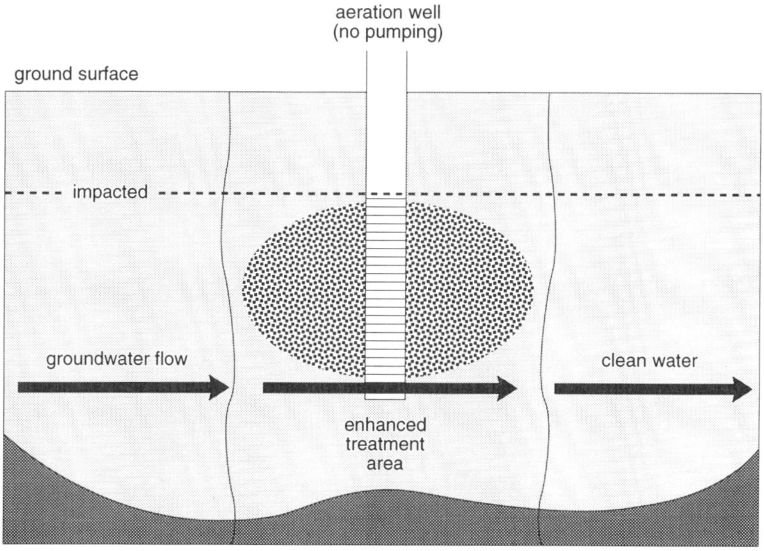

Figure 3. Conceptual illustration of a bioventing system..

4.4.1. Liability. The issue of liability itself has several directions: the liability of the original polluter, the liability of the new owner or developer, and the liability of the contractors who implement the remedy. In each case, liability can also be renamed "fear of the unknown". The original polluter often will not accept liability for fear of criminal and civil penalties. To render some brownfields properties attractive to developers, liability must be mitigated to belong only to the original polluter.

4.4.2. Redefining the Scope of Remedies. The objective of all cleanup actions is to protect human health and the environment. This cannot be redefined; however the approach we take to determine when we achieve this goal should be reconsidered. One example is the determination of contaminant cleanup levels. The use of risk-based decision-making has assisted in moving regulations away from generic target levels to site-specific standards that consider the future land use. This step alone required several years of pressure from the regulated community combined with the realization of the limited resources available to remediate sites. It is currently integrated into some regulated programs in most states.

The next step in evaluating our cleanup success is to reconsider the bioavailability of compounds and the actual toxicity of compounds. Briefly, the bioavailability issue (Loehr, 1996) has emerged as an explanation for residual contamination following active biotreatment. In essence, this argument pleads that contamination that is measurable in soil is bound to the soil in such a way that it is unavailable to be degraded. Considering that it is unavailable to microorganisms, it should not be a hazard for higher beings. This argument is the focus of a great deal of basic and applied research. Currently, most of the database concerns polyaromatic hydrocarbons (PAH), but the precedent will affect many other organic compounds.

4.4.3. Treatment Periods. In many cases, the remedy for a site is chosen based on the speed at which the remedy renders the site clean. To promote redevelopment of brownfields sites, the treatment period may need to be reconsidered. For example, instead of waiting for the entire site to reach the designated target level, the plan would specify attaining a proportion of treatment that would protect a more immediate and realistic exposure route (e.g., construction workers potentially exposed for only days or weeks, not children exposed for 70 years). A different approach would allow construction to begin at one area while problem areas are treated elsewhere. The benefits of considering these alternatives are less removal and disposing of contamination in off-site facilities, and less time elapsed prior to redevelopment.

4.4.4. Partnerships. The key to all progress in brownfields will undoubtedly be the spirit of partnership. The USEPA and many state agencies have recognized the power of partnerships and have created "voluntary" programs that allow owners and developers alike to work with regulators in a team effort to release a site. It is surprising the increase in progress resulting from deleting the classic adversarial attitudes of these two parties. Similarly, brownfields redevelopment will require partnering with local stakeholders (residents, business people, potential employees). When stakeholders feel that they will receive a benefit, almost any proposal for a site becomes attractive. One excellent example is the East Fork Poplar Creek Project in Oak Ridge, Tennessee. This federal site was primarily mercury contaminated sediment in a creek running through the metropolis. By applying site-specific risk analysis, involving stakeholders, and segregating property according to the extent of characterization, all parties benefitted by avoiding $1 billion in site remediation costs (Department of Energy, 1994). Stakeholders and regulators were convinced that the species of mercury present across much of the site was not a hazard. Furthermore, when presented with the options, these parties understood that spending the money to excavate all of the targeted sediment did not prove to have a significant cost benefit. Ultimately, the remedy required very small quantities of sediment to be excavated. The remediation was scheduled to be complete within 5 years of the record of decision (Department of Energy, 1994).

5. CONCLUDING STATEMENT

Brownfields are a negative influence on our society. Their physical appearance is contagious. Their contamination may be hazardous to citizens, and their blight has cost needless loss of precious greenfields. Bioremediation in its many different forms provides the balance between protection of human health and economic health. This balance is critical to widespread redevelopment of brownfields sites. Bioremediation is also one of very few alternatives that can truly bring brownfields back into productive use without delay. However, the most essential realization in addressing brownfields is that no single party or remedial tool can mitigate these sites unilaterally. It will require the commitment and resources of all factions of our society.

REFERENCES

Abou-Rizk, J.A., M.E. Leavitt, and D.A. Graves. 1995. "In Situ Aquifer Bioremediation of Organics Including Cyanide and Carbon Disulfide", *in* R. Hinchee ed., *In Situ and On-Site Bioreclamation*. Battelle Press, Columbus, Ohio.

Abou-Rizk, J.A. and M.E. Leavitt. in press. "Natural Attenuation and A Microbial Fence: Re-engineering Corrective Action", in Proceedings of In Situ and On-Site Bioremediation, Battelle Press.

Butler, EL, RC Prince, GS Douglas, T Aczel, CS Hsu, MT Bronson, JR Clark, JE Linstrom, WG Steinhauer. "Hopane, a New Chemical Tool for Measuring Oil Biodegradation" in: RE Hinchee and RF Olfenbuttel, eds., *On-Site Bioreclamation*, Battelle Press.1994. pp 515–522.

Department of Energy. 1994. Feasibility Study for the East Fork Poplar Creek – Sewer Line Beltway. DOE/OR/02-1185&D2

Lee, M.D. , JM Thomas, RC Borden, PB Bedient, JT Wilson, CH Ward. 1988. Biorestoration of Aquifers Contaminated with Organic Compounds. *CRC Critical Reviews in Environmental Control,* Vol 18(1) pp 29–89. CRC Press, Inc.

Loehr, Raymond C. 1996. The Environmental Impact of Soil Contamination: Bioavailability, Risk Assessment, and Policy Implications. Reason Foundation and the National Environmental Policy Institute. Policy Study No. 211.

Vogel, T., and P. McCarthy. 1985. Biotransformation of tetrachloroethylene to trichloroethylene, dichloroethylene, vinyl chloride and carbon dioxide under methanogenic conditions. *Appl. Environ. Microbiol.*, 49:1080–1083.

Wackett, L.P. and D.T. Gibson. 1988. Degradation of trichloroethylene by toluene dioxygenase in whole cell studies with P. putida F1. *Appl. Env. Microbiol.* 54:1703–8.

Wackett, L.P., G.A. Brusseau, S.R. Householder, R.S. Hanson. 1989. Survey of microbialoxygenases: Trichloroethylene degradation by propane oxidizing bacteria. *Appl. Environ. Microbiol.* 55:2960–64.

Wilson, J.T. and B.H. Wilson. 1985. Biotransformation of trichloroethylene in soil. *Appl. Env. Microbiol.* 49:242–43.

28

BIOTREATABILITY KINETICS

A Critical Component in the Scale-up of Wastewater Treatment Systems

C. P. Leslie Grady, Jr., Shawn M. Sock, and Robert M. Cowan

Environmental Systems Engineering
L. G. Rich Environmental Research Laboratory
Clemson University
Clemson, South Carolina 29634-0919

1. INTRODUCTION

The design of a biological wastewater treatment facility requires many decisions, and the procedures for making most of them are well established from decades of experience, particularly for domestic wastewater treatment. Consequently, scale-up of those facilities is seldom an issue. Heuristic guidelines help to ensure that the final facility will function in the manner planned. The treatment of industrial wastewaters, or ex situ treatment of contaminated groundwaters, is more complicated and subject to more uncertainty. This difference follows from the nature of the organic contaminants being destroyed. Domestic wastewater contains primarily biogenic organic material which is easily biodegradable. Consequently, the kinetics of biodegradation is seldom an issue and design is primarily based on stoichiometry (Daigger and Grady, 1995). Industrial wastewaters and contaminated groundwaters, on the other hand, routinely contain synthetic organic chemicals (SOCs), and their removal to very low levels is usually the objective of the treatment process. In this situation, the kinetics of biodegradation is often the issue around which the entire design process is centered, requiring the translation of information collected in the lab to use in a full-scale facility. Since the lab data may have been collected in systems a few liters in size, whereas the full-scale facility may be several thousand cubic meters in size, the question of how best to move from one to the other is quite important. In other words, how do we use lab data to ensure that a full-scale system will work as planned? This chapter will address this question by reviewing an example.

The need to remove specific SOCs to very low levels provides a major challenge to design engineers. From a fundamental scientific perspective, there are many questions associated with the biodegradation of an individual organic chemical in a complex mixture. That knowledge base is still evolving, but luckily, it appears that for systems designed

Biotechnology in the Sustainable Environment, edited by Sayler *et al.*
Plenum Press, New York, 1997

with low net specific growth rates, it is possible to use single substrate biodegradation kinetics to depict a compound's removal (Ellis et al., 1995). This simplifies the design task to a great degree. Complicating the task, however, is the uncertainty associated with the input concentrations of the SOCs of interest, particularly in industrial wastewaters. Thus, from a pragmatic point of view, the challenge is to make the system robust enough to handle frequent input perturbations, without making it so large and so complex that it will be uneconomical or difficult to operate. The designer is also faced with long-term uncertainty about both the input levels of the SOCs and the output criteria that they will face in the future. As a result of all of this uncertainty, designers tend to be very conservative in their approach, particularly when they consider the consequences to the environment, their clients, and themselves of an inadequate system.

Conservatism, in turn, influences design. For example, many systems employ completely mixed bioreactors because of their ability to dampen input variability by immediate dilution throughout the entire volume. Such a configuration also maintains all organic constituents at low concentration, helping to ensure that simultaneous biodegradation occurs. There are disadvantages associated with the configuration, however, among them being the fact that the concentrations in the bioreactor are also the concentrations in the effluent. As a consequence, when the bioreactor concentration rises in response to a perturbation, the effluent concentration also rises. Another tendency is to design systems with long hydraulic retention times (HRTs), i.e., with large system volumes relative to the flow rate, in order to maximize dilution of input perturbations. This increases system cost. Finally, many systems are designed with long solids retention times (SRTs). This has two positive impacts, but also a negative one. Because the bacterial specific growth rate is inversely proportional to the SRT, maintenance of a long SRT maintains the bacterial growth rate low, thereby providing a low substrate concentration and ensuring that simultaneous substrate removal will occur. It also maintains a large mass of biomass in the system, which helps to minimize the impact of shock loads. At the same time, because the bacteria grow at a constant low specific rate, a large fraction of them is inactive and unable to respond quickly when perturbations occur.

In concert with conservative design, many engineers tend to take an empirical approach to problem solving. That is, they rely upon their experience. This is to be expected, given the consequences associated with failure of a design, but it limits innovation. The typical approach is to select several alternative systems that experience indicates will be capable of meeting the treatment requirements. These are then subjected to a detailed screening procedure, which may or may not involve kinetic studies. The outcome of the screening exercise is the selection of a limited number for detailed study during pilot-scale testing, after which the "best" is selected for full-scale implementation.

While the approach described above has been successful to a large degree, it has its disadvantages, chief among which is the discouragement of innovation. Because of the cost associated with pilot-scale studies, the number of alternatives that can be tried is limited, and the tendency is to stick with those that have been used before. Even that does not guarantee success, however, since the approach has also led to processes that did not meet expectation. There is a better way. It relies on a combination of approaches, beginning at the lab-scale and progressing through pilot-scale to full-scale design. The difference, however, lies in the number of alternatives that can be screened economically. This approach is based on the well-proven chemical engineering technique of reactor engineering. Basic studies in the lab establish the kinetics of biodegradation and those kinetic expressions are then used in mathematical models to study a broad range of process alternatives through computer simulation. Once the model has been formulated, a large number of alternative systems can be screened rapidly and economically, allowing the designer to use imagination and innovation with no risk and little cost. This provides a much broader base of alternatives, helping to ensure that the ones chosen for pilot-scale trials are those most likely to work well. This approach has been used successfully for a number of years for the design of biological nutrient removal systems, primarily because of the availability of consensus-based models for the processes (Henze et al., 1987, 1995). While similar models are not yet available for SOC removal, their development is a main focus of the environmental biotechnology research program at Clemson University. This chapter will demonstrate the advantages of the approach through presentation of a case study. The example comes from a study we performed on the removal of 1,4-dioxane from esterification wastewaters. More details can be found in Sock (1993) and Cowan et al., (1996).

2. THE PROBLEM

1,4-Dioxane is a cyclic diether that is a commonly used industrial solvent and is a byproduct of organic synthesis processes involving ethoxylation with ethylene oxide or the use of ethylene glycol as a feed stock. Because of the stability of the two ether bonds in its saturated heterocyclic structure, 1,4-dioxane is very difficult to biodegrade. In addition, because of its low volatility, it is not amenable to removal by stripping. It also has a very low octanol:water partition coefficient, and thus is not amenable to removal by sorption either. As a consequence, most of it passes through wastewater treatment processes that receive it. Some of it, however, weakly sorbs to the biomass so that if the waste biomass is disposed of in a landfill, the potential exists for desorption with subsequent groundwater contamination. Although such a practice has been discontinued, this was done in the past, leading to some contamination. The major concern with discharge of 1,4-dioxane is that it is a suspected human carcinogen and will probably be regulated as such in the future. In the meantime, NPDES permits require it to be removed to low levels from wastewaters and various regulatory agencies are also requiring its removal from contaminated groundwater. Because of the potential problems associated with it, corporations with 1,4-dioxane in their wastewaters are investigating modifications to their production processes to eliminate it. Until that can be accomplished, however, they want to remove it from their wastewaters through treatment, not just to meet regulatory requirements but because of their corporate environmental policies. Therefore a study was undertaken at Clemson University to investigate alternative means of accomplishing this.

Even though 1,4-dioxane is known to be difficult to biodegrade, there was a limited amount of evidence in the literature when this study was started to suggest that it could be transformed biologically, particularly by cultures capable of degrading tetrahydrofuran (THF) and morpholine (Bernhardt and Diekmann, 1991; Dmitrenko et al., 1987). In addition, there was evidence of seasonal removal by the activated sludge systems at some industrial facilities that appeared to be due to biological activity. Consequently, biodegradation was one process option chosen for study and this paper is limited to it. A major question during our study was whether bacteria were capable of using 1,4-dioxane as a sole carbon and energy source or whether transformation occurred only by gratuitous metabolism. Since the completion of our work reports have appeared in which the biodegradability of 1,4-dioxane by mineralization was demonstrated (Burback and Perry, 1993; Parales et al., 1994, and Roy et al., 1994).

Several objectives were established for the portion of the study considered here. First, we needed to determine if 1,4-dioxane could serve as the sole carbon and energy source for bacterial growth. If it could, we needed to establish an enrichment culture that could be used for kinetic studies and as a seed culture for future experimental investigations. After determining the kinetics of biodegradation it was necessary to conduct reactor engineering studies using computer simulation to establish possible reasons for the lack of significant 1,4-dioxane biodegradation in existing wastewater treatment facilities. That knowledge was then used to guide other reactor engineering studies from which to propose process flow schemes that could be used to remove 1,4-dioxane from wastewaters biologically. Others (CH2M HILL, 1994) tested a limited set of those flow schemes at pilot scale. By presenting this case study we hope to demonstrate how experimental investigations of biodegradation kinetics can be combined with computer simulation studies to solve difficult environmental problems and facilitate the scale-up of processes based on environmental biotechnology.

3. MINERALIZATION OF 1,4-DIOXANE

The question of whether an SOC can serve as sole carbon and energy source for microbial growth is very significant to the design of a biological process for removing it from a wastewater. If an SOC can serve as a growth substrate, then its biodegradation will be growth associated and the mass of bacteria in the treatment system capable of degrading it will be proportional to its concentration in the wastewater. Under those circumstances, a treatment system can be designed using the conventional equations for biodegradation (Grady and Daigger, 1996). Conversely, if an SOC is transformed by gratuitous metabolism, the mass of enzyme(s) involved in the transformation will be independent of the mass of SOC being transformed and will be determined by the mass of the inducer for the enzyme(s) entering the bioreactor. In that case, the inducer must be identified and a strategy must be devised for controlling its input, thereby controlling the concentration of the bacteria producing the enzyme(s). This is a difficult task and the techniques for establishing a design based on gratuitous metabolism are not well established. Consequently, the entire direction of the design task depended on establishing whether 1,4-dioxane could serve as the sole carbon and energy source for microbial growth.

Two bioreactors were started using activated sludge from a number of wastewater treatment plants receiving 1,4-dioxane. One of the bioreactors employed a suspended growth culture and was operated as a sequencing batch reactor. The other employed a submerged attached growth culture and was operated in a continuous manner. Both bioreactors received a complex organic feed containing many biogenic organic compounds plus 10 mg/L each of 1,4-dioxane and THF. Complete removal of both SOCs (the detection limit on 1,4-dioxane was 1.0 milligrams per liter or mg/L) was achieved in the attached growth reactor by the fourth week of operation so their concentrations were doubled three times, until by the seventh week the attached growth bioreactor was receiving a feed containing 80 mg/L of each. Nevertheless, by week ten complete removal was again achieved and the bioreactor operated stably until week 23. The suspended growth bioreactor required a longer time for development of the community, perhaps because microbial retention was less effective, and 20 weeks were required before complete removal was achieved when the feed concentration was 80 mg/L. During week 23 the 1,4-dioxane concentration was increased to 130 mg/L and the THF concentration to 108 mg/L while the sugars and half of the proteins and organic acids were removed from the biogenic portion of the feed. Nevertheless, the attached growth reactor continued to achieve complete removal of both compounds. The suspended growth reactor showed a transient increase in

1,4-dioxane concentration, but returned to complete removal by week 25. During week 25 all biogenic organic compounds were removed from the feed to the attached growth bioreactor and removal of the two SOCs continued unabated. During week 27 the 1,4-dioxane concentration was increased to 150 mg/L, while the THF concentration in the feed was reduced to 90 mg/L, with no measurable effect on their removal. Finally, on week 29 all THF was eliminated from the feed and the attached growth bioreactor was operated on 1,4-dioxane alone for five more weeks. Following a transient rise in concentration to 10 mg/L during week 30, the effluent 1,4-dioxane concentration dropped below the detection limit again and remained there until the bioreactor was shut down at the end of week 34. The suspended growth bioreactor did not do as well and following the removal of the biogenic substrates during week 27, significant discharge of 1,4-dioxane occurred until the bioreactor was shut down during week 34.

The ability of the attached growth bioreactor to remove 1,4-dioxane for five weeks while it was supplied as the sole substrate provided strong evidence that 1,4-dioxane could serve as a growth substrate. However, the evidence was not conclusive because of the large amount of biomass retained in the system. In order to obtain more conclusive evidence, suspended growth batch reactors were seeded from the attached growth continuous bioreactor during the thirteenth week and studies on them were run in parallel with the continuous bioreactors. These cultures consistently removed 1,4-dioxane without having other carbon or energy sources present and without the accumulation of soluble microbial products. Because of the suspended nature of the culture, it could not have been maintained if it had not been growing on the 1,4-dioxane. Thus it was concluded that 1,4-dioxane could serve as a sole carbon and energy source for microbial growth. This conclusion was confirmed by kinetic experiments based on microbial growth.

The culture degrading 1,4-dioxane was a complex bacterial community with several bacterial genera present. However, despite repeated attempts, we were unable to isolate a pure culture capable of growth on 1,4-dioxane alone, nor were we able to reconstruct the mixed culture from the isolated organisms (Morin, 1995).

4. KINETICS OF 1,4-DIOXANE BIODEGRADATION

4.1. Determination of Biodegradation Kinetics

Because the biodegradation of 1,4-dioxane was growth-associated, the kinetics of its biodegradation could be determined by biomass growth, substrate utilization, or oxygen consumption (Dang et al., 1989). Kinetic experiments were conducted using both of the latter two techniques and the results were in agreement. However, because growth was relatively slow, more use was made of substrate depletion experiments than of respirometry. From these studies it was clear that 1,4-dioxane biodegradation followed Monod kinetics with no indication of either substrate or product inhibition at 1,4-dioxane concentrations up to 2,100 mg/L. Consequently, the rate of biomass growth (r_X) could be represented as:

$$r_X = \mu \cdot X \tag{1}$$

and the rate of 1,4-dioxane removal (r_S) as:

$$r_S = -q \cdot X \tag{2}$$

where:

$$\mu = \mu_{max} \cdot S/(K_S + S) \tag{3}$$

and

$$q = \mu/Y \tag{4}$$

In these equations, X is the 1,4-dioxane-degrading biomass concentration (mg/L), μ is the specific growth rate coefficient (h^{-1}), μ_{max} is the maximum specific growth rate coefficient (h^{-1}), K_S is the half-saturation coefficient (mg 1,4-dioxane/L), S is the 1,4-dioxane concentration (mg/L), q is the specific substrate removal rate (mg 1,4-dioxane/(mg biomass·h)), and Y is the biomass yield (mg biomass formed/mg 1,4-dioxane used). Consequently, quantification of the biodegradation kinetics required estimation of the coefficients μ_{max}, K_S, and Y.

During preliminary studies it was discovered that growth of the culture on 1,4-dioxane appeared to be particularly sensitive to temperature. Consequently, a series of substrate removal experiments was conducted at a variety of temperatures to quantify the temperature effects. The values of the kinetic coefficients are shown in Table 1 and the relationship between the substrate concentration and the specific growth rate resulting from the interactions between μ_{max} and K_S (i.e., Eq. 3) are shown in Figure 1.

4.2. Implications of Kinetic Studies

The strong temperature dependency of the coefficients in Table 1 suggested a possible explanation for the failure of conventional wastewater treatment systems to achieve significant removal of 1,4-dioxane. It seemed likely that the 1,4-dioxane-degrading bacteria simply could not grow rapidly enough to stay in the bioreactor, particularly in the winter. Using the coefficients for 25°C, the lowest temperature studied, the minimum SRT for biomass growth was calculated to be on the order of 20 days. At any SRT less than that, the culture would be washed out of the bioreactor because it could not grow fast enough to maintain itself. The temperature of most activated sludge systems drops below 20°C in the winter, and at that temperature the minimum SRT would be even longer. Since it is uncommon for activated sludge systems to have SRTs in excess of 20 days, it is very likely that the primary reason little removal of 1,4-dioxane is seen in most activated sludge systems is that the SRT is simply not long enough to maintain a culture capable of its biodegradation. The kinetic coefficients reported in Table 1 are after improvement of the culture following long-term continuous culture. The kinetics of the original culture were even slower than values in Table 1 suggest (Sock, 1993), making it even more likely that the failure of activated sludge systems to remove 1,4-dioxane is kinetically-based and is not

Table 1. Effect of temperature on 1,4-dioxane biodegradation kinetics

	Temperature			
Parameter	40°C	35°C	30°C	25°C
μ_{max}, h^{-1}	0.022	0.043	0.014	0.010
K_S, mg/L	0.30	1.04	9.93	13.51
Y, mg/mg	0.70	0.64	0.23	0.33

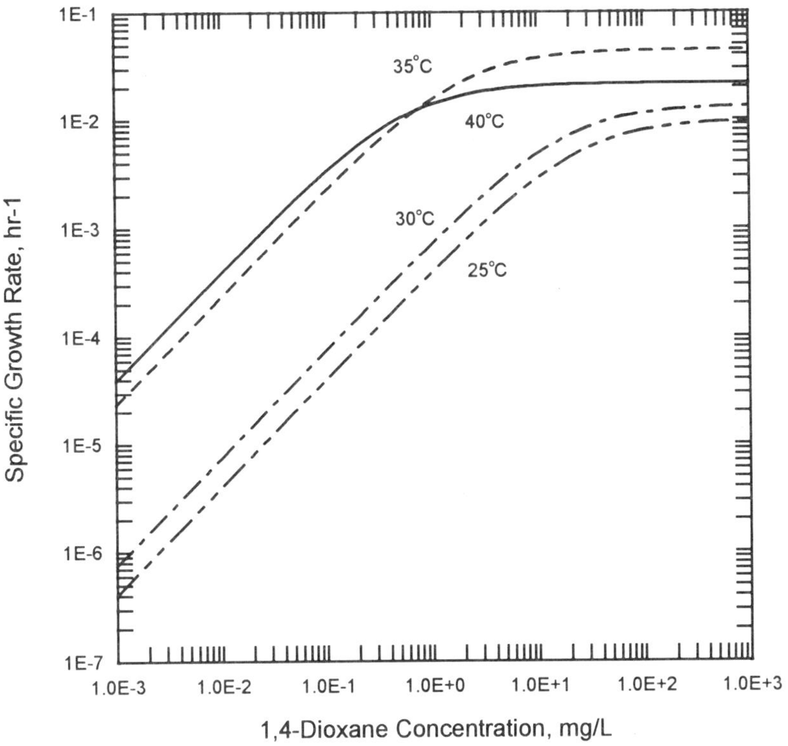

Figure 1. Effect of temperature on 1,4-dioxane biodegradation kinetics.

due to the inability of bacteria to degrade it or to any particular inhibitory or toxic characteristics.

Because the kinetics of 1,4-dioxane biodegradation are strongly dependent on temperature, the obvious question is whether it would be possible to achieve biodegradation to a sufficiently low concentration of 1,4-dioxane in a completely-mixed bioreactor by elevating the wastewater temperature. The reported coefficients can be used to determine this by using them to calculate the minimum attainable substrate concentration, S_{min}. It is the concentration of substrate in a bioreactor with an infinitely long SRT. Under that circumstance, no biomass is wasted and biomass is lost only by decay and other maintenance-related mechanisms (Grady and Lim, 1980). In other words, biomass growth is as slow as it can possibly be, and thus the substrate concentration is as low as it can get while maintaining a steady state in a completely mixed bioreactor. The value of S_{min} at each temperature was calculated as:

$$S_{min} = K_S \cdot b/(\mu_{max} - b) \tag{5}$$

where b is the decay coefficient, which reflects maintenance energy needs. Typical values of b for slowly growing cultures were assumed and the resulting minimum 1,4-dioxane concentrations were calculated. Figure 2 shows the results. A horizontal line is drawn at a concentration of 0.10 mg/L because it represents the highest likely allowable effluent concentration. Even at that concentration, however, it can be seen that the temperature would have to be elevated to at least 36°C to achieve it at an infinite SRT. Since any practical

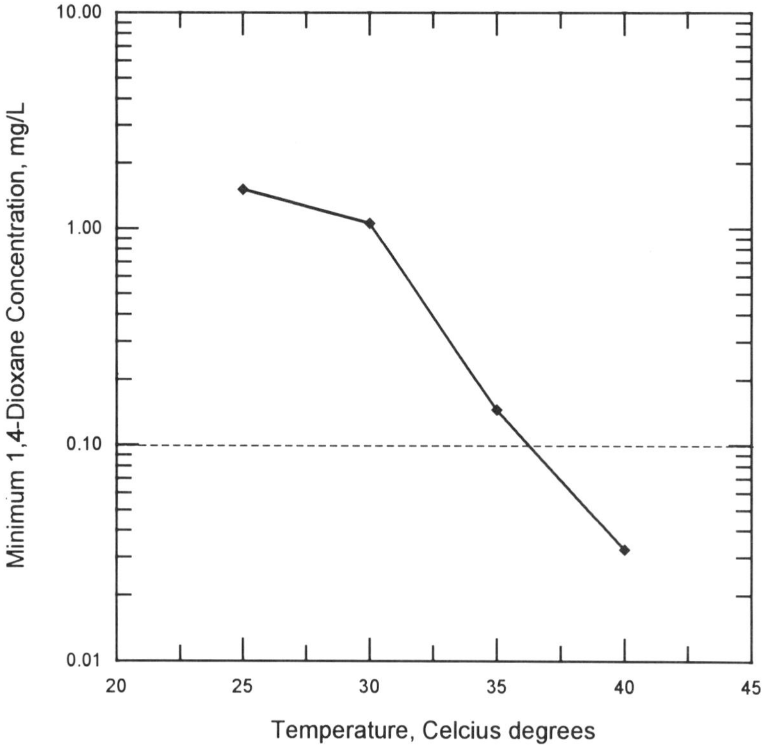

Figure 2. Effect of temperature on the minimum attainable 1,4-dioxane concentration from a single completely-mixed bioreactor.

bioreactor will have an SRT on the order of 15 days or so, the results shown in Figure 2 demonstrated that a single completely-mixed bioreactor is not a feasible configuration. Thus, other configurations had to be investigated.

5. INVESTIGATION OF ALTERNATIVE BIOREACTOR CONFIGURATIONS

Having determined that a typical completely-mixed activated sludge system would be incapable of meeting expected maximum allowable effluent concentrations of 1,4-dioxane, theoretical modeling studies were conducted with the known kinetics of biodegradation to investigate potential alternative configurations. Two situations will be considered here: the use of completely-mixed tanks in series as the bioreactor configuration and the use of pretreatment of segregated 1,4-dioxane-containing streams.

5.1. Tanks-in-Series Configuration

An activated sludge system containing completely-mixed tanks-in-series with biomass recycle from a single final settler to the first bioreactor has several potential advantages. First, it will produce a lower effluent substrate concentration than a single bioreactor with the same SRT and HRT (Grady and Lim, 1980). In fact, there is no minimum attainable substrate concentration from such a configuration. Second, it will allow less substrate to be discharged in response to influent perturbations than a single completely-mixed bioreactor (Grady, 1971; Santiago and Grady, 1990). This follows from the time delays inherent in the configuration and the fact that the effluent concentrations reflect the concentrations in the last bioreactor of the chain. Such a configuration is particularly well suited to 1,4-dioxane biodegradation because 1,4-dioxane is not subject to substrate inhibition within the concentration ranges likely to be encountered (Sock, 1993).

To determine the impacts of a tanks-in-series configuration, the simplest possible system was investigated through simulation; one containing only two tanks in series with biomass recycle from the final settler to the first tank. To start with, a very short SRT of four days was assumed with a bioreactor temperature of 35°C so that a worst possible case could be visualized. If it could meet desired output concentrations, then more conservative designs would do even better and would provide a margin of safety. The first study investigated the effect of the relative sizes of the two bioreactors. The results are shown in Figure 3 where it can be seen that two reactors in series are clearly better than one, producing an effluent that is almost three orders of magnitude lower than that from a single bioreactor. Furthermore, even at an SRT of four days, the effluent 1,4-dioxane concentration would be lower than 1.0 micrograms per liter (μg/L) as long as 35 to 85% of the system volume was in the first tank. Consequently, using two or more bioreactors in series would make it possible to meet expected 1,4-dioxane limits without difficulty.

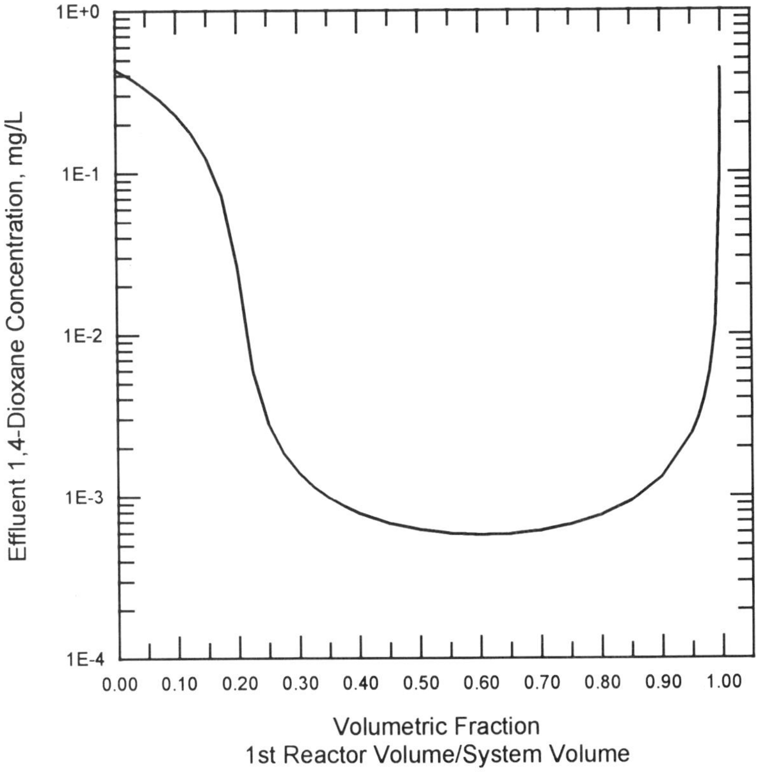

Figure 3. Effect of the relative sizes of two bioreactors in series on the effluent 1,4-dioxane concentration when the SRT = 4 days. Theoretical results from modeling studies using the kinetic parameters in Table 1 for 35°C.

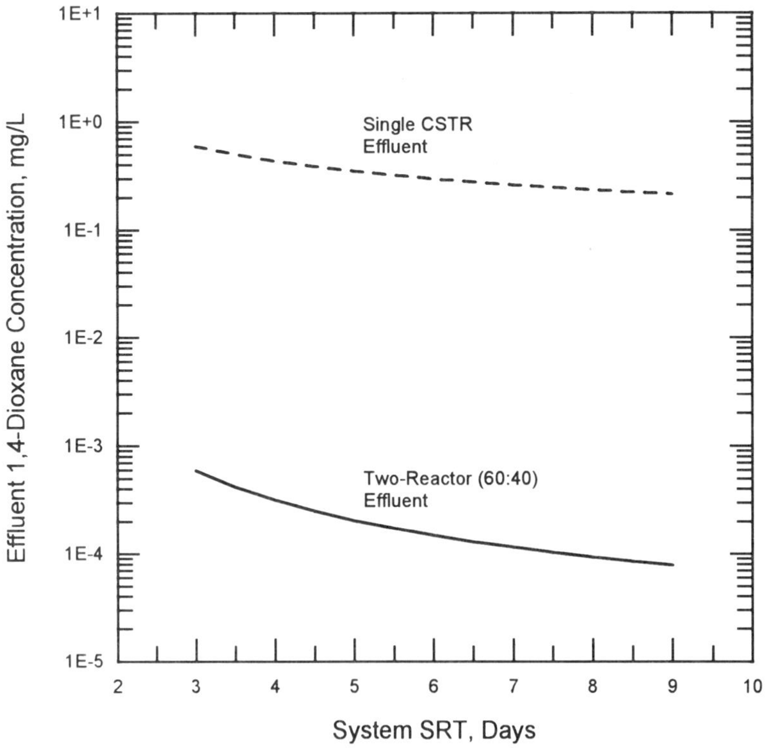

Figure 4. Effect of SRT on the effluent 1,4-dioxane concentration from two bioreactors in series when 60% of the system volume is in the first bioreactor. Theoretical results from modeling studies using the kinetic parameters in Table 1 for 35°C.

An SRT of four days was chosen for the first set of simulations in order to investigate a severe situation. It is unlikely that such a short SRT would be used in practice, however, because it would not be resilient enough to shock loads (Grady, 1971; Santiago and Grady, 1990). Consequently, the modeling study was extended to other SRT values. Because the combination of reactor volumes giving the lowest effluent 1,4-dioxane concentration was 60% in the first and 40% in the second, that combination was chosen for the SRT study. The results from that set of simulations is shown in Figure 4. There it can be seen that extending the SRT to nine days would allow the effluent 1,4-dioxane concentration to be reduced to less than 0.1 μg/L for that configuration. Thus, it was apparent that a tanks-in-series bioreactor configuration should be used for best removal of 1,4-dioxane and that benefits would accrue from the use of longer SRTs.

5.2. Pretreatment of 1,4-Dioxane Containing Streams

If most of the 1,4-dioxane in a wastewater originates in a few waste streams within the production facility, then it may be possible to segregate those streams and pretreat them separately prior to blending them with the principal wastewater flow for treatment in the main treatment plant. This concept is illustrated in Figure 5. Because it is impossible to segregate all 1,4-dioxane-containing wastewater streams, 1,4-dioxane would still enter the main bioreactor, but the mass flow rate would be reduced. Theoretical considerations suggest that two major benefits could accrue from such a practice.

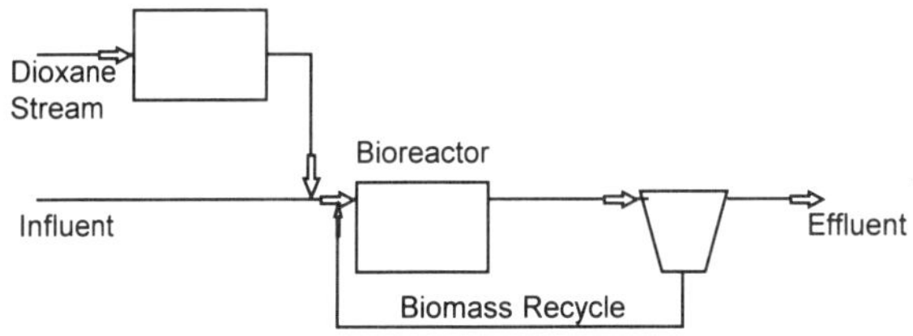

Figure 5. Schematic diagram of a treatment system incorporating pretreatment of 1,4-dioxane.

The first comes simply from the fact that pretreatment would reduce the mass of 1,4-dioxane entering the activated sludge system. Theoretically, this would have no beneficial effect on the activated sludge system if it contained only a single completely mixed bioreactor. In such a system, the effluent substrate concentration is independent of the influent concentration and is determined solely by the system SRT (Grady and Lim, 1980). However, pretreatment would have a very large benefit if the activated sludge system were configured in a tank-in-series arrangement. At the SRTs likely to be used in any real-world activated sludge system, the 1,4-dioxane concentration in all of the bioreactors would be low relative to K_S, making the 1,4-dioxane removal rate first order with respect to the 1,4-dioxane concentration. Under that circumstance, the fractional removal across the system will be independent of the influent concentration. This means that the effluent concentration would be reduced in direct proportion to the reduction in the influent concentration. Consequently, the more 1,4-dioxane that can be removed by pretreatment, the lower the 1,4-dioxane concentration from the activated sludge system. This suggests that pretreatment in combination with a tanks-in-series configuration for the activated sludge system would have very real benefits.

The benefit discussed in the preceding paragraph will accrue regardless of the type of pretreatment system employed. The second benefit, however, will only accrue if the pretreatment system is biological and the excess biomass from it is discharged to the activated sludge system. The simplest form of pretreatment would be to use a simple continuous stirred tank reactor (CSTR), with all biomass growth being discharged from it to the activated sludge system. Under that condition, the activated sludge system becomes a bioreactor that receives in the influent significant quantities of biomass capable of degrading a slowly degradable compound. The magnitude of the benefit to such a bioreactor depends on the mass input rate of capable biomass relative to the mass rate of capable biomass growth on the specific substrate actually applied to it (Grady and Daigger, 1996). An example of the magnitude of the benefit associated with such biomass addition can be seen by considering the reduction in S_{min} from a completely mixed activated sludge system, even though, as we have seen, such a configuration would not be used in this circumstance. In this case,

$$S_{min}^* = (1.0 - \alpha)S_{min} \tag{6}$$

where S_{min}^* is the minimum 1,4-dioxane concentration attainable with pretreatment, α is the fraction of the 1,4-dioxane removed by pretreatment, and S_{min} is the minimum 1,4-dioxane concentration attainable in the absence of pretreatment, as given by Eq. 5. Consideration of Eq. 6 shows that considerable additional benefit accrues from using a biological pretreatment system with discharge of the biomass to the main treatment system and this became an important consideration during process design.

6. PROPOSED TREATMENT SYSTEM

Several of the sponsors of our work are polyester fiber manufacturers which use activated sludge for wastewater treatment. They have taken various approaches to meet current standards for 1,4-dioxane removal, but they desire to achieve even better treatment for the reasons outlined earlier. One characteristic of the esterification process for polyester fiber production is that over 90% of the 1,4-dioxane comes from low-flow, high-concentration streams that can be segregated easily. The remainder is in low concentration streams and must be dealt with in the activated sludge process. As a consequence, based on the reactor engineering studies presented above, a treatment system was proposed that made use of biological pretreatment for the stream containing the bulk of the 1,4-dioxane, with discharge of the effluent and its associated biomass to an activated sludge system containing bioreactors in series.

The waste streams to be pretreated typically contain 1,4-dioxane at concentrations up to around 2,000 mg/L, which our kinetic studies showed had no inhibitory effects on the degrading biomass. However, the streams also contain around 25,000 mg/L of chemical oxygen demand (COD) due to acetaldehyde and ethylene glycol, making it impossible to pretreat them with only a simple aerobic CSTR. Consequently, based on other research done in our lab (Cowan et al., 1995), a two-stage pretreatment system was proposed, with an anaerobic first stage for removal of everything except the 1,4-dioxane. This would be followed by an aerobic second stage in the configuration of a completely-mixed bioreactor with biomass recycle. All biomass wastage would be discharged to the downstream activated sludge system. Because the 1,4-dioxane-containing streams were warm, both systems could be operated at 35°C without heat addition. In concert with the findings from the reactor engineering studies, the proposed activated sludge system for treatment of the main wastewater flow and the pretreated esterification wastewater was configured as three tanks-in-series with 50% of the system volume in tank 1 and 25% each in tanks 2 and 3 (CH2M HILL, 1994). It would be expected to operate at ambient temperature because of the large flow rates involved.

CH2M HILL (1994) tested the proposed flow scheme, as well as several others, in pilot scale on-site at a polyester fiber manufacturing facility. The system routinely reduced the 1,4-dioxane concentration to less than 40 μg/L while operating on real wastewater with all of the variability associated with it. The sponsor is currently considering installation of the process.

Although the modeling study had suggested that the output 1,4-dioxane concentration would be lower than that observed in pilot-scale, it must be borne in mind that the modeling was done with kinetic parameters obtained in the lab on a culture that was being exposed only to 1,4-dioxane. It is very likely that in the real wastewater, other constituents influenced the kinetics, making biodegradation less effective than when 1,4-dioxane was the only carbon source present. Our knowledge of microbial kinetics in complex chemical environments is not sufficiently advanced to allow highly accurate predictions of effluent concentrations of individual contaminants. It is sufficient, however, to allow the relative efficacy of alternative systems to be compared. The important point is that the reactor engineering studies allowed an understanding of the conditions required for effective biodegradation and led to a treatment system that gave consistently good removal of a constituent that had heretofore been considered to be nonbiodegradable. Furthermore, knowledge of the biokinetics allowed the screening of multiple alternatives, thereby allowing the pilot studies to be focused on only the most promising.

7. SUMMARY

The case study presented herein has demonstrated clearly that SOCs that are biodegraded very slowly can be effectively removed in properly configured treatment systems using biological processes. Identification of the best system configuration requires knowledge of the kinetics of biodegradation in a form that can be incorporated into mathematical models. That means that parameters such as μ_{max}, K_S, and Y must be quantified in such a way that they are representative of the culture in the treatment system. This can be done by lab-scale studies using a variety of techniques (Grady and Daigger, 1996). Lab-scale studies also offer an opportunity to prove the feasibility of a concept, or conversely, to show that a concept won't work, without huge investments of time and money. Reactor engineering via modeling provides a very economical way to screen a large number of alternatives once the concept has been confirmed. Currently, microcomputer-based generalized programs are available that can be used quickly and effectively to simulate the performance of biological nutrient removal systems for domestic wastewater treatment. A major focus of the biotechnology research effort at Clemson University is the development of a similar program for predicting the fate of SOCs in treatment systems because the availability of such a program would do much to encourage the use of reactor engineering during biological process design. Finally, the most promising alternatives must be tested at pilot-scale on real wastewater as an essential part of scale-up. This is necessary to ensure that the system is sufficiently robust to handle the unexpected events that are common during wastewater treatment, but which are very difficult to incorporate into models. By preceding pilot-scale studies by biokinetic studies, however, the pilot-scale studies can be focused on the alternatives most likely to succeed, thereby reducing costs and increasing effectiveness.

8. ACKNOWLEDGMENTS

The work at Clemson University reported herein was funded by a grant from the South Carolina Hazardous Waste Management Research Fund and a consortium of six companies concerned about the removal of 1,4-dioxane from wastewater and groundwater, including: The Dow Chemical Company, DuPont de Neumures, Hoechst Celanese Corporation, Eastman Chemical, Owens-Corning, and Wellman Industries. The views and opinions expressed are those of the authors and not necessarily those of the funding sources.

Robert M. Cowan is currently an Assistant Professor of Environmental Engineering in the Department of Environmental Sciences and Bioresource Engineering at Rutgers University. Shawn M. Sock is currently an environmental engineer with CH2M HILL in Atlanta, GA.

9. NOMENCLATURE

9.1. Abbreviations

CSTR - Continuous Stirred Tank Reactor
HRT - Hydraulic Retention Time
SOC - Synthetic Organic Chemical
SRT - Solids Retention Time
THF - Tetrahydrofuran

9.2. Symbols

b - Decay coefficient, T^{-1}
K_S - Half-saturation coefficient, ML^{-3}
q - Specific substrate removal rate, T^{-1}
r_S - Substrate removal rate, $ML^{-3}T^{-1}$
r_X - Biomass growth rate, $ML^{-3}T^{-1}$
S - Substrate concentration, ML^{-3}
S_{min} - Minimum attainable substrate concentration from a CSTR
S_{min}* - Minimum attainable substrate concentration from a CSTR with pretreatment
X - Biomass concentration, ML^{-3}
Y - Biomass yield, M biomass (M substrate)$^{-1}$
α - Fraction of substrate removed by pretreatment
μ - Biomass specific growth rate, T^{-1}
μ_{max} - Biomass maximum specific growth rate coefficient, T^{-1}

REFERENCES

Bernhardt, D. and Diekmann, H. (1991). "Degradation of 1,4-dioxane, tetrahydrofuran, and other cyclic ethers by an environmental *Rhodococcus* strain," *Applied Microbiology and Biotechnology*, **36**, 120–123.

Burback, B. L. and Perry, J. J. (1993). "Biodegradation and biotransformation of groundwater pollutant mixtures by *Mycobacterium vaccae*," *Applied and Environmental Microbiology*, **59**, 1025–1029.

CH2M HILL (1994), *1,4-Dioxane Treatment Guidance Document*, Report to Hoechst Celanese Corporation, Charlotte, NC.

Cowan, R. M., Krishnan, H. and Grady, C. P. L. Jr. (1995). "Anaerobic pretreatment of a synthetic industrial wastewater in a fluidized bed biological reactor," *Proceedings of the 49th Industrial Waste Conference, May 1994, Purdue University*, R. F. Wukasch, Editor, Lewis Publishers, Chelsea, MI. pp 543–554.

Cowan, R. M., Sock, S. M., and Grady, C. P. L. Jr. (1996). "Evaluation of biodegradation as a means of removing 1,4-dioxane from industrial wastewaters and contaminated groundwaters," (in preparation).

Daigger, G. T. and Grady, C. P. L. Jr. (1995). "The use of models in biological process design," *Proceedings of the Water Environment Federation 68th Annual Conference and Exposition, Volume 1, Wastewater Treatment Research and Municipal Wastewater Treatment*, pp 501–510.

Dang, J. S., Harvey, D. M., Jobbágy, A., and Grady, C. P. L. Jr. (1989). "Evaluation of biodegradation kinetics with respirometric data," *Research Journal of the Water Pollution Control Federation*, **61**, 1711–1721.

Dmitrenko, G. N., Gvozdyak, P. I., and Uoda, V. M. (1987). "Selection of destructor microorganisms for heterocyclic xenobiotics," *Soviet Journal of Water Chemistry and Technology*, **9**, 77–81.

Ellis, T. G., Smets, B. F., and Grady, C. P. L. Jr. (1995). "Influence of simultaneous multiple substrate biodegradation on the kinetic parameters for individual substrates," *Proceedings of the Water Environment Federation 68th Annual Conference and Exposition, Volume 1, Wastewater Treatment Research and Municipal Wastewater Treatment*, pp 167–178.

Grady, C. P. L. Jr. (1971). "A theoretical study of activated sludge transient response", *Proceedings of the 26th Industrial Waste Conference, May 1971, Purdue University*, pp 318–335.

Grady, C. P. L. Jr. and Daigger, G. T. (1996). *Biological Wastewater Treatment: Theory and Applications*, 2nd Edition, Marcel Dekker, Inc., New York, NY, (in preparation).

Grady, C. P. L. Jr. and Lim, H. C. (1980). *Biological Wastewater Treatment: Theory and Applications*, Marcel Dekker, Inc., New York, NY.

Henze, M., Grady, C. P. L. Jr., Gujer, W., Marais, G. v. R., and Matsuo, T. (1987). *Activated Sludge Model No. 1*, Scientific and Technical Report No. 1, International Association on Water Pollution Research and Control, London, England.

Henze, M., Gujer, W., Marais, G. v. R., Matsuo, T., Mino, T., and Wentzel, M. C. (1995). *Activated Sludge Model No. 2*, Scientific and Technical Report No. 3, International Association on Water Quality, London, England.

Morin, M. D. (1995). *Degradative Characterization of a Mixed Bacterial Culture Capable of Mineralizing 1,4-Dioxane and Preliminary Identification of Its Isolates*, Master of Science Thesis, Microbiology, Clemson University, Clemson, SC.

Parales, R. E., Adamus, J. E., White, N., and May, H. D. (1994). "Degradation of 1,4-dioxane by an actinomycete in pure culture," *Applied and Environmental Microbiology*, **60**, 4527–4530.

Roy, D., Anagnostu, G., and Chaphalka, P. (1994). "Biodegradation of dioxane and diglyme in industrial waste," *Journal of Environmental Science and Health*, **A29**, 129–147.

Santiago, I. and Grady C. P. L. Jr. (1990). "Simulation studies of the transient response of activated sludge systems to biodegradable inhibitory shock loads", Proceedings of the 44th Industrial Waste Conference, May 1989. Purdue University, J. M. Bell, Editor, Lewis Publishers, Chelsea, MI, pp 191–198.

Sock, S. M. (1993). *A Comprehensive Evaluation of Biodegradation as a Treatment Alternative for Removal of 1,4-dioxane*, Master of Science Thesis, Environmental Systems Engineering, Clemson University, Clemson, SC.

29

MOLECULAR ANALYSIS AND CONTROL OF ACTIVATED SLUDGE

C. A. Lajoie,[1] A. C. Layton,[1] R. D. Stapleton,[1] I. R. Gregory,[1] A. J. Meyers,[2] and G. S. Sayler[1]

[1]University of Tennessee
Center for Environmental Biotechnology
676 Dabney Hall
Knoxville, Tennessee 37996-1605
[2]Tennessee Eastman Division
Eastman Chemical Company, P.O. Box 511
Kingsport, Tennessee 37662

1. INTRODUCTION

1.1. The Activated Sludge Process

Activated sludge is a common biological treatment method for both industrial and domestic wastewaters. It is often used for the treatment of waste streams that contain readily biodegradable substances that otherwise would exert a significant oxygen demand on receiving waters. Activated sludge is also used for the treatment of waste streams that contain potential environmental contaminants that are toxic to human or aquatic life.

The major components of activated sludge plants are the aeration basins, clarifiers and sludge dewatering equipment (Figure 1). In the aeration basin soluble and particulate contaminants are converted to carbon dioxide, water and microbial biomass. Some contaminants may be removed by adsorption to microbial flocs. The microbial biomass is removed from the waste stream by gravity settling in clarifiers. The clarified effluent is discharged to adjacent receiving waters such as streams or rivers. A portion of the settled sludge is recycled to the aeration basin. Sludge recycling effectively maintains a high microbial biomass in the aeration basin in the presence of high hydraulic loads, and selects for microbial populations which grow as settleable flocs (Figure 2). The remaining sludge is dewatered, and then incinerated, composted or landfilled. Municipal wastewater treatment plants also have settling basins upstream of the aeration basin. This is referred to as primary treatment. Secondary treatment consists of the aeration basins and clarifiers. Tertiary treatment consists of additional capabilities for nitrogen and phosphorus removal.

Activated sludge process control is directed toward maintaining conditions in the plant necessary for rendering the wastewater suitable for discharge to local receiving wa-

Biotechnology in the Sustainable Environment, edited by Sayler *et al.*
Plenum Press, New York, 1997

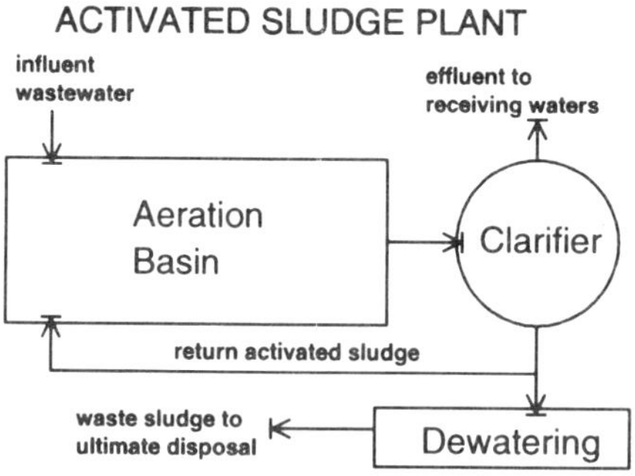

Figure 1. Schematic diagram of activated sludge secondary treatment processes.

ters. Discharge of a low-quality effluent may result in environmental problems in the receiving waters. There are many factors in the design and control of activated sludge plants that affect effluent quality. Operational flexibility is achieved primarily through changes in the rate of return of activated sludge to the aeration basin. An increased return activated sludge flow rate increases the amount of biomass in the aeration basin (mixed-liquor suspended solids), thereby decreasing the ratio of incoming organic substrates to cell mass (food/mass ratio). This is most commonly expressed as sludge age. Changes in sludge age can have considerable effects on both sludge and effluent quality. Other parameters that can be adjusted include aeration rate, flow patterns, and nutrient additions.

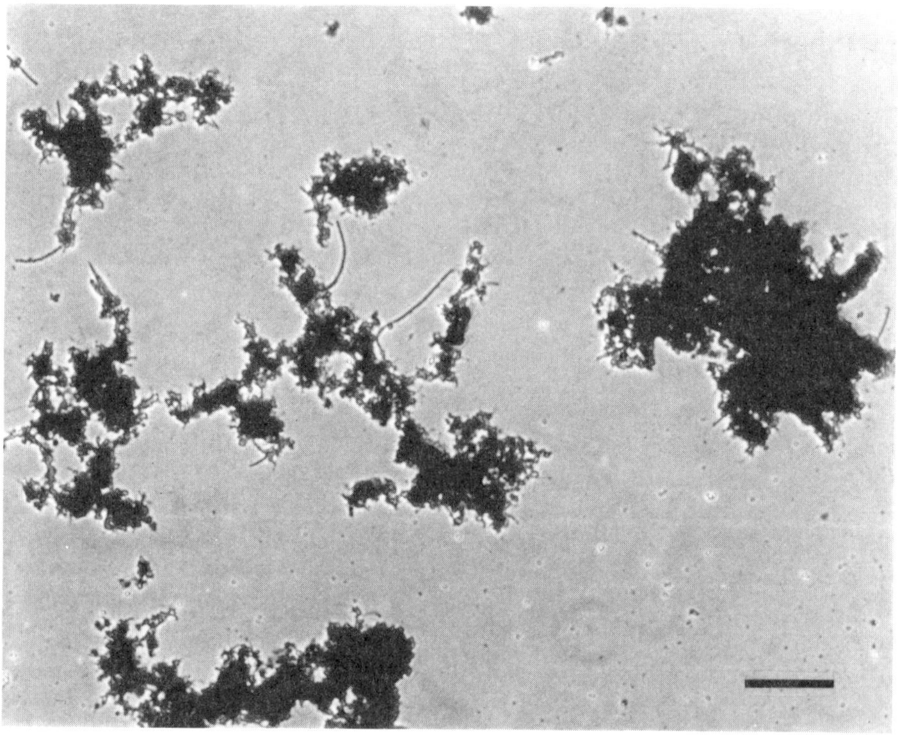

Figure 2. Microscopic appearance of activated sludge flocs (200X phase contrast; bar = 50 μm).

It is common for industrial activated sludge plants to have more operational flexibility than municipal plants.

Although the activated sludge process has been in use for many years, and is the most widely used secondary wastewater treatment process in the world, upsets in plant performance remain quite common (Jenkins et al., 1993). Activated sludge process problems can be divided into two categories. General plant upsets can result in problems with sludge settling, compaction and dewatering, and consequent problems in effluent quality and sludge disposal. These upsets have been variously categorized as filamentous bulking, non-filamentous, zoogloeal or viscous bulking, pin-floc, foaming, dispersed growth, and rising sludge (Wanner, 1994). The causative factors are not completely understood, but tend to relate to process configuration, wastewater loading, macronutrient concentrations, wastewater composition, and/or dissolved oxygen concentrations, and their selective effects on the microbial community. The second class of problems relates more to biodegradation of specific wastewater components and their potential effects on the quality of receiving waters or human health. An example of this was the foaming observed in receiving waters before the advent of biodegradable surfactants (Mitchell, 1974). A more recent case involves the accumulation in sludge of nonylphenols derived from the partial degradation of nonylphenolethoxylate surfactants (Giger et al., 1981, 1984). Nonylphenols are toxic to aquatic life (Patoczka and Pulliam, 1990; Dorn et al., 1993). Nonylphenols also are reported to have estrogenic activity, mimicking the effect of normal estrogens in humans and other species, resulting in reproductive abnormalities (Hileman, 1994; Soto et al., 1991; White et al., 1994).

Although a considerable amount of effort has been dedicated to activated sludge process engineering analysis, surprisingly little is known about the community composition, degradative activities, and metabolic strategies of the component microorganisms. This is due to the complexity of the activated sludge community, the difficulty of simulating the activated sludge environment at the test tube or bench-scale level, and the limitations of traditional microbiological methods. Microscopic identification of filamentous organisms and observations of general floc characteristics have been useful in diagnosing the causes of some process problems (Jenkins et al., 1993). However, upsets in process operations still remain quite common in both domestic and industrial activated sludge plants. Corrective actions such as chlorination of sludge to kill filamentous organisms or synthetic polymer addition to enhance settling in clarifiers or sludge dewatering are relatively crude and often expensive. The fate of specific industrial chemicals cannot always be adequately predicted. Nor can the responsible degradative populations be monitored. These uncertainties can result in expensive corrective actions, regulatory problems, and redesign of consumer products. Continued population growth, industrial expansion, and more stringent effluent quality standards are expected to put increasing demands on existing wastewater treatment plants. The acceptable margin of error in plant operations may decrease accordingly. A deeper understanding of activated sludge microbial communities may result in more effective and dynamic process control.

Recent developments in molecular methods for analysis of microbial communities have sparked a renewed interest in the microbiology of activated sludge. Whereas traditional approaches may have reached the point of diminishing returns, molecular analysis has the potential to increase our understanding of the activated sludge process, and consequently, improve process control. In this chapter the underlying basis of molecular approaches to activated sludge analysis will be discussed. The potential applications of molecular diagnostics will be examined in the context of traditional approaches to activated sludge community analysis.

1.2. The Activated Sludge Community

Activated sludge consists of a diverse community of protozoa, invertebrates, fungi and bacteria. The microbial community is limited by organic carbon, and consequently, competition for organic substrates is the primary selective pressure. The two major groups comprising activated sludge are the floc-forming (non-filamentous) and filamentous bacteria. Whereas the floc-forming organisms produce the exopolysaccharide responsible for floc formation, it has been suggested that the filamentous organisms within the flocs provide secondary structure (Sezgin et al., 1978). The examples of floc-forming and filamentous organisms that are commonly described are *Zoogloea ramigera* and *Sphaerotilus natans*, respectively.

Activated sludge process problems can often be traced to the microbial community. For example, excessive growth of filamentous bacteria is responsible for filamentous bulking. Due to the extended nature of the filaments protruding from the flocs, the sludge does not settle readily nor compact well in the clarifier. Eventually the sludge blanket in the clarifier overflows resulting in a marked deterioration in effluent quality. In viscous or zoogloeal bulking, excessive exopolysaccharide production by the floc-forming organisms results in a highly water-retentive sludge which settles and compacts poorly and is difficult to dewater. Pin-flocs are very small flocs which result from a weak floc structure. This can result from toxic shock or mechanical shearing of the flocs. The small flocs do not settle effectively in the clarifier, resulting in a turbid effluent. In dispersed growth the microorganisms grow as individual cells rather than as flocs and are lost from the system (Jenkins, 1992). The settleability of the activated sludge is commonly determined by placing a sample in a graduated cylinder and allowing it to settle for 30 minutes. The volume of settled sludge, expressed as milliliter per gram (ml/g) dry weight suspended solids, is refered to as the sludge volume index (SVI). Problems with sludge quality are often indicated by the visual appearance of the settled activated sludge in the SVI determination (Figure 3).

Diagnosis of activated sludge problems consists of determining the conditions selecting for the responsible microbial populations and making process changes that select against these populations. This can only be done effectively if the responsible organisms can be identified and their growth requirements or competitive metabolic strategies are known.

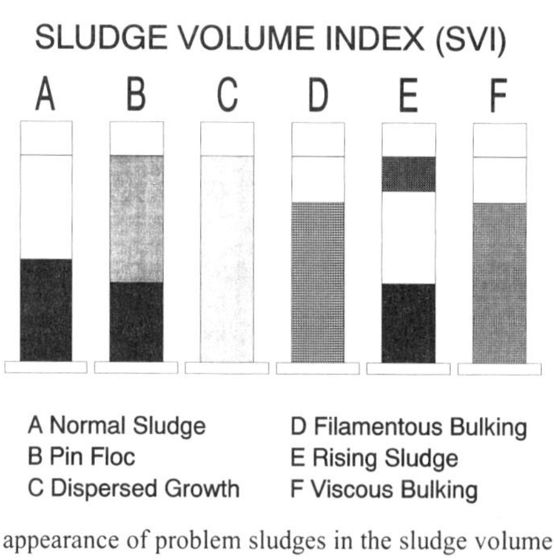

Figure 3. The appearance of problem sludges in the sludge volume index (SVI) test.

2. CURRENT APPROACHES TO ACTIVATED SLUDGE ANALYSIS

2.1. Traditional Microbial Systematics

Historically, the classification and identification of microorganisms has been very difficult. Unlike higher plants or animals, microorganisms have few distinctive morphological characteristics. Identification has been based primarily on shape (e.g., rod, coccus), cell wall type (gram-negative or gram-positive), and a host of biochemical tests (e.g., oxidase-positive, oxidase-negative) (Unz, 1984). Unlike higher organisms, the classification of microorganisms has not been effectively tied to their evolutionary relatedness as this is very uncertain when biochemical tests are used as the basis of species identification (Woese, 1987).

There are a number of serious limitations inherent in traditional approaches to microbial systematics. The series of biochemical tests needed to identify an unknown isolate are extensive, tedious, and often unreliable. A practical problem in the identification of organisms in activated sludge is that the majority of the biochemical tests can only be performed on pure cultures. This requires isolation and culture of the individual strains present in activated sludge, which is often not possible. The growth requirements of activated sludge microorganisms are not always known, and as indicated earlier, test tube simulation of the activated sludge environment is difficult. Although inoculation of activated sludge into many types of non-selective medium or plating of sludge on solidified medium will result in the growth of a variety of organisms, it is difficult to verify that these isolates are representative of the most numerous species in the treatment plant. It has been estimated that only 5–15% of the organisms present in activated sludge are culturable (Wagner et al., 1993). As the activated sludge environment is generally deficient in organic substrates, and the dissolved oxygen concentration is maintained at a moderate level, it is possible that many of the prevalent organisms are slow-growing, microaerophilic species. These types of organisms are difficult to culture in the laboratory, and consequently, many common species have probably never been characterized. A related difficulty is that effective process control requires real time or near real time analysis of problems with the activated sludge community. Isolation, cultivation and identification of the causative organisms by these traditional methods is frequently too slow to be effective as a diagnostic approach to process control.

The difficulties inherent in microbial identification are exemplified by the case of *Zoogloea ramigera*. Due to the lack of other distinctive traits, finger-like floc formation (zoogloeas) has been the primary basis for identifying *Zoogloea*. The original type strain for this genus was *Zoogloea* Itzigsohn 1868. This strain was not adequately preserved and was subsequently replaced by a new representative *Z. ramigera* I-16-M. This strain, however, does not produce the characteristic zoogloeas originally described by Itzigsohn. A new strain (*Z. ramigera* 106) was, therefore, proposed as the new type strain for this genus. There are three *Z. ramigera* strains commonly used in activated sludge research. Although they were all isolated from wastewaters and produce flocs, they possess other distinct phenotypic differences. As many different bacteria produce flocs depending on the culture medium, this is not a very distinctive trait on which to base a genus description (Shin et al., 1993). It is common for any floc-forming bacteria isolated from sludge to be referred to as *Zoogloea*. The term is now often used to refer to the flocs themselves or specific types of flocs (finger-like or amorphous), rather than to the actual organisms that produce the flocs. For many organisms commonly observed in activated sludge there are no type strains that can be used for comparative purposes, as these strains have never been successfully cultivated (Wanner and Grau, 1989).

2.2. Microscopic Analysis of Activated Sludge

A practical approach for diagnosing activated sludge problems via examination of the microbial populations began with a paper published by Eikelboom in 1975 entitled "Filamentous Organisms Observed in Activated Sludge" (Eikelboom, 1975). From observations of more than 1,000 wastewater treatment plant samples, 26 different filamentous organisms were described. This work was significant in that an identification key was developed in which all common filamentous organisms were included irrespective of whether the genera or species were known. Filament types are identified based on morphological and staining characteristics observable with a microscope. Although some strains have been given Latin names, most are given numerical codes associated with morphological or staining characteristics. Once a consistent method of identification was accepted by activated sludge researchers a sufficient data base ensued which allowed the development of correlations between filament types and plant operations conditions and problems (Strom and Jenkins, 1984, Blackbeard et al., 1986). This has provided the basis for in situ microscopic observation and analysis of activated sludge upsets.

This approach to diagnosing treatment plant problems has proven very effective for some types of major activated sludge problems. For example, if the major filament type observed during an episode of filamentous bulking is *Sphearotilus natans* or type 1701, it is likely that the problem is resulting from low dissolved oxygen in the aeration basin. Type 021N and *Thiothrix* spp. are characteristic of wastewaters in which nitrogen and phosphorus are growth-limiting. The presence of excessive zoogloeas or finger-shaped flocs is indicative of a high substrate to biomass ratio (high F/M). These three problems can be corrected by increasing aeration, addition of nitrogen and/or phosphorus, or increasing the rate of return of clarifier settled sludge to the aeration basin (raising sludge inventory), respectively (Jenkins et al., 1993; van Niekerk et al., 1987; Richard et al., 1985).

The tools needed to identify filamentous organisms in activated sludge are a good quality microscope, stains and dyes, and an identification key. Identification of filamentous organisms and descriptions of floc structure can be determined routinely by trained plant personnel. Although this diagnostic approach has proven effective in many cases, there are some difficulties and uncertainties in the method. Visual determinations of whether filaments have sheaths or cell septa can be difficult. Some of the staining characteristics may not be conserved between different treatment plants, especially in the case of industrial wastewaters. For example, in some environments a filament type may be gram-negative, whereas in another plant, the same filament will appear gram-positive. Without extensive experience in filament identification, this may lead to misidentification of the filament type and a consequent inaccurate diagnosis of the underlying problem. At minimum it has the potential to discourage efforts to adopt microscopic observation as a process control aid. Since most of the diagnostic types have never been isolated, in spite of previous efforts to do so, there are few type strains which can be used for comparison (Wanner and Grau, 1989). Whether those strains which exhibit variability are in actuality members of the same species or genus, or indicative of the same activated sludge environments, has not been adequately established. In some cases, the conditions that favor growth of some types of filaments are either not known or are attributable to more than one factor. Perhaps the greatest impediment to use of microscopic examination as a process control aid is that most plant personnel are not familiar with the use of microscopes. In spite of these limitations, microscopic observation is the most effective microbiologically based diagnostic procedure in use today.

The use of microscopy in activated sludge process control may have reached the point of limiting returns. Whereas this technique has proven useful for correcting major plant upsets, suboptimum plant performance still remains a problem. Routine polymer addition is often used to enhance settling or dewatering, and this can become a considerable expense over time. In many cases plant operations are essentially a series of trial-and-error corrections. In the case of variable waste streams, plant performance can be erratic and unpredictable, with no hard data on which to base decision-making. Unfortunately, microscopic observation is not always sufficiently responsive, predictive or sensitive to avoid these problems.

Ideal process control should be routine, quantitative and automated. Microscopic observation has little possibility of achieving these goals because it is a somewhat subjective method requiring special training. Although filament prevalence can be approximated, the appropriate process control modifications that follow tend to be empirical. Attempts have been made to interface graphic analysis programs with microscopes, but none have proven highly effective.

Probably the most significant components of the activated sludge community are the single-cell microorganisms which produce the exopolysaccharides that hold the floc together. Although microscopic observations can be made of general floc structure, the organisms comprising the floc are typically morphologically indistinguishable, being almost universally gram-negative rods. The number of species comprising a typical floc is still unknown. Considerable diagnostic information could be obtained via an understanding of the floc-forming organisms and suitable methods for their routine examination.

3. MOLECULAR ANALYSIS OF ACTIVATED SLUDGE

3.1. Microbial Systematics Using 16S rRNA

Modern systematics is based on the nucleotide sequence of the ribonucleic acid (RNA) found in the 16S subunit of the ribosome (16S rRNA). All living things possess ribosomes which are responsible for protein synthesis. All RNA is composed of four nucleotides; adenine (A), cytosine (C), guanine (G) and uracil (U). The sequence of these four bases in 16S rRNA is used for the classification and identification of microorganisms. The 16S rRNA molecule is approximately 1,500 bases in length. Some regions of 16S rRNA are highly conserved, whereas other regions are more variable. Differences between two species in a region that tend to be conserved indicates that divergence occurred in the distant past. Two species that differ only in a highly variable region are considered to be more closely related. In this manner, 16S rRNA is used as a molecular chronometer for establishing relationships between species. Currently, microbial systematics is undergoing a revolution in which reclassification of major groups of microorganisms is occurring. Due to the length of the sequences (1,500 nucleotides) this is generally done using computer assistance with mathematical methods such as distance matrix tree analysis (Woese, 1987).

An example of a distance matrix tree for some bacteria isolated from activated sludge is shown in Figure 4. Phylogenetic tree analysis confirms that isolates classified as *Sphaerotilus natans*, based on morphological characteristics, are closely related and may represent a single species within the *beta* subclass of the *Eubacteria* (Cortstgens and Muyzer, 1993). However, isolates classified as *Zoogloea* do not represent a single species (Shin et al., 1993). *Zoogloea* sp. ATCC 19324 and *Z. ramigera* ATCC 19544 are essentially identical and are in the *beta* subclass of the *Eubacteria*. *Z. ramigera* ATCC 25935

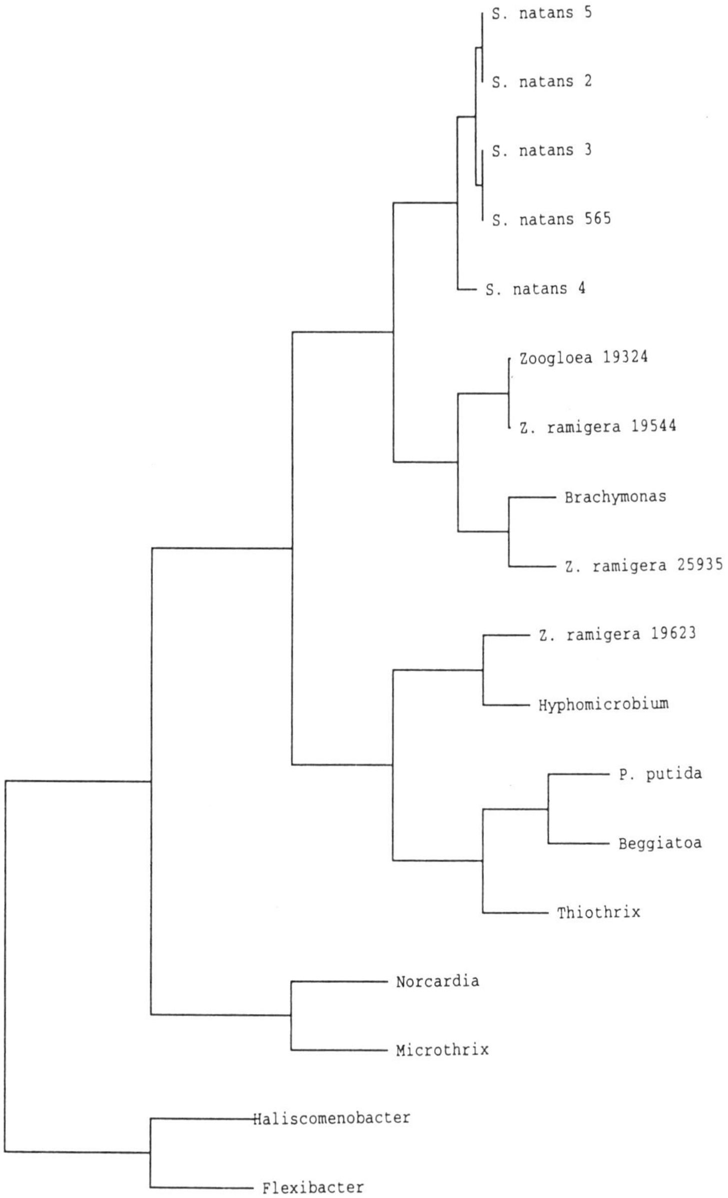

Figure 4. Phylogram showing relatedness of bacteria present in activated sludge with *Brachymonas denitrificans* and *Pseudomonas putida* used as reference isolates for the *beta* and *gamma* subclasses of *Eubacteria*, respectively. Isolate abbreviations with Genbank accession numbers in parenthesis are as follows: S. natans 2, *Sphaerotilus natans* ATCC 15291 (L33976); S. natans 5, *Sphaerotilus natans* ATCC 13338 (L33980); S. natans 3, *Sphaerotilus natans* ATCC 29329 (L33977); S. natans 565, *Sphaerotilus natans* str. 565 (Z18534); S. natans 4, *Sphaerotilus natans* ATCC 29330 (L33978); Z. ramigera 19324, *Zoogloea* sp. ATCC 19324 (D14257); Z. ramigera 19544, *Zoogloea ramigera* ATCC 19544 (X74913); Z. ramigera 25935, *Zóogloea ramigera* ATCC 25935 (X74194); Brachymonas, *Brachymonas denitrificans* (D14320); Z. ramigera 19623, *Zoogloea ramigera* ATCC 19623 (X74915); Hyphomicrobium, *Hyphomicrobium vulgare* (X53182); P. putida, *Pseudomonas putida* (L37365); Beggiatoa, *Beggiatoa* sp. 1401–13 (L40997); Thiothrix, *Thiothrix ramosa* (U32940); Norcardia, *Norcardia asteroides* str. N19 (X53205); Microthrix, *Microthrix parvicella* (X89774); Haliscomenobacter, *Haliscomenobacter hydrossis* (M58790); Flexibacter, *Flexibacter litoralis* ATCC 23117 (M58784). 16S rDNA sequences were obtained from Genbank and analyzed using the Software Package of the University of Wisconsin Genetic Computer Group (GCG) (Deveraux et al., 1984). The program PILEUP was used for alignment of DNA sequences. Distance matrix analysis using the Jukes-Cantor correction method was performed using the program DISTANCES. The phylogram was created using the UPGMA method in the program GROWTREE.

also is in the *beta* subclass, but is on a different branch, whereas *Z. ramigera* ATCC 19623 is a member of the *alpha* subclass of the *Eubacteria*. Based on 16S rRNA and quinone analysis Shin et al., (1993) have suggested that *Zoogloea* sp. ATCC 19324 and *Z. ramigera* ATCC 19544 remain in the *Zoogloea* genus and the other two isolates be reclassified. The difficulties in classifying bacteria which form zoogloeal clusters by classic taxonomic methods is due to a lack of distinctive morphological, staining and biochemical characteristics. This is less of a problem with many of the filamentous bacteria observed in activated sludge. Other filamentous bacteria found in activated sludge, including *Hyphomicrobium, Beggiatoa, Thiothrix, Nocardia, Microthrix, and Flexibacter* represent a wide range of subclasses of *Eubacteria* (Figure 4).

The use of 16S rRNA sequences as the basis for microbial classification and identification has considerable significance for the analysis of activated sludge microbial communities. Microorganisms can be identified or categorized without the need for isolation or cultivation, and in the absence of distinct biochemical or morphological traits (Gobel, 1995; Wagner et al., 1993). Probes can be made for important groups of organisms irrespective of whether they have been isolated, cultivated or given Latin names. There is continuing development of the relevant technologies for simplifying and automating molecular methods of analysis (Amann et al., 1990a; Amann et al., 1990b; Wagner et al., 1994a). All of these factors suggest that molecular analysis of activated sludge microbial populations is the diagnostic method of the future.

3.2. 16S rRNA Oligonucleotide Probes

DNA-DNA or DNA-RNA probing is based on the fact that the nucleotide bases pair in a complementary fashion. Adenine is complementary to uracil or thymine, and guanine is complementary to cytosine. If the target nucleotide sequence is known, a complementary sequence can be synthesized which will hybridize with it at the appropriate temperature. For example, if the target sequence is CTAGCAAAA the complementary sequence would be GATCGTTTT. The probe can be labeled such that hybridization with an unknown sample will indicate the presence of the target. Common labels include ^{32}P (radioactive), fluorescein (fluorescent) and digoxigenin (antibody coupled enzyme reaction producing a colored product) (Stahl and Amann, 1991). Probing can be done by extracting activated sludge RNA, immobilizing it on a filter, and hybridizing it with the probe. Since 16S rRNA oligonucleotide probes are small, intact cells in activated sludge fixed on microscope slides can be probed with fluorescently labeled probes. Fluorescence from target cells probing positive can be visualized directly via fluorescence microscopy (Amann et al., 1990b).

Probes are designed based on the RNA sequence of the desired target organism. For example, probes have been designed for the three *Z. ramigera* strains discussed previously. Sequencing of the 16S rRNA indicated that there are regions in the RNA molecule where the three strains differ. Probes were designed that are complementary to the regions of each strain that are distinctive (Rossello-Mora et al., 1995). Fluorescein labeling of the probes allows the detection of these strains in an environment in which they are otherwise indistinguishable from other species with similar morphology (Figure 5).

Samples of mixed liquor from sewage treatment plants experiencing different loading rates were probed with the *Zoogloea* spp. probes (Table 1). Significant numbers of cells hybridized with the probe designated ZRA (*Zoogloea ramigera* ATCC19544), but few cells hybridized to the two other *Zoogloea* spp. probes. The presence of *Zoogloea ramigera* ATCC 19544 correlated with loading, but probe-positive populations were not

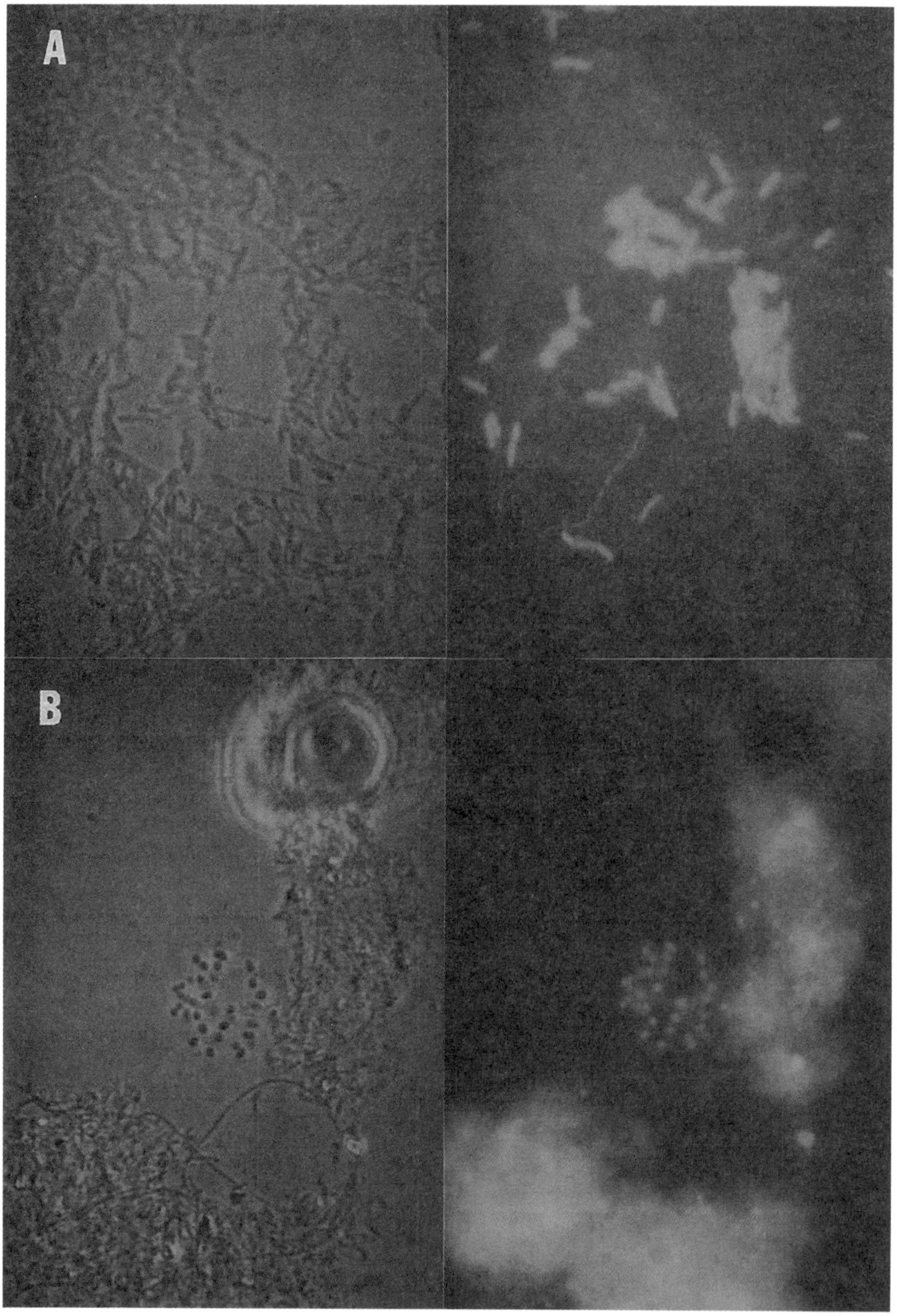

Figure 5. A. Probing of a mixture of *E. coli* and *Z. ramigera* ATCC 25935 with the fluorescein labeled probe ZBE for *Z. ramigera*. **B.** Probing of a zoogloeal cluster in activated sludge with the fluorescein labeled probe EUB for *Eubacteria*. For each panel, identical fields were viewed by phase-contrast microscopy (left) and epifluorescence microscopy (right) at 1,000×.

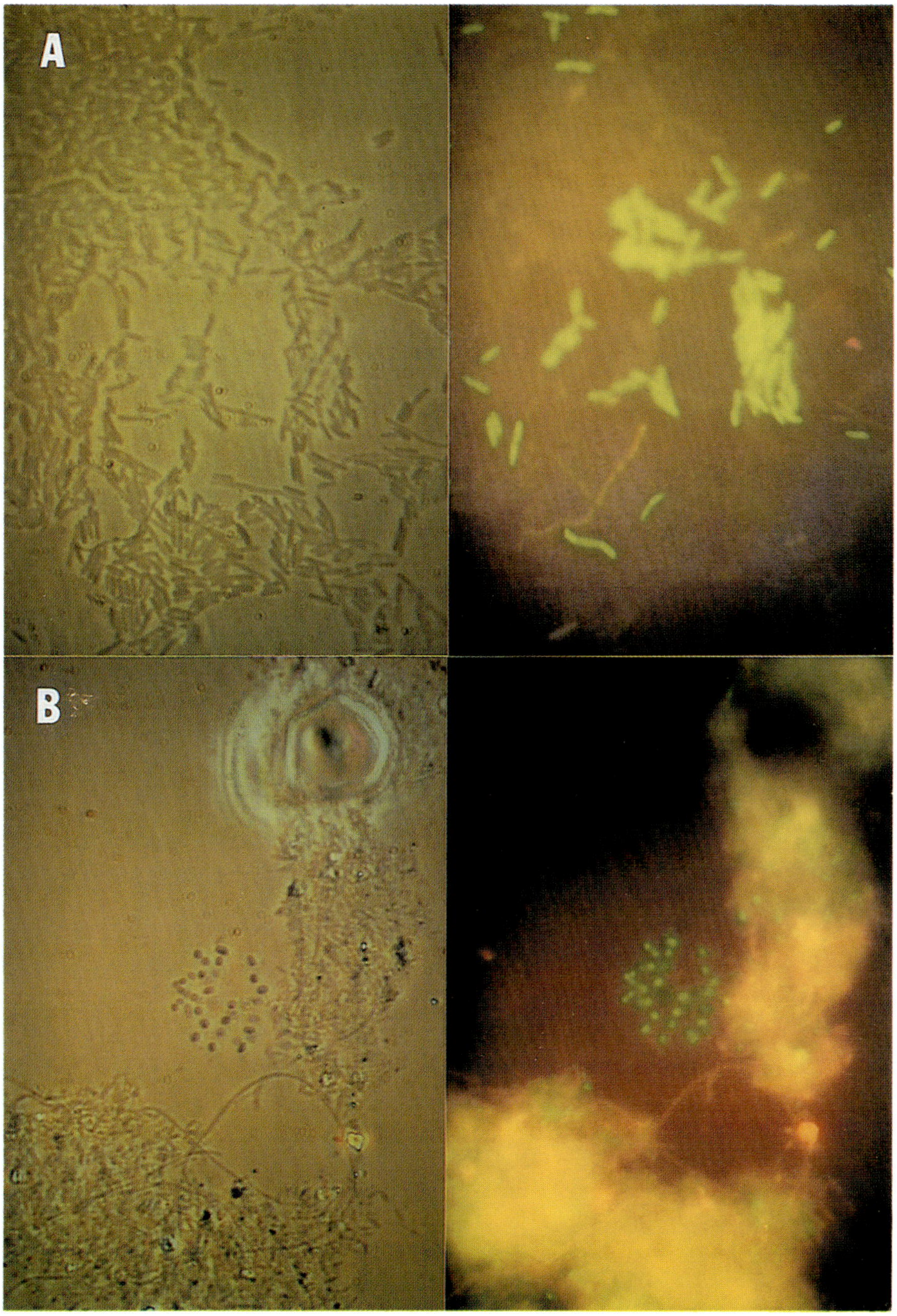

Figure 5. A. Probing of a mixture of *E. coli* and *Z. ramigera* ATCC 25935 with the fluorescein labeled probe ZBE for *Z. ramigera*. **B.** Probing of a zoogloeal cluster in activated sludge with the fluorescein labeled probe EUB for *Eubacteria*. For each panel, identical fields were viewed by phase-contrast microscopy (left) and epifluorescence microscopy (right) at 1,000×.

Table 1. Probing of activated sludge with 16S rRNA oligonucleotide probes

Target, probe sequence (5′→3′), and abbreviation	Results, observations, and comments
Case 1 *Zoogloea* spp. probes; loading	
Samples obtained from 11 different sewage treatment plants, including three high capacity plants, three medium capacity plants and four low-capacity plants. Some plants had primary and secondary aeration basins (Rossello-Mora et al., 1995).	
Zoogloea ramigera ATCC19544 CTGCCGTACTCTAGTTAT (ZRA)	Hybridization occurred in most samples from high and medium capacity plants. More positive signals in primary basins than in secondary basins. Up to 10% of the total number of cells probed positive. Typical finger-like zoogloeal structures observed in only two samples. No positive responses in low capacity activated sludge plants.
Zoogloea ramigera ATCC25935 TGCCAAACTCTAGCCTTG (ZBE)	No cells probed positive.
Zoogloea ramigera ATCC19623 CTTCCATACTCTAGGTAC (ZAL)	Low numbers of cells detected in five plants. Probe positive cells did not occur as aggregates (zoogloeal clusters).
Case 2 probes for filamentous bacteria; bulking	
Grab samples of mixed liquor were collected from the aeration basins of five municipal wastewater treatment plants and three industrial wastewater treatment plants (Hoechst, animal waste processing, and municipal wastewater mixed with sewage of a brewery and galvanization factory), and a sequencing batch reactor treating dairy waste (Wagner et al., 1994a).	
Haliscomenobacter hydrossis GCCTACCTCAACCTGATT (HHY)	Positive response in 6 out of 10 plants examined. Thin filaments common within flocs, that were otherwise difficult to observe via phase contrast. HHY positive filaments were also CF positive.
Thiothrix nivea CTCCTCTCCCACATTCTA (TNI)	Positive response in 4 out of 10 plants examined. Simultaneous occurrence with, but distinct from, type 21N in 3 plants. Also hybridized to large coccoid cells (possibly gonidia). Sulfur storage test positive. Dominant during bulking in an industrial wastewater plant. TNI positives were also GAM positive.
Eikelboom Type 021N TCCCTCTCCCAAATTCTA (21N)	Positive response in 3 out of 10 plants examined. Simultaneous occurrence with, but distinct from, type TNI in 3 plants. Sulfur storage test positive. Present as long trichomes (no rosettes). 21N positives were also GAM positive.
Leucothrix mucor CCCCTCTCCCAAACTCTA (LMU)	No positive responses.
Sphaerotilus natans CATCCCCCTCTACCGTAC (SNA)	Positive responses in 7 of 10 plants examined. Various morphotypes of filaments resembling trichomes of *S. natans*, *L. cholodnii* and type 1701 probed positive. Rod shaped single cells (possibly swarmers or sheath - types) also probed positive. SNA positives were also BET positive.
Leptothrix discophora CTCTGCCGCACTCCAGCT (LDI)	Positive response in 6 out of 10 plants examined. Two distinct morphotypes probed positive. Both contained PHB. LDI positives were also BET positive.
Case 3 heiracheal probes; loading	
Grab samples of mixed liquor were collected from two successive activated sludge aeration basins of a large municipal sewage treatment plant. The first and second basins are high and low load stages, respectively (Wagner et al., 1993).	
Eubacteria GCTGCCTCCCGTAGGAGT (EUB)	In high and low load basins 89% and 70%, respectively, of DAPI positives also probed positive with EUB. *Proteobacteria* accounted for 60 to 75% of DAPI positives.

Table 1. (*Continued*)

Target, probe sequence (5′→3′), and abbreviation	Results, observations, and comments
Alpha subclass of *Proteobacteria* CGTTCG(C/T)TCTGAGCCAG (ALF)	In high and low load basins 10% and 30%, respectively, of EUB positives were also ALF positive. ALF positives were dominant in low load basin. ALF hybridized to clusters of coccoid rods in the low load basin.
Beta subclass of *Proteobacteria* GCCTTCCCACTTCGTTT (BET)	In high and low load basins 42% and 37%, respectively, of EUB positives were also BET positive. BET positives were dominant in high and low load basin. BET hybridized to rods in the high load basin.
Gamma subclass of *Proteobacteria* GCCTTCCCACATCGTTT (GAM)	In high and low load basins 34% and 7%, respectively, of EUB positives were also GAM positive. GAM positives were dominant in high load basin. GAM hybridized to filaments (type 1863) in the high load basin.
Eikelboom type 1863	Probed positive with GAM in high load aeration basin.
Case 4 *Nitrosomonas* spp., *Zoogloea* spp., and heiracheal probes; nitrification	
Grab samples of mixed liquor were collected from the aeration basins of eight municipal sewage treatment plants and an animal waste processing plant. The animal waste plant was the most intensely analyzed. Some municipal plants were two stage, with nitrification occurring in the second stage (Wagner et al., 1995).	
Nitrosomonas europaea *Nitrosomonas eutropha* CCCCTCTGCTGCACTCTA (NEU)	Dense spherical and rod shaped clusters of rod shaped cells probed positive in plants with stable nitrification. No hybridization in plants without nitrifying activity. In samples from animal waste processing plant up to 20% of total bacteria probed positive. All NEU positive cells were also BET positive.
Beta subclass of *Proteobacteria* GCCTTCCCACTTCGTTT (BET)	Greater than 70% of the total bacteria probed positive in the animal waste processing plant. NEU and ZRA positive cells constituted one half of all BET positives.
Gamma subclass of *Proteobacteria* GCCTTCCCACATCGTTT (GAM)	Less than 10% of the cells probed positive with the GAM or ALF probes.
Alpha subclass of *Proteobacteria* CGTTCG(C/T)TCTGAGCCAG (ALF)	Less than 10% of the cells probed positive with the GAM or ALF probes.
Gram-positive with high G+C DNA content TATAGTTACCACCGCCGT (HGC)	No cells probed positive.
Cytophaga-Flavobacterium cluster TGGTCCGTGTCTCAGTAC (CF)	No cells probed positive.
Zoogloea ramigera ATCC19544 CTGCCGTACTCTAGTTAT (ZRA)	20% of the total bacteria probed positive in the animal waste processing plant. Zoogloeal matrices apparent by light microscopy.
Zoogloea ramigera ATCC25935 TGCCAAACTCTAGCCTTG (ZBE)	No cells probed positive
Zoogloea ramigera ATCC19623 CTTCCATACTCTAGGTAC (ZAL)	No cells probed positive.
Case 5 *Acinetobacter* spp. and heiracheal probes; phosphorus removal (EBPR)	
Grab samples of mixed liquor were collected from the aerobic and anaerobic basins of a municipal wastewater treatment plant using the Phoredox process for phosphorus removal (EBPR– enhanced biological phosphorus removal). Samples from this plant were compared to samples from the aerobic and anaerobic basins of two sewage treatment plants plant without EBPR (Wagner et al., 1994b).	
Eubacteria GCTGCCTCCCGTAGGAGT (EUB)	In aerobic and anaerobic basins of EBPR plant 78% and 83%, respectively, of DAPI positives probed positive with EUB.

Table 1. (*Continued*)

Target, probe sequence (5′→3′), and abbreviation	Results, observations, and comments
Alpha subclass of *Proteobacteria* CGTTCG(C/T)TCTGAGCCAG (ALF)	In aerobic and anaerobic basins of EBPR plant 9% and 11%, respectively, of EUB positives were also ALF positive.
Beta subclass of *Proteobacteria* GCCTTCCCACTTCGTTT (BET)	Dominant in anaerobic and aerobic basins. In aerobic and anaerobic basins of EBPR plant 26% and 24%, respectively, of EUB positives were also BET positive. BET positives contained PHB, but no polyphosphate granules.
Gamma subclass of *Proteobacteria* GCCTTCCCACATCGTTT (GAM)	In aerobic and anaerobic basins of EBPR plant 10% and 5%, respectively, of EUB positives were also GAM positive. 60% of GAM positives were ACA positive.
Cytophaga–Flavobacterium cluster TGGTCCGTGTCTCAGTAC (CF)	In aerobic and anaerobic basins of EBPR plant 8% and 9%, respectively, of EUB positives were also CF positive.
Gram-positive with high G+C DNA content TATAGTTACCACCGCCGT (HGC)	Dominant in anaerobic and aerobic basins. In aerobic and anaerobic basins of EBPR plant 19% and 24%, respectively, of EUB positives were also HGC positive. HGC positives contained polyphosphate inclusions. Appear to be important in P removal. In 2 municipal plants without EBPR 7% and 1% of EUB positives probed positive with HGC. *Microthrix parvicella* probed positive with HGC.
Acinetobacter spp. ATCCTCTCCCATACTCTA (ACA)	Less than 10% of all bacteria. Observed as rods and cocci commonly occurring in clusters or chains. In aerobic and anaerobic basins of EBPR plant 7% and 3%, respectively, of EUB positives were also ACA positive. Appear to be unimportant in P removal. In 2 municipal plants without EBPR 3% and 8% of EUB positives probed positive with ACA. Eikelboom type 1863 probed positive with GAM and ACA.
Eikelboom type 1863	Probed positive with GAM and ACA probes in EBPR plant. Present as free floating filaments between flocs. Mostly gram-negative cells with some gram positive cells.
Microthrix parvicella	Probed positive with HGC probe in EBPR plant. Gram-positive filaments containing large amounts of phosphate. Probe penetration difficult.

numerous, with a maximum of 10% of the cells in the mixed liquor probing positive. In two cases typical finger-like zoogloeas were observed. In an animal waste processing plant, zoogloeal matrices were observed, and 20% of the total cells probed positive with the ZRA probe. No hybridization occurred with the other *Zoogloea spp.* probes.

Microscopic examination of activated sludge in an industrial wastewater treatment plant experiencing dewatering problems indicated the presence of numerous amorphous zoogloeal type structures (Figure 6). India ink exclusion tests suggested the presence of excess exopolysaccharide production (Figure 7). Probing of the sludge with a fluorescently labeled 16S rRNA *Eubacteria* probe indicated that probe penetration into the zoogloeas could be achieved, and the responsible organisms were members of the *Eubacteria* (Figure 5). Dot blot hybridization experiments with three different 16S rRNA oligonucleotide probes specific for previously characterized *Zoogloea ramigera* strains and in situ hybridizations with the ZBE probe did not indicate the presence of these strains in the sludge. Apparently, in this case, the microorganisms comprising the zoogloeas are not closely related to known *Zoogloea ramigera* strains.

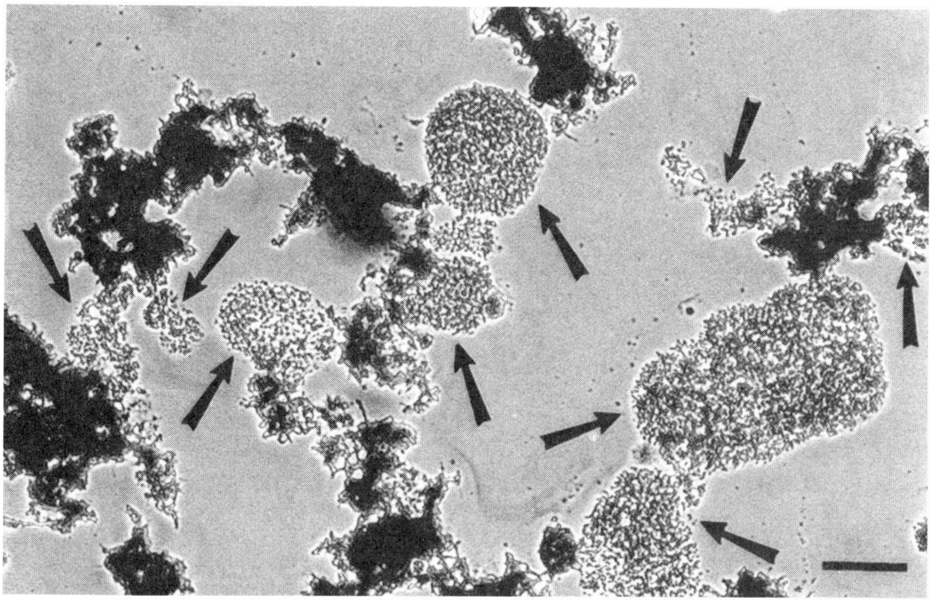

Figure 6. Microscopic appearance of amorphous zoogloeal clusters in activated sludge (arrows indicate zoogloeal clusters, 200× magnification; bar = 50 μm).

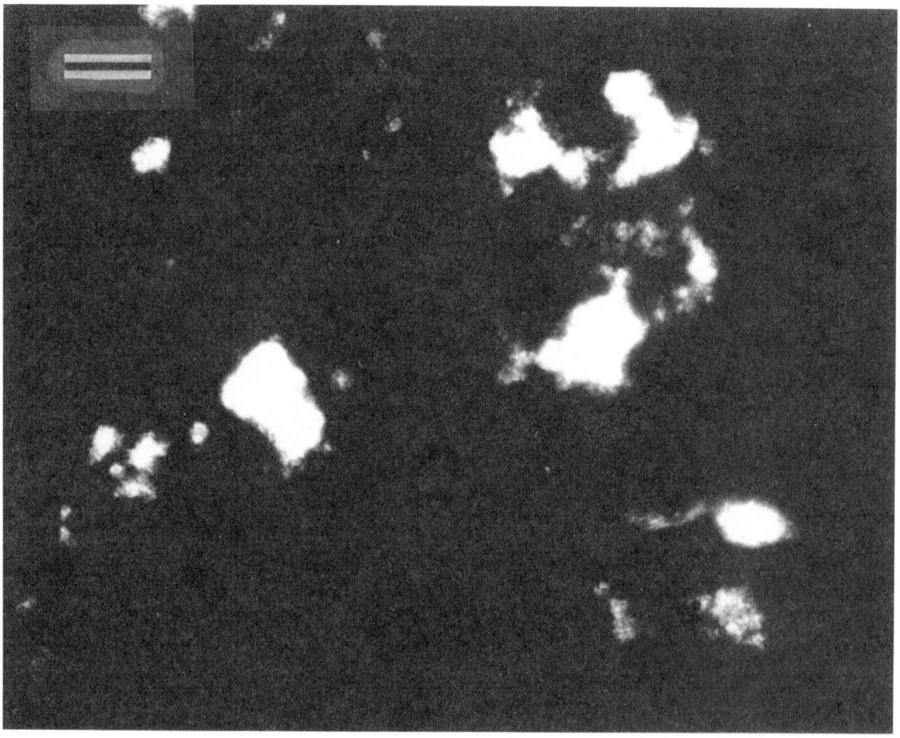

Figure 7. Reverse-staining of activated sludge with India ink demonstrating the presence of "excess" exopolysaccharide (200× magnification; bar = 50 μm).

The use of 16S rRNA as the basis of microbial systematics allows considerable flexibility in probe design. Probes can be designed for individual species or major groups of species. For example, all prokaryotic organisms are divided into two groups; *Archae* or *Eubacteria*. These groups differ in regions of the 16S rRNA molecule that tend to be highly conserved. Probes designed for this region can distinguish members of these two groups (Burggraf et al., 1994). A *Eubacteria* probe would target all members of such genera as diverse as *Pseudomonas*, *Alcaligenes*, *Rhodobacter*, and *Bacillus*, as well as the three *Z. ramigera* strains. Probes based on major phylogenetic groups are sometimes referred to as heiracheal probes. The first level is *Eukaryotes*, *Eubacteria* and *Archae* (higher organisms, true bacteria and primitive "bacteria", respectively). Within the *Eubacteria* are the *alpha, beta* and *gamma* subclasses of the *Proteobacteria*. This is essentially a "top-down" approach based on microbial classification. Probes can also be made more specific by targeting more variable regions of the 16S rRNA. Genus and group-specific probes have been developed for sulfate-reducing bacteria (Devereux et al., 1992). The three *Z. ramigera* probes are examples of very specific species level probes. This versatility in probe design is based on comparative sequence analysis between the target microorganisms and all other known sequences. The degree of specificity required depends on the application.

A summary of activated sludge studies in which heiracheal and species specific probes have been used is presented in Table 1. The specific probes used in these studies are based on what is commonly believed to be present in activated sludge as suggested by more traditional methods. Several probes have been designed for filamentous microorganisms. Results of probing of activated sludge samples indicate that these probes yield results consistent with identification by classical stain and dye coupled microscopic examination (Wagner et al., 1994a). *Sphaerotilus natans* and Eikelboom type 1701 probed positive with the same probe (SNA) indicating that these types are closely related. Both tend to be found in environments with low oxygen concentrations. The results indicate that some advantages of 16S rRNA probes over traditional methods of microscopic examination is that they provide unambiguous identification, are not dependent on variable morphological or staining characteristics, and can be of value in observing filaments otherwise obscured by the floc structure.

Heirachael probing, based on broad phylogenetic groups, indicates that in the range of 70–90% of the observable microorganisms in activated sludge probe positive with the *Eubacteria* probe EUB (Table 1). Whether the remaining organisms are members of the *Archae*, have low RNA content due to low metabolic activity, or are not accessible to the EUB probe is unknown. *Proteobacteria* accounted for 60 to 75% of the *Eubacteria*, with members of the *beta* subclass (BET) prevalent in both high- and low-load aeration basins. ALF positive strains (*alpha* subclass) were more common in the low load basin, whereas GAM positives (*gamma* subclass) were more common in the high-load basin (Wagner et al., 1993). At present there is insufficient data to indicate if this is a general trend.

A probe has been developed for some of the more commonly isolated nitrifying bacteria (NEU). Dense spherical and bacillary clusters of rod-shaped cells probed positive in plants exhibiting stable nitrification. No hybridization was observed in plants lacking nitrifying activity. Up to 20% of the total bacteria probed positive with the NEU probe in an animal waste processing plant. All NEU positives were also BET-positive, indicating that the nitrifiers were all members of the *beta* subclass of the *Eubacteria*. In the same plant zoogloeal matrices were evident, and 20% of the total population probed positive with the ZRA probe for *Zoogloea ramigera* (Wagner et al., 1995).

Microbial populations in phosphorus removal plants have been examined using both heirachael type probing and an *Acinetobacter* spp. probe. Results indicated that *Acineto-*

bacter spp. tend to be in the range of only 3–8% of the microbial community in phosphorus removal plants. The group referred to as the gram-positive with high G+C content (HGC) were more prevalent, and it was suggested that these are the organisms important in phosphorus removal (Wagner et al., 1994b).

It is impossible with the limited range of studies so far performed to make broad generalizations about the microbial communities in activated sludge. It appears that members of the *alpha*, *beta*, and *gamma* subclasses of the *Proteobacteria* tend to be the dominant organisms, but more specific classifications have not as yet been achieved.

3.3. 16S rRNA Community Analysis

Whereas a number of probes have been designed for microorganisms believed to be present in activated sludge, it is possible to analyze microbial communities without making any assumptions as to the organisms likely to be present. This is accomplished via the polymerase chain reaction (PCR). Short stretches of synthetic DNA (primers) can be made which are complementary to the highly conserved regions at the ends of the 16S rDNA molecule. The enzyme *Taq* polymerase begins copying the DNA at the regions of double-stranded DNA formed by hybridization of the primers to the 16S rDNA, and makes a copy of the entire intervening 16S rDNA molecule. Each copy serves as a template for further copies. The number of 16S rDNA molecules increases exponentially with each round of synthesis (Giovannoni, 1991). In this manner, a small amount of sludge DNA, containing all DNA-coding functions for the entire microbial community, as well as 16S rRNA-coding regions (16S rDNA), is amplified such that it contains almost exclusively 16S rDNA sequences. As with oligonucleotide probes, the primer specificity can be adjusted by choosing sequences that amplify only certain groups of organisms. For example primers can be made to a 16S rDNA region unique to nitrifying bacteria (Voytek and Ward, 1995; Hiorns et al., 1995). The amplified DNA in this case would only contain 16S rDNA from nitrifying bacteria.

The PCR-generated 16S rDNA sequences are then cloned into plasmids and inserted into *E. coli* host strains to create a 16S rDNA library. A representative number of the cloned 16S rDNA molecules are sequenced (Lane, 1991). The sequences are then compared with each other and all other known sequences in a computer database to determine the relatedness between species within the activated sludge and other previously sequenced species, respectively (Figure 8). Analyses can be performed on either full-length sequences (1500 bases) or partial sequences (250–400 bases). If the sequences are similar to a previously characterized species, the unknown species in the activated sludge are thereby identified. If the sequences are not similar, their relationship to other species can be determined by making a distance matrix tree (Woese, 1987).

3.4. Designing Diagnostic Probes

Probes can be made for specific groups of organisms in activated sludge from sequences in the PCR generated 16S rDNA library derived originally from that sludge. Probing of the original sludge sample indicates if the PCR generated 16S rDNA library is indeed representative of the sludge microbial community (Altwegg, 1995). Probes can be fluorescently labeled for in situ hybridization and microscopic analysis to determine if morphologically distinct organisms (filaments) or floc types (zoogloeas) correspond to specific 16S rDNA sequences. Similarly, all the major filaments commonly observed in activated sludge can be identified from sludge samples without the need for pure culture.

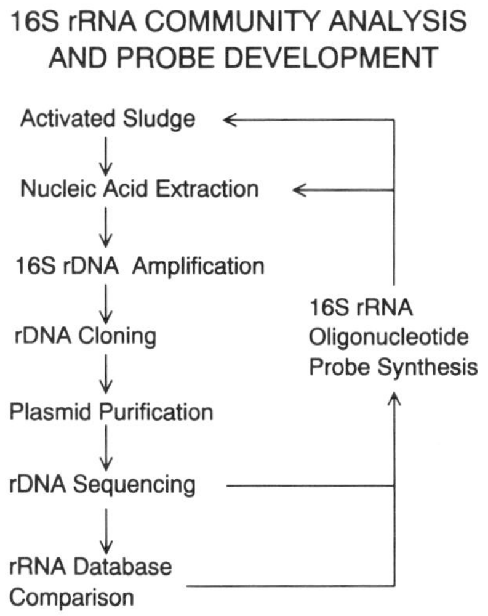

Figure 8. Characterization of a microbial community by analysis of a 16S rDNA library.

Fluorescent probes developed from the sequences could be used as an adjunct to traditional microscopic approaches to verify the identity of problem organisms in sludge (Wagner et al., 1994a).

In some cases comparison of sequences amplified from sludge may match sequences found in existing sequence databases such as the Illinois 16S Ribosomal RNA Data Base Project or GenBank (Larsen et al., 1993) and the probable cause of an activated sludge upset inferred from the types of organisms identified. Probes designed specifically from the activated sludge sequences, or previously designed probes for this group of organisms, may be used routinely to assist in avoiding a recurrence of the same problem. In other cases, the sequences obtained may not match any previously known sequences. This is not unexpected as not all known species have been sequenced, and more importantly, many common organisms in activated sludge have probably never been isolated. This does not prevent the development of suitable probes for the unknown organisms, and these probes can assist in efforts to isolate the problem strains (Kane et al., 1993).

In some cases diagnostic information from previously unknown sequences can only be obtained by comparison between plants or between periods of favorable and unfavorable sludge quality within the same plant. This points to the major current limitation of 16S rRNA analysis as a diagnostic tool for activated sludge. As this is a relatively new technology, more information on the microbial communities of activated sludge must be obtained to develop consistent correlations between individual species and causative factors in plant upsets. Computer accessible databases specifically for this purpose would assist in collating activated sludge 16S rRNA sequences and associated sludge quality problems. Eventually it is expected that probe arrays will be developed for the detection of many of the activated sludge organisms associated with particular types of process upsets.

3.5. Gene Probes

Whereas 16S rRNA probes are effective for identifying microbial populations, gene probes can be used for determining if particular genetic capabilities are present in the microbial community. Although there is some overlap, 16S rRNA probes are expected to be

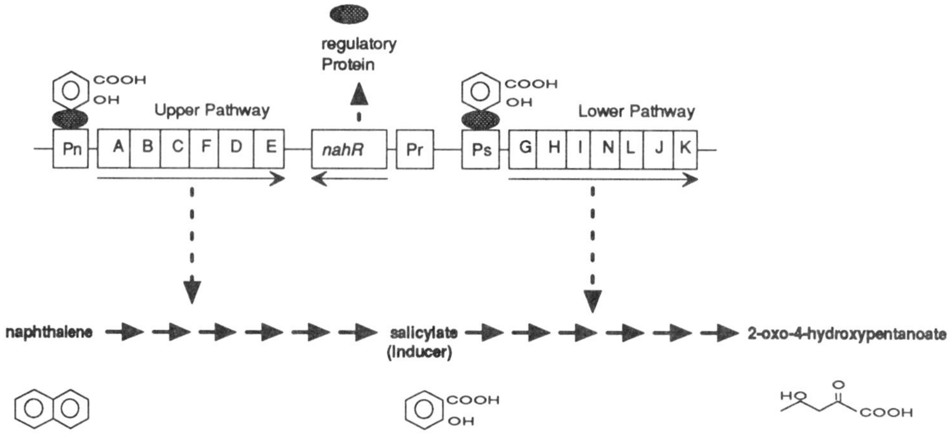

Figure 9. The plasmid encoded *nah* operon for naphthalene degradation.

more useful for problems with sludge quality, whereas gene probes are more suitable for evaluating the fate of specific problem chemicals in activated sludge. In both cases probes hybridize with potential target sequences via complementary sequences. Genes probes are generally designed to target DNA sequences coding for a particular enzymatic activity, whereas 16S rRNA probes target ribosomal RNA.

An example of a gene probe is the *nah* probe developed for naphthalene degradation. Genes coding for naphthalene degradation are located on the 83 kilobase NAH7 plasmid. The naphthalene degradative operon consists of an upper pathway and a lower pathway (Figure 9). The upper pathway (*nah*ABCDEF) encodes enzymes for the transformation of naphthalene to salicylate. The lower pathway (*nah*GHINLJK) encodes enzymes for the conversion of salicylate to acetaldehyde and pyruvate (Fleming et al., 1993).

A probe based on the upper pathway of the *nah* operon was used to investigate naphthalene degradation in activated sludge treatment of a petrochemical wastewater (Blackburn et al., 1987). Biodegradation was the primary mechanism of naphthalene removal when *nah* positive strains were present at high concentrations. Decreases in the naphthalene degradative populations resulted in increased naphthalene loss via air-stripping. Enumeration of naphthalene degradative populations via probing was much more sensitive than traditional vapor plate methods.

Gene probes can be used for optimizing conditions for degradation of particular chemicals that are a problem due to their volatility, toxicity or recalcitrance. They may also be useful for determining if the potential for biodegradation of new consumer products exists in municipal wastewater treatment facilities.

4. CONCLUSIONS

Research on the microbial communities in activated sludge has been slow in recent years due to the limitations of traditional techniques for microorganism identification and enumeration. The advent of modern molecular approaches has led to a resurgence of interest in activated sludge community dynamics. The ability to identify, enumerate, and detect individual populations in activated sludge using 16S rRNA techniques far surpasses anything previously available. The detection of specific organisms or metabolic capabilities using molecular probes has applications both in plant operations and the design of consumer products destined for disposal in municipal waste treatment plants. Results to date have in some cases supported previous conclusions concerning activated sludge microbial populations. In other cases, existing beliefs appear to be oversimplified. More importantly, it is possible to obtain a view of activated sludge that was previously inaccessible. A greater understanding of activated sludge microbiology is expected to lead to improvements in analysis and control of activated sludge treatment processes.

ACKNOWLEDGMENTS

We thank Don Taylor from Eastman Chemical Company, Tennessee Eastman Division, for providing activated sludge photomicrographs.

REFERENCES

Altwegg, M., 1995, General problems associated with diagnostic applications of amplification methods, *J. Microbiol. Meth.* 23:21–30.

Amann, R.I., Binder, B.J., Olsen, R.J., Chisholm, S.W., Devereux, R., and Stahl, D.A., 1990a, Combination of 16S rRNA-targeted oligonucleotide probes with flow cytometry for analyzing mixed microbial populations, *Appl. Environ. Microbiol.* 56:1919–1925.

Amann, R.I., Krumholz, L., and Stahl, D.A., 1990b, Fluorescent-oligonucleotide probing of whole cells for determinative, phylogenetic, and environmental studies in microbiology, *J. Bacteriol.* 172:762–770.

Blackbeard, J.R., Ekama, G.A., and Marais, G.v.R., 1986, A survey of filamentous bulking and foaming in activated sludge plants in South Africa, *Water Pollut. Contr.* 85:90–100.

Blackburn, J.W., Jain, R.K., and Sayler, G.S., 1987, Molecular microbial ecology of a naphthalene-degrading genotype in activated sludge, *Environ. Sci. Technol.* 21:884–890.

Burggraf, S., Mayer, T., Amann, R., Schadhauser, S., Woese, C.R., and Stetter, K.O., 1994, Identifying members of the domain *Archaea* with rRNA-targeted oligonucleotide probes, *Appl. Environ. Microbiol.* 60:3112–3119.

Cortstgens, P. and Muyzer, G., 1993, Phylogenetic analysis of the metal-oxidizing bacteria *Leptothrix discophera* and *Sphaerotilus natans* using 16S rDNA sequencing data, *System. Appl. Micobiol.* 16:219–223.

Deveraux, J., Haeberle, P., and Smithes, O., 1984, A comprehensive set of sequence analysis programs for the VAX, *Nucl. Acid. Res.* 12:387–395.

Devereux, R., Kane, M.D., Winfrey, J., and Stahl, D.A., 1992, Genus- and group-specific hybridization probes for determinative and environmental studies of sulfate-reducing bacteria, *System. Appl. Microbiol.* 15:601–609.

Dorn, P.B., Salanitro, J.P., Evans, S.H., and Kravetz, L., 1993, Assessing the aquatic hazard of some branched and linear nonionic surfactants by biodegradation and toxicity, *Environ. Toxicol. Chem.* 12:1751–1762.

Eikelboom, D.H., 1975, Filamentous organisms observed in activated sludge, *Water Res.* 9:365–388.

Fleming, J.T., Sanseverino, J., and Sayler, G.S., 1993, Quantitative relationship between naphthalene catabolic gene frequency and expression in predicting PAH degradation in soils at town gas manufacturing sites, *Environ. Sci. Technol.* 27:1068–1074.

Giger, W., Stephanou, E., and Schaffner, C., 1981, Persistent organic chemicals in sewage effluents. I. Identification of nonylphenols and nonylphenolethoxylates by glass capillary gas chromatography/mass spectrometry, *Chemosphere* 10:1253–1263.

Giger, W., Brunner, P.H., and Schaffner, C., 1984, 4-Nonylphenol in sewage sludge: accumulation of toxic metabolites from nonionic surfactants. *Science* 225:623–625.

Giovannoni, S., 1991, The polymerase chain reaction, in: *Nucleic Acid Techniques in Bacterial Systematics* (E. Stackebrandt and M. Goodfellow, eds.), John Wiley & Sons, Chichester, pp.177–201.

Gobel, U.B., 1995, Phylogenetic amplification for the detection of uncultured bacteria and the analysis of complex microbiota, *J. Microbiol. Meth.* 23:117–128.

Hileman, B., 1994, Environmental estrogens linked to reproductive abnormalities, cancer, *Chemical and Engineering News,* January pp. 19–23.

Hiorns, W.D,, Hastings, R.C., Head, I.M., McCarthy, A.J., Saunders, J.R., Pickup, R.W., and Hall, G.H., 1995, Amplification of 16S ribosomal RNA genes of autotrophic ammonia-oxidizing bacteria demonstrates the ubiquity of nitrosospiras in the environment, *Microbiology* 141:2793–2800.

Jenkins, D., 1992, Toward a comprehensive model of activated sludge bulking and foaming, *Wat. Sci. Tech.* 25:215–230.

Jenkins, D., Richard, M.G., and Daigger, G.T., 1993, *Manual on the Causes and Control of Activated Sludge Bulking and Foaming*, 2nd ed., Lewis Publishers, Chelsea.

Kane, M.D., Poulsen, L.K., and Stahl, D.A., 1993, Monitoring the enrichment and isolation of sulfate-reducing bacteria by using oligonucleotide hybridization probes designed from environmentally derived 16S rRNA sequences, *Appl. Environ. Microbiol.* 59:682–686.

Lane, D.J., 1991, 16S/23S rRNA sequencing, in: *Nucleic Acid Techniques in Bacterial Systematics* (E. Stackebrandt and M. Goodfellow, eds.), John Wiley & Sons, Chichester, pp.115–148.

Larsen, N., Olsen, G.J., Maidak, B.L., McCaughey, M.J., Overbeek, R., Macke, T.J., Marsh, T.L., and Woese, C.R., 1993, The ribosomal database project, *Nucleic Acids Res.* 21:3021–3023.

Mitchell, R., 1974. *Introduction to Environmental Microbiology*, Prentice-Hall, Englewood Cliffs.

Patoczka, J. and Pulliam, G.W., 1990, Biodegradation and secondary effluent toxicity of ethoxylated surfactants, *Wat. Res.* 24:965–972.

Richard, M.G., Shimizu, G.P., and Jenkins, D., 1985, The growth physiology of the filamentous organism type 021N and its significance to activated sludge bulking, *J. Wat. Pollut. Control Fed.* 57:1152–1160.

Rossello-Mora, R., Wagner, M., Amann, R., and Schleiffer, K.-H., 1995, The abundance of *Zoogloea ramigera* in sewage treatment plants, *Appl. Environ. Microbiol.* 61:702–707.

Sezgin M., Jenkins, D., and Parker, D.S., 1978, A unified theory of filamentous activated sludge bulking, *J. Wat. Pollut. Control Fed.* 50:362–381.

Shin, Y.K., Hiraishi, A., and Sugiyama, J., 1993, Molecular systematics of the genus *Zoogloea* and amendation of the genus, *Int. J. Syst. Bacteriol.* 43:826–831.

Soto, A.M., Justicia, H., Wray, J.M., and Sonnenschein, C., 1991, p-Nonylphenol: An estrogenic xenobiotic released from "modified" polystyrene, *Environ. Health Perspect.* 92:167–173.

Stahl, D.A., and Amann, R., 1991, Development and application of nucleic acid probes, in: *Nucleic Acid Techniques in Bacterial Systematics* (E. Stackebrandt and M. Goodfellow, eds.), John Wiley & Sons, Chichester, pp. 205–242.

Strom, P.F., and Jenkins, D., 1984, Identification and significance of filamentous microorganisms in activated sludge, *J. Wat. Pollut. Control Fed.* 56:449–459.

Unz, R.F., 1984, Genus IV. *Zoogloea* Itzigsohn, in: *Bergey's Manual of Systematic Bacteriology*, Volume 1 (N.R. Krieg and J.G. Holt, eds.), Williams & Wilkins, Baltimore, pp.214–219.

van Niekerk, A.M., Jenkins, D., and Richard, M.G., 1987, The competitive growth of *Zoogloea ramigera* and type 021N in activated sludge and pure culture-A model for low F:M bulking, *J. Wat. Pollut. Control Fed.* 59:262–273.

Voytek, M.A., and Ward, B.B., 1995, Detection of ammonium-oxidizing bacteria of the *beta*-subclass of the class *Proteobacteria* in aquatic samples with the PCR, *Appl. Environ. Microbiol.* 61:1444–1450.

Wagner, M., Amann, R., Lemmer, H., and Schleiffer, K.-H., 1993, Probing activated sludge with oligonucleotide specific for proteobacteria: Inadequacy of culture-dependent methods for describing microbial community structure, *Appl. Environ. Microbiol.* 59:1520–1525.

Wagner, M., Amann, R., Kampfer, P., Assmus, B., Hartmann, A., Hutzler, P., Springer, N., and Schleiffer, K.-H., 1994a, Identification and in situ detection of gram-negative filamentous bacteria in activated sludge, *System. Appl. Microbiol.* 17:405–417.

Wagner, M., Erhart, R., Manz, W., Amann, R., Lemmer, H., Wedi, D., and Schleiffer, K.-H., 1994b, Development of an rRNA-targeted oligonucleotide probe specific for the genus *Acinetobacter* and its application for in situ monitoring in activated sludge, *Appl. Environ. Microbiol.* 60:792–800.

Wagner, M., Rath, G., Amann, R., Koops, H.-P., and Schleiffer, K.-H. 1995. In situ identification of ammonia-oxidizing bacteria, *System. Appl. Microbiol.* 18: 251–264.

Wanner, J., 1994, *Activated Sludge Bulking and Foaming Control*. Technomic Publishing, Penn.

Wanner, J., and Grau, P., 1989, Identification of filamentous microorganisms from activated sludge: A compromise between wishes, needs and possibilities, *Wat. Res.* 23:883–891.

White, R., Jobling, S., Hoare, S.A., Sumpter, T.P., and Parker, M.G., 1994, Environmental persistent alkylphenolic compounds are estrogenic, *Endocrinology* 134:175–182.

Woese, C.R., 1987, Bacterial evolution, *Micro. Rev.* 51:221–271.

30

ANAEROBIC BIOTECHNOLOGY FOR SUSTAINABLE WASTE TREATMENT

W. Verstraete, T. Tanghe, A. De Smul, and H. Grootaerd

University of Gent
Lab. Microbial Ecology
Coupure L 653, B-9000 Gent, Belgium

1. ABSTRACT

Anaerobic digestion of dissolved, suspended and solid organics has rapidly evolved in the last decades, but nevertheless still faces several scientific unknowns. For treating wastewaters, a novel and highly performing new system has been introduced in the last decade, i.e., the upflow anaerobic sludge blanket system (UASB). This reactor concept requires anaerobic consortia to grow in a dense and eco-physiologically well organized way. The application of this bioprocess in worldwide wastewater treatment is indicated. Challenges for the next decade are: the design of adequate bio-supportive supplements for high-rate anaerobic reactors, the development of proper measurement and control equipment, the direct anaerobic treatment of domestic sewage in tropical countries, the combination of anaerobic wastewater treatment with N and P nutrient removal and finally the total reclamation of treated wastewater.

For the treatment of organic suspensions, there is currently a tendency to evolve from the conventional mesophyllic continuously stirred tank system to the thermophilic configuration, as the latter permits higher conversion rates and better hygienization levels. Integration of ultrafiltration in anaerobic slurry digestion allows to operate at higher volumetric loading rates and at shorter hydraulic residence times. Challenges for the near future are the improved use of auxin-rich digested liquors in agriculture or forestry and of phosphate recovered from sewage sludge digesters for recycling in the chemical industry.

With respect to organic solids, the recent trend in society towards source separated collection of biowaste has opened a broad range of new application areas for solid state fermentation. Several full-scale systems are currently in operation in Europe. This technology opens new perspectives for recycling various fractions of domestic wastes and furthermore has important implications for the producers of consumer goods. Challenges are the rerouting of the fermentation so that less odorous compounds are formed, the implementation of anaerobically-aerobically treated compost in soil bioremediation and the more accurate estimation of the positive significance of anaerobic digestion of biowaste in the global greenhouse effect. In recent years, anaerobic biocatalysts have become of inter-

Biotechnology in the Sustainable Environment, edited by Sayler *et al.*
Plenum Press, New York, 1997

est for the treatment of xenobiotic pollutants in wastestreams and polluted sites. Novel challenges are the facilitated production of such biocatalysts, their effective use in the environmental technology and the evaluation of their effects in pragmatic ecotoxicology. Finally, the recent progress of implementing sulphate reducing bacteria to recover sulphur from SO_2-wastegases and heavy metals from groundwater is indicated.

2. INTRODUCTION

The Brundtland report and the Rio Conference in 1992 (United Nations, 1992; Larsen and Gujer, 1996a) have forced technologists to rethink their procedures and processes in terms of sustainability. However, the report also pointed out that protection of ecosystems has no chance in areas of extreme poverty and that economic prosperity depends on the conservation of an essential physical basis for our activities. This clearly sets the stage for duality: indeed society has to produce goods and assure services but at the same time, it generates wastes and entropy which endanger the prospects of future generations.

Environmental technologists have in the last decades been very concerned with straightforward technological-economic challenges such as drinking water production, wastewater treatment, refuse handling and treatment, soil and sediment clean-up and wastegas purification. Only recently, they have started to look at their activities from the point of view of sustainability and they have had to admit that in many cases, they were far from holistic (Verstraete and Top, 1992). Typical examples of non-sustainable approaches are current practices in aerobic wastewater treatment and refuse landfilling.

In the field of biotechnology, the anaerobic bacteria have specific characteristics. Indeed, only a minor part of the energy, potentially available in the substrate, is consumed by the organisms bringing about the conversion. Hence, they act as true catalysts: they bring about the treatment but do not change themselves very much in mass or volume. This principle makes anaerobic biotechnology particularly suited in the context of sustainable development.

The aim of this paper is to review some anaerobic process technologies and indicate some areas of research and development in order to increase their applicability in the context of sustainability. For fundamental aspects of anaerobic digestion, the reader is referred to Verstraete et al., (1996). The subsequent considerations deal with actual processes and the research and development challenges they hold for the next century.

3. THE TREATMENT OF WASTEWATERS

3.1. State of the Art

One of the most important events in wastewater treatment in the last decades has been the development of the so-called Upflow Anaerobic Sludge Blanket or UASB reactor (Lettinga et al., 1980; Verstraete et al., 1996). The principle of the reactor is quite straightforward: the wastewater is pumped upwards through a reactor under strictly anaerobic conditions at a rate between 0.5 and 1.5 meters per hour; inside the reactor a selection process occurs which can result in the growth of anaerobic microorganisms in a kind of conglomerate (granule) varying between 0.5–5 millimeters in diameter. These granules are powerful biocatalysts and convert the biodegradable organic matter in the influent in a rapid (space loadings varying from 10–20 kilograms chemical oxygen demand, or COD, per cubic meter reactor per day) and complete way to biogas. Actually, the granular biomass is such a valuable biocatalyst that it is to our knowledge the only mixed culture which is at present commercially handled worldwide at a respectful price of the order of about 1–2 US dollars (USD) per kg dry weight. The sludge is separated from the water and the gas phase by means of an internal settler. Generally, effluents approaching discharge standards are thus obtained for wastewaters from breweries and soft drink plants, from potato processing plants and from certain paper recycling plants. For concentrated wastewaters, an aerobic treatment has to succeed the anaerobic treatment; yet, it is smaller and less energy consuming. In Table 1, an example is given of the costs of direct aerobic treatment and anaerobic plus aerobic treatment respectively. Clearly, the anaerobic system consumes less energy and while producing much less surplus sludge.

Accepting that anaerobic digestion generally cannot provide a complete treatment (i.e., various minerals are left), understanding of process fundamentals has overcome drawbacks such as (Verstraete et al., 1996):

- The presumed low stability of anaerobic treatment—anaerobic digestion systems are highly stable, provided they are designed, operated and controlled properly;
- The slow start-up of reactors—full-scale installations can now be started up within a few weeks, sometimes even days, due to a better understanding of the growth conditions of anaerobic bacteria and to large quantities of highly active anaerobic sludge from existing full-scale installations that can be used as inoculum;
- Production of malodorous nuisance problems—problems in this respect can be prevented by using relatively simple means, i.e., physical-chemical methods, or biofilters; and
- Relatively high susceptibility of methanogens and acetogens to toxic compounds—despite this, the potential of anaerobic consortia to adapt and to effect conversions of unwanted chemicals have been found to be much larger than perceived before.

3.2. Challenges

At present, several hundreds of UASB reactors have been installed worldwide, particularly to treat industrial wastewaters with a COD exceeding 2.0 grams per liter (g/l).

Table 1. Comparison of the costs of an anaerobic/aerobic treatment of wastewater and a direct aerobic treatment

	Anaerobic + aerobic (USD/inhabitant equivalent/yr)	Aerobic (USD/inhabitant equivalent/yr)
Investment*	6.8	8.0
Operation		
Aeration	0.6	3.3
Sludge dewatering	0.2	1.7
Sludge disposal	0.7	6.5
Others	4.0	4.0
Total	12.3	23.5
Biogas use	−2.4	0

*The overall treatment concerns 52,000 inhabitant equivalents; 1 I.E. = 135 grams COD per day.
Source: Derycke et al., 1993.

They generally are implemented when the wastewater is rich in carbohydrates and relatively poor in other contaminants. A number of aspects are currently in need of further development:

- The growth of the sludge bed is dependent on a number of factors. There is substantial evidence that a fraction of the minimum (10%) of the COD must be high energy substrates in order for the acidogenic bacteria to grow and produce exocellular polymers (Vanderhaegen et al., 1991), and this is not always the case in practice. Moreover, there are indications that surfactants can be of specific use in the induction of granulation, as they lower the surface tension (Daffonchio et al., 1995; Thaveesri et al., 1995). In this respect, in order to make UASB technology a more reliable technology, it appears worthwhile to develop a type of bio-supportive supplement for these reactors for periods of start-up or of low-quality input wastewater, where the biocatalyst may be maintained in a proper state.
- Although the anaerobic digester systems are relative robust, they must however be protected against shock loadings of, for instance, substrate, soda, and peroxides, etc. Proper design of the equalisation basin can generally ascertain the overall safety of the process. Yet, on-line warning systems capable of detecting an oncoming calamity early would be an important improvement. Current developments (Grijspeerdt et al., 1995) indicate that integrated process control of these reactor systems will become available in the near future.
- The COD of domestic sewage typically varies between 0.3 and 1 g/l and generally the temperature of these wastewaters is rather low. Hence, the biogas produced does not suffice to heat them to the proper 25–30°C minimum temperature. Nevertheless, direct anaerobic treatment of sewage has become a reality and several full scale systems are currently operational in tropical countries such as Columbia, Brazil and India (Van Haandel and Lettinga, 1994). Yet, there is room for improvement. Indeed, the particulate matter in domestic wastewater tends to accumulate in the upflow reactor and thus occupies the space of the methanogenic biocatalyst. Moreover, the treated water still contains sulphide and ammonium; the first can give odor nuisance and the second may contribute toward fish mortality in receiving waters. For tropical countries, direct anaerobic treatment of domestic sewage is a channel that must be explored further in terms of additional sun-heat input (e.g., by means of solar heating systems), and additional nutrient removal (e.g., by means of a nutrient immobilizing straw-biofilter as reported by Avnimelech et al., 1993).
- In industrialized countries, domestic wastewater has to be purified not only in terms of its carbonaceous load, but also in terms of its nitrogen and phosphorous content. These treatment processes require a minimum COD: nitrogen (N) ratio of 4 and a minimum COD: phosphorous (P) ratio of 20. Hence, when the water is rich in these nutrients, anaerobic pretreatment is not directly compatible with advanced treatment since it takes away the necessary reducing equivalents. Particularly N removal is a burden in this respect. Yet, it must be stressed that the current approach of removing nitrogen by the sequence of nitrification (requiring energy) and subsequent denitrification (requiring reducing equivalents which otherwise could be recovered as methane gas) is not holistic. One could visualize of systems by which the so-called "Anthropogenic Nutrient Solution" fraction (e.g., urine), containing some 75 % of the N discharged per inhabitant per day, is collected in time pulses by the sewerage system (Larsen and Gujer, 1996b) . The latter con-

centrated flows would then permit ammonia recovery by stripping. Concomitantly, the remainder of the domestic sewage would be more amenable to anaerobic treatment. Finally, there is evidence that it must be possible to anaerobically oxidize ammonium to nitrogen gas, but this process awaits further technical development (Muller et al., 1995; Van de Graaf et al., 1995)

- The most important recovery product in wastewater treatment is the water itself. As a matter of fact, in the coming decades, population growth will inevitably push many developing nations into conditions of chronic water scarcity. By the year 2000, two-thirds of the countries of the world are expected to fall into the category of low to very low water availability. In areas of conflict, water is of growing importance as a strategic weapon (Clarke, 1992). At present, several industries are already implementing wastewater reclamation by using anaerobic technology as illustrated in Figure 1 (Verstraete, 1995). A combination of the biodegradative possibilities of anaerobic and aerobic microorganisms, along with physico-chemical techniques, offers effective perspectives for overall water recycling.

4. DIGESTION OF ORGANIC SLURRIES

Organic slurries containing particulate organic matter such as animal manures and primary and secondary sewage sludges are normally digested in completely mixed reactors at volumetric loading rates which are quite low, such as 1 to 3 kilograms (kg) COD per cubic meter (m^3) per day. Little increase in rate or performance has been achieved over the last decades.

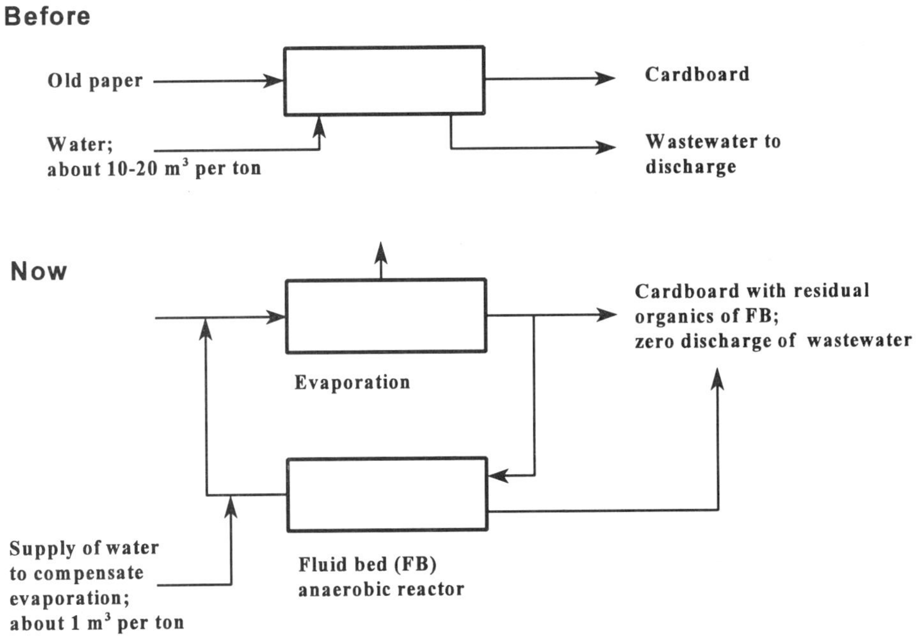

Figure 1. Zero effluent of cardboard production by recycling the water over an anaerobic fluid bed reactor (Barascud, 1993).

4.1. State of the Art

New insights obtained during the last decade in thermophilic digestion resulted in the construction of large scale thermophilic digesters to treat farm manure. The digesters co-treat certain organic wastes from the agro and food industries and ensure both good stabilization and a fair degree of hygienization. Normally, a reduction in fecal enterococci from 10^4–10^5 colony forming units per milliliter (CFU/ml) may be achieved (Bendixen, 1994). This is the most important factor, as it is by no other means economically feasible to provide a fair hygienization for farm wastes.

The developments in thermophilic slurry digestion open new possibilities in the field of sewage sludge digestion. Conventionally, sanitary engineers preferred mesophyllic digestion for reasons of reliability. Yet, regulations in terms of pathogen reduction have become more stringent. In the US, for example, biosolids destined for land should have $< 2.0 \times 10^6$ faecal coliforms per gram total solids. Most US mesophilic digesters achieve the standard (97%), although usually just barely (Stukenberg et al., 1994). In the near future, it should be possible, both in terms of thermal and microbial engineering, to re-examine the digesters to operate in a thermophilic mode and thus achieve higher sanitation standards.

4.2. Challenges

It has been demonstrated that digested residues can contain plant hormones (Kostenberg et al., 1995). Hence, slurries thus enriched should be upgraded for special purposes, such as the growth of specially demanding plants. Of particular interest could be the use of such digested wastes to produce plant plates capable to facilitate reforestation.

Anaerobic sludge digestion of the primary and secondary sludge produced in domestic wastewater treatment allows the attainment of up to 80% energy independency for a domestic wastewater treatment plant, by using the methane gas produced. Yet, the conventional digester for organic slurries is a completely mixed reactor, with no solids recycle. Consequently, the solids residence time (SRT) is identical to the hydraulic residence time (HRT). This limits the volumetric capacity of the digester, since a relatively long SRT (15 to 20 days) is required for effective solids destruction. Improvements to the conventional reactor design generally involve methods to selectively retain the solids in the digester. Solids-liquid separation can be effectively done by membrane separation techniques. Subsequent recycling of the solids into the digester enables operation at a higher active biomass concentration, allowing the loading of organics can be increased. Since the HRT is decoupled from the SRT, the volumetric throughput of the digester can be increased. It also allows the maintenance of a constant sludge age in the digester.

The coupling of anaerobic digestion with either ultrafiltration or microfiltration is therefore an area of development which offers interesting prospects (Ross et al., 1994; Nagano et al., 1992; Harada et al., 1994). Besides the possibility of having a more complete degradation of the solids, it also gives permeates which are free of solids. Since phosphates generally are associated with the particulates, this technology thus allows to produce waters poor in P and inversely, concentrates enriched in insoluble and sorbed phophate. As a matter of fact, the re-use of wastewater phosphate is at present limited to the use of phosphate-rich sludge in agriculture. This is not always possible or desirable in case the sludge is contaminated. Hence, in terms of sustainability, it is essential that new approaches are developed in which the wastewater phosphate concentrated may be by means of digestion coupled with membrane technology. The phosphate may then be upgraded to a product capable of being recycled in the detergent or food industry.

5. SOLID STATE FERMENTATION OF BIOWASTES

For the fermentation of solid wastes, i.e., domestic household waste, most of the COD is in the form of insoluble volatile solids (VS) at a concentration of at least 150 kg COD per ton. Hence, an intensive hydrolysis/acidification phase is required in solid state fermentation (20–40% total solids) where the solid waste is converted into methanogenic substrates. This can be achieved very effectively, particularly at temperatures of 50–55°C. In this way, some 100–150 m^3 of biogas can be recovered per ton original organic solids (30–40% dry weight) (Baeten and Verstraete, 1993).

5.1. State of the Art

The anaerobic fermentation of domestic refuse has been of interest in the US for about two decades (Wujich & Jewell, 1980). However, effective full-scale technical developments in solid state fermentation (20–40% total solids) were realized only in the last few years in Europe (Baeten & Verstraete, 1993; Gellens et al., 1995). A variety of reactor designs, which vary considerably in terms of operational temperature and residence time, have come into practice in recent years. Figure 2 gives a typical flow scheme of a full-scale solid state digester and shows that anaerobic digestion can be a most valuable technology to convert source-separated municipal solid waste into two valuable end products, energy (biogas) and a peat like compost (so-called humotex in this process). By this method, a final product is obtained which has been subjected to both anaerobic and aerobic microbial biodegradation processes; the resulting so-called double processed

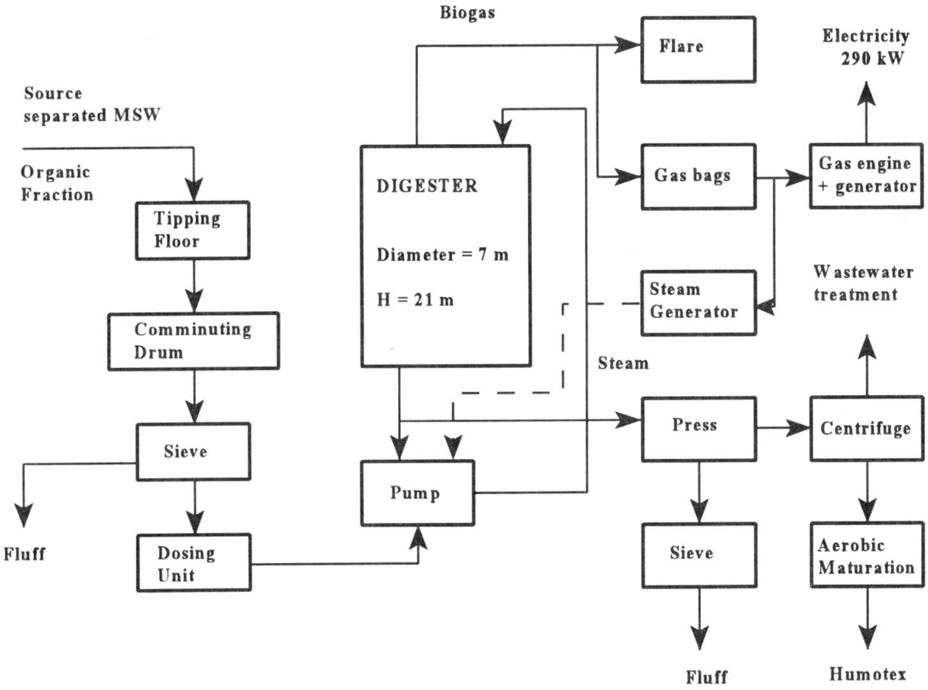

Figure 2. Flowsheet of the full-scale DRANCO plant in Brecht, Belgium (Gellens et al., 1995). The plant serves a population of ± 81,000 inhabitants and treats 10,500 tons of biowaste per year.

Table 2. Investment (1000 USD) and operation (USD/ton waste) costs for the biological treatment of municipal biowaste. The data relate to installations with a capacity of 25,000 ton/year

	Investment	Operation
DRANCO	11,000	70
BIOCEL	5,600	48
BTA	8,700	57
Aerobic Composting	6,000–8,000	60–75

Note: Specific costs according to Edelmann & Engeli (1992).
Total costs (capital + operation) for
- anaerobic systems 108-150 USD/ton
- aerobic systems 135-172 USD/ton

(DP) compost has improved hygienic qualities relative to conventional single processed (SP) compost (Gellens et al., 1995). Moreover, as indicated in Table 2, the overall costs of the anaerobic digestion of biowastes are comparable to that of conventional aerobic composting.

5.2. Challenges

The reservoir of carbon as carbon dioxide (CO_2) in the atmosphere is no longer at steady state and is growing from year to year. Between 1860 (prior to the industrial revolution) and 1980, atmospheric CO_2 rose by approximately 70 parts per million (ppm) from pre-industrial 270 ppm to 340 ppm (La Marche et al., 1984). The increasing CO_2 concentration of the atmosphere, which is occurring at a rate of about 1 ppm per year, is undeniably giving rise to major concerns about the equilibria of our ecosystems by the year 2030. Slowing-down the greenhouse effect is a global responsibility and is supported by the international bodies implementing the Agenda 21 of the United Nations Conference on Environment and Development (UNCED) of June 1992 in Rio de Janeiro. The concern about the contribution of CO_2 emission to the greenhouse effect has recently been concertized by a proposal of the European Union to impose a CO_2 levy. The proposed levies are considerable and amount to about 15 European currency units (ECU) per ton mineral oil (e.g., gasoline, domestic fuel and heavy fuels) and 3 ECU per ton of CO_2 emitted by combusting cokes, natural gas, tar and peat (Vande Woestyne et al., 1994).

If the industrialized countries continue to opt for mass incineration of their organic residues, they avoid the opportunity to act, when there is still time, to curb their CO_2 emission. They should also certainly avoid further dumping of organic wastes in uncontrolled landfills, since this leads to CO_2 and methane (CH_4) emissions. Although it is known that CH_4 traps heat more effectively than CO_2 does, the biggest uncertainty arises through the greenhouse weighting of methane. Literature data are very divergent. According to Atlas and Bartha (1993), methane traps heat 4 to 5 times more effectively than CO_2, while Wallis (1994) has reported a factor of 35 times more effective. This means that even in relatively small amounts, methane can contribute significantly to the greenhouse effect.

A more environmentally friendly alternative for the disposal of biowaste is to convert it to stable organic matter using aerobic or anaerobic composting processes. When anaerobic residues are fermented in a digester, the labile fraction is converted in a contained system to biogas. The latter is combusted to CO_2 thereby saving on fossil fuel energy. The remainder of the organic matter is composed of lignin-linked organics which decompose very slowly. Actually, like humus in soil, one can approximate their decomposition rate at 1–2% per year, which corresponds with a 50% disappearance time (DT_{50}) of 35–70 years. In other words, by subjecting organic residues to anaerobic digestion and thus converting them for a major part (60 to 70% reduction) to stable humus, one can build a significant sink of organic carbon in the biosphere, releasing CO_2 over a long period of time.

The amount of organic matter produced in industrialized societies per capita per year which is treatable by anaerobic digestion can be estimated as 100 kg biowaste at 30% dry matter, i.e., 30 kg dry weight biowaste and 27 kg dry weight sewage sludge production. Consider the case of the solid biowastes only, where they are treated and converted anaerobically. This represents about 30 kg organic matter corresponding to approximately 30 kg CO_2 per capita per year which is emitted if this route is chosen. If however, this biowaste is landfilled, the greenhouse effect of the landfill gas (CH_4 and CO_2), expressed as the amount of CO_2-equivalent greenhouse gas released, would be in the order of 315 kg. The projected decrease of CO_2 for the industrialized countries is 5% (UNCED, 1992) or about 1 ton CO_2 per capita per year. This means that by proper treatment of the biowaste produced, a significant decrease of CO_2 emission can be obtained without affecting society's level of comfort. However, to make such an approach conceptually attractive, the humus produced should be used properly and an overall biosphere sustainability management scheme should be adopted. Figure 3 illustrates the concept of inter-region recycling of organics (Vande Woestyne et al., 1994); the latter is slowly gaining support from political and commercial authorities.

For the sake of completeness, it should be pointed out that aerobic composting and incineration also abate the uncontrolled emission of CH_4 gases in the atmosphere. However, aerobic composting does not allow the production of a highly hygienic residue (Baeten and Verstraete, 1993). Incineration, on the other hand, gives rise to a direct conversion of all organics to CO_2. In contrast herewith, in the case of both aerobic and anaerobic bioconversions, about 50% of the CO_2 is rapidly released while the rest is produced with a lag-time of approximately 50–100 years.

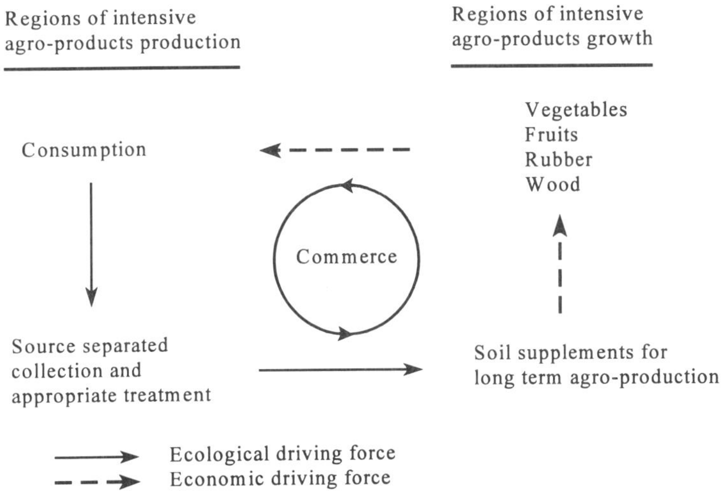

Figure 3. Flow-sheet of inter-regional recycling of organic matter (Vande Woestyne et al., 1994).

The material leaving the anaerobic digester contains considerable amounts of amines, ammonium and sulphides. Although the digester is completely closed and free of odor, the handling of the digested residue can give rise to odor problems. Rechanneling anaerobic digestion pathways so that malodorous compounds can be avoided or further degraded would be a major step forward. The very offensive dimethyl sulfide appears to be due to acetogenic bacteria such as *Acetobacterium woodii* and *Clostridium thermoaceticum* (Kene and Hines, 1995). Rerouting of terminal anaerobic digestion not only offers important hygienic possibilities for hospitals and care-taking homes, but also for anaerobic digestion of biowastes.

The so-called double processed (anaerobically and subsequently aerobically) DP-compost is quite distinct in properties from the single processed SP-compost (Table 3). It contains less salts and is also enriched in bacteria which remove the remainder of CH_4, metal sulphides and ammonium. These methanotrophs, sulfur oxidizers and nitrifiers have important potentials for soil remediation. As a matter of fact, field experiences with DP-compost tend to indicate that this product is highly conducive to the enhancement of pollutant biodegradation and the binding of pollutants to soils and sediments (unpublished results).

It is clear that a variety of aspects need further research in order to better delineate the role of anaerobic digestion in improving the long-term sustainability of the biosphere. Urgently needed are:

- Accurate estimates of the mass balances of organics which can be collected and treated by the anaerobic route;
- Environmental impact analyses of the anaerobic route versus the conventional alternatives in terms of the greenhouse effect;
- Potential financial support which can be supplied via the CO_2 levies as proposed by the European Union; and
- Removal of psychological barriers in favor of incentives in the industrial and non-industrialized countries and promotion of inter-region recycling of organics and the concomitant contributions to biosphere sustainability.

Table 3. Characteristics of DP-compost relative to a reference conventional aerobic biowaste compost. Both were produced on the basis of biowaste containing soiled paper

Characteristics	DP biowaste compost	Conventional SP compost
Dry matter (%)	52.6	66.6
Volatile solids (% VS/DM)	37.2	30.9
pH	8.4	8.4
EC (μS/cm)	1153	2720
P (g/ton WW)	1746	3144
NH_4^+–N (g/ton WW)	890	257
NO_3^-–N (g/ton WW)	95	337
Total N (g/ton WW)	7200	10000
Heavy metals	< German RAL standards	< German RAL standards
	Moderate for P, K, Ca & Mg	Moderate for P, K, Ca & Mg
Fertilization potential	None	Present
Indicator micro-organisms		
Weed seeds	None	Few
Phytotoxicity	None	Moderate
Application	Soil conditioner, potting substrate	Soil conditioner, potting substrate

WW = Wet weight

6. ENGINEERING OF ANAEROBIC BIOCATALYSTS FOR XENOBIOTIC REMOVAL

In recent years, it has become increasingly clear that anaerobes can be used quite advantageously to convert a variety of xenobiotics. It was well known for a long time that anaerobes can use nitro-compounds as electron acceptors and thus convert these compounds to more reduced and often more polymerized compounds (Field et al., 1995). Reductive dechlorination has been shown to be an exergonic process with deltaG values of -150 to -190 kilojoules reaction, with hydrogen (H_2) as the electron donor (Dolfing and Janssen, 1994). Mohn and Tiedje (1991) showed that dehalogenation is coupled to ATP formation and Holliger et al., (1993) showed that it leads to cell growth. Recently, cloning of the 4-chlorobenzoate dehalogenase enzyme complex of the aerobic bacterium *Pseudomonas* sp. and transfer into a denitrifying bacterium has resulted in a constructed strain able to metabolise 4-chlorobenzoate under denitrifying conditions (Coschigano et al., 1994). Many genera are able to dehalogenate, including *Pseudomonas* (Fulthorpe et al., 1993), in addition to anaerobes such as *Clostridium* (Madsen and Licht, 1992) and sulphate reducing bacteria (Mohn and Tiedje, 1991; Pavlostathis and Zuang, 1991).

6.1. State of the Art

Few studies have reported on the degradation of halogenated compounds in operating methanogenic reactor systems. Removal of adsorbable organic halogen (AOX) by a UASB treating kraft-mill bleach wastewater was 27 to 65%, depending on the residence time (Parker et al., 1993). Mohn and Kennedy (1992) reported incomplete degradation of tri- and dichlorophenols by anaerobic sludge. Hendriksen and Ahring (1992) reported the UASB degradation of pentachlorophenol to, mainly, di- and monochlorophenols. However, shortly after this, Hendriksen et al., (1992) reported almost complete (94%) dehalogenation of pentachlorophenol in an UASB. Glucose addition was necessary for sludge growth and increased the total dechlorination by a factor of 5, possibly by acting as an electron donor.

Spontaneous adaption of an UASB reactor to a new xenobiotic compound is a slow process, although it can be accelerated. Ahring et al., (1992) inoculated UASB reactors with either a mono culture of *Desulfomonile tiedjei*, which was able to dechlorinate 3-chlorobenzoate (3-CB), or a three-membered consortium consisting of *D. tiedjei*, a benzoate degrader and a H_2-utilizing methanogen. No degradation occurred in a non-inoculated control reactor started up with the same granular sludge, but inoculated UASB reactors rapidly transformed 3-CB. The degradation was stable for several months, and with antibody staining, it was shown that *D. tiedjei* was incorporated in the methanogenic granules. This demonstrates that inoculation of UASB reactors with specific degraders can be an effective means of establishing the consortium needed for degradation.

The use of anaerobic technology has also been reported in soil and sediment cleaning. The firm Umweltschutz Nord (Ganderkesee, Germany) reports that it has incorporated an anaerobic tunnel (TERRANOX®system) in its soil treatment system. Soil of a former trinitrotoluene (TNT) production plant was treated by a two-step anaerobic/aerobic biological process. In the first anaerobic step of the biotreatment, TNT and aminonitrotoluenes are reduced in the anaerobic reactor by adding nutrients and co-substrates for the autochthonous anaerobic bacteria. The final reduction product, triaminotoluene, is highly reactive and polymerizes under aerobic conditions. Thus, the second aerobic treatment process provides irreversible binding of triaminotoluene to the soil fraction and mineralization of the fermentation products from the anaerobic treatment (Stolpmann et al., 1995; Daun et al., 1995). Instead of complete degradation, the biotreatment process results in humification of the contaminants in the soil which can be a very cost-effective process for cleaning up TNT-contaminated sites. This approach has also been successfully used to decontaminate soils polluted with chloroethene and BTEX aromatics (mainly toluene, ethylbenzene and xylene). At the end of the combined anaerobic and aerobic treatment phases (six and two weeks, respectively), volatile chlorinated hydrocarbons were reduced from 2975 micrograms per kilogram (mg/kg) to 4 mg/kg and BTEX aromatics were reduced from 102,970 mg/kg to 27 mg/kg (both reported as soil dry weight) (Meyer et al., 1993).

6.2. Challenges

It actually often takes months before a community of microorganisms, obtained from sewage sludge or soil, adapts to the metabolism of xenobiotics. Important for the understanding of the long start-up times is the finding that the ability of a bacterial community to dehalogenate is often associated with elevated amounts of plasmids (Fulthorpe et al., 1993). Adaptation of the association is probably based on the proliferation of the right plasmids and of the desired species, taking many generation times. There is clearly a need for better insight in genetic evolution, plasmid transfer and species interaction in anaerobic communities dealing with xenobiotics (Fetzner and Lingens, 1994; Top and Forney, 1994).

Recently, Jain et al., (1995) reported the fact that they have developed a culturable and stable granular anaerobic microbial consortia capable of mineralizing polychlorinated biphenyls (PCBs). The fate of PCBs in sediments and soils incubated with this consortia has been investigated respectively by Natarajan et al., (1995) and Jain et al., (1995). In both cases a notable observation was the occurrence of dechlorination predominantly higher chlorinated compounds (i.e., having more than five sites where chlorine binds with biphenyl) and relative changes on tri and tetra chlorinated congeners. These changes indicate that the progressive dechlorination of PCBs occurred due to the biological activity of the anaerobic microbial granules. Further research by Natarajan et al., (1996b) showed extensive dechlorination of less chlorinated congeners and the appearance of biphenyl, which indicates that the anaerobic consortia may be capable of total dechlorination of PCBs. Complementary results demonstrate the enhanced dechlorination of PCBs in naturally contaminated river sediments by augmentation with exogenously developed anaerobic microbial consortia (Natarajan et al., 1996a). The degradation and mineralization of biphenyl by the same anaerobic microbial consortium/granules has recently been observed, which completes the total microbial breakdown of PCBs (personal communication, Jain M.K., 1996). This opens interesting perspectives for the industrial production of these consortia and the subsequent implantation of these consortia in polluted environments such as river sediments.

Pesticide legislation worldwide accepts the fact that for many highly toxic chemicals, complete mineralization is only to a minor extent taking place. This is due to the binding of the xenobiotic to the soil humus matrix. In this respect, anaerobic binding of pollutants has been reported (Bilbao et al., 1996) and certainly opens perspectives. However, it is clear that together with such integration of the xenobiotic into the soil organic complex, appropriate questions of ecotoxicity and its determination must be asked and answered (Verstraete et al., 1995).

7. WASTEGAS TREATMENT

In the second half of the 20th century, the growing concern about the possible worldwide effect of air pollution by industry and traffic on the general quality of life has resulted in ever-stricter norms for wastegases containing sulphur oxides (SO_2) and nitrogen oxides (No_x). In most processes these standards (i.e., emissions below 40 kilogram per hour for NO_x–N and below 60 kilogram per hour for SO_2, according to European legislation) cannot be reached without superposition of a treatment of the generated wastegas to modifications in the process itself (Boersema et al., 1986).

7.1. State of the Art

The different recognized wastegas treatment technologies for NO_x control are mainly chemical and catalytic (Bosch and Janssen, 1988). The conventional SCR (Selective Catalytic Reduction) process, in which No_x is transformed into N_2 at 300–400°C on a fixed vanadium oxide/tungsten oxide/titanium dioxide catalyst ($V_xO_y/W_xO_y/TiO_2$) bed with residual O_2 and injected NH_3 , gives the best performances (>90%) but is expensive (4,000 USD to 10,000 USD per ton of NO_x removed). Therefore, in the last five years, innovations in NO_x control technology tend to combine SCR with the relatively low-cost (500 USD to 1,500 USD) and flexible urea-based SNCR (Selective Non-Catalytic Reduction) process. The SCNR process, which by itself is only able to reduce the NO_x level by up to 75%, consists of introducing reduced nitrogen species such as commercially available urea fertilizers into the furnace (700–1000°C) of the process (Christiansen, 1995). Competitive biotechnological anaerobic alternatives for NO_x control are not yet available, probably because of the poor water solubility of NO, i.e., 0.07 liter per liter water under normal conditions; NO is the main component of NO_x emissions.

Up to the beginning of the nineties, flue gas desulfurization (FGD) was mostly achieved by the very efficient (>95%) and regenerative LFSO (Limestone/Gypsum Forced Oxidation)-process (Kwong and Meissner, 1995). In the latter, SO_2 is first transformed to $SO_3^{2-}/SO_4^{2-}/HSO_3^-$ by scrubbing the waste gas with an alkaline aqueous solution that can be reused after removal of the formed sulphur oxides as gypsum. Recent studies on lab- and pilot-scale of the flue gas from a 600 megawatt power station in Geertruidenberg (The Netherlands) by the two Dutch companies Hoogovens TS E&E B.V. and Paques N.V. confirm the possibility of a biotechnological process, called Biostar, with approximately 30% less total costs and with a performance equivalent to the LSFO process (Hoogovens TS E&E, 1994). In this process, valuable sulphur is formed through a cascade of an anaerobic and aerobic treatment of the $SO_3^{2-}/SO_4^{2-}/HSO_3^-$ -rich scrubwater. By adding a relatively inexpensive stoichiometric amount of a COD source (e.g., ethanol or synthesis gas) to the scrubwater in the anaerobic reactor, the sulphur oxides are converted by SRB (Sulphate Reducing Bacteria) to sulphide that is consequently aerobically transformed to high quality sulphur under oxygen limiting conditions by colorless sulphur bacteria. Because of the significantly lower capital cost and cost of bulk reagents, in combination with the economically more attractive end product sulphur, the biotechnological solution for FGD is believed to out compete the more expensive conventional FGD-technology in the near future, or to be at least a potential post-treatment in the event of more stringent legislation.

7.2. Challenges

Challenges of wastegas treatment are:

- Integrating the knowledge of denitrification and nitrification of waste water in NO_x removal technologies to design a competitive but economically more attractive process than the combined SCR/SNCR process, and
- Transferring the understanding of the promising biotechnological FGD-technology to a successful treatment of sulphate-rich wastewaters derived from several industries like petrochemistry, metallurgy and mining.

8. VARIOUS

An interesting case study of the successful full-scale application of anaerobic biotechnology for sustainable waste treatment is the remediation of groundwater contaminated with the heavy metals Zn (50 milligrams per liter, or mg/l) and Cd (0.1 mg/l) and with sulphate (500 mg/l), of the zinc refinery Budelco B.V. in the Netherlands. Only a SRB-based water treatment process, compared with an ion exchange and liquid membrane permeation process, could fulfill the Zn (0.3 mg/l), Cd (0.01 mg/l) and the sulphate (200 mg/l) standards while pumping groundwater at a rate of 5,000 cubic meters per hour (Yspeert and Buisman, 1995). In the anaerobic sulphate reducing reactor, the heavy metals are precipitated with the produced sulphide, while in the aerobic compartment, the excess sulphide is converted into elemental sulphur. The collected solids, metal sulphides and sulphur, are brought back to the zinc refinery plant. This rather special case of anaerobic biotechnology can be considered indicative of many other potentials that are awaiting a chance and challenge to be explored.

9. SUMMARY

Even without detailed knowledge of the specific characteristics of anaerobic bacteria, it should be clear from the various examples that anaerobic biotechnology in many cases offers valid alternatives for waste treatment compared to other existing chemical or biotechnological treatment methods. Moreover, applying anaerobic biotechnology for waste treatment has the advantage of resulting in a more sustainable waste treatment, because contrary to most existing technologies, anaerobic biotechnology allows the recovery of energy from waste (e.g., UASB reactor for wastewater treatment, anaerobic digestion of solid waste), reduces the numbers of pathogenic bacteria in waste (e.g., thermophilic digestion of farm waste), and recovers metals from wastewater (e.g., Zn).

However, one should recognize that additional research is needed to further practical applications of anaerobic biotechnology. This research should concentrate on a better understanding of the basic principles of anaerobic processes, but should also pay attention to the development of relatively simple and user-friendly methods to efficiently monitor and control these anaerobic processes. The combined effect of such research efforts would result in an easier start-up and improved operational stability of anaerobic processes, which would finally allow the to maximize the benefits of anaerobic biotechnological processes, and thus result in more sustainable ways to treat waste.

REFERENCES

Ahring, B.K., Christiansen, N., Mathrani, I.,Hendriksen, H.V., Macario, A.J.L. and Conway de Macario,E., 1992. Introduction of a *de novo* bioremediation ability, aryl reductive dechlorination, into anaerobic granular sludge by inoculation of sludge with *Desulfomonile tiedjei*, *Appl. Env. Microbiol.* 58: 3677–3682.

Atlas, R.M. and Bartha, R., 1993. *Microbial Ecology : Fundamentals and applications*, 3rd Ed. Benjamin/Cummings Publ. Com. pp. 563.

Avnimelech, Y., Diab, S. and Kochba, M., 1993. Development and evaluation of a biofilter for turbid and nitrogen rich irrigation water, *Wat. Res.* 27: 785–790.

Baeten, D. and Verstraete, W., 1993. In-reactor anaerobic digestion of MSW-organics, In : *Science and engineering of compost - Design, environmental, microbiological and utilization aspects* (H.A.J. Hoitink and H.M. Keener, Eds.), Ohio State University, pp. 111–130.

Barascud, M.-C., 1993. Intégration d'un réacteur anaérobie dans les circuits d'eaux d'une machine à papier. Thèse de docteur en génie des procédés, Institut national polytechnique de Grenoble, pp. 187.

Bendixen, H.J., 1994. Safeguards against pathogens in Danish biogas plants, In : *Proceedings of the 7th International Symposium on Anaerobic Digestion*, 23–27 January 1994, Cape Town, South Africa, pp. 629–638.

Bilbao, V., De Rore, H., Dries, J., Devliegher, W., Top, T. and Verstraete,W., 1996. Soil remediation through binding of pollutants to the soil matrix, Submitted for publication in *Biodegradation*.

Boersema, J.J., De Groot, W.T. and Peereboom Copius, J.W., 1986. Basisboek milieukunde, Boom Meppel, Amsterdam.

Bosch, H. and Janssen, F., 1988. Formation and control of nitrogen oxides. *Catalysis Today* 2: 369–532.

Christiansen, P.B., 1995. NO_x control technology advances rapidly, *Waste Management Technology* 19: 40–43.

Clarke, 1992. *Water: The International Crisis*, Earthscan Publications Ltd., London.

Coschigano, P.W., Häggblom, M.M. and Young, L.Y., 1994. Metabolism of both 4-chlorobenzoate and toluene under denitrifying conditions by a constructed bacterial strain, *Appl. Env. Microbiol.* 60: 989–995.

Daffonchio, D., Thaveesri, J. and Verstraete, W., 1995. Contact angle measurement of cell hydrophobicity of granular sludge from upflow anaerobic sludge bed reactors, *Appl. Env. Microbiol.* 61: 3676–3680.

Daun, G., Lenke, H., Desiere, F., Stolpmann, H., Warrelmann, J., Reuss, M. and Knackmuss, H.-J., 1995. Biological treatment of TNT-contaminated soil by a two-stage anaerobic/aerobic process, in: *Contaminated soil '95* (W.J. van den Brink, R. Bosman and F. Arendt, eds.), Kluwer Acad. Pub., the Netherlands, pp. 337–346.

Dolfing, J. and Janssen, D.B., 1994. Estimates of Gibbs free energy of formation of chlorinated aliphatic compounds, *Biodegradation* 5: 21–28.

Fetzner, S. and Lingens, F., 1994. Bacterial dehalogenases: biochemistry, genetics, and biotechnological applications, *Microbiological Reviews* 58: 641–685.

Field, J.A., Stams, A.J.M., Kato, M. and Schraa, G., 1995. Enhanced biodegradation of aromatic pollutants in cocultures of anaerobic and aerobic consortia, *Antonie van Leeuwenhoek* 67: 47–77

Fulthorpe, R.R., Liss, S.N. and Allen, D.G., 1993. Characterization of bacteria isolated from a bleached pulp mill wastewater treatment system, *Canadian Journal of Microbiology* 39: 13–24.

Gellens, V., Boelens, J. and Verstraete, W., 1995. Source separation, selective collection and in reactor digestion of biowaste, *Antonie van Leeuwenhoek* 67: 79–89.

Grijspeerdt, K., Muller, A., Aivasides, A., Wandrey, C., Guwy, A., Hawkes, F., Hawkes, D., Van Der Schueren, D., Verstraete, W., di Pinto A. & Rozzi,A., 1995. Development of an integrated process control system for a multi-stage wastewater treatment plant - Part I : sensor technology. Poster presented at the 9th Forum for Applied Biotechnology, In : *Proceedings part II of the Ninth Forum for Applied Biotechnology*, Gent, 27–29 September 1995. pp. 2459–2467.

Harada, H., Momomoi, K., Yamazaki, S. and Takizawa, S., 1994. Application of anaerobic-uf membrane reactor for a wastewater containing high strength particulate organics, *Wat. Sci. Technol.* 30: 307–319.

Hendriksen, H.V. and Ahring, B.K., 1992. Metabolism and kinetics of pentachlorophenol transformation in anaerobic granular sludge, *Appl. Microb. and Biotechn.* 37: 662–666.

Hendriksen, H.V., Larsen, S. and Ahring, B.K., 1992. Influence of supplement carbon source on anaerobic dechlorination of pentachlorophenol in granular sludge, *Appl. Env. Microbiol.* 58: 365–370.

Hoogovens Technical Services Energy & Environment B.V. 1994. Biotechnological flue gas desulfurisation: a new biotechnological process for gas cleaning. News bulletin Hoogovens Group 2, January, 1–17.

Jain, M.K., Natarajan, M.R., Nye, J., and Wu, W.-M., 1995. Development of culturable anaerobic microbial consortium for bioremediation of polychlorinated biphenyls, Presented at 1995 PCB Seminar, August 29–31, 1995, Electric Power Research Institute Sponsored, at Boston, MA.

Kene, R.P. and Hines, M.E., 1995. Microbial formation of dimethyl sulfide in anoxic spagnum peat, *Appl. Env. Microbiol.* 61: 2720–2726.

Kostenberg, D., Marchaim, U., Watad, A.A. and Epstein, E., 1995. Biosynthesis of plant hormones during anaerobic digestion of instant coffee waste, *Plant Growth Regulation* 17: 127–132.

Kwong, V. and Meissner, R., 1995. Rounding up sulphur, *Chem. Engineering*, 74–83.

La Marche, V.C., Greybill, D.A. Jr., Fritts, H.-C. and Rose, M.R., 1984, Increasing atmospheric carbon dioxide : Three ring evidence for growth enhancement in natural enhancement in natural vegetation, *Science* 225: 1019–1021.

Larsen, T.A. and Gujer, W., 1996a. Fundamentals of sustainable urban water management, T.A. Larsen, Institute of Hydromechanics and Water Resources Management, Swiss Federal Institute of Technology, personal communication.

Larsen, T.A. and Gujer, W., 1996b. Separate management of anthropogenic nutrient solutions, Accepted for the IAWQ conference in Singapore 1996.

Lettinga, G., Van Velsen, A.F.M., Hobma, W., de Zeeuw, J. and Klapwijk, A., 1980, Use of the upflow sludge blanket (USB) reactor concept for biological waste water treatment, especially for anaerobic treatment, *Biotech. and Bioeng.* 22: 699–734.

Madsen, T. and Licht, D., 1992. Isolation and characterization of an anaerobic chlorophenol transforming bacterium, *Appl. Env. Microbiol.* 58: 2874–2878.

Meyer, O., Refae, R.I., Warrelmann, J. and von Reis, H., 1993. Development of techniques for the bioremediation of soil, air and groundwater polluted with chlorinated hydrocarbons: the demonstration project at the model site in Eppelheim, *Microb. Releases*. 2: 11–22.

Mohn, W.W. and Kennedy, K.J., 1992. Limited degradation of chlorophenols by anaerobic sludge granules, *Appl. Env. Microbiol.* 58: 2131–2136.

Mohn, W.W. and Tiedje, J.M., 1991. Evidence for chemiosmotic coupling of reductive dechlorination and ATP synthesis in *Desulfomonile tiedjei*, *Arch. Microbiol.* 157: 1–6.

Muller, E.B., Stouthamer, A.H. and Van Verseveld, H.W., 1995. Simultaneous NH_3 and N_2 production at reduced O_2 tensions by sewage sludge subcultured with chemolithotrophic medium, *Biodegradation* 6: 339–349.

Nagano, A., Arikawa, E. and Kobayashi, H., 1992. The treatment of liquor wastewater containing high-strength suspended solids by membrane bioreactor systems, *Wat. Sci. Technol.* 26: 887–895.

Natarajan, M.R., Nye, J., Wu, W.-M., Wang, H.Y. and Jain, M.K., 1996a. Reductive dechlorination of PCB-contaminated River Raisin sediments by anaerobic microbial granules, *Appl. Env. Microbiol.* (Submitted, under review).

Natarajan, M.R., Wu, W., Nye, J., Wang, H., Bhatnagar, L. and Jain, M.K., 1996b. Dechlorination of polychlorinated biphenyl (PCB) congeners by an anaerobic microbial consortium, *Environ. Sci. and Technol.* (Submitted, under review).

Natarajan, M.R., Wu, W., Rajan, R., Nye, J., Wang, H. and Jain, M.K., 1995,. Dechlorination of PCBs by anaerobic microbial granules, *Organohalogen compounds* 24: 33–36.

Parker, W.H., Hall, E.R. and Farquhar, G.J., 1993. Assessment of design and operating parameters for high rate anaerobic dechlorination of segregated kraft mill bleach plant effluents, *Wat. Env. Res.* 65: 264–270.

Pavlostathis, S.G. and Zuang, P., 1991. Transformation of trichloroethylene by sulphate-reducing cultures enriched from contaminated subsurface soil, *Appl. Microbiol. Biotechnol.* 36: 416–420.

Ross, W.R., Strohwald, N.K.H., Gobler, C.J. and Sanetra, J., 1994. Membrane-assisted anaerobic treatment of industrial effluents: the South African ADUF process, In: *Proceedings of the 7th International Symposium on Anaerobic Digestion*, IAWQ, RSA Litho Ltd., South Africa, pp. 550–559.

Stolpmann, H., Lenke, H., Warrelmann, J., Heuermann, E., Freuchtnicht, A., Daun, G. and Knackmuss, H.-J., 1995. Bioremediation of TNT-contaminated soil by the TERRANOX® system, in: *Biological unit processes for hazardous waste treatment*, 3(9) (R.E. Hinchee, R.S. Skeen and G.D. Sayles, eds.), Battelle Press, Ohia, USA, pp. 283–287.

Stukenberg, J.R., Shimp, G., Sandino, J., Clark, J.H. and Crosse, J.T., 1994. Compliance outlook : meeting 40 CFT part 503 class B pathogen reduction criteria with anaerobic digestion, *Wat. Env. Res.* 66: 255–263.

Thaveesri, J., Daffonchio, D., Liessens, B., Vander'meeren, P. and Verstraete, W., 1995. Granulation and sludge bed stability in upflow anaerobic sludge bed reactors in relation to surface thermodynamics, *Appl. Env. Microbiol.* 61: 3681–3686.

Top, E. and Forney, L.Y., 1994. The significance of degradative plasmids in soil bioremediation, Med. Fac. Landbouww. Univ. Gent 59/4a, 1847–1855.

UNCED, 1992. Agenda 21, Rio de Janeiro. Bescherming van de atmosfeer. *In*: Bossenverklaring, biodiversiteitsverdrag en klimaat-verdrag, Den Haag, pp. 152–164.

United Nations, 1992, Agenda 21 : Programme of Action for Sustainable Development, ISBN 92–1–100509–4.

Van De Graaf, A. A., Mulder, A., De Bruijn, P., Jetten, M.S.M., Robertson, L.A. and Kuenen, J.G., 1995. Anaerobic oxidation of ammonium is a biological process, *Appl. Env. Microbiol.* 61: 1246–1251.

Vanderhaegen, B., Ysebaert, E., Favere, K., Van Wambeke, M., Peeters, T., Panic, V., Vandenlangenbergh, V. and Verstraete, W., 1991. Acidogenesis in relation to in-reactor granule yield, *Wat. Sci. Techn.* 25: 21–30.

Vande Woestyne, M., Gellens, V., Anasi, I. and Verstraete, W., 1994. Anaerobic digestion and inter-regional recycling of organic soil supplements, In: *Sustainable rural environment and energy networ (SNER) - Biogas technology as an environmental solution to pollution*, Fourth DAO/SREN Workshop, Migal, Israel, 14–17 June 1994. (U. Marchaim and G. Ney, Eds.) REUR Technical Series Number 33, ISSN 1024–2368.

Van Haandel, A.C. and Lettinga, G., 1994. *Anaerobic sewage treatment - A practical guide for regions with a hot climate*, John Wiley & Sons, Ltd. Chichester. ISBN 0–471–95121–8.

Verstraete, W., 1995, Role of biotechnology in water-cycle management, In : *OECD Documents. Bioremediation : The Tokyo '94 Workshop*, 27–30 November 1994, Tokyo, Japan, OECD, Paris, ISBN 92–64–14634–2, pp. 455–467.

Verstraete, W. and Top, E., 1992. Holistic environmental biotechnology, In : *Microbial control of pollution* (J. Fry, G. Gadd, R. Herbert and I. Watson-Craik eds.), Cambridge Univ. Press., pp. 1–18.

Verstraete, W., Top, E., Vanneck, P., De Rore, H. and Genouw, G., 1995. Lessons from the soil, In : *Proceedings of the Fifthe International FZK/TNO Conference on Contaminated Soil*, Contaminated Soil '95, Volume I (W.J. van den Brink, R. Bosman and F. Arendt, eds.), 30 October - 3 November 1995, Maastricht, The Netherlands.. pp. 15–24.

Verstraete, W., de Beer, D., Pena, M., Lettinga, G. and Lens, P., 1996. Anaerobic bioprocessing of organic wastes, Accepted for publication in World Journal of Microbiology and Biotechnology.

Wallis , M., 1994. Waste incineration reassessed, *Warmer Bull.* 41: 18–19.

Wujich, W.J. and Jewell, W.J., 1980. Non-acetoclastic methanogenesis from acetate-acetate oxidation by thermophilic syntrophic co-culture, *Arch. Microbiol.* 138: 263–272.

Yspeert, P. and Buisman, C., 1995. Full-scale biological treatment of groundwater contaminated with heavy metals and sulfate, In : *OECD Documents. Bioremediation : The Tokyo '94 Workshop*, 27–30 November 1994, Tokyo, Japan, OECD, Paris, ISBN 92–64–14634–2, pp. 335–343.

31

ADVANCES IN BIOLOGICAL NUTRIENT REMOVAL FROM WASTEWATER

R. N. Dawson

Stanley Associates Engineering Ltd.
Suite 402
Victoria, British Columbia V8W 3B9

1. INTRODUCTION

Removal of nitrogen and phosphorous from wastewater is becoming a common requirement for municipalities throughout North America, Europe and Australia. This is particularly true for inland river systems which are tributaries to freshwater lakes and for coastal estuary situations. Phosphorous is usually the limiting nutrient for controlling eutrophication in freshwater and therefore effluent standards of less than 1.0 milligrams per liter (mg/L) are in place for such water bodies as the Great Lakes. For mountain lakes with long retention times effluent total phosphorous levels of 0.2 mg/L, (which represents best available technology) are in place, e.g., Okanagan Lake, British Columbia, Canada and Flathead Lake, Montana, USA.

For inland river systems nitrogen is not usually a trigger for eutrophication. However, toxicity problems because of ammonia concentrations in the effluent result in a requirement for complete nitrification to achieve ammonia (NH_3) levels of less than 1.0 mg/L, often on a seasonal basis. Another reason for reducing ammonia levels is that instream nitrification causes an additional dissolved oxygen (DO) demand on the river, leading to significant DO depletion which in turn endangers fish and other aquatic organisms. These situations are typical for large cities located on river systems on the Prairies, e.g., Calgary, Alberta, Canada. Often denitrification is included in the treatment plant design to reduce overall aeration and power requirements. As well denitrification improves settleability of the sludge in suspended growth biological treatment systems.

For estuarine situations such as the Chesapeake Bay area in the US, nitrogen is the macro-nutrient causing eutrophication of the saline estuary and near shore embayed waters and therefore nitrogen removal down to 3 to 5 mg/L is required.

Retrofitting of existing sewage treatment plants for nitrogen and phosphorous removal has been occurring with increasing frequency over the last 10 years. Recent advances in biological nutrient removal (BNR) have significantly reduced the costs of both retrofits and new nutrient removal facilities particularly where implementation of primary sludge fermentation provides simple carbon compounds to accelerate and stabilize nitrogen and phosphorous removal.

2. HISTORICAL BACKGROUND OF NITROGEN AND PHOSPHOROUS REMOVAL

When phosphorous removal initially became a common requirement for sewage treatment plants, precipitation of phosphorous by addition of divalent metallic salts of iron and aluminum (e.g., alum and ferric chloride) to the effluent from the aeration section of activated sludge plants was utilized to achieve 1.0 mg/L total phosphorous. At some plants, lime was added to primary settling tanks to achieve the phosphorous precipitation. Because of the low cost of retrofitting the chemical addition and the stability of the precipitated phosphorous in sludge the process has remained very popular. However, the disadvantages of the phosphorous precipitation are high operational costs for chemical addition and increase in sludge quantities of 20 to 25% above the quantities usually associated with secondary biological treatment. Additional capital costs for sludge stabilization, dewatering, and disposal facilities are required, as well as increased operational costs for sludge management. For example, for a 150 million gallons per day (mgd) plant these annual additional costs for phosphorous removal by precipitation can be as high as $1,400,000 (US)/annum. To achieve phosphorous level of 0.2 to 0.3 mg/L effluent filtration is necessary.

Nitrogen is present in sewage as organic nitrogen (up to 30 to 50 mg/L) which rapidly degrades to ammonia in the sewers and secondary sewage treatment plant through the activity of anaerobic and aerobic bacteria. For a conventional secondary wastewater treatment plant in North America total nitrogen will be present in the raw wastewater at about 25 mg/L, consisting of 15 to 20 mg/L of ammonia and the remainder as organic nitrogen and nitrates of about 5 to 8 mg/L depending upon the degree of ammonia conversion in the aeration tank.

Ammonia conversion to nitrates is achieved aerobically in a two step bacteriological process involving two types of bacteria *Nitrosomonas* and *Nitrobacter* which sequentially convert ammonia to nitrite and nitrite to nitrate according to the following reactions.

$$55NH_3 + 76O_2 + 109HCO_3^- \xrightarrow{Nitrosomonas} C_5H_7O_2N + 54NO_2^- + 57H_2O + 104H_2CO_3 \quad (1)$$

$$400NO_2^- + NH_4 + 4H_2CO_3 + HCO_3^- + 195O_2 \xrightarrow{Nitrobacter} C_5H_7O_2N + 3H_2O + 400NO_3^- \quad (2)$$

In the process about 4.3 mg of O_2 per mg of ammonia is required and about 8.64 mg of HCO_3^- alkalinity is used up. *Nitrobacter* reacts much more quickly than *Nitrosomonas* so little nitrate is ever detected. Nitrifying bacteria are autotrophic organisms which are relatively slow growing in comparison to carbon degrading bacteria, and therefore the sludge age for nitrification is usually between 10 to 20 days in comparison to 3 to 6 days for carbon removal in conventional activated sludge.

Extended aeration activated sludge processes usually achieve complete or significant nitrification because of their long hydraulic retention time of greater than 20 hours and long sludge-age or solids retention time (SRT).

Denitrification or conversion of nitrate to elemental nitrogen gas can be carried out by a large number of facultative heterotrophic bacteria present in activated sludge under anaerobic or anoxic conditions provided sufficient organic carbon is available for cell synthesis. Energy for growth is obtained by the conversion of nitrate to nitrogen gas. The carbon source can either be organic carbon in the wastewater or any externally applied organic carbon such as methanol.

Denitrification occurs in a two stage process as follows:

$$NO_3^- \longrightarrow NO_2^- \longrightarrow NO^- \longrightarrow N_2O\uparrow \longrightarrow N_2\uparrow \qquad (3)$$

A typical overall equation for methanol as the carbon source consists of:

$$3NO_3^- + 1.08CH_3OH{=}H^+ \longrightarrow 0.065C_5H_7O_2N + 47N_2 + 0.76CO_2 + 2.44H_2O \qquad (4)$$

Biological nitrification and denitrification can be achieved in either suspended growth (activated sludge) or fixed film (trickling filter) biological treatment systems.

In the early 1960s nitrogen conversion was carried out in two stage- separate sludge suspended growth systems such as that shown in Figure 1. The idea was to separate out carbon removal/ nitrification into a long sludge age, aerobic system where nitrifiers were plentiful from a shorter sludge age anoxic, denitrification system to which an external carbon source was applied. However, nitrogen removal researchers in the US and Europe developed single sludge-recycle systems such as that shown in Figure 1 where a highly nitrified stream was recycled to an initial anoxic reaction zone in the bioreactor where the organic carbon in the raw wastewater acted as the carbon source.

Also shown in Figure 1 is alum addition for chemical precipitation of phosphorous.

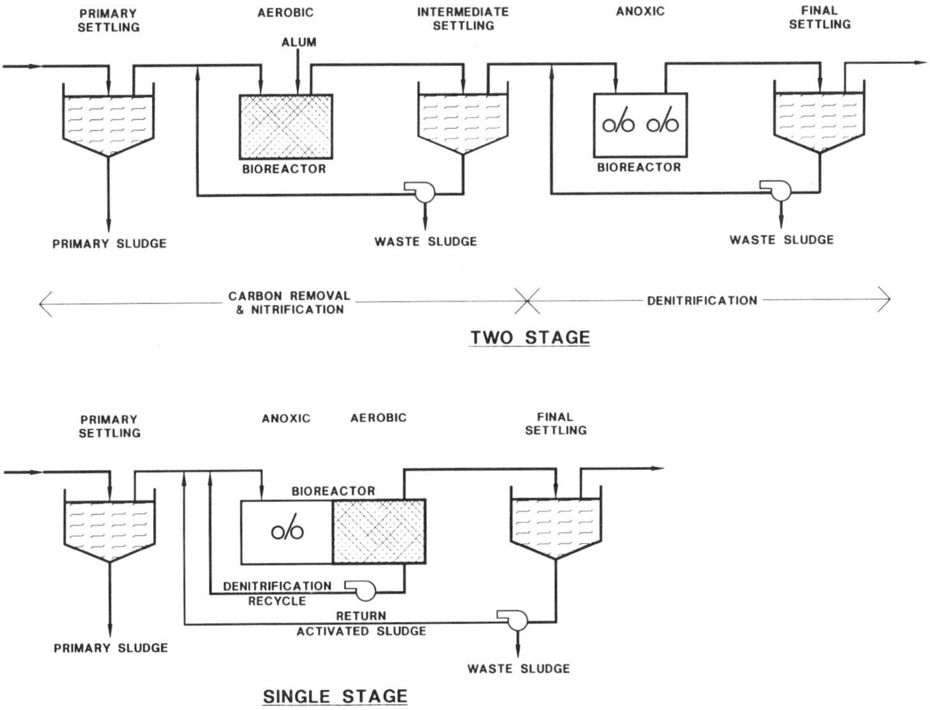

Figure 1. Historical nitrogen and phosphorous removal processes.

To achieve both nitrogen and phosphorous removal in these single sludge reactors, bioreactors with hydraulic retention times of 16 to 24 hours were designed with sludge ages of 10 to 20 days in the late 1960s and early 1970s.

3. BIOLOGICAL NUTRIENT REMOVAL

Biological Nutrient Removal (BNR) is the term used for a number of biological process configurations which achieve nitrogen and phosphorous removal from wastewater in a sequence of anaerobic, anoxic and aerobic reactor cells. Nitrogen removal is achieved by nitrification and denitrification as discussed above in a single sludge reactor system with influent carbon used as the carbon source. Biological phosphorous removal is achieved by introducing wastewater and activated sludge into an anaerobic reactor cell (no oxygen, no nitrates). If single carbon compounds such as volatile fatty acids (VFAs) are present in the wastewater or added to the anaerobic cell in a ratio of 4:1 VFA : phosphorous, then specific bacteria such as *Acinetobacter* initiate a biochemical pathway which allows them to store carbon intracellularly as polyhydroxy butyrates (PHBs). In the anaerobic zone, phosphorous is released from these cells into the surrounding wastewater as part of the energy cycle. However, when these bacteria pass into the aerobic zones of the bioreactor they rapidly grow and take up phosphorous from the reactor contents beyond their metabolic needs—reducing soluble phosphorous levels to as low as 0.05 mg/L in the effluent. When these cells are wasted from the process as waste activated sludge (WAS), then total phosphorous can be removed down to less than 0.4 mg/L without filtration and to less than 0.2 mg/L in conjunction with filtration. The key to the biological phosphorous removal process is the presence of the simple carbon compounds which initiate the biochemical process, and accelerate the phosphorous uptake and nitrogen removal rates.

Typical BNR processes are shown in Figure 2.

4. DEVELOPMENT OF THE BNR PROCESS

The development of the single-sludge biological nitrogen removal process was an improvement over the 2- and 3-step sludge processes of the early 1960s. This development took place when it was recognized that a single sludge mass could perform carbonaceous removal, nitrification and denitrification when subjected to sequential unaerated and aerated conditions. The two principal single-sludge processes were the post-denitrification Wuhrmann process and the pre-nitrification Ludzack-Ettinger process (See Figure 2). In 1974 Barnard endeavored to combine the positive features of these two processes in the 4-stage Bardenpho process (Barnard, 1975).

In the late 1950s and early 1960s, enhanced biological phosphorous removal was reported to occur on an erratic basis in various parts of the world. Levin and Shapiro (1965) observed up to 80% and 96% biological phosphorous removal, at plants in Washington, D.C. and San Antonio, Texas. However, very little was understood about the nature of the biological phosphorous removal mechanism and the conditions necessary to reliably trigger it in the activated sludge process.

In the 1970s a number of researchers (Barnard, 1975; Fuhs and Chen; 1975) attributed the "Luxury Uptake" mechanism for phosphorous removal to the consequence of subjecting the sludge to an anaerobic stress condition (defined as the absence of both nitrate and dissolved oxygen) of such an intensity at some point in the process that phospho-

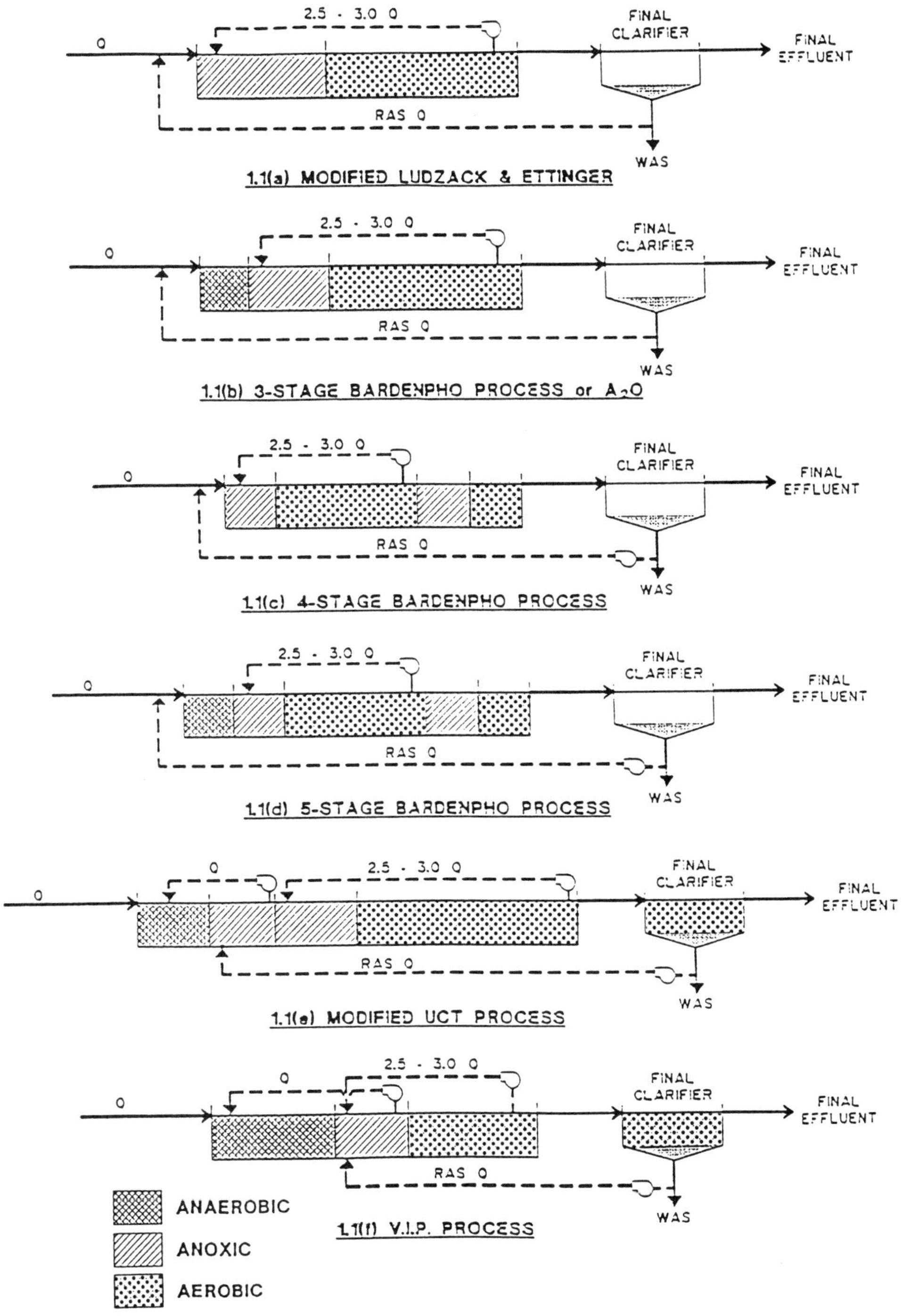

Figure 2. Typical biological nutrient removal process configurations.

rous is released by the organisms into solution. The 5-stage Modified Bardenpho (Phoredox) process was developed out of this hypothesis (See Figure 2). In this process, the required anaerobic stress condition is intended to be created in the anaerobic zone at the head end of the plant.

In the late 1970s, approximately 20 plants were built in South Africa using the modified Bardenpho process configuration, and these plants are currently operating with varying degrees of success. A number of treatment plants in the US were also retrofitted to operate in the Bardenpho configuration. Paepke (1983) conducted an extensive review of 11 of the South African Bardenpho plants and found at that time that only one was achieving consistently good phosphorous removal, and three more plants achieved good removal only after improving operational and control procedures. Incomplete denitrification, unsuitably weak wastewater characteristics, and the presence of dissolved oxygen and nitrate entering the anaerobic zone of the process were identified as the major factors adversely affecting phosphorous removal in the poorly performing plants.

During the late 1970s, researchers at the University of Cape Town (UCT) uncovered a number of major limitations in the Bardenpho design concept. In general, the plants were designed on an empirical design formula based on the nominal detention time of five reaction zones. Little or no consideration was given to the strength of chemical characteristics of the incoming wastewater at a particular plant location. The Bardenpho process had very little operating flexibility and, once commissioned, phosphorous removal was found to be largely dependent on the influent sewage characteristics, and was adversely affected by components of the mechanical plant that introduced dissolved oxygen into the anaerobic zone. In addition, low effluent chemical oxygen demand (COD) and high total Kjeldahl nitrogen (TKN) values in the influent result in inordinately large concentrations of nitrate being recycled to the anaerobic zone in the return sludge, which is known to be detrimental to the phosphorous removal mechanism. Finally, it was found that in the Bardenpho process, the denitrification rate in the secondary anoxic zone be omitted and the anoxic capacity be more efficiently applied in the primary anoxic zone where reaction rates are higher due to the presence of substrate in the incoming wastewater. This 3-stage Bardenpho configuration (see Figure 2) is the basis of the high rate A^2/O process currently being marketed in the US.

In the late 1970s and early 1980s, researchers began to explore the biochemical nature of the phosphorous removal mechanism and the reason why the anaerobic-aerobic sequence was essential for the mechanism to operate. Research workers of the City of Johannesburg, South Africa proposed that simple carbonaceous substrates (principally VFAs) were stored by the organisms involved in enhanced phosphorous removal in the anaerobic zone of the process. The stored substrates helped these organisms survive in the anaerobic zone and provided the energy required for phosphorous storage in the subsequent aerobic zone. It was suggested that increasing the availability of simple carbonaceous substrates in the anaerobic zone enhanced this organism growth. It was shown that nitrate entering the anaerobic zone had detrimental effect on phosphorous removal because the substrate required in the removal mechanism is utilized preferentially in the denitrification reaction.

Furthermore, the UCT group proposed a simple refinement to the Bardenpho process (the configuration became known as the UCT process shown in Figure 2); this facilitates a large degree of operational flexibility and ensures that no nitrate enters the anaerobic zone under a wide variety of influent sewage characteristics. For these reasons, the UCT process proved to be capable of consistent phosphorous removal from wastewaters with a wide variation in influent characteristics without a major detrimental effect on the nitrogen removal of the process. In a further refinement of the UCT process proposed in 1981 by Siebritz et al., the anoxic zone was split in two. This modification facilitated an even greater degree of operational flexibility by completely separating the nitrogen and phosphorous removal cycle of the process. The process is commonly referred to as the Modified UCT Process.

In 1985 it was demonstrated at the pilot-scale that enhanced biological phosphorous removal could be achieved in a high-rate UCT process configuration having little or no nitrogen removal. This process has become known as the Virginia Initiative Plant (VIP).

5. ENHANCED BIOLOGICAL NUTRIENT REMOVAL — UTILIZING PRIMARY SLUDGE FERMENTATION

A number of biological nutrient removal (BNR) plants were designed and constructed in the US and Canada in the late 1970s and early 1980s based upon experience with approximately 20 single sludge modified Bardenpho process plants constructed in South Africa. The US plants have met with varied success in phosphorous removal since they depended upon the development of volatile fatty acids in the sewers before arriving in the plants. In Canada it was recognized that domestic wastewaters in Canadian communities are relatively dilute and because of low temperature wastewater there was little likelihood that VFA concentrations would be sufficiently high to stimulate biological phosphorous removal. In 1975 when application of the first BNR plant was contemplated at Kelowna, British Columbia bench and pilot plant investigations were carried out. Also, full scale optimization studies of a five-stage modified Bardenpho process were implemented at a 27.7 million liters per day (ML/day) plant at Kelowna. As a result of this work it was determined that fermentation of sludge settled out in the primary clarifiers generated volatile fatty acids such as acetic acids which when added to the anaerobic cells of BNR processes (e.g., Bardenpho, modified UCT, VIP) and significantly improved biological phosphorous removal. Since that time a number of process improvements have been made in fermenter design to increase VFA production which have been applied at plants in Canada at Penticton, British Columbia; Calgary, Alberta, and in Denmark at Friedericksvaerk and in the US at Kalispell, Montana.

Bench scale research at University of British Columbia in 1975 and followup pilot scale and full scale plant optimization at Kelowna, British Columbia (Oldham) demonstrated the significance of primary sludge fermentation in enhancing biological phosphorous removal in cold dilute wastewater. A static flow-through fermenter/thickener as shown on Figure 3 at Kelowna and was successful in enabling the Kelowna plant to reduce effluent phosphorous from 1.0 mg/L to less than 0.2 mg/L (Oldham, 1985). A 4-year pilot-scale study (Rabinowitz and Oldham, 1986) into the use of primary sludge germination to generate the substrates (VFAs) required for phosphorous removal resulted in the development of a simple primary clarifier / fermenter operating system. This system can be used with any biological nutrient removal process. For example, the use of this system improved the phosphorous removal characteristics of a UCT pilot process by approximately 50% and resulted in a consistent phosphorous removal exceeding 90%. Using organically weak domestic wastewater, a median effluent total phosphorous concentration of 0.3 mg/L was maintained over a 1-year study period without chemical addition or effluent filtration (Rabinowitz, 1987).

Other work carried out recently at University of British Columbia includes the development of a control strategy for the optimization of activated sludge process perform-

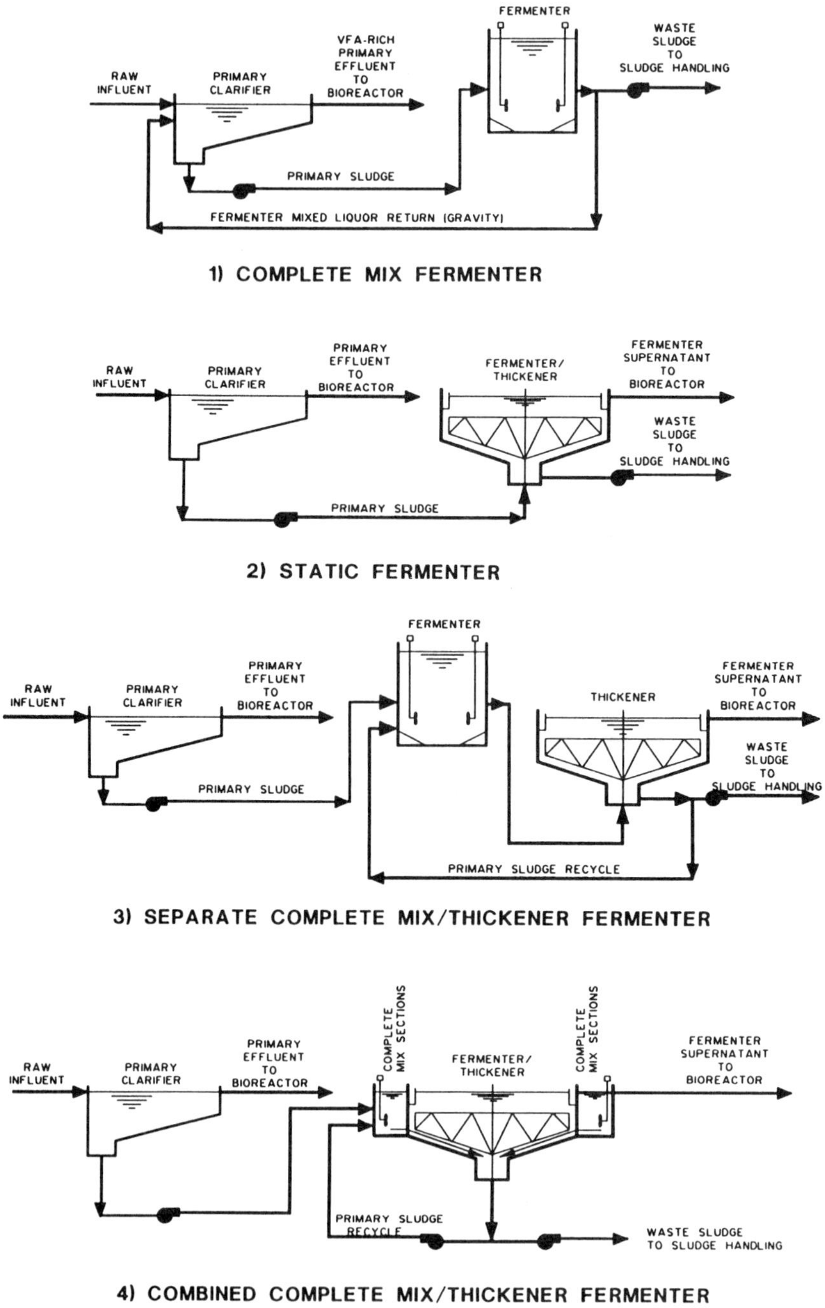

Figure 3. Primary sludge fermenter configurations.

ance, the effect of temperature on phosphorous removal characteristics and the development of an integrated biochemical model for the enhanced biological phosphorous removal mechanism.

A complete mix primary sludge fermenter system shown in Figure 3 was subsequently designed and installed at the Penticton, British Columbia (18 ML/day) and the

Friedericksvaerk, Denmark BNR plant in 1990. Although the complete mix fermenter system is considered to produce more VFA per kilogram (kg) of primary sludge than the static fermenter at Kelowna, the VFAs cannot be added directly to the anaerobic zone of a BNR unit, and there could be some loss of VFAs. Separate, complete mix / thickener, primary sludge fermenters have been designed and are being installed at Kalispell, Montana (12 ML/day) and Calgary, Alberta (100 ML/day). Such a fermenter system, as shown in Figure 3, provides optimum production of VFA in complete mix reactor section and subsequently the settling out of the sludge and decanting of VFAs allows direct application of VFAs to the anaerobic section of the bioreactor where it is needed to stimulate the biological phosphorous process. A further innovation of the primary sludge fermenter which has advantages for retrofitting to existing tankage in activated sludge plants such as anaerobic digesters is the combined, complete-mix/thickener primary sludge fermenter shown in Figure 3. Such a system was designed for Solrod, Denmark as a retrofit into an existing digester during the recent plant upgrade from secondary treatment to BNR.

In addition to these alternatives, at Westbank, British Columbia, a 0.75 ML/day BNR plant, a two-tank, in-series static fermenter was utilized which provides a simple fermenter system for small BNR plants.

The advantages and disadvantages of the various fermenter options discussed above are shown in Table 1.

6. RETROFITTING EXISTING PLANTS — USING BNR TECHNOLOGY AND PRIMARY SLUDGE

6.1. Fermentation

When nitrogen and phosphorous removal to low levels is required at existing plants, retrofitting BNR Technology using primary sludge fermentation (PSF) makes economic sense for the following reasons:

- chemical addition for phosphorous removal is avoided;
- sludge production is reduced by up to 25%;
- energy savings are produced when denitrification is included—up to 20%;
- operating cost savings pay back capital costs of retrofit within 5 to 8 years; and

Table 1. Advantages and disadvantages of alternate fermenter designs

Fermenter type	Advantage	Disadvantage
Static—flow through	Simple operation Allows direct VFA addition—bioreactor	Lowest VFA generation rate VFA optimization difficult
Complete mix/recycle to primary	Optimum VFA production	No direct addition to VFA VFA loss possible
Complete mix/thickener	Optimum VFA production Direct addition possible Good control	High cost—2 stage reactor Complex operation
Complete mix/thickener combination	Optimum VFA production Direct addition possible Easy to retrofit Reduced cost	Complex operation

- primary sludge fermentation result in increased reaction rates for phosphorous removal, denitrification and overall nitrification.

Initially plants were designed as BNR plants with hydraulic retention times of 20 to 24 hours without primary sludge fermentation. Because of implementation of PSF and optimization of the enhanced BNR process, these facilities can now be retrofit with hydraulic retention times as low as 0.5 hours to achieve effluent total nitrogen of 3 to 5 mg/L and total phosphorous of 0.2 to 0.5 mg/L. Case histories of plants which demonstrate the capability of enhanced BNR technology follow.

6.2. Kelowna, British Columbia — Bardenpho Process/Static Primary Sludge Fermenter

Kelowna, British Columbia is a city of approximately 60,000 persons situated on Okanagan Lake in central British Columbia. The Okanagan Lake has been experiencing eutrophication partially due to the discharge of nutrients in the early 1960s, and nitrogen was implicated in stimulating near-shore growths of Asian water milfoil, a rooted aquatic plant.

In 1975, a 22.7 ML/day 5-stage Bardenpho plant was constructed based upon South African experience. Conservative design criteria were selected because of the dilute wastewater and low temperature of wastewater in winter months. The plant was initially designed with a 20 hour hydraulic retention time bioreactor subdivided into 21 cells as shown in Figure 4. Initially difficulty was experienced in achieving a 1 mg/L effluent phosphorous concentration until pilot scale testing and research at the full scale plant carried out by W. Oldham of University of British Columbia successfully optimized the process. A major factor in achieving consistently low ortho-phosphorous levels of approximately 0.1 mg/L in the effluent has been the operation of the static, flow-through primary sludge fermenter.

Sludge from the primary clarifiers is introduced into an old anaerobic digester at a rate of approximately 5–10% of average plant flow. By installing a slow-moving rake, the old digester was converted to a thickener with a sludge age of about 20 days. In Table 2, typical volatile fatty acid concentrations produced in the Kelowna fermenter are shown.

Initially the decant from the fermenter was directed through the primary clarifiers and then to the two halves of the bioreactor. During the first five years of operation and concentrated research, phosphorous levels in the final filtered effluent were maintained between 1.0 and 1.5 mg/L. In early 1988, the fermenter supernatant was directed to the anaerobic zone of the bioreactor rather than through the primary clarifiers. As a result, the effluent phosphorous concentration was gradually reduced to around 1.0 to 0.2 mg/L as the impact of VFAs changed the biological population and accelerated nitrogen removal rates. At the same time hydraulic retention times were reduced to 8 to 10 hours and the process configuration was changed to a 3-stage Bardenpho process. These changes uprated the capacity of

Table 2. Kelowna, British Columbia — static flow-through fermenter typical operating results

Date	VFA (mg/L as acetate)	Soluble COD
June 6, 1992	235	500
January 13, 1992	212	505
May 4, 1992	192	527
May 11, 1992	273	541

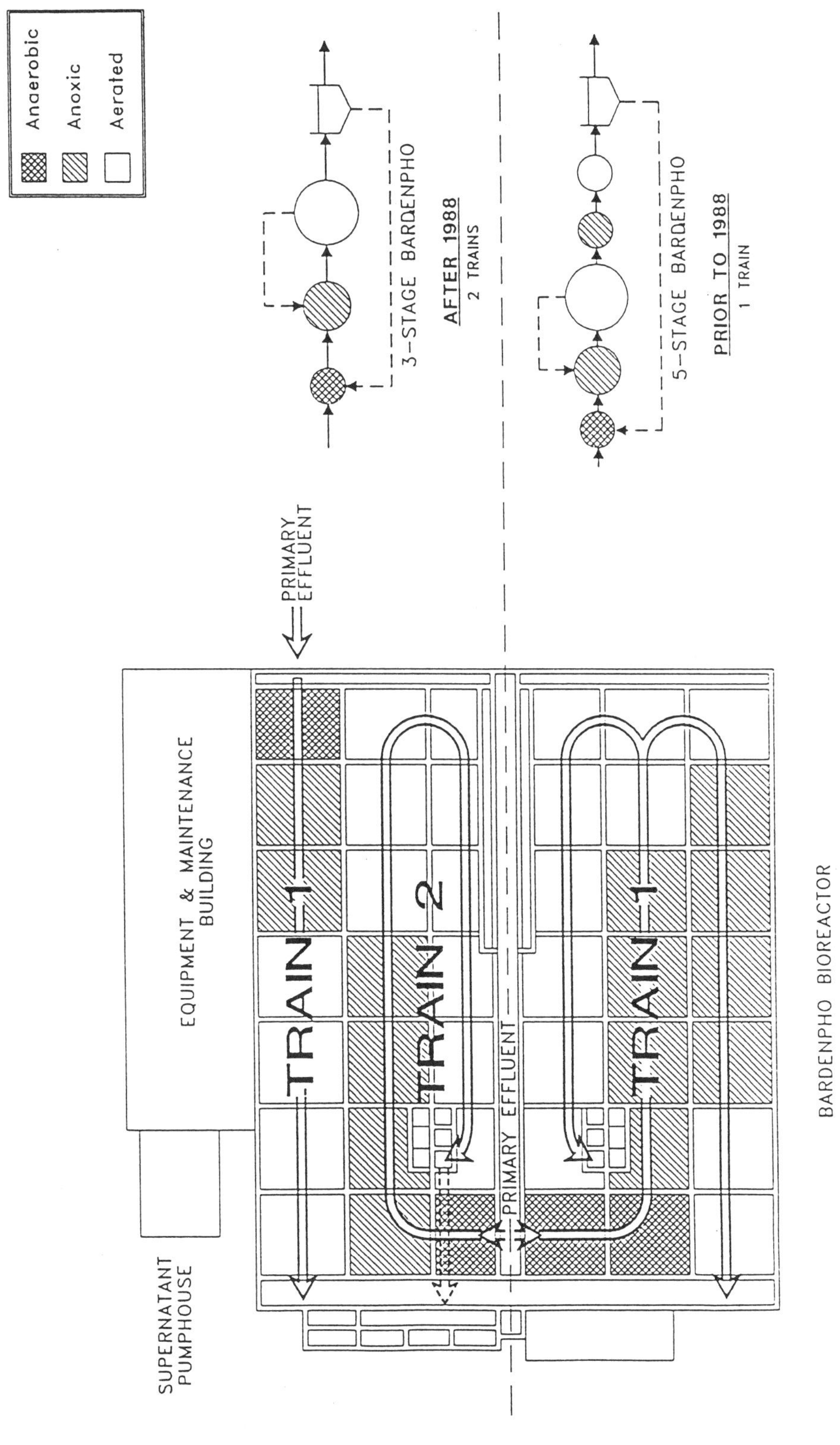

Figure 4. Kelowna wastewater treatment plant Bardenpho reactor.

the plant to 45 ML/day. Further improvements to the aeration/mixing equipment, adjustments to the cell baffles and addition of secondary settling capacity have uprated the facility to approximately 60 ML/day utilizing the initial bioreactor.

6.3. Penticton, British Columbia — Modified UCT Process/Complete Mix Primary Sludge Fermenter

At Penticton, British Columbia an 18 ML/day BNR plant has been operational since 1990. The bioreactor is an eleven cell modified UCT process constructed in two mirror images as shown in Figure 5 and the schematic in Figure 6. Initially the bioreactor had a hydraulic retention time (HRT) of 14 to 16 hours. Each half of the plant can be operated as a separate system for research purposes. A complete mix primary sludge fermenter is dedicated to each half of the bioreactor. Because of variable wastewater temperatures (24°C and 7°C winter) and seasonally fluctuating hydraulic and organic loadings, the plant was designed so that the number of cells operating as the anaerobic, anoxic, and aerobic sections of the process can be varied. The use of gated channels for plant feed, overflow and internal recycles permits this flexibility and the use of low head axial flow pumps achieve the recycles in an energy efficient manner. A schematic diagram of the complete mix fermenter system is included in Figure 3. The fermenter is designed with a variable operating depth to allow control of hydraulic retention time. Sludge age is controlled by wasting fermented solids to the anaerobic digesters.

The key design criteria for the fermenter at Penticton are summarized in Table 3. The hydraulic retention time can also be varied by changing the primary sludge underflow rate as well as by varying the tank operating depth. This operating flexibility was built into the fermenter design to cope with fluctuating loads at the Penticton plant and because the design of fermenters at the time of the project was more of an art than a science. Within the design HRTs and SRTs, the fermenter has successfully stayed in the acid generation phase of the sludge fermentation process.

The objective of the fermenter is to produce approximately 400 to 500 mg/L of VFA in the overflow to the primary clarifier or 30 to 40 mg/L of VFA added in the bioreactor feed to the anaerobic zone of the UCT process. To date the plant operators have maintained too long an HRT and SRT in the fermenter with the result that VFA production and soluble COD production have suffered. Typically when the VFA levels have fallen below 350 mg/L, the performance of the bioreactor in terms of phosphorous removal has suffered.

In the Kelowna treatment plant, the cells are all equal in size, and originally there were two 21 cell modules (which are now configured as four 3-stage Bardenpho process

Table 3. Penticton, British Columbia — primary sludge fermenter design criteria summary

Description	Criteria
Hydraulic loading rate	5–15% plant flows
Hydraulic retention time	5–21 hours
Solids residence time	4–12 days
Operating temperature	10–24°C
Solids concentration	1.5–2.0% solids
Mixing energy	3–8 watts/m^3
Internal recycle pump rate	variable
Recycle rate to primary clarifier	4–14% plant flow

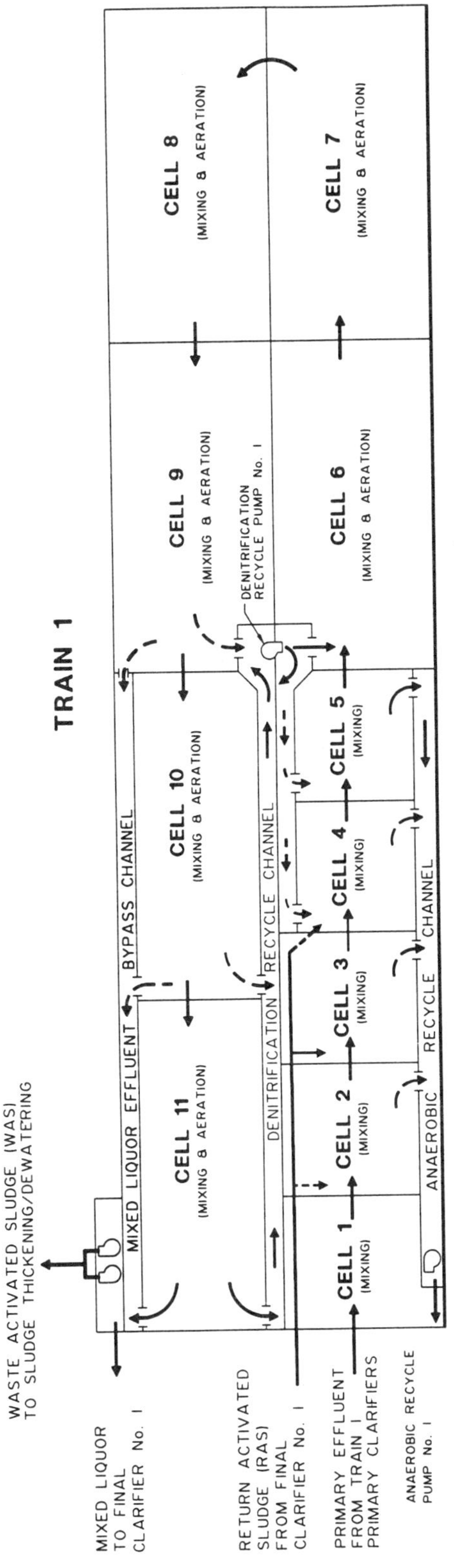

CELL No.	MODE OF OPERATION	VOLUME FRACTION
1	ANAEROBIC	4%
2	ANAEROBIC or Ist ANOXIC	4%
3	ANAEROBIC or Ist ANOXIC	4%
4	ANAEROBIC Ist OR 2nd ANOXIC	4%
5	Ist OR 2nd ANOXIC	4%
6	2nd ANOXIC or AEROBIC	10%
7	2nd ANOXIC or AEROBIC	10%
8	AEROBIC	15%
9	AEROBIC	15%
10	AEROBIC	15%
11	AEROBIC	15%

Figure 5. The City of Penticton advanced wastewater treatment plant bioreactor layout.

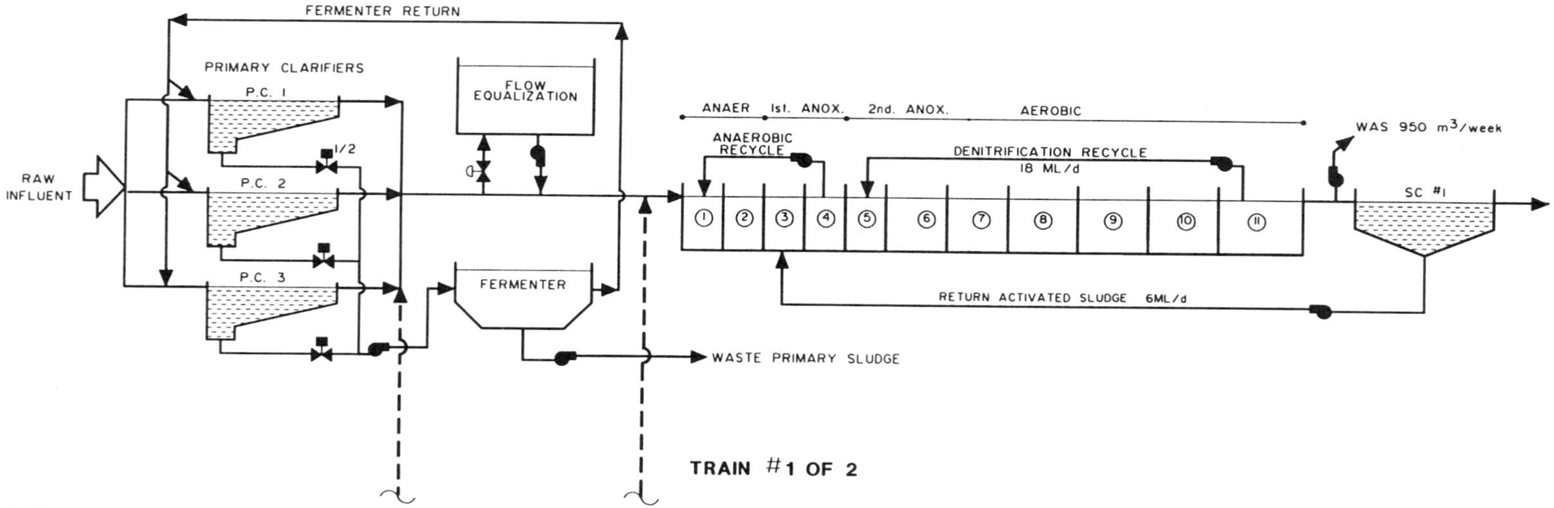

TRAIN #1 OF 2

MODIFIED UCT

CELL No.	MODE OF OPERATION	VOLUME FRACTION
1	ANAEROBIC	4%
2	ANAEROBIC or 1st ANOXIC	4%
3	ANAEROBIC or 1st ANOXIC	4%
4	ANAEROBIC 1st OR 2nd ANOXIC	4%
5	1st OR 2nd ANOXIC	4%
6	2nd ANOXIC or AEROBIC	10%
7	2nd ANOXIC or AEROBIC	10%
8	AEROBIC	15%
9	AEROBIC	15%
10	AEROBIC	15%
11	AEROBIC	15%

Figure 6. Process flow schematic for the City of Penticton advanced wastewater treatment plant.

streams). In Penticton, there are also two bioreactors, the first five cells are each 4% of the total volume, the next 2 cells are each 10% of the total volume and the final 4 cells are each 15% of total volume. Because the population in the Penticton area rises during the summer months due to tourism, the design allowed for a 40% reduction in operating HRT.

The Penticton plant operators perform routine grab sample analysis of all the cells in the bioreactor. This data indicates that nitrification is essentially complete long before the last cell of the bioreactor, and excess phosphorous uptake is complete in the first 25% of the aeration zone (Figure 7). The excess bioreactor capacity is a form of extended aeration and has been shown to produce unwanted effects like cell lysis, causing the re-release of ammonia and phosphorous.

In May 1992, Stanley Associates Engineering Ltd. embarked on an applied research project to optimize the fermenter and bioreactor operations at Penticton. Up to that time overall plant performance with respect to phosphorous removal to less than 0.3 mg/L had been erratic. As a first step in system optimization, a series of carefully controlled experiments have been initiated to determine the effect of fermenter HRT and SRT on VFA production.

The two fermenters at Penticton were set up with SRT levels of about 14 days and with HRT of 28 hours in one and 20 hours in the other. Volatile fatty acids production and soluble COD levels in the fermenter with the lower HRT increased by 15% and 19% respectively. At the same time as this VFA production increase was observed, the plant effluent phosphorous has stabilized at around 0.2 mg/L. Further experimentation confirmed that at SRTs below 6 days and HRTs below 10 hours, VFA production begins to fail.

Based on the observations indicated above in Figure 7, it was suggested that the total bioreactor HRT could be reduced to a nominal 8 hours and the Penticton WWTP would still maintain effluent quality. During the Fall of 1993, all wastewater flows were introduced into one bioreactor, and the nominal HRT for the process was reduced to 8.2 hours. Figure 8 illustrates the bioreactor's performance once the plant stabilized.

Based on the positive results illustrated in Figure 8, it was decided to reduce the HRT further by removing an additional aeration cell from operation. The nominal HRT for the process was then reduced to 6.7 hours. Excellent effluent results were produced and the bioreactor was maintained at this HRT for a period of 3 months (Figure 9). Nitrification was inhibited by the aeration system being unable to meet the aeration demand at high mixed liquor suspended solids (MLSS) concentrations.

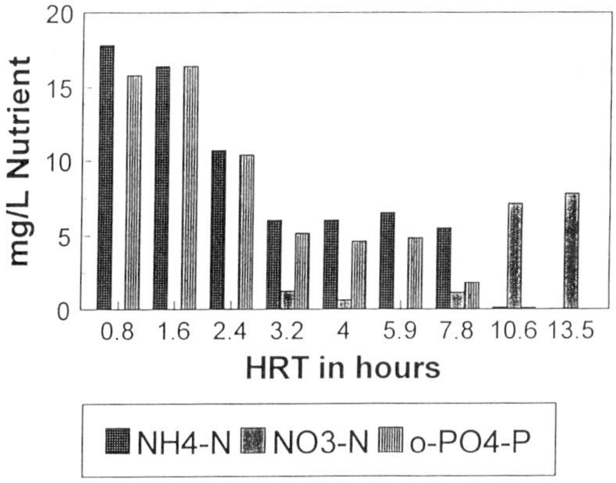

Figure 7. Penticton wastewater treatment plant bioreactor nutrient scan 13.5 hour HRT.

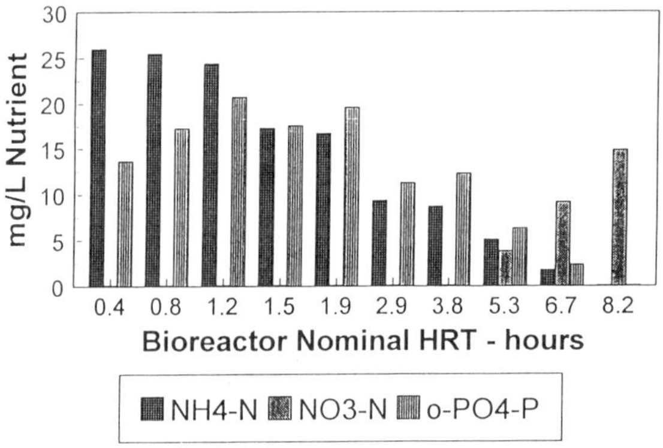

Figure 8. Penticton wastewater treatment plant bioreactor nutrient scan 8.6 hour HRT.

The effluent quality was not consistent in this period at the stringent total effluent phosphorous criterion of less than 0.5 mg/L. The reasons for this were not fully determined, however, the short bioreactor HRT did not allow for any buffering of influent wastewater variation due to recycle streams, and the high MLSS concentration resulted in high sludge blankets and some secondary release of phosphorous effluent. At the end of the trial period, the nominal HRT was increased to 8.2 hours to ensure compliance with B.C. Ministry of Environment effluent criterion. Effectively the optimization program on the reactor and fermenter increased the capacity by 40%, creating a 25 ML/day.

Through the period of record from August 1991 to April 1992, VFA concentrations dropped from around 600 mg/L to 350 mg/L, which roughly corresponds to loading reductions on the Penticton plant. A good correlation was evident between the VFA concentration and ortho-phosphorous release in the anaerobic cell, i.e., 19 mg/L to 12 mg/L. However, because the bioreactor operation had not been optimized during this period, the effluent phosphorous and VFA concentration could not be correlated.

Continuing work at the plant has further stabilized the effluent to 5 mg/L total nitrogen and 0.2 mg/L total phosphorous.

Table 4 shows the effect of temperature on the performance of the fermenters. From data collected during the study, it was apparent that operating temperature of the fermenter influenced VFA production. Data in Table 4 shows that VFA and soluble COD production

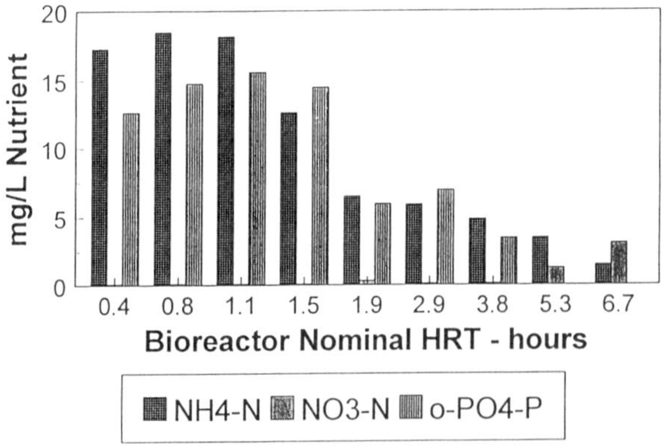

Figure 9. Penticton wastewater treatment plant bioreactor nutrient scan 6.5 hour HRT.

Table 4. Penticton, British Columbia — typical fermenter operating conditions effect of temperature

Item	November 19, 1991	August 6, 1991
Ambient temperature °C	15	28
Average HRT (hrs)	34	36.4
Average SRT (days)	11.9	12.1
Soluble COD in (mg/L)	463	348
Soluble COD out (mg/L)	987	1173
VFA produced (mg/L as Ac)	351	575
TSS influent (mg/L)	245	154
TSS fermenter (%)	1.41	1.62

in November (ambient temperature 15°C) were significantly less than production during August (when the ambient temperature was about 28°C).

6.4. Comparative Fermenter Performance Static versus Complete Mix

A review of pilot scale complete mix fermenter performance at the University of British Columbia in comparison to the full scale static fermenter at Kelowna indicated that a complete mix system would produce substantially more volatile fatty acids than a static system. Table 5 shows a comparison of VFA production results for the full scale Kelowna static fermenter and the Penticton complete mix fermenter. From this data it is apparent that the complete mix fermenter produces about 50% greater volatile fatty acids than a static fermenter operating under similar ambient conditions.

7. FURTHER ADVANCES IN BNR TREATMENT

Fermenter technology has significantly improved the biological nutrient removal process in terms of stabilizing the process and speeding up reaction rates. Retrofit of BNR to existing secondary plants can be made relatively economically with the exception of the fermenter facilities and the increase in final settling tank capacity which is usually required to ensure good phosphorous removal. Surface settling rates for final clarifiers for

Table 5. Penticton and Kelowna — comparative fermenter performance

Item	Penticton—complete mix	Kelowna—static
Hydraulic load (m^3/d)	528	1123
HRT (hours)	32.1	7.7
Nominal SRT (days)	12.6	18.8
Waste (m^3/d)	28	19.3
VFA (mg/L as Ac)	394	228
VFA (kg/d as Ac)	197	251.6
VFA produced as acetate per fermenter volume per day (kg as Ac/m^3d)*	0.37	0.22
Soluble COD produced (mg/L)	992	534
Soluble COD produced (kg/d)	496	589
Soluble COD per fermenter volume per day (kg as Ac/m^3d)*	0.94	0.52

*The Penticton fermenter had not been optimized at this time as shown by the high HRT of 32 hours and SRT of about 13 days.

the BNR process should be in the range of 15 to 18 cubic meters per square meter per day (m^3/m^2/day) as opposed to 20 to 24 m^3/m^2/day for secondary plants. Actually, final clarifiers or solids separation is the unit process—which limits the BNR process from being carried out at higher MLSS levels, i.e., high biological rates. Advances in emerging technologies which will allow reduction in size of BNR facilities as well as capital costs include:

- BNR technology combined with enriched oxygen aerobic bioreactors—smaller bioreactors.
- Complete mix fermenters / centrifuge solids separation smaller fermenter facilities.
- On-line process monitoring for phosphorous, ammonia and nitrate.
- High rate solids separation to replace final clarifiers—e.g., cross flow microfiltration or membrane separation.
- Submerged attached growth technology to increase nitrification rate, e.g., Ringlace®.
- BNR technology combined with fixed film processes.

Many of these emerging technologies are being researched at the demonstration and full scale throughout US and Canada. A good example of a full scale research project is the Nutrient Removal project at New York City where fermenter technology is being researched in conjunction with fixed film, attached submerged growth, and on-line process monitoring.

8. CONCLUSIONS

BNR processes have made significant advances in the last 10 years. Costs for retrofitting existing facilities and constructing "green field" plants have significantly decreased for BNR plants, particularly because of the use of fermenter technology. The combination of fixed film technology with suspended growth BNR processes utilizing primary sludge fermentation and high solids (high rate aerobic) processes will make the choice of BNR technology more attractive to pollution control agencies as effluent quality standards become more stringent. Within the next few years, BNR processes will probably represent Best Practical Technology for North America.

REFERENCES

Barnard, J. L. 1975. Biological phosphorous removal without the addition of chemicals. Water Research, 9:485.

Fuhs, G. W. and M. Chen. 1975. Microbiological basis of phosphate removal in the activated sludge process for the treatment of wastewater. Microbiol, Ecol. 2:119–138

Levin, G. V. and J. Shapiro. 1965. Metabolic uptake of phosphorous by wastewater organisms. J WPCF, 37:800–821

Oldham, W. K. 1985. Full-scale optimization of biological phosphorous removal at Kelowna, Canada. Water Science and Technology. 17:243–257

Paepke, B. H. 1983. Performance and operational aspects of biological phosphate removal plants in South Africa. Prog. Wat. Technol., 15(3/4):219–232

Rabinowitz, B. and W. K. Oldham. 1986. Excess phosphorous removal in activated sludge process using primary sludge fermentation. Can. Jour. of Civil. Eng., 13:345–351

Rabinowitz et al., 1987. A novel operational mode for a primary sludge fermentation for use with the enhanced phosphorous removal process. Advances in Water Pollution Control (Ed. R. Ramadori) Proceedings of IAWPRC conference, Rome

32

BIOTECHNOLOGY IN THE SUSTAINABLE ENVIRONMENT

A Review

J. J. Gauthier

The University of Alabama at Birmingham
Department of Biology, UAB Station
Birmingham, Alabama 35294-1170

At the beginning of this symposium, the participants were asked to reflect upon where we are now and where we are we heading in the development of biotechnology in the sustainable environment. As the meeting progressed, it became clear that achieving the goal of a sustainable environment depends on continued public concern about the environment, policy issues dealt with by governments and application of the tools of science of environmental engineering.

Bioremediation, whether it be by natural attenuation, bioaugmentation, or use of genetically engineered microorganisms (GEMs), is only one of four aspects of long-term environmental protection. Biological treatment technologies and process control are important *abatement* tools needed for maintaining a sustainable environment. *Pollution prevention* can be achieved by waste minimization, and by development of new processes that are environmentally benign. Long-term *sustainability* will depend not only upon the application of new advances in science, but also upon the establishment of realistic, cost effective policies by legislators and regulatory agencies.

1. POLICY ISSUES: PUBLIC CONCERN AND THE ECONOMY

Based on surveys of sustainability in twenty-six developed countries, it was concluded that public concerns about the health of current and future generations and about the sustainability of both economic and ecological systems related to the environment, has not been diminished by current global economic situations (D. Miller). There is worldwide recognition that, as the views of governments and regulatory agencies change and industries recognize the link between quality and stewardship, green technologies offer new opportunities for economic growth. Cost reduction, market demand for green products, and reduced liability will drive the application of green technologies.

Bioremediation in the U.S. is estimated to become a $500 million business by the year 2000. In 1995, the National Science and Technology Council identified agriculture, environmental biotechnology, manufacturing and processing, marine biotechnology and aquaculture as areas of opportunity that will create jobs while improving and sustaining the environment (J. Grimes). Several federal agencies currently support environmental research, mainly on bioremediation, because of the need for cost-effective approaches to cleanup. The major research needs are in the areas of microbial community structure, degradative pathways for specific pollutants in aerobic and anaerobic systems, bioaugmentation, microbial genetics, applications of biosensors, model evaluation, and development of methods for determining the efficacy of bioremediation technologies.

Many nations are dealing with environmental cleanup issues. In Japan, soil and groundwater pollution are significant problems, especially halogenated organic compounds such as trichloroethylene and tetrachloroethylene (O. Yagi). Work is in progress to develop methods of evaluation for soil cleanup technologies. Russia also faces significant problems as the shift to a market-based economy places increased pressure on natural resources and environmental quality. There is a need for bioremediation to clean up existing pollution and for applications of biotechnology to reduce pollution in the future (A. Boronin). High priorities for Russian researchers include biotechnologies for wastewater treatment and bioremediation of oil spills, chemical weapons, radionuclides and heavy metal pollution. Legislation and regulations concerning release of GEMs into the environment is now under consideration by the Russian government.

There is a clear recognition that environmental biotechnology should be exported to assist other nations to include waste treatment technology as an ongoing part of their development (T. McIntyre). Mexico is a potential market for cost-effective products of the new technology (J. Solliero). Because new international agreements have required modification of environmental legislation initiated in the 1970s, and Mexico is limited in its capacities to develop its own environmental solutions, imported biotechnologies give hope for conserving natural resources and improving environmental quality.

2. POLICY ISSUES: RISK ASSESSMENT AND LEGISLATION

In the early 1970s, waste treatment strategies were based on the assumption that the mere presence of a pollutant creates a risk to health and/or to the environment. Subsequent research has shown that organic and inorganic chemicals bind to soil or sediment and become less available for uptake by living organisms. This understanding has shifted the focus of regulations from dependence on measuring minute concentrations of chemicals to determining the actual impact of these chemicals on health and the environment (D. Ritter). In a 1995 survey, many states indicated a strong interest in a risk-based approach to cleanup which focuses resources on sites that pose the greatest health and environmental risks and consequently, several states are developing guidelines for intrinsic bioremediation (K. Devine). Superfund reauthorization proposes to establish a risk range for cleanup, and the US Environmental Protection Agency (USEPA) plans to establish risk-based levels in redefining hazardous waste. Research, development and commercial implementation of bioremediation will be directly affected by risk-related issues.

In many regions of the country, especially in urban areas, properties abandoned due to known or perceived contamination represent a liability due to unknown cleanup requirements. The USEPA's Brownfields Initiative is an attempt to bring state and federal agencies together to promote cooperative interactions in an attempt to solve this problem

(M. Leavitt). Application of environmentally acceptable endpoints should facilitate a more cost-effective recovery of potentially valuable property.

3. RISK ASSESSMENT AND MONITORING

As state and federal regulatory agencies become more receptive to risk-based cleanup, there will be a need to develop more cost effective, sensitive and quantifiable methods to reduce uncertainty (R. Zimmer). The USEPA is responsible for assessing the biodegradability of both new and existing chemicals under the Toxic Substances Control Act (TSCA) (R. Boething). Because biodegradability data are not available for many of the thousands of compounds submitted each year, the USEPA has developed mathematical models to estimate degradability. In the future, these same models could be used in the design of new chemicals with greater degradability. Uncertainty can be further reduced by application of new technologies to quantify molecular endpoints for risk assessment and to monitor microbial activities in bioprocesses. Microfabrication technology is being used to construct genosensors, which consist of a miniature arrays of DNA probes bound to surfaces (K. Beattie). In addition, bioreporter bacteria can be used to monitor genetic expression in the environment, to track movement of microorganisms through soil and to detect the presence of contaminants (R. Burlage).

From the time that the first GEMs were constructed, there has been concern about the potential risks of releasing them into the environment. To meet their obligations under TSCA, the USEPA has developed a review process for evaluating the risks associated with releasing genetically engineered organisms. *Pseudomonas fluorescens* HK44 is an organism which produces visible light in the presence of 2- and 3-ring polyaromatic hydrocarbons. The review of plans to release this organism in a field study was described (P. Sayre) and it was concluded that such a study poses low risk.

4. BIOREMEDIATION

The two major challenges to the use of microorganisms for removing pollutants from the environment are the recalcitrance of some compounds and the overall cost.

Consequently, much current research is focused on enhancing the biodegradation of xenobiotic compounds in a cost-effective manner. For example, the use of selected plants to remove metals from water and soil (phytoremediation) represents a potentially cost-effective method of bioremediation (B. Ensley). Hydroponically cultivated plants can transport heavy metals from water into their roots, whereas other plants can extract radionuclides and heavy metals from soil, These heavy metals and radionuclides can the be recovered by ashing.

It is known that polychlorinated biphenyls (PCBs) are dechlorinated by natural microbial populations in the environment at a slow rate. Addition of 2,6-dibromobiphenyl successfully increased the rate of dechlorination of Arochlor 1260, thereby decreasing bioaccumulation in the food chain and reducing the risk of carcinogenicity and toxicity to human populations (D. Bedard). Soils contaminated with PCBs often have low numbers of naturally occurring PCB degraders and the PCBs adsorbed to soil particles may reduce their susceptibility to degradation. Bioaugmentation with addition of surfactants can enhance the rate of PCB removal. PCB-degrading GEMs capable of using the surfactants as a growth substrate are being tested (M. Beck). Laboratory studies have been completed, pilot studies are in progress, and field studies are planned for the near future.

In Japan, trichloroethylene is a common contaminant of soil and groundwater. When this compound is degraded by organisms that grow on methane or aromatic compounds as the primary carbon source, a careful balance must be maintained between the trichloroethylene concentration and that of the growth substrate, which is inhibitory to trichloroethylene degradation. Japanese investigators have obtained a mutant strain of a phenol-degrading *Moraxella* species that utilizes the non-toxic compound, malate, as the growth substrate and degrades trichloroethylene in the absence of phenol (T. Imamura).

A challenge faced in bioremediation of groundwater is the use of modeling to describe specific environments and to predict treatment outcomes. A heterogeneous test site at Columbus Air Force Base has been extensively characterized, and the data obtained provides a scientific basis for predicting biodegradation and natural attenuation in groundwater (T. Stauffer).

5. WASTEWATER TREATMENT

Biological wastewater treatment facilities have played a major role in preventing water pollution and environmental degradation. A significant step toward being able to more economically prevent pollution in the future is to predict the distribution and persistence of new chemicals during wastewater treatment. Current biodegradation test procedures do not provide an accurate measure of degradation in a natural setting. Application of new test protocols, which simulate realistic conditions and provide more appropriate methods of data analysis, will overcome these limitations (T. Federle).

Synthetic compounds that degrade slowly in conventional wastewater treatment systems may be treatable to a larger degree when facilities are specifically configured to meet the physiological requirements of the active microbial consortium. For example, 1,4-dioxane can be degraded in a staged system when the dioxane-containing wastewater is pretreated in an aerobic reactor and excess biomass from this reactor provides a continual seed to subsequent stages (L. Grady).

Process control of activated sludge has traditionally been based on gross indicators of sludge condition (e.g., SVI, TSS) and effluent quality (e.g., BOD). Microscopy has been limited to identifying relative numbers of protozoa as indicators of sludge health, and detecting the presence of filamentous organisms associated with sludge bulking. The usefulness of microscopic observations has been limited by inability to easily identify specific genera or to correlate problems with the presence of particular organisms. Molecular analysis based on 16S rRNA probes provides an excellent method for specifically identifying microorganisms associated with wastewater treatment problems, without having to employ culturing techniques (C. Lajoie). Molecular techniques such as denaturing gradient gel electrophoresis and rRNA probe procedures coupled to polymerase chain reaction (PCR), are being investigated as tools to monitor changing microbial populations (D. Crawford). Future process control and diagnosis of problems associated with specific processes may be based on such molecular analyses.

During the past few years, effluent limits for nitrogen and phosphorus have become increasingly stringent. Studies of nitrogen removal via nitrification/denitrification revealed that when an anaerobic reactor precedes an aerobic system, phosphorus in recycled water becomes concentrated in the biomass, thus reducing the phosphorus effluent concentration. Further studies showed that volatile fatty acids, produced during fermentation of primary sludge, stimulates uptake of phosphorus in the anaerobic reactor. These advances in understanding of the microbial ecology of treatment systems have resulted in more stable and efficient nitrogen and phosphorus removal (R. Dawson).

Anaerobic digestion of wastes has not been favored in the US due to the potential for odor problems. Improvements in the technology made during the last decade offer promise of a more cost-effective approach that can be utilized by less industrialized nations (W. Verstraete). Although mesophilic systems currently predominate, thermophilic processes offer the advantage of a faster rate of treatment. As source separation of solid wastes becomes more popular, opportunities to recycle the organic material and to degrade xenobiotic compounds by composting become environmentally beneficial alternatives. Elimination of odor problems and removal of certain air pollutants are becoming more feasible through use of biofilters, bioscrubbers and bio-trickling filters (H. Bohn). Advantages of these technologies include absence of fuel or chemical consumption, prevention of secondary pollution, reduced energy requirements, and adaptability to a wide variety of pollutants.

6. VERSATILITY IN THE MICROBIAL WORLD AND NEW APPLICATIONS

Concern about the future health of human beings and of the environment has encouraged industries to seek "greener alternatives" in manufacturing processes. Enzymes have long been used in industrial processes and their applications can be expanded to perform many tasks that are currently accomplished with chemicals, at less cost to the industry and to the environment (G. Nedwick). Obtaining enzymes which function in biosynthetic processes to replace harmful chemicals is achievable through selective enrichment of microorganisms and applications of recombinant DNA technology. Another approach is to design biodegradable substrates. One example is the use of fungal peroxidases, which oxidize natural substrates such as lignin, and are also capable of oxidizing various xenobiotic compounds. By characterizing the active site of peroxidases and delineating the mechanism of substrate degradation, synthetic compounds such as dyes and other commercially important compounds degradable by these enzymes can be designed (R. Crawford).

During the course of evolutionary history, microorganisms have developed the capacity to produce enzymes which degrade most, if not all, naturally occurring compounds. Although 1,2-dichloroethane is not known to occur naturally, this compound can be biodegraded, suggesting that new enzymatic capabilities may rapidly evolve in response to the availability of new synthetic substrates (D. Janssen).

Studying the mechanism of enzymatic dehalogenation of this compound, may provide insights into construction of enzymes which degrade other xenobiotics. Specificity may be modified by site-directed mutagenesis, allowing the construction of enzymes with increased capacity to degrade xenobiotics. Degradation of xenobiotics can be achieved by expressing these enzymes in a suitable host organism and constructing recombinant organisms that carry out sequences of reactions to metabolize recalcitrant compounds and detoxify reactive intermediates.

7. ENVIRONMENTAL TECHNOLOGY: WHERE ARE WE GOING?

There are many reasons to be optimistic that we can achieve a sustainable environment. As governments throughout the world recognize the need for environmental protection and shift from a concentration-based to a risk-based approach to environmental regulations, resources can be focused on problems most harmful to human health. To supplement and support more rational legislation, a variety of tools are becoming available to provide more cost-effective cleanup and abatement. GEMs constructed to produce enzymes with new capabilities can be used to degrade synthetic compounds and to monitor the progress of treatment processes. A better understanding of the nutritional and physical conditions required for environmental cleanup and waste removal can lead to more efficient remediation and waste treatment processes. As we look to the future, we see industrial processes based on more benign technologies, which will lead to less use of polluting chemicals and reduced costs. The development of biodegradable products will also become an important part of environmental protection. It is the responsibility of everyone in the environmental industry to inform the public of advances being made and the need for continued effort in this direction. It must be remembered, however, that the new paradigm in legislation and all of the advances in biotechnology will be of little long-term value without population control, protection of species diversity, prevention of soil erosion, recognition of the limited carrying capacity of air, water and soil, and adequate global sharing of the planet's resources.

INDEX